Theilheimer's
Synthetic Methods
of Organic Chemistry

Vol. 60

Vol. 60 2001

Theilheimer's
Synthetic Methods
of Organic Chemistry

Editor **Alan F. Finch, Cambridge**

Assistant Editor Gillian Tozer-Hotchkiss, Derwent Information
Editorial Consultant William Theilheimer, Nutley, N.J.

Basel · Freiburg
Paris · London
New York · New Delhi
Bangkok · Singapore
Tokyo · Sydney

KARGER

Deutsche Ausgaben			Vol. 19	1965	
Vol. 1	1946	1. Auflage	Vol. 20	1966	with Reaction Titles Vol. 16-20 and Cumulative Index
	1948	2., unveränderte Auflage			
	1950	3., unveränderte Auflage	Vol. 21	1967	
Vol. 2	1948		Vol. 22	1968	
Vol. 3	1949	with English Index key	Vol. 23	1969	
	1953	2., unveränderte Auflage	Vol. 24	1970	
	1966	3., unveränderte Auflage	Vol. 25	1971	with Reaction Titles Vol. 21-25 and Cumulative Index
	1975	4., unveränderte Auflage			
Vol. 4	1950	with English Index key	Vol. 26	1972	
	1966	2., unveränderte Auflage	Vol. 27	1973	
			Vol. 28	1974	
English Editions			Vol. 29	1975	
Vol. 1	1948	Interscience Publishers	Vol. 30	1976	with Reaction Titles Vol. 26-30 and Cumulative Index
	1975	(Karger) Second Edition	Vol. 31	1977	
Vol. 2	1949	Interscience Publishers	Vol. 32	1978	
	1975	(Karger) Second Edition	Vol. 33	1979	
Vol. 5	1951	with Reaction Titles Vol. 1-5 and Cumulative Index	Vol. 34	1980	
			Vol. 35	1981	with Reaction Titles Vol. 31-35 and Cumulative Index
	1966	Second Edition			
Vol. 6	1952		Vol. 36	1982	
	1975	Second Edition	Vol. 37	1983	
Vol. 7	1953		Vol. 38	1984	
	1975	Second Edition	Vol. 39	1985	
Vol. 8	1954		Vol. 40	1986	with Reaction Titles Vol. 36-40 and Cumulative Index
	1975	Second Edition	Vol. 41	1987	
Vol. 9	1955		Vol. 42	1988	
Vol. 10	1956	with Reaction Titles Vol. 6-10 and Cumulative Index	Vol. 43	1989	
			Vol. 44	1990	
	1975	Second Edition	Vol. 45	1991	with Reaction Titles Vol. 41-45 and Cumulative Index
Vol. 11	1957				
	1975	Second Edition	Vol. 46	1992	
Vol. 12	1958		Vol. 47	1993	
	1975	Second Edition	Vol. 48	1994	
Vol. 13	1959		Vol. 49	1995	
	1975	Second Edition	Vol. 50	1996	with Reaction Titles Vol. 46-50
Vol. 14	1960		Vol. 51	1997	
	1975	Second Edition	Vol. 52	1997	
Vol. 15	1961	with Reaction Titles Vol. 11-15 and Cumulative Index	Vol. 53	1998	
			Vol. 54	1998	
Vol. 16	1962		Vol. 55	1999	
Vol. 17	1963		Vol. 56	1999	
Vol. 18	1964		Vol. 57	2000	
			Vol. 58	2000	
			Vol. 59	2001	

Library of Congress, Cataloging-in-Publication Data

Theilheimer's synthetic methods of organic chemistry = Synthetische Methoden der organischen Chemie. – Vol. 60 (2001) - Basel; New York: Karger, © 1982 -
v.
Continues: Synthetic methods of organic chemistry.
Editor: Alan F. Finch.
1. Chemistry, Organic – yearbooks   I. Finch, Alan F.   II. Theilheimer, William, 1914–
ISBN 3-8055-7266-2

All rights reserved.
No part of this publication may be translated into other languages, reproduced or utilized in any form or by any means, electronic or mechanical, including photocopying, recording, microcopying, or by any information storage and retrieval system, without permission in writing from the publisher.

© Copyright 2001 by S. Karger AG, Basel (Switzerland), and Derwent Information Ltd., London
Distributed by S. Karger AG, Allschwilerstrasse 10, P.O. Box, CH-4009 Basel (Switzerland)
Printed in Switzerland on acid-free paper by Schüler AG, Biel
ISBN 3-8055-7266-2

Contents

Preface to Volume 60	VI
Advice to the User	VII
General Remarks	VII
Method of Classification	VIII
High-Coverage Searches	X
Further Trends and Developments in Synthetic Organic Chemistry 2001	XI
Systematic Survey	XXIII
Abbreviations and Symbols	XXV
Reactions	1
Reviews	262
Subject Index	268
Supplementary References	313

Preface

This is the second of the biannual volumes of *Theilheimer* for 2001 containing abstracts of new synthetic methods and supplementary data mainly from papers published in the literature during October 2000 to March 2001.

For browsing purposes, abstracts are displayed according to the Systematic Classification (symbol notation) so that reactions of the same type and associated data appear together. For example, all deprotections appear in the early symbols (under HO↓↑, HN↓↑, HS↓↑); reduction of oxo compds., imines and carbon-carbon multiple bonds under the HC⇊ sections; C-defunctionalization under the HC sections; oxy-functionalization under the OC sections; aminations, nitrations, peptide coupling etc. under the NC sections; halogenation under the HalC sections; sulfurations under the SC sections; selenation, stannylation, phosphorylation, etc. under the RemC sections; syntheses involving C-C bond formation in the latter half of the book under the CC sections; and data on resolutions (Res) at the end. A list of reaction symbols and references thereto is given in the Systematic Survey (p. XXIII).

The displayed data are supported by the customary in-depth Subject Index and access to supplementary data can be made in the usual manner via the Supplementary Reference Index, e.g. the reader interested in updates to palladium-catalyzed, asym. α-allylation (Synth. Meth. *48*, 772) will note from p. 318 that additional references can be found on p. 190 of this volume.

As usual, the volume contains a 'Reviews' section (p. 262), covering reviews published up to and including September 2001, and a 'Trends' section (p. XI) incorporating key developments in the synthetic chemistry up to and including September 2001. Most of these latter references will appear as abstracts in the next volume.

I would like to express my gratitude to Dr. Theilheimer for his encouragement in the preparation of these yearbooks, and to my colleagues at Thomson Scientific, involved in the production of the Journal of Synthetic Methods. A special thankyou goes to my assistant editor, Gillian Tozer-Hotchkiss, and to Kath Ince, Jill Entwistle and Jack Ranner for handling the electronic processing by which these volumes are published.

October 2001 *A.F. Finch*, Editor

Advice to the User

General Remarks

New methods for the synthesis of organic compounds and improvements of known methods are being recorded continuously in this series.

Reactions are classified on a simple though purely formal basis by symbols, which can be arranged systematically. Thus searches can be performed without knowledge of the current trivial or author names (e.g. 'Oxidation' and 'Friedel-Crafts reaction').

Users accustomed to the common notations will find these in the subject index. By consulting this index, use of the classification system may be avoided. It is thought that the volumes should be kept close at hand. The books should provide a quick survey, and obviate the immediate need for an elaborate library search. Syntheses are therefore recorded in the index by starting materials and end products, along with the systematic arrangement for the methods. This makes possible a sub-classification within the reaction symbols by reagents, a further methodical criterion. Complex compounds are indexed with cross reference under the related simpler compounds. General terms, such as synthesis, replacement, heterocyclics, may also be brought to the attention of the reader.

A brief review (*Trends* section), stresses highlights of general interest and calls attention to key methods too recent to be included in the body of the text.

The abstracts are limited to the information needed for an appraisal of the applicability of a desired synthesis. In order to carry out a particular synthesis it is therefore advisable to have recourse to the original papers or, at least, to an abstract journal. In order to avoid repetition, selections are made on the basis of most detailed description and best yields whenever the same method is used in similar cases. Continuations of papers already included will not be abstracted, unless they contain essentially new information. They may, however, be quoted at the place corresponding to the abstracted papers. These supplementary references (see page 313) make it possible to keep abstracts of previous volumes up-to-date.

Syntheses that are divided into their various steps and recorded in different places can be followed with the help of the notations such as *startg. m. f.* (starting material for the preparation of ...).

Method of Classification

Reaction Symbols. As summarized in the Systematic Survey (p. XXIII), reactions are classified firstly according to the bond formed in the synthesis, secondly according to the reaction type, and thirdly according to the bond broken or the element eliminated. This classification is summarized in the reaction symbol, e.g.

$$\underset{\underset{\text{Bond formed}}{\nearrow} \underset{\text{Reaction type}}{\uparrow} \underset{\text{Bond broken or element eliminated}}{\nwarrow}}{\text{OC} \Uparrow \text{N}}$$

The first part of the symbol refers to the chemical bond formed during the reaction, expressed as a combination of the symbols for the two elements bonded together, e.g. HN, NC, CC. The order of the elements is as follows:

H, O, N, Hal (Halogen), S, Rem (Remaining elements), and C.

Thus, for the formation of a hydrogen-nitrogen bond, the notation is HN, not NH.

If two or more bonds are formed in a reaction, the 'principle of the latest position' applies. Thus, for the reduction

$$\text{RCH=O} + \text{H}_2 \longrightarrow \text{R-CH(H)-OH}$$

in which both hydrogen-oxygen and hydrogen-carbon bonds are formed, the symbol is HC⇓OC and not HO⇓OC.

The second part of the symbol refers to the reaction type. Four types are distinguished: addition (⇓), rearrangement (∩), exchange (⇅), and elimination (⇑), e.g.

RCH=CH$_2$ + H$_2$O ⟶ R-CH(OH)-CH$_3$		OC⇓CC
(thiophene-allyl) ⟶ (thienyl-ethenyl)		CC∩SC
R-Cl + CN$^-$ ⟶ R-CN [+Cl$^-$]		CC⇅Hal
R-CH(Br)-CH$_3$ ⟶ RCH=CH$_2$ [+HBr]		CC⇑Hal

Monomolecular reactions are either rearrangements (∩), where the molecular weight of the starting material and product are the same, or eliminations (⇑), where an organic or inorganic fragment is lost; bimolecular and multicomponent reactions are either additions (⇓), such as intermolecular

Diels-Alder reactions, Michael addition and 1,4-addition of organometallics, or exchanges (↓↑), such as substitutions and condensations, where an organic or inorganic fragment is lost.

The last part of the symbol refers to the essential bond broken or, in the case of exchange reactions and eliminations, to a characteristic fragment which is lost. While the addition symbol is normally followed by the two elements denoting the bond broken, in the case of valency expansion, where no bonds are broken, the last part of the symbol indicates the atom at which the addition occurs, e.g.

| R₂S | ⟶ | R₂SO | | OS⇅S |
| RONO | ⟶ | RONO₂ | | ON⇅N |

For addition, exchanges, and eliminations, the 'principle of the latest position' again applies if more than one bond is broken. However, for rearrangements, the most descriptive bond-breakage is used instead. Thus, for the thio-Claisen rearrangement depicted above, the symbol is CC∩SC, and not CC∩CC.

Deoxygenations, quaternizations, stable radical formations, and certain rare reaction types are included as the last few methods in the yearbook. The reaction symbols for these incorporate the special symbols El (electron pair), Het (heteropolar bond), Rad (radical), Res (resolutions), and Oth (other reaction types), e.g.

| R₂S=O | ⟶ | R₂S | | ElS⇅O |
| R₃N | + R'Cl ⟶ | R₃N⁺R' Cl⁻ | | Het⇅N |

The following rules simplify the use of the reaction symbols:

1. The chemical bond is rigidly classified according to the structural formula without taking the reaction mechanism into consideration.

2. Double or triple bonds are treated as being equivalent to two or three single bonds, respectively.

3. Only stable organic compounds are usually considered: intermediates such as Grignard compounds and sodiomalonic esters, and inorganic reactants, such as nitric acid, are therefore not expressed in the reaction symbols.

Reagents. A further subdivision, not included in the reaction symbols, is based on the reagents used. The sequence of the reagents usually follows that of the periodic system. Reagents made up of several components are arranged according to the element significant for the reaction (e.g. $KMnO_4$ under Mn, NaClO under Cl). When a constituent of the reagent forms part of the product, the remainder of the reagent, which acts as a 'carrier' of this constituent, is the criterion for the classification; for example, phosphorus is the carrier in a chlorination with PCl_5 and sodium in a nitrosation with $NaNO_2$.

High-Coverage Searches

A search through *Synthetic Methods* provides a selection of key references from the journal literature. For greater coverage, as for bibliographies, a supplementary search through the following publications is suggested:

Derwent Reaction Service[1]. Designed for both current awareness and retrospective retrieval. Its monthly publication, the *Derwent Journal of Synthetic Methods*, covers the journal and patent literature, and provides 3,000 abstracts of recently published papers annually.

Access is available in-house via RX-JSM to over 100,000 reactions, including the data in all the abstracts in *Synthetic Methods*, while online access to data from 1980 is provided on STN as DJSMONLINE.

Science Citation Index[2]. For which *Synthetic Methods* serves as a source of starting references. This is particularly useful for accessing papers quoting details of a particular method which has been included in these volumes from a preliminary communication.

Chemical Abstracts Service[3]. References may not be included in *Synthetic Methods* (1) to reactions which are routinely performed by well known procedures; (2) to subjects which can be easily located in handbooks and indexes of abstracts journals, such as the ring system of heterocyclics or the metal in case of organometallic compounds, and (3) to inadequately described procedures, especially if yields are not indicated.

References to less accessible publications such as those in the Chinese or Japanese language are usually only included if the method in question is not described elsewhere.

---

[1] Derwent Information Ltd., 14 Great Queen Street, London WC2B 5DF, England; http://www.derwent.com
[2] Institute for Scientific Information, Philadelphia, Pa., USA.
[3] Chemical Abstracts Service, Columbus, Ohio, USA.

Further Trends and Developments in Synthetic Organic Chemistry 2001

The chemistry of dendrimers ('dendritic chemistry') is fast developing into an applied science, as did fullerene chemistry in the nineties. A timely review outlines the current state-of-the-art, with special reference to the real and potential application of these uniform, globular, highly branched molecules in such fields as surface chemistry, biomimetics, membrane technology, and especially drug delivery[1]. In synthetic organic chemistry, outlets are manifold. High-loading dendritic supports are now well-established in solid-phase synthesis[2], as are readily recyclable transition metal-functionalized dendrimers[3], both aspects having been recently reviewed in depth. The latter - seen as bridging the gap between homogeneous and heterogeneous catalysts - embrace two key facets: catalysis with peripherally functionalized dendrimers, wherein cooperative effects of multiple metal centres are a particular manifestation; and catalysis with core-functionalized dendrimers where all the benefits of site isolation are tapped. The design of amphiphilic dendrimers as singlet oxygen sensitizers is a more recent innovation[4], together with the emergence of robust, chiral dendritic ligands for asymmetric synthesis. Note, for example, a chiral, readily retrievable, dendritic N-sulfonyl-1,2-diamine which induces high asymmetric induction in catalytic transfer-hydrogenation of ketones[5].

In asymmetric synthesis, the quest for new chiral ligands is fundamental to all aspects of asymmetric catalysis, none more evident than in the asymmetric synthesis of sec. alcohols from aldehydes[6] and in palladium-catalyzed asymmetric allylation[7]. In the latter arena, we see, for the first time, the concept of *multichirality* with the emergence of 2'-subst. 1,1'-P,N-ferrocenyl ligands having both *central* and *planar* chirality, as well as exhibiting *axial* chirality on coordination with the central metal[8]. 2,2'-Biphospholes, featuring both axial and central chirality (at phosphorus), are equally of note in the same context, these undergoing spontaneous resolution on crystallization to yield each diastereoisomer independently[9]. Planar chiral η^5-cyclopentadienyl-rhenium(I) tricarbonyl complexes are an improvement on ferrocene analogs in asymmetric synthesis of sec. benzyl alcohols[10], while new chiral imidazolidin-2-ylidene ligands are a feature of a new asymmetric oxindole ring closure by Pd-catalyzed intramolecular α-arylation[11]. The synergistic effect of two different chiral ligands has also been duly noted, as illustrated in a Ti(IV)-catalyzed asymmetric ene reaction where (S)-

BINOL and (R)-5,5′,6,6′,7,7′,8,8′-octafluoro-BINOL combine as a pseudoracemic system to afford higher enantioselectivity and yield than in the same reaction with each ligand in isolation[12]. As underlined in the *Trends* section of Volume **59**, there has been a significant development in asymmetric induction with relatively simple chiral amines, such as L-proline, as catalysts in their own right. Such *'organo-catalysis'* [in the *absence* of a binding metal] has now been extended to asymmetric Michael addition[13], Robinson annulation[14], Friedel-Crafts alkylation[15], Diels-Alder cycloaddition[16], and the asymmetric synthesis of cyanohydrin esters from ketones[17] (the latter two reactions featuring *Cinchona* derivs.). In asymmetric aldol condensation, chiral calcium alkoxides are readily available[18], benign alternatives to the more familiar chiral transition metal complexes, while an axially chiral aluminum salen complex is the reagent of choice in the synthesis of chiral α-amino-β-hydroxycarboxylic acids via asymmetric aldol-type condensation with 5-alkoxyoxazoles[19]. Well-established chiral copper-(I) and -(II) bis(2-oxazoline) complexes, however, are favoured for asymmetric nitro-Mannich (aza-Henry) reaction, which are equally facile with free aliphatic nitro compds.[20] and their silyl nitronates[21]. The precise placement of coordination sites in the substrate(s) may be as critically important as the design of the chiral ligand and choice of central metal. Thus, in a recent zirconium(IV) BINOLate-catalyzed asymmetric synthesis of homoallylamines from aldimines and allylstannanes, hydroxyl groups are specifically incorporated in *both* substrates in order to secure the required delivery and face-selective coupling of nucleophile with electrophile within the coordination sphere of the chirally modified central metal[22]. However, perhaps the most significant development in the field of asymmetric catalysis is in the application of dialkylzincs as more versatile nucleophiles than malonates (and related CH-acidic compds.) in asymmetric allylic substitution. Here, modular chiral pyridyldipeptides serve as bifunctional ligands, one centre of which serves to bind the copper(I) catalyst while the second coordinates to the dialkylzinc to facilitate high face selectivity (up to 90% e.e. for generation of chiral quaternary carbon centres)[23].

Regiospecific oxidative functionalization of hydrocarbons and remote, unactivated carbon centres is a recurring theme in modern organic synthesis, which has drawn much from intensive studies of natural, enzymatic processes. Thus, the identification of *two* iron atoms in the active site of methane monooxygenase[24] has inspired the design of relatively simple di-iron complexes in search of the 'holy grail' of hydrocarbon oxidation with aerial oxygen at room temperature[25]. Rhodium(I)-catalyzed activation of hydrocarbons has been a feature of both carbonylation and dehydrogenation in dense carbon dioxide[26], as well as in the direct borylation of aromatic hydrocarbons[27], while intramolecular amination of an unactivated hydrocarbon centre can be effected with $Rh_2(OAc)_4$ in the presence of phenyl iodosoacetate[28]. A mild radical nitration of hydrocarbons with nitric acid is also reported with N-hydroxyphthalimide as catalyst[29]. No less important is the search for new methodologies based on carbon-carbon single +bond activation. Noteworthy in this respect are tandem aldol transfer reactions[30], an interesting allyl alcohol reorganisation based on In(III)-mediated retro-ene reaction[31], and an asymmetric synthesis of aryl α-hydroxyketones with kinetic resolution via an enzymatic C-C cleavage of benzoins[32]. An electron transfer-initiated C-debenzylation has also been reported as the prelude to an ionic lactolide ring closure[33], and an enzyme-catalyzed retro-Claisen condensation with a hydrolase has been applied to the desymmetrization of cyclic β-diketones[34].

Adapting enzymes for syntheses in organic media is another recurring theme. One device is to stabilize the enzymes in the form of cross-linked enzyme crystals (CLECS)[35], being not unlike zeolites in having a uniform pore size. Readily prepared enzyme-coated microcrystals also display enhanced catalytic activity and stability in organic media[36]. Pig liver esterase, however, may be simply activated by prior treatment with an organic polymer[37], while a recombinant form of the same enzyme displays higher enantioselectivity in ester hydrolysis than the natural enzyme which normally contains contaminants[38]. Inexpensive, non-toxic, recyclable liquid petroleum gas is a useful alternative to traditional organic solvents for asymmetric reduction of ketones with yeast[39].

The development of alternative, eco-friendly media has been a major focus in recent years, thanks largely to the emergence of supercritical[40] or dense phase carbon dioxide and ionic liquids which support a wide spectrum of reactivity. Conducting nucleophilic substitution in these solvents is a new feature: in ionic liquids, the medium itself effectively serves as the phase transfer catalyst, as in the nucleophilic substitution of halogen by cyano groups with KCN[41]. Here, work-up simply involves product extraction into supercritical carbon dioxide and removal of potassium halide with water prior to recycling the ionic liquid. Such functional group displacement in supercritical CO_2, however, requires a readily retrievable silica gel-supported quaternary ammonium salt as phase transfer catalyst with, preferably, water as co-solvent[42]. Grafting ionic liquids onto reactants is a further device which avoids handling solvents altogether, thereby facilitating work-up by simple solvent washes[43]. The incorporation of 'fluorous' functionality (phase tags) into reactants and reagents is a further ploy to ease the burden of work-up. Thus, lipase-catalyzed resolution of sec. alcohols by asymmetric transesterification can be effected by using a fluorous ester as acylating agent, the chiral products (the slower-reacting enantiomeric alcohol and new fluorous ester) being retrieved by partitioning between organic and fluorous phases, respectively[44]. The inventory of easily recyclable fluorous reagents and catalysts is also increasing in size, one newcomer being a 'lightly' fluorous sulfoxide as an alternative to DMSO for the Swern oxidation[45]. The procedure is particularly worthwhile in that the liberated fluorous sulfide, unlike dimethyl sulfide, does not smell. However, the most significant advance in fluorous chemistry has the potential of rivalling polymer-based combinatorial synthesis: dubbed 'fluorous mixture synthesis', it facilitates the simultaneous, rapid and clean synthesis of a mixture of like-compounds in the same solution by differential fluorous tagging of the starting materials for simple separation of the final products[46]. Water, of course, is the friendliest medium of all, supporting a complete spectrum of reactivity. Note, for example: a new radical cross-coupling in aq. THF using allylgallium compds.[47]; Rh(I)-catalyzed Heck or reductive Heck arylation of styrenes with arylboronic acids in water itself[48]; and Rh(I)-catalyzed 1,4-addition of stannanes[49] under the same conditions. A variety of techniques is also at hand to encourage water-insoluble organic substrates to react in water. Perhaps the most memorable in recent months is a new, boron-*catalyzed* aldol-type condensation with enoxysilanes. Here, reaction takes place in water within the colloidal particles so that the intermediate boron enolates are protected from hydrolysis by surfactant molecules[50].

In the expansive field of palladium-catalyzed synthesis, a number of P-ligands has evolved to enhance coupling of unactivated and/or deactivated aryl chlorides. These include: air-stable phosphinous acids as phosphine oxide precursors[51]; highly active phosphine-coordinated o-palladacyclics, which facilitate Suzuki coupling with turn-over numbers as high as 99, 000[52]; and new ferrocenyldiphosphines for Pd-catalyzed carbonylation[53]. Catalytic activity may also be enhanced by the addition of simple amines. Thus, Heck arylation with t-Bu$_3$P as ligand can now be conducted *at room temperature* in the presence of dicyclohexylmethylamine[54], while diethylamine facilitates Stille coupling with tetraalkylstannanes as an alternative to added copper(I) salts[55].

Coupling of aryl iodides with enesilanols has also been realized with NaH or KH, thereby avoiding the use of fluoride ion[56]. Palladium-catalyzed N-arylation is now well established, but mild alternatives are nonetheless welcome. Here, one is attracted to a room temperature version of the Goldberg synthesis with 1.2 mol% CuI and a 1,2-diamine as ligand[57]. A mild di- and tri-arylamine synthesis with air-stable $Cu(PPh_3)_3Br$ is also available[58], as well as an oxidative Cu(II)-catalyzed N- and O-arylation with arylboronic acids in place of the traditional ar. halides[59].

While developments in palladium-catalyzed synthesis have followed largely in the footsteps of allylation via π-allyl complexes, there has always been a limitation in this methodology in that complete regio- and stereo-selectivity can rarely be guaranteed. Not so under rhodium catalysis. It has now been shown that allylic substitution can proceed with complete retention of regio- and stereo-selectivity of the leaving group by displacement in the presence of Wilkinson's catalyst modified with trimethyl phosphite[60]. The principle of tuning rhodium complexes by the simple device of adding a halide ion or protic acid has also been demonstrated by the enhancement of asymmetric Rh(I)-catalyzed nucleophilic ring opening of bridged 2,5-dihydrofurans[61]. Rhodium-catalyzed hydroformylation has benefited by fine-tuning of phosphine or phosphite ligands, the most recent new-comers being *hybrid* phosphite-O-acyl phosphites for securing linear aldehydes from *internal* olefins by *isomerizing* hydroformylation[62]. Adapting chiral rhodium complexes for asymmetric hetero-Diels-Alder reaction has likewise proved profitable, modular chiral Rh(I)-carboxamidates effecting high enantioselectivity at the lowest catalyst/substrate ratio as yet recorded (0.0001!!) for this reaction[63].

Under ruthenium catalysis, there are four principle lines of research: optimization of chiral Ru-complexes for asymmetric homogeneous hydrogenation and transfer-hydrogenation; diversification of Ru-carbene ligands for syntheses via olefin or alkyne metathesis; the development of new atom-economical procedures, such as cycloisomerization [notably of dienes, diynes and enynes] and inter- and intra-molecular cycloaddition[64]; and, lastly, ventures into proximity coordination effects, as realised to excellent effect by Murai et al. in the field of Ru- or Rh-catalyzed C-H bond activation (cf. **48**, 681s**60**). Illustrating each in turn, note the availability of recyclable ruthenium complexes based on diguanidinium- and PEG-modified versions of Noyori's BINAP ligand which, in the asymmetric hydrogenation of ketones, induce the highest reported levels of enantioselectivity in as little as 0.01 mol%[65]. Among new metathetic procedures, there have been particular advances in domino processes for delivering complex functionality[66]; however, in this context, attention might also be drawn to less fashionable, but sometimes more adaptable and simpler, molybdenum carbene complexes, notably for improving asymmetric olefin metathesis[67]. Note also, as an aside, the first alkane σ-bond cross-metathesis using a silica-supported tantalum(V) alkyl alkylidene complex[68]. Atom economy, of course, brings to mind the contributions of B.M. Trost et al., such as recent applications of Ru-catalyzed intramolecular [5+2]-cycloaddition[69] and an interesting regio- and stereo-specific addition of protected allylamines to functionalized alkynes via ruthenacyclopentenes[70]; the first Ru-catalyzed pyridine ring closure by [2+2+2]-cycloaddition has also been reported[71]. With regard to the fourth aspect of proximity effects, suffice to illustrate the principle with a recent Ru-catalyzed decarbalkoxylation of benzoic acid 2-pyridylmethyl esters, where the nitrogen atom is critical in facilitating the initiating oxidative insertion of ruthenium into the C-O ester bond prior to decarbonylation[72].

Among new cobalt-catalyzed procedures, note especially the first metal-catalyzed intramolecular [2+2]-cycloaddition as an alternative to the traditional photochemical route[73], as well as a novel Co-catalyzed reductive route to enolates via hydrometalation[74]. A new Ir-catalyzed enyne cycloisomerization is additionally available[75], while a novel, basic binuclear Ir(II) hydride complex has been designed for the reversible deprotonation of C-H acidic compds. under neutral conditions[76]. Note also that carbothiolation of alkynes is more efficient under Pt-catalysis than Pd-catalysis[77].

Innovations in the arena of radical synthesis are perhaps less prolific now than in the nineties. However, new radical initiators are always welcome, especially as alternatives to toxic tin reagents. Triethylaluminum comes to mind for Giese-type addition to 1,1-nitroethylenes[78], while radical synthesis of ketones from alkyl iodides is now possible via methyl α-alkoximinosulfones, reaction being sustained by liberated methyl radical which effectively traps iodine atom[79]. Isomerization of allyl silyl ethers to enoxysilanes can also be achieved by radical means (through the agency of a thiol as polarity reversal catalyst), thereby providing an alternative to the traditional metal-catalyzed routes[80]. Perhaps more notable is a new asymmetric synthesis of N-protected α-amino esters via radical 1,4-addition, the key step involving a face-selective hydrogen atom transfer in the complexation sphere of a chiral magnesium bis(2-

oxazoline)[81]. New procedures based on a radical-ionic pathway have already been illustrated[33], but the reverse mechanism is less familiar. Note, however, the novel [anionic] Michael addition in tandem with a radical ring closure, made possible by single electron transfer, wherein the initially formed enolate is oxidized to the ketone radical with ferrocenium ion prior to radical cyclization onto an appended unsaturation[82]. Tandem routes based on radical reduction with SmI_2 have also diversified, as reflected in a new tetrasubst. furan synthesis via intramolecular oxypalladation-Heck-type arylation[83], and in the novel intramolecular Barbier-type reaction in tandem with Grob fragmentation to produce medium-ring carbocyclic enones[84].

Pursuing the 'tandem' theme for generating polyfunctionalization, note especially the one-step asymmetric synthesis of δ,ε-ethylene-β,β'-dihydroxyketones via intramolecular silylformylation-allylsilylation[85], as well as polycyclic ring formation via [3+2]-cycloaddition in tandem with carbene insertion into the C-H bond of enolethers[86]. Sequencing aldol condensation with etherification and the Meerwein-Ponndorf-Verley reduction has been reported for preparing 1,3-diol monoethers[87], while a three-component synthesis of N-protected α-cyclopropylamines from alkynes involves sequencing hydrozirconation with 1,2-addition to a protected aldimine, and terminating with cyclopropanation using *methylene chloride*[88]. Further multicomponent routes based on isonitriles have been elaborated, largely inspired by the Ugi condensation and with obvious potential in combinatiorial chemistry[89]. The Petasis reaction is finding more and more outlets[90], as also are transition metal-catalyzed 3-component couplings, such as a newly devised route to β-amino-α,β-ethylene-sulfones and -phosphonates (under Rh-catalysis)[91] and an iridium-catalyzed 3-component route to N-allylaldimines[92].

Polymer-based syntheses are still very much in vogue as research into new resins and resin linkers intensifies, together with developments in on-resin functionalization and special cleavage devices. The concept of 'resin-capture-release' has been reviewed[93], and extended to a new macrocyclization by initial immobilization of a functionalized boronic acid with a Dowex resin, followed by intramolecular Suzuki-Miyaura coupling with simultaneous release of the resin[94]. Notable among recently reported, readily retrievable polymer-based reagents are chiral, polymer-based Cu(II)-bis(Δ^2-oxazoline) complexes (for heterogeneous aldol-type condensations[95]) and polymer-based, metal-free oxoammonium resins (for TEMPO-type oxidation[94]). Immobilization of expensive transition metal oxidants by sol-gel entrapment is an alternative 'trick' to facilitate recycling and minimize leaching of the metal[97]. Heterogenization of Lewis acids by anchoring on a quaternary ammonium-functionalized mesoporous silica support enhances stability and selectivities, as well as aiding work-up[98].

A few random highlights include: the concept of generating chirally modified and activated Lewis acids *in situ* by pretreatment of an achiral Lewis acid with a chiral Lewis base[99]; novel syntheses with eco-friendly H_2O_2, such as a clean, heterogeneous Baeyer-Villiger oxidation catalyzed by a tin-doped zeolite[100], and biomimetic, non-heme, iron-catalyzed epoxidation and dihydroxylation of olefins[101]; a simple route to metal carbene complexes from S-ylids[102]; a novel oxidative glycosylation with glycals[103]; the design of a microreactor for the rapid, high-yielding solution-phase synthesis of peptides[104]; kinetic resolution of 2-aminoalcohols via aziridinium salts by asymmetric nucleophilic displacement of the hydroxyl group by chlorine (with N-chlorosuccinimide/(S)-BINAP)[105]; and, finally, the emergence of a new carbon material (carbon 'nanofilaments') as a stable, heterogeneous support to rival graphite[106].

We have seen here, and in previous *Trends* sections, that as much attention is being paid to new isolation and separation techniques as to the design and application of new synthetic methods. Another remarkable advance has been made with the concept of 'precipitons'[107]. This is a simple device by which a substrate is tagged with a particular residue which, in one geoisomeric form, ensures solubility, but at the same time renders the product insoluble (and readily retrievable) if converted to the geoisomeric form prior to work-up. The stilbene residue is ideal (the *cis*-tagged substrate being soluble, while the *trans*-form renders the product insoluble)[108]. The impact on multi-step synthesis may not prove decisive, but the influence on large-scale synthesis, particularly at the industrial level, may be enormous.

[1] D.C. Tulley, J.M.J. Fréchet, Chem. Commun. *2001*, 1229-39; review of dendritic encapsulation of function, applying nature's site isolation principle from biomimetics to materials science s. S Hecht, J.M.J. Fréchet, Angew. Chem. Int. Ed. Engl. *40*, 74-91 (2001); phosphorus dendrimers as tools for drug delivery s. R. Göller, A.-M. Caminade et al., Tetrahedron Lett. *42*, 3587-90 (2001); centrally dye-labelled dendrimers s. J. Zhang et al, ibid. 3599-601.
[2] *50*, 555s60.
[3] *53*, 59s60.
[4] S. Hecht, J.M.J. Fréchet, J. Am. Chem. Soc. *123*, 6959-60 (2001).
[5] Y.-C. Chen, J.-G. Deng, Chem. Commun. *2001*, 1488-9.
[6] Review s. L. Pu, H.-B. Yu, Chem. Rev. *101*, 757-824 (2001); asymmetric synthesis of sec.

alcohols with trialkylalanes in place of the conventional dialkylzincs s. J.-S. You et al., Chem. Commun. *2001*, 1546-7; by asymmetric induction with helical hydrocarbons s. I. Sato, K. Soai et al., Angew. Chem. Int. Ed. Engl. *40*, 1096-8 (2001).

[7] Update s. *48*, 772s60

[8] W.-P. Deng, X.-L. Hou et al., J. Am. Chem. Soc. *123*, 6508-19 (2001).

[9] O. Tissot, M. Gouygou et al., Angew. Chem. Int. Ed. Engl. *40*, 1076-8 (2001).

[10] C. Bolm et al., Angew. Chem. Int. Ed. Engl. *40*, 1488-90 (2001).

[11] S. Lee, J.F. Hartwig et al., J. Org. Chem. *66*, 3402-15 (2001).

[12] S. Pandiaraju, A.K. Yudin et al., J. Am. Chem. Soc. *123*, 3850-1 (2001).

[13] With (S)-1-(2-pyrrolidinylmethyl)pyrrolidine s. J.M. Betancort, C.F. Barbas, III et al., Tetrahedron Lett. *42*, 4441-4 (2001).

[14] With (S)-proline s. D. Rajagopal, S. Swaminathan et al., Tetrahedron Lett. *42*, 4887-90 (2001).

[15] With chiral 4-imidazolidone salts s. N.A. Paras, D.W.C. MacMillan, J. Am. Chem. Soc. *123*, 4370-1 (2001).

[16] With recyclable, perfluorinated cinchona derivs. s. F. Fache, O. Piva, Tetrahedron Lett. *42*, 5655-7 (2001).

[17] With modified cinchona alkaloids s. S.-K. Tian, L. Deng, J. Am. Chem. Soc. *123*, 6195-6 (2001).

[18] T. Suzuki, R. Noyori et al., Tetrahedron Lett. *42*, 4669-71 (2001).

[19] D.A. Evans et al., Angew. Chem. Int. Ed. Engl. *40*, 1884-8 (2001).

[20] N. Nishiwaki, K.A. Jørgensen et al., Angew. Chem. Int. Ed. Engl. *40*, 2992-5 (2001).

[21] K.R. Knudsen, K.A. Jørgensen et al., J. Am. Chem. Soc. *123*, 5843-4 (2001).

[22] T. Gastner, S. Kobayashi et al., Angew. Chem. Int. Ed. Engl. *40*, 1896-8 (2001).

[23] C.A. Luchaco-Cullis, A.H. Hoveyda et al., Angew. Chem. Int. Ed. Engl. *40*, 1456-60 (2001).

[24] Review of methane hydroxylation by soluble methane monooxygenase s. M. Merkx, S.J. Lippard et al., Angew. Chem. Int. Ed. Engl. *40*, 2782-807 (2001).

[25] R.L. Rawls, Chem. Eng. News *79*, No.24, 23-5 (2001).

[26] J.-C. Choi, T. Sakakura et al., J. Org. Chem. *66*, 5262-3 (2001).

[27] S. Shimada, T.B. Marder et al., Angew. Chem. Int. Ed. Engl. *40*, 2168-71 (2001).

[28] C.G. Espino, J. Du Bois et al., J. Am. Chem. Soc. *123*, 6935-6 (2001); reviews of syntheses with hypervalent iodine compds. s. *47*, 262s60.

[29] S. Isozaki, Y. Ishii et al., Chem. Commun. *2001*, 1352-3.

[30] Recent examples of catalytic aldol-Tishchenko reactions s. I. Simpura, V. Nevalainen, Tetrahedron Lett. *42*, 3905-7 (2001); C. Schneider, M. Hansch, Chem. Commun. *2001*, 1218-9.

[31] T.-P. Loh et al., Angew. Chem. Int. Ed. Engl. *40*, 2921-2 (2001).

[32] A.S. Demir, M. Muller et al., J. Chem. Soc. Perkin Trans. 1 *2001*, 633-5.

[33] V.S. Kumar, P.E. Floreancig, J. Am. Chem. Soc. *123*, 3842-3 (2001).

[34] G. Grogan, N.J. Turner et al., Angew. Chem. Int. Ed. Engl. *40*, 111-4 (2001).

[35] Review of cross-linked enzyme and protein crystals s. A.L. Margolin, M.A. Navia, Angew. Chem. Int. Ed. Engl. *40*, 2205-22 (2001).

[36] M. Kreiner, M.C. Parker et al., Chem. Commun. *2001*, 1096-7.

[37] H.-J. Gais et al., J. Org. Chem. *66*, 3384-96 (2001).

[38] A. Musidlowska, U.T. Bornscheuer et al., Angew. Chem. Int. Ed. Engl. *40*, 2851-3 (2001).

[39] M.K. Johns, A.J. Smallridge et al., Tetrahedron Lett. *42*, 4261-2 (2001).
[40] Review of syntheses in supercritical fluids s. *57*, 106s60.
[41] C. Wheeler, C.A. Eckert et al., Chem. Commun. *2001*, 887-8.
[42] J. DeSimone, P. Tundo et al., J. Org. Chem. *66*, 4017-9 (2001).
[43] B. Hungerhoff, F. Theil et al., Angew. Chem. Int. Ed. Engl. *40*, 2492-4 (2001).
[45] D. Crich, S. Neelamkavil, J. Am. Chem. Soc. *123*, 7449-50 (2001).
[46] D.P. Curran et al., Science *291*, 1766 (2001).
[47] S. Usugi, K. Oshima et al., Tetrahedron Lett. *42*, 4535-8 (2001).
[48] M. Lautens et al., J. Am. Chem. Soc. *123*, 5358-9 (2001).
[49] S. Venkatraman, C.-J. Li et al., Tetrahedron Lett. *42*, 4459-62 (2001); Rh(I)-catalyzed 1,4-addition of arylboronic acids in water s. R. Itooka, N. Miyaura et al., Chem. Lett. *2001*, 722-3.
[50] Y. Mori, S. Kobayashi et al, Angew. Chem. Int. Ed. Engl. *40*, 2815-8.
[51] G.Y. Li, Angew. Chem. Int. Ed. Engl. *40*, 1513-5 (2001).
[52] R.B. Bedford, C.S.J. Cazin, Chem. Commun. *2001*, 1540-1.
[53] W. Mägerlein, M. Beller et al., Angew. Chem. Int. Ed. Engl. *40*, 2856-9 (2001).
[54] A.F. Littke, G.C. Fu, J. Am. Chem. Soc. *123*, 6989-700 (2001).
[55] M.T. Burros, C.D. Maycock et al., Chem. Commun. *2001*, 1662-3.
[56] S.E. Denmark, R.F. Sweis, J. Am. Chem. Soc. *123*, 6439-40 (2001).
[57] A. Klapars, S.L. Buchwald et al., J. Am. Chem. Soc. *123*, 7727-9 (2001).
[58] R. Gujadhur, D. Venkataraman et al., Tetrahedron Lett. *42*, 4791-3 (2001).
[59] P.Y.S. Lam, P.K. Jadhav et al., Tetrahedron Lett. *42*, 3415-8 (2001).
[60] Application to tandem C-allylation-Pauson Khand reaction s. P.A. Andrews, J.E. Robinson, J. Am. Chem. Soc. *123*, 4609-10 (2001).
[61] M. Lautens, K. Fagnou, J. Am. Chem. Soc. *123*, 7170-1 (2001).
[62] D. Selent, A. Börner et al., Angew. Chem. Int. Ed. Engl. *40*, 1696-8 (2001); Pd-catalyzed isomerizing carbonylation of internal olefins with chelating bis(phosphaadamantyl)diphosphines as ligand s. R.I. Bugh, E. Drent et al., Chem. Commun. *2001*, 1476-7.
[63] M.P. Doyle et al., J. Am. Chem. Soc. *123*, 5366-7 (2001).
[64] Review of ruthenium-catalyzed C-C forming reactions (non-metathetic) s. B.M. Trost et al., Chem. Rev. *101*, 2067-96 (2001).
[65] P. Guerreiro, J.-P. Genêt et al., Tetrahedron Lett. *42*, 3423-6 (2001); for an appreciation of Noyori's contributions to the field of asymmetric catalysis s. A.M. Rouhi, Chem. Eng. News *79*, No.30, 33-4 (2001).
[66] Recent example of a ring-closing metathesis-ring opening metathesis-cross metathesis s. A. Ruckert, S. Blechert et al., Tetrahedron Lett. *42*, 5245-7 (2001).
[67] With an axially chiral Mo-carbene complex s. S.L. Aeilts, A.H. Hoveyda, R.R. Schrock et al., Angew. Chem. Int. Ed. Engl. *40*, 1452-6 (2001).
[68] C. Coperet, J. Thivolle-Cazat et al., Angew. Chem. Int. Ed. Engl. *40*, 2331-4 (2001).
[69] B.M. Trost, H.C. Shen, Angew. Chem. Int. Ed. Engl. *40*, 2313-6 (2001).
[70] B.M. Trost, J.-P. Surivet, Angew. Chem. Int. Ed. Engl. *40*, 1468-71 (2001).
[71] Y. Yamamoto, K. Itoh et al., J. Am. Chem. Soc. *123*, 6189-90 (2001).
[72] N. Chatani, S. Murai et al., J. Am. Chem. Soc. *123*, 4849-50 (2001); proximity-induced dehydrogenative *o*-silylation of 2-aryl-2-oxazolines s. F. Kakiuchi, S. Murai et al., Chem.

Lett. *2001*, 422-3.
[73] T.-G. Baik, M.J. Krische et al., J. Am. Chem. Soc. *123*, 6716-7 (2001).
[74] Application to intramolecular reductive aldol condensation and Michael addition s. T.-G. Baik, M.J. Krische et al., J. Am. Chem. Soc. *123*, 5112-3 (2001).
[75] N. Chatani, S. Murai et al., J. Org. Chem. *66*, 4433-6 (2001).
[76] Application to Michael addition and cleavage of aromatic glycols s. Z. Hou, Y. Wakatsuki et al., J. Am. Chem. Soc. *123*, 5812-3 (2001).
[77] K. Sugoh, H. Kurosawa et al., J. Am. Chem. Soc. *123*, 5108-9 (2001).
[78] J.-Y Liu, C.-F. Yao et al., Tetrahedron Lett. *42*, 3613-5 (2001); reviews on radical addition to alkenes s. *39*, 646s60.
[79] S. Kim et al., Angew. Chem. Int. Ed. Engl. *40*, 2524-6 (2001).
[80] A.J. Fielding, B.P. Roberts, Tetrahedron Lett. *42*, 4061-4 (2001).
[81] M.P. Sibi et al., Angew. Chem. Int. Ed. Engl. *40*, 1293-6 (2001).
[82] U. Jahn, Chem. Commun. *2001*, 1600-1.
[83] J.M. Aurrecoechea, E. Perez, Tetrahedron Lett. *42*, 3839-41 (2001).
[84] G.A. Molander et al., J. Org. Chem. *66*, 4511-6 (2001).
[85] S.J. O'Malley, J.L. Leighton, Angew. Chem. Int. Ed. Engl. *40*, 2915-7 (2001).
[86] N. Iwasawa et al., J. Am. Chem. Soc. *123*, 5814-5 (2001).
[87] N. Aremo, T. Hase, Tetrahedron Lett. *42*, 3637-9 (2001).
[88] P. Wipf et al., J. Am. Chem. Soc. *123*, 5122-3 (2001).
[89] Review of multicomponent reactions with isocyanides s. *43*, 700s60; benzimidazoles by Ugi condensation-ring closure, offering 4 points of diversity, s. P. Tempest, C. Hulme et al., Tetrahedron Lett. *42*, 4959-62 (2001); with aminoisocyanides s. V. Nair et al., Chem. Lett. *2001*, 738-9.
[90] Recent example of ring closure via Petasis reaction s. F. Berrée, B. Carboni et al., Tetrahedron Lett. *42*, 3591-4 (2001).
[91] Y.-S. Lin, H. Alper et al., J. Am. Chem. Soc. *123*, 7719-20 (2001).
[92] S. Sakaguchi, Y. Ishii et al., Angew. Chem. Int. Ed. Engl. *40*, 2534-6 (2001).
[93] *50*, 555s60.
[94] V. Lobregat, M. Vaultier et al., Chem. Commun. *2001*, 817-8.
[95] S. Orlandi, P. Salvadori et al., Angew. Chem. Int. Ed. Engl. *40*, 2519-21 (2001).
[96] S. Weik, J. Rademann et al., Angew. Chem. Int. Ed. Engl. *40*, 1436-9 (2001); review of oxidations with tetrapropylammonium per-ruthenate s. *43*, 219s60.
[97] Sol-gel entrapped chromium(VI) s. M. Gruttadauria et al., Tetrahedron Lett. *42*, 5199-201 (2001); sol-gel entrapped per-ruthenate s. M. Pagliaro, R. Ciriminna, ibid. 4511-4.
[98] T.M. Jyothi, M.V. Landau et al., Chem. Commun. *2001*, 992-3; s.a. Angew. Chem. Int. Ed. Engl. *40*, 2881-4 (2001).
[99] S.E. Denmark, T. Wynn, J. Am. Chem. Soc. *123*, 6199-200 (2001).
[100] A. Corma et al., Nature *412*, 423 (2001).
[101] Epoxidation s. M.C. White, E.N. Jacobsen et al., J. Am. Chem. Soc. *123*, 7194-5 (2001); dihydroxylation of alkenes s. M. Costas, L. Que, Jr. et al., ibid. 6722-3.
[102] M. Gandelman, D. Milstein et al., J. Am. Chem. Soc. *123*, 5372-3 (2001); review of syntheses with carbene complexes s. *47*, 955s60.
[103] L. Shi, D.Y. Gin et al., J. Am. Chem. Soc. *123*, 6939-40 (2001).

[104] P. Watts, S.J. Haswell et al., Chem. Commun. *2001*, 990-1; reviews of peptide synthesis s. *19*, 33s*60*.
[105] G. Sekar, H. Nishiyama, Chem. Commun. *2001*, 1314-5.
[106] G. Mestl, R. Schlögl et al., Angew. Chem. Int. Ed. Engl. *40*, 2066-8 (2001).
[107] Overview s. S. Borman, Chem. Eng. News *79*, No.23, 49-50 (2001).
[108] Application to the Baylis-Hillman reaction s. T. Bosanac, C.S. Wilcox, Chem. Commun. *2001*, 1618-9; to the synthesis of β-keto esters s. Tetrahedron Lett. *42*, 4309-12 (2001); to isoxazoline synthesis s. Angew. Chem. Int. Ed. Engl. *40*, 1875-9 (2001).

Systematic Survey

Reaction symbol	Page				
HO∩OC	1	ORem⇓HRem	32	NS↓↑Hal	70
HO↓↑S	2	ORem↓↑H	33	NS↓↑C	70
HO↓↑Rem	2	ORem↓↑O	33	NC⇓ON	71
HO↓↑C	3	ORem↓↑N	33	NC⇓OC	71
HN↓↑O	8	ORem↓↑Hal	34	NC⇓NC	73
HN↓↑N	9	ORem↓↑Rem	34	NC⇓CC	74
HN↓↑S	9	ORem↓↑C	34	NC∩HN	76
HN↓↑Rem	10	OC⇓HC	35	NC∩OC	77
HN↓↑C	10	OC⇓OO	35	NC∩CC	77
HS↓↑S	11	OC⇓OC	35	NC↓↑H	78
HS↓↑C	11	OC⇓NC	37	NC↓↑O	81
HC⇓OC	12	OC⇓CC	37	NC↓↑N	88
HC⇓NC	19	OC∩HO	45	NC↓↑Hal	90
HC⇓CC	19	OC∩HC	46	NC↓↑S	95
HC↓↑O	24	OC∩ON	47	NC↓↑Rem	96
HC↓↑N	25	OC∩NC	47	NC↓↑C	96
HC↓↑Hal	25	OC∩CC	47	NC↑↑H	98
HC↓↑S	26	OC↓↑H	48	NC↑↑O	99
HC↓↑Rem	27	OC↓↑O	49	NC↑↑N	100
HC↓↑C	27	OC↓↑N	53	NC↑↑Hal	100
HC↑↑O	27	OC↓↑Hal	55	HalC⇓OC	101
HC↑↑C	28	OC↓↑S	56	HalC⇓CC	102
ON⇓HN	28	OC↓↑Rem	58	HalC∩NHal	103
ON⇓N	28	OC↓↑C	59	HalC↓↑H	103
ON↓↑H	29	OC↑↑H	62	HalC↓↑O	106
ON↑↑O	29	OC↑↑O	67	HalC↓↑N	107
OS⇓S	30	OC↑↑N	67	HalC↓↑Hal	108
OS↓↑N	30	OC↑↑Hal	67	HalC↓↑S	108
OS↓↑Hal	31	OC↑↑S	68	HalC↓↑Rem	109
ORem⇓HO	32	OC↑↑C	68	HalC↓↑C	110
		NN↓↑H	69	SS↓↑H	110
		NHal↓↑H	70	SS↓↑S	110

Reaction symbol	Page				
		SC⇑C	119	CC↓↑O	177
		RemC⇓NC	120	CC↓↑N	191
SS↓↑C	110	RemC⇓HalC	121	CC↓↑Hal	197
SS⇑C	111	RemC⇓CC	121	CC↓↑S	218
SRem⇓HRem	111	RemC∩ORem	123	CC↓↑Rem	221
SRem↓↑H	111	RemC↓↑H	123	CC↓↑C	242
SRem↓↑O	112	RemC↓↑O	124	CC⇑H	247
SRem↓↑Hal	112	RemC↓↑Hal	124	CC⇑O	248
SRem↓↑Rem	112	RemC↓↑Rem	125	CC⇑N	250
SC⇓CC	112	RemC↓↑C	126	CC⇑Hal	250
SC∩HC	113	CC⇓OC	126	CC⇑S	254
SC∩ON	113	CC⇓NC	138	CC⇑Rem	255
SC∩CC	113	CC⇓RemC	143	CC⇑C	257
SC↓↑O	114	CC⇓CC	143	ElN⇑O	258
SC↓↑N	116	CC∩HC	169	ElN⇑N	259
SC↓↑Hal	116	CC∩OC	172	ElS⇑O	259
SC↓↑S	117	CC∩NC	174	ElRem⇑S	259
SC↓↑Rem	118	CC∩HalC	174	ElRem⇑Rem	259
SC↓↑C	118	CC∩RemC	174	Res	260
SC⇑O	119	CC∩CC	175	Oth	261
SC⇑S	119	CC↓↑H	175		

Abbreviations and Symbols

abs.	absolute
alc.	alcoholic
aq.	aqueous
ar.	aromatic
atm.	atmosphere(s)
compd(s).	compound(s)
deriv(s).	derivative(s)
e.e.	enantiomeric excess
eq(s).	equivalent(s)
E.	Example
F.e.s.	Further example(s) see
M	molar
prepn.	preparation
prim.	primary
s60	supplementary reference in Volume 60
sec.	secondary
startg. m.f.	starting material for (the preparation of ...)
subst.	substituted
sym.	symmetrical
tert.	tertiary
v.i.	via intermediates
w.a.r.	without additional reagents
Y *	Yield
⚡	Electrolysis
⚏	Irradiation
○	Ring closure
◔	Ring contraction
◐	Ring expansion
◡	Ring opening
Ⓗ	Ring hydrogenation
←	'see title or reagent on the left half of the page'

* Yields in parentheses refer to the immediately preceeding step of a multi-step reaction

Derwent
Journal of Synthetic Methods

Every month this service covers everything new and important in synthetic organic chemistry...

Reactions selected from worldwide journal and patent literature

- Each reaction rigorously checked to ensure it only appears once
- Each reaction clearly illustrated with its own scheme
- Comprehensive indices, with an annual cumulation
- Retrospective retrieval through online access to over 100,000 reactions, including Theilheimer's 'Synthetic Methods'
- Structure-searchable either online on STN or in-house as RX-JSM.

Available in a variety of formats to suit your needs.

For further details contact Sales Support,
Derwent Information Ltd, 14 Great Queen Street, London WC2B 5DF UK
Telephone +44 (0)20 7424 2347 Fax +44 (0)20 7344 2972 Email salesup@derwent.co.uk

Formation of H-O Bond

Rearrangement

Oxygen/Carbon Type

HO ∩ OC

Chiral lithium amides R_2NLi
2-Ethylenealcohols from epoxides by asym. ring opening   ▽₀↗ → C(OH)C=C
s. *4*, 3s*53, 54, 58*; chiral 4-aminocyclopent-2-en-1-ols s. S. Barrett, P. O'Brien et al., Tetrahedron *56*, 9633-40 (2000); desymmetrization of a *meso*-cyclohexene oxide with a norephedrine-derived base s. B. Colman, P. O'Brien et al., Tetrahedron:Asym. *10*, 4175-82 (1999).

Lithium diisopropylamide $i\text{-}Pr_2NLi$
γ-Acylation of β,γ-ethylenephosphonic acid esters   ↻
via 5-α-phosphonyl-Δ²-isoxazolines and (E)-α,β-ethylene-γ-phosphonylketoximes

1.

A stirred soln. of startg. isoxazoline [prepared according to *21*, 436] in dry THF at -78° under N₂ treated dropwise by syringe with 1 eq. 1 M LDA in heptane/THF/ethylbenzene, stirred for 1 h, allowed to warm to room temp., after 0.5 h treated with 2 eqs. glacial acetic acid, then brine, worked up, the resulting crude oil taken up in DMF, treated with TiCl₃ (8.6% soln. in 28% HCl), stirred for ca. 7 h at room temp., and water added → diethyl (E)-4-oxo-2-hexenylphosphonate. Y 78% (64% overall from the allylic phosphonate). For β-unsubst. derivs. the products had the (E)-configuration exclusively, whereas for β-subst. derivs. (Z)-isomers predominated in a 1.7:1 Z/E ratio. F.e. and conversion to **(E)-2,4-dienones** s. S.Y. Lee, D.Y. Oh et al., J. Org. Chem. *65*, 256-7 (2000); poly[het]aryl ketoximes with methanolic NaOMe or LAH s. A. Corsaro, V. Pistarà et al., Synthesis *2000*, 1469-73.

Via intermediates   *v.i.*
Cyclic α,β-ethylene-α-hydroxyketones from α,β-epoxyketones   ▽₀↗ → C(OH)=C
via quaternization of cyclic α,β-ethylene-α-(2-pyridyloxy)ketones

2.

Diosphenols. 2 Drops of 30% KH oil suspension added under N₂ to a stirred soln. of 2 eqs. 6-methyl-2-pyridone in dry HMPA (1.3 parts)/dry dibutyl ether (10 parts), a soln. of isophorone oxide in dibutyl ether (2 parts) added rapidly, and heated at reflux for 6 h → intermediate 2-pyridyloxyketone (Y 70%), added at 0° under N₂ to a stirred soln. of 1.1 eqs. methyl triflate in dry methylene chloride, and kept at the same temp. for 30 min then at room temp. for 2.5 h → intermediate N-methylpyridinium salt (Y 91%), added to 1:1 1 M aq. Na₂CO₃/acetone, and stirred overnight → product (Y 86%). cf. *30*, 524. Other N-hindered *o*-hydroxy-N-heteroarenes, such as 2-hydroxyquinoline or 6-phenyl-2-pyridone, may be used as the hydroxide equivalent, but offer little advantage (save crystallinity of the diosphenol ether) over readily available 6-methyl-2-pyridone. F.e. incl. cyclopentane analogs s. A.A. Ponaras, M.Y. Meah, Tetrahedron Lett. *41*, 9031-5 (2000).

Exchange

Sulfur ↑   HO ↕ S

Cerium(III) chloride/sodium iodide $CeCl_3/NaI$
O-Detosylation under neutral conditions   OTs → OH

with SmI_2 cf. *21, 9s57*; general procedure with $CeCl_3/NaI$ in acetonitrile s. G.S. Reddy, G.H. Mohan, D.S. Iyengar, Synth. Commun. *30*, 3829-32 (2000).

Titanium trichloride/lithium $TiCl_3/Li$
Low-valent titanium-mediated O- and N-detosylation

Phenols. A mixture of 4 eqs. $TiCl_3$ and 14 eqs. Li in dry THF refluxed at 80° for 3 h under argon, cooled to 25°, a soln. of *p*-(*p*-tosyloxy)toluene in THF added, and stirring continued at 25° for 18 h → *p*-cresol. Y 75% (only 32% using $TiCl_3$/Mg at reflux for 18 h). The method is simple and general, alkenes, ar. chlorides, THP- and TBDPS-ethers being compatible with the reaction conditions. F.e. and N-detosylation s. S.K. Nayak, Synthesis *2000*, 1575-8.

Remaining Elements ↑   HO ↕ Rem

Potassium tert-*butoxide*   KOBu-t
Protection of carboxyl groups as tris(2,6-diphenylbenzyl)silyl esters   COOSi≤ → COOH

Bowl-shaped tris(2,6-diphenylbenzyl)silyl (TDS) esters of carboxylic acids are unusually stable towards a variety of reactive nucleophiles and bases (e.g. BuLi, MeLi, LDA, MeMgBr), and, in particular, serve to shield the α-protons [in a molecular 'pocket'] so that attack by nucleophiles and α-deprotonation are inhibited. However, **removal of the protective group** can be easily effected with py·HF or as follows. **E:** A soln. of 4-phenylbutyric acid TDS ester in DMSO treated with 10 eqs. KOBu-*t* at 25° for 1 h → 4-phenylbutyric acid. Y 86%. The corresponding prim. alcohol was obtained in >99% yield by reductive deprotection with *i*-Bu_2AlH. Protection is easily carried out by treatment of the carboxylic acid with tris(2,6-diphenylbenzyl)silyl bromide in the presence of AgOTf. F.e.s. A. Iwasaki, K. Maruoka et al., J. Am. Chem. Soc. *122*, 10238-9 (2000).

Boron chloride BCl_3
Preferential O-de-*tert*-butyldimethylsilylation $OSiMe_2Bu$-t → OH
at primary sites with Bu_4NF/BF_3 cf. *51, 3*; at the 6-position of per-O-silylated carbohydrates with BCl_3 in THF s. Y.-Y. Yang, C.-H. Lin et al., Synlett *2000*, 1634-6.

2,8,9-Trialkyl-2,5,8,9-tetraaza-1-phosphabicyclo[3.3.3]undecanes
Superbase-catalyzed O-de-*tert*-butyldimethylsilylation ← OSiMe$_2$Bu-*t* → OH

n-C$_8$H$_{17}$OSiMe$_2$Bu-t ──────→ n-C$_8$H$_{17}$OH

Startg. silyl ether added via syringe to a soln. of 0.2 eq. 2,8,9-trimethyl-2,5,8,9-tetraaza-1-phosphabicyclo[3.3.3]undecane in *DMSO* under N$_2$, stirred at 80° for 34 h, cooled to room temp., and poured into water → product. Y 94% (1:1 mixture of geoisomers). Reaction is generally applicable to the cleavage of prim., sec., tert. and phenolic *tert*-butyldimethylsilyl ethers, desilylation of *tert*-butyldiphenylsilyl ethers being less efficient (22-45%). It is possible that Si-O cleavage is effected by catalytically generated solvent anion, or by prior activation of the Si-O bond via P-Si interaction, followed by nucleophilic attack of the solvent anion. F.e. and in acetonitrile, also with the 2,8,9-triisopropyl deriv. for cleavage of TBDMS ethers of 1-octanol, 2-phenoxyethanol and racemic α-phenylethanol, s. Z. Yu, J.G. Verkade, J. Org. Chem. **65**, 2065-8 (2000).

Tetra-n-butylammonium fluoride/acetic acid Bu$_4$NF/AcOH
O-De-*tert*-butyldiphenylsilylation OSiPh$_2$Bu-*t* → OH
preferential cleavage cf. *40*, 1; with retention of *tert*-butyldimethylsilyl ethers in the presence of AcOH and water s. S. Higashibayashi, K. Hashimoto et al., Synlett *2000*, 1306-8.

Carbon ↑   HO ↓↑ C

Irradiation ⇝
Photochemical cleavage of phenacyl esters COOCH$_2$COAr → COOH
s. *29*, 2s*54*; of 2,5-dimethylphenacyl esters by *intramolecular* H-abstraction (so that no photosensitizer or external H-atom donor is required) s. P. Klán, D. Heger et al., Org. Lett. **2**, 1569-71 (2000).

Microwaves s. under *InI$_3$ and I$_2$* ←

Sodium hydroxide NaOH
Dicarboxylic acid monoesters from diesters COOR → COOH
Partial hydrolysis

An aq. soln. of NaOH (0.25 *N*) added in small portions to a mixture of startg. diester in *THF/water* (1:10) at 0° until TLC indicated consumption of the substrate, stirred at the same temp. for 30-60 min, then acidified with 1 *N* aq. HCl → product. Y 99%. The procedure is clean and straightforward, and work-up easy. O,O-Isopropylidene derivs. and epoxides remain unchanged. F.e. and partial hydrolysis of dimethyl esters s. S. Niwayama, J. Org. Chem. **65**, 5834-6 (2000).

n-Butyllithium/pyrrolidine BuLi/R$_2$NH
Protection of glycols as 2-(carbalkoxymethyl)-1,3-dioxolanes C
Removal of the protective group s. *17*, 234s*60*

Sodium carbonate Na$_2$CO$_3$
Cyclic α,β-ethylene-α-hydroxy- ←
from α,β-ethylene-α-(2-pyridiniooxy)-ketones s. *60*, 2

Lithium bromide LiBr
Cleavage of tetrahydropyran-2-yl ethers s. *56*, 73s*60* OThp → OH

Ammonia NH₃
O-Deacylation OAc → OH
s. *24*, 387; recovery of chiral α-hydroxyphosphonic acid esters with 25% aq. NH₃/alcohols or THF s. D.G. Piotrowska, A.E. Wroblewski et al., Synth. Commun. *30*, 3935-40 (2000).

Ammonia/pyridine NH₃/C₅H₅N
Protection of phosphoryl hydroxyl groups P-OR → P-OH
as 2-cyanoethyl esters cf. *17*, 169s29; as 2-cyano-1-*tert*-butylethyl or 2-cyano-1-(1,1-diethyl-3-butenyl)ethyl esters during oligonucleotide synthesis by the *in situ*-phosphoramidite method (cf. *17*, 169s59, 60), and cleavage with 1:1 concd. aq. NH₃/py s. A. Kitamura, T. Yoshida et al., Chem. Lett. *2000*, 1134-5.

Piperidine R₂NH
Protection of hydroxyl groups as allyl carbonates OCOOCH₂CH=CH₂ → OH
and removal of the protective group
with Ni(CO)₄ cf. *29*, 28; protection of phenolic α-amino acids for loading onto the Wang resin, also selective removal of the protective group with 20% piperidine in DMF (with retention of N-Aloc groups), s. A.D. Morley, Tetrahedron Lett. *41*, 7401-4 (2000).

Zinc/acetic acid Zn/AcOH
Ethylene derivs. from 1,2-alkoxybromides ←
with Zn cf. *1*, 775; with liberation of chiral 3-hydroxy-2-azetidinones s. S.N. Joshi, B.M. Bhawal et al., Tetrahedron:Asym. *11*, 1477-85 (2000).

Magnesium methoxide Mg(OMe)₂
O-Deacetylation OAc → OH
s. *42*, 3s53; of carbohydrate derivs. with retention of N-tetrachlorophthaloyl groups s. Z.H. Qin, Z.J. Li et al., Chin. Chem. Lett. *11*, 941-4 (2000).

Diisobutylaluminum hydride i-Bu₂AlH
Regiospecific O-debenzylation OBn → OH

7.

of per-O-benzylcyclodextrins. 133 eqs. Diisobutylaluminum hydride (1.5 *M* in toluene) added to a stirred soln. of perbenzylated β-cyclodextrin in anhydrous toluene at room temp. under argon, and heated at 30° for 2 h → product. Y 83%. Only one of 33 possible regioisomers was formed. F.e. and mono-6-O-debenzylation with 0.5 *M* diisobutylaluminum hydride s. A.J. Pearce, P. Sinay, Angew. Chem. Int. Ed. Engl. *39*, 3610-2 (2000); regiospecific mono-O-debenzylation of perbenzylated mono- and di-saccharides s. Compt. Rend. Acad. Sci. Ser. 2. *1999*, 441-8.

Indium triiodide/silica gel/microwaves ←
Carboxylic acids from esters COOR → COOH
under microwave irradiation on a solid support in the absence of solvent

8. PhCH₂COOMe ⟶ PhCH₂COOH

Silica gel (HF 254) added to a soln. of InI₃ (prepared by heating In and 1.5 eq. I₂ under reflux in THF), the mixture stirred for 15 min (for uniform mixing), THF removed under reduced pressure, methyl phenylacetate added with stirring to the resulting bed of InI₃-impregnated silica, followed by a few drops of water, and the mixture (as a solid mass) irradiated in a domestic microwave oven (BPL, India) at 480 W for 25 min (with continued moistening of the silica gel bed with a few drops of water every five min) → phenylacetic acid. Y 92%. The method is general under nearly neutral conditions, efficient, high-yielding, and tolerant of ketone, hydroxyl, methoxyl and double bond moieties. Reaction was very slow and incomplete (even after 12-15 h) by conventional heating in THF. F.e.s. B.C. Ranu, P. Dutta, A. Sarkar, Synth. Commun. *30*, 4167-71 (2000).

Cerous chloride $CeCl_3$
Cleavage of tetrahydropyran-2-yl ethers   OThp → OH
with CAN cf. *51*, 2; with $CeCl_3 \cdot 7H_2O$ in methanol, selectivity, s. G.S. Reddy, D.S. Iyengar et al., Synth. Commun. *30*, 4107-11 (2000).

Lipase   ←
Carboxylic acids from esters   COOR → COOH
s. *28*, 13; with highly active, recyclable lipase entrapped in hydrophobic sol-gels under aq. heterogeneous conditions s. M.T. Reetz et al., Synthesis *2000*, 781-3.

Parallel kinetic resolution of sec. alcohols and prim. amines by asym. aminolysis   ←

9.

A mixture of 3 mmol 1-phenylethyl acetate, 3 mmol 2-aminoheptane, 350 mg 4 Å molecular sieves, and 500 mg lipase B (from *Candida antarctica*) in dry dioxane stirred (250 r.p.m.) at 30° under N_2 for 17 h, the enzyme filtered off and washed, the organic phase extracted with 2 *N* HCl and water, evaporated, and the crude mixture purified by silica gel chromatography → (S)-1-phenylethyl acetate (Y >75%; e.e. 92%), (R)-1-phenylethanol (Y >75%; e.e. 99%), (S)-2-aminoheptane (Y >85%; e.e. 93%) and (R)-2-acetylaminoheptane (Y >85%; e.e. 97%). A catalytic cycle is proposed wherein enantiodiscrimination of each substrate takes place in different steps, presumably at the same catalytic site. All four products can be isolated easily in high yield and enantiomeric excesses. F.e.s. E. Garcia-Urdiales, V. Gotor et al., Tetrahedron:Asym. *11*, 1459-63 (2000).

Lipase or immobilized lipase   ←
Asym. enzymatic hydrolysis of carboxylic acid esters   COOR → COOH + HOR
update s. *28*, 13s*59*; asym. hydrolysis of cyanohydrin acetates with kinetic resolution in propanol/toluene s. U. Hanefeld et al., Synlett *2000*, 1775-6; kinetic resolution of 3,4-dihydro-2(1*H*)-pyrimidinone-5-carboxylic acid esters s. B. Schnell, C.O. Kappe et al., J. Chem. Soc. Perkin Trans. 1 *2000*, 4382-9; of 1-(pyridyl)ethanol and 1-phenylethanol esters, effect of ester group on enantioselectivity, s. F. Bellezza, A. Cipiciani et al., ibid. 4439-44; of 3-nitrocyclopent(or hex)-2-en-1-yl acetates s. Tetrahedron:Asym. *11*, 2259-62 (2000); of P-chiral phosphonyl- and phosphoryl-acetates s. ibid. *9*, 2641-50 (1998); of N-subst. 4-benzoyloxy-3-carbomethoxy-piperidines s. M. Roberti, D. Simoni et al., ibid. *11*, 4397-405 (2000); of methyl 2-aryloxy-propionates in water or aq. organic media s. A. Cipiciani et al., ibid. *10*, 4599-605 (1999); desymmetrization of *cis*-1,3-cyclohexanedicarboxylic acid esters s. N.W. Boaz, ibid. 813-6; of 1,4-dihydropyridine-3,5-dicarboxylic acid esters s. A. Sobolev, M.C.R. Franssen et al., ibid. *11*, 4559-69 (2000); desymmetrization of 1,3-diol acetates on a *lipase-based HPLC stationary phase* cf. C. Bertucci, P. Salvadori et al., ibid. *10*, 4455-62 (1999).

Phosphotriesterase   ←
Kinetic resolution of phosphoric acid esters via asym. hydrolysis   P-OR → P-OH

10.

2.6 nM Phosphotriesterase mutant (G60A) added with stirring at room temp. to a soln. of 8 mM startg. racemic phosphoric acid ester in acetonitrile (20%)/pH 9 CHES buffer containing 0.1 mM

Co²⁺, and worked up after 40 min by extraction with chloroform → (Rp)-unhydrolyzed enantiomer. Y 99% (at 50% conversion; e.e. 98%). The yields, enantiomeric purity and turnover numbers are superior to those reported for non-enzymatic methods. The face selectivity was reversed with mutant I106G/F132G/H257Y. F.e.s. F.Y. Wu, F.M. Raushel, J. Am. Chem. Soc. *122*, 10206-7 (2000).

Lipase or microorganisms ←
Diols from cyclic carbonic acid esters with kinetic resolution
with lipase cf. *53*, 4; isolation of C₂-symmetric diols with a strain of *Pseudomonas diminuta* s. K. Matsumoto et al., Tetrahedron:Asym. *11*, 1965-73 (2000); chiral long-chain glycols with lipase (PPL) s. Tetrahedron *56*, 9281-8 (2000).

Trifluoroacetic acid/thioanisole/trimethylsilyl bromide ←
Protection of *vic*-hydroxyl groups as cyclic orthoformic acid esters
of carbohydrate *vic*-diols cf. *28*, 268; of 3,4-dihydroxyphenylalanine as the cyclic ethyl orthoformate for solid-phase peptide synthesis by the Fmoc method, and removal of the protective group with CF₃COOH/PhSMe/Me₃SiBr, s. B.-H. Hu, P.B. Messersmith, Tetrahedron Lett. *41*, 5795-8 (2000); protection of catechol groups as 2,2-diphenyl-1,3-benzodioxoles, and removal of the protective group by Pd(OH)₂/H₂ s. C. Cren-Olivé, C. Rolando et al., ibid. 5847-51.

Trifluoroacetic acid/thioanisole/trimethylsilyl triflate ←
Cleavage of polymer-based benzyl ethers $OCH_2Ar \rightarrow OH$
with CF₃COOH cf. *37*, 152s*55*, *56*; liberation of phenols with 10% trifluoroacetic acid in dichloromethane containing excess of PhSMe/Me₃SiOTf, also identification of the phenols from single resin beads by Fourier transform ion cyclotron resonance mass spectrometry of the dansylates prepared *in situ* s. M.H. Todd, C. Abell, Tetrahedron Lett. *41*, 8183-7 (2000).

Allyl bromide $CH_2=CHCH_2Br$
Cleavage of acetals in aq. organic media $C(OR)_2 \rightarrow CO$

11.

A soln. of [4-(3,3-dimethoxypropyl)phenyl]dimethoxymethane in 3:1 water/THF treated with 3 eqs. allyl bromide at room temp., and stirred for 4 h → 4-(3,3-dimethoxypropyl)benzaldehyde. Y 94%. Allyl bromide in aq. medium serves as a convenient, easily handled, controllable *precursor of HBr*. The method is applicable to the cleavage of dimethyl acetals derived from aromatic aldehydes, cinnamaldehyde and aromatic or [isocyclic or acyclic] aliphatic ketones, whereas acetals of aliphatic aldehydes are inert. In contrast, optimization of the conditions using aq. HBr itself was difficult. F.e.s. J.S. Kwon, Y.S. Cho et al., Bull. Korean Chem. Soc. *21*, 457-8 (2000).

N-Iodosuccinimide/trifluoromethanesulfonic acid NIS/CF_3SO_3H
Selective cleavage of *p*-methoxybenzyl ethers $OCH_2Ar \rightarrow OH$
with simultaneous formation of glycosides from thioglycosides and retention of benzyl ethers s. *39*, 189s*60*

Trimethylsilyl bromide s. under CF₃COOH Me_3SiBr

Bismuth(III) nitrate or chloride $Bi(NO_3)_3$ or $BiCl_3$
Selective cleavage of acetals under mild, neutral conditions $C(OR)_2 \rightarrow CO$

12.

0.5 eq. BiCl₃ added to startg. acetal *in methanol*, stirred for 30 min, methanol removed under reduced pressure, and the mixture quenched by the addition of ice → product. Y 85%. The method is simple, efficient and selective, leaving tetrahydropyran-2-yl ethers, Si-ethers, benzyl ethers and styrenes unaffected. Methanol is a better solvent for the cleavage than dichloromethane. The reagent is inexpensive, commercially available and non-toxic. F.e. incl. cleavage of acyclic acetals s. G. Sabitha et al., Chem. Lett. *2000*, 1074-5; cleavage of α,β-unsatd. acetals and ketals with $Bi(NO_3)_3 \cdot 5H_2O$ cf. K.J. Eash, R.S. Mohan et al., J. Org. Chem. *65*, 8399-401 (2000).

Sodium dithionite s. under $NaBrO_3$ $\qquad Na_2S_2O_4$
Trimethylsilyl triflate s. under CF_3COOH $\qquad Me_3SiOTf$
Trifluoromethanesulfonic acid s. under NIS $\qquad CF_3SO_3H$

Iodine/microwaves $\qquad \leftarrow$
Cleavage of tetrahydropyran-2-yl ethers $\qquad OThp \rightarrow OH$
with I_2 s. *56*, 9s*57*; with a little I_2 in methanol under microwave irradiation s. N. Deka, J.C. Sarma, Synth. Commun. *30*, 4435-41 (2000).

Sodium bromate/sodium dithionite $\qquad NaBrO_3/Na_2S_2O_4$
Selective O-debenzylation $\qquad OBn \rightarrow OH$
s. *58*, 10; with retention of O-benzyloxycarbonyl groups s. M. Adinolfi, A. Iadonisi et al., Synlett *2000*, 1277-8.

Tetra-n-butylammonium fluoride $\qquad Bu_4NF$
Protection of carboxyl groups $\qquad COOCH_2CH(R)Si\lessdot \rightarrow COOH$
as 2-(trimethylsilyl)ethyl esters
Removal of the protective group
with Bu_4NF cf. *33*, 8; more rapid and cleaner cleavage of 2-phenyl-2-(trimethylsilyl)ethyl esters, being orthogonal to N-Fmoc, N-Cbz, N-Boc and N-Aloc groups, as well as benzyl and allyl esters, s. M. Wagner, H. Kunz, Synlett *2000*, 400-2.

Palladium $\qquad Pd$
Protection of hydroxyl groups as *p*-(trifluoromethyl)benzyl ethers $\qquad OCH_2Ar \rightarrow OH$
Reductive removal of the protective group

13.

A soln. of startg. protected hydroxyester in THF hydrogenated over palladium black for 24 h under 5 kg/cm² $H_2 \rightarrow$ product. Y 100%. Unlike the O-benzyl group, which is cleaved by DDQ and may even be partially oxidized in air (to the O-benzoyl deriv.), the *p*-(trifluoromethyl)benzyl group is resistant to DDQ and stable in air (making the final purification easier in the application to the synthesis of *Helicobacter pylori* lipid A (and an analog)). However, the O-benzyl group is *more* susceptible to hydrogenolysis (under 2 kg/cm² H_2). F.e.s. Y. Sakai, S. Kusumoto et al., Tetrahedron Lett. *41*, 6843-7 (2000).

Palladium (nanoparticles)
O-Debenzylation
with Pd-C cf. *1*, 13; rapid solution-phase and *on-resin* O-debenzylation of benzyl-protected carbohydrates with Pd nanoparticles (ca. 17 Å mean diameter) s. O. Kanie, C.-H. Wong et al., Angew. Chem. Int. Ed. Engl. *39*, 4545-7 (2000).

Palladium-carbon/ethylenediamine or squaric acid derivs. $\qquad \leftarrow$
O-Debenzylation s. *51*, 32s*60*

Palladium hydroxide $\qquad Pd(OH)_2$
Protection of catechol groups as 2,2-diphenyl-1,3-benzodioxoles $\qquad \subset$
Removal of the protective group s. *28*, 268s*60*

Formation of H-N Bond

Exchange ⇅

Oxygen ↑   HN ⇵ O

Lithium/ethylenediamine/n-propylamine   ←
Prim. amines from oximes s. *5*, 32s*60*   C=NOH → CHNH$_2$

Indium/acetic acid   In/AcOH
Acylamines from oximes   C=NOH → CHNHAc

14. i-Pr-C(=O)-C(=NOH)-CO$_2$Me + Ac$_2$O → i-Pr-C(=O)-C(NHAc)-CO$_2$Me

4 eqs. Indium powder (100 mesh) added to a soln. of startg. oxime, 2.5 eqs. Ac$_2$O and 4 eqs. acetic acid in THF, the mixture heated at reflux for 18 h, satd. aq. Na-bicarbonate soln. added, and stirred for 1 h → product. Y 85%. The method is simple, selective, and general, and the startg. materials are readily available. Significantly, carbonyl groups remained unaffected (reduction with hydride transfer reagents yielding the 2-aminoalcohols). There was no need for exclusion of air or use of dry solvents, and work-up was simple. F.e.s. J.R. Harrison, C.J. Moody, M.R. Pitts, Synlett *2000*, 1601-2.

Indium/ammonium chloride   In/NH$_4$Cl
Ar. amines from nitro compds.   NO$_2$ → NH$_2$
s. *56*, 13; reduction of polycyclic nitroarenes and heteroaromatic nitro compds s. B.K. Banik et al., Synth. Commun. *30*, 3745-54 (2000).

Borane-tetrahydrofuran/polymer-based (R)-2-hydroxymethyl-4-sulfonylpiperazine   ←
Prim. amines from alkoximes by asym. reduction   C=NOR → CHNH$_2$
with a chiral polymer-based borane-2-aminoalcohol complex cf. *25*, 15s*41*; with borane-tetrahydrofuran/polymer-based (R)-2-hydroxymethyl-4-sulfonylpiperazine s. S. Itsuno, T. Matsumoto et al., J. Org. Chem. *65*, 5879-81 (2000).

Borane-trimethylamine/palladous hydroxide-carbon   BH$_3$·NMe$_3$/Pd(OH)$_2$-C
Palladium-catalyzed activation of borane-amine adducts   NO$_2$ → NH$_2$
Ar. amines from nitro compds. under neutral conditions

15. NC-C$_6$H$_4$-NO$_2$ → NC-C$_6$H$_4$-NH$_2$

The scope of reductions with commercially available, easily handled, hydrolytically and thermally stable borane-amines has been increased by activation *in situ* via palladium catalysis. Thus, reduction of ar. nitro compds., which are normally inert towards borane-amines alone, may be effected in high yield. E: A stirred mixture of startg. nitroarene and 1.2 eqs. BH$_3$·NMe$_3$ in methanol in an open vessel treated with Pearlman's catalyst (50% wet; 2.5 dry wt%; 1 mol%), heated to reflux until consumption of the borane complete (1.5 h), cooled to room temp., and filtered through Celite → *p*-aminobenzonitrile. Y >99%. The method is operationally simple and since it is applicable to borane-amines derived from volatile amines, work-up is straightforward. Benzylic functionalities are tolerated as are electron-donating and -withdrawing groups (such as ester, cyano, hydroxyl, methoxy, acetamido or fluorine); however, ar. chlorine was also reduced. F.e.s. M. Couturier et al., Tetrahedron Lett. *42*, 2285-8 (2001).

Samarium diiodide SmI_2
Cleavage of hydroxamic acid benzyl esters $CON(R)OCH_2Ar \rightarrow CONHR$
s. *51*, 9; N-subst. amides by cleavage of traceless alkoxylamine-type polymer linkers s. R.M. Myers, C. Abell et al., Org. Lett. *2*, 1349-52 (2000).

Iron/ammonium chloride Fe/NH_4Cl
Amines from nitro compds. $NO_2 \rightarrow NH_2$
ar. amines s. *18*, 26; *48*, 17; 2-aryl-*prim*-amines with retention of ar. chlorine and methylenedioxy groups s. C.Z. Ran, M.H. Xie et al., Chin. Chem. Lett. *11*, 855-6 (2000).

Nickel Ni
Amines from nitro compds.
with Raney nickel cf. *2*, 48; under milder conditions with easily regenerable Urushibara nickel s. X. Liu et al., Org. Prep. Proced. Int. *32*, 485-8 (2000).

Palladium hydroxide-carbon s. under $BH_3 \cdot NMe_3$ $Pd(OH)_2\text{-}C$

Platinum-carbon/ammonium formate $Pt\text{-}C/HCOONH_4$
Ar. amines from nitro compds. via catalytic transfer-hydrogenation
with hypophosphorous acid as H-donor and NH_4VO_3 as catalyst cf. *2*, 49s56; with $HCOONH_4$ or HCOOH as H-donor, selectivity, s. D.C. Gowda, B. Mahesh, Synth. Commun. *30*, 3639-44 (2000).

Nitrogen ↑ HN ↓↑ N

Aluminum amalgam Al, Hg
Prim. amines from 1*H*-naphtho[1,8-*de*]-1,2,3-triazin-2-ium N-ylids s. *24*, 852s60

Sodium tetrahydridoborate/lithium chloride $NaBH_4/LiCl$
Amines from azides $N_3 \rightarrow NH_2$
with $NaBH_4/[C_{16}H_{33}PBu_3]Br$ cf. *23*, 27s38; general procedure with 1:1 $NaBH_4/LiCl$, selectivity, s. S.R. Ram, D.S. Iyengar et al., Synth. Commun. *30*, 4495-500 (2000).

Sodium tetrahydridoborate/cobalt(II) chloride $NaBH_4/CoCl_2$
Carboxylic acid amides from azides $CON_3 \rightarrow CONH_2$
in methanol cf. *48*, 18; with $CoCl_2 \cdot 6H_2O$ *in water*, also aliphatic and aromatic amines from azides, and sulfonamides from sulfonyl azides, selectivity, s. F. Fringuelli, F. Pizzo et al., Synthesis *2000*, 646-50.

Ferrous sulfate/ammonia $FeSO_4/NH_3$
Amines from azides $N_3 \rightarrow NH_2$
with Fe/NH_4Cl cf. *52*, 13s54; with $FeSO_4$/aq. NH_3 (or aq. $MeNH_2$) in methanol, also lactams from azidocarboxylic acid esters, incl. pyrrolo[2,1-*c*][1,4]benzodiazepinones, s. A. Kamal et al., Tetrahedron Lett. *41*, 7743-6 (2000).

Cobalt(II) chloride s. under $NaBH_4$ $CoCl_2$

Nickel(II) chloride/lithium/4,4'-di-tert-butylbiphenyl ←
Prim. amines from hydrazines $RNHNH_2 \rightarrow RNH_2$
with Raney nickel cf. *20*, 356; with $NiCl_2 \cdot 2H_2O/Li/4,4'$-*di-tert*-butylbiphenyl (water of hydration being H-donor), also with 4-vinylbiphenyldivinylbenzene copolymer as catalytic electron transfer agent, **and from azo compds.**, s. F. Alonso, M. Yus et al., Tetrahedron *56*, 8673-8 (2000).

Sulfur ↑ HN ↓↑ S

Electrolysis
Electrochemical N-desulfonylation $NSO_2R \rightarrow NH$
s. *27*, 18s58; cathodic cleavage of N-dimethylsulfamyl protective groups from indole nitrogen, selectivity, s. M. Langeron, R.H. Dodd et al., Tetrahedron Lett. *41*, 9403-6 (2000).

Titanium trichloride/lithium $TiCl_3/Li$
N-Detosylation s. *60*, 3 $NTs \rightarrow NH$

Remaining Elements ↑   HN ↓↑ Rem

Hydrogen chloride   HCl
Amines from phosphoromonoamidates   NPO(OR)$_2$ → NH
s. *37*, 447; prim. *tert*-alkylamines s. T. Gadja, A. Zwierzak et al., Pol. J. Chem. *74*, 1385-7 (2000).

Carbon ↑   HN ↓↑ C

Irradiation   *hν*
Removal of photo-labile N-carbalkoxy protective groups   NCOOR → NH
s. *44*, 23s*56*; removal of N-carbo-2,2′-bis(2-nitrophenyl)ethoxy and N-carbo-6-nitroveratryloxy groups from solid-supported oligodeoxynucleotide heteropolymers (carrying base-sensitive S-pivaloylthioethyl phosphotriester linkages) with simultaneous cleavage of the resin s. K. Alvarez, J.-J. Vasseur et al., J. Org. Chem. *64*, 6319-28 (1999).

Electrolysis/tris(2,2′-bipyridyl)nickel(II) fluoroborate   ⊖/Ni(bipy)$_3$(BF$_4$)$_2$
Cathodic N-decarballyloxylation   NCOOCH$_2$CH=CH$_2$ → NH

16.

0.1 eq. Ni(bipy)$_3$(BF$_4$)$_2$ added to a mixture of startg. allyl carbamate and Bu$_4$NBF$_4$ in dry DMF, and electrolyzed at 20° in a single compartment cell fitted with zinc/stainless steel electrodes at a constant current of 60 mA until 3.4 F/mol passed → product. Y 99%. The mild, neutral conditions prevent allylamine formation and racemization while avoiding the need for stoichiometric amounts of reducing agents. Ketones, acetals, nitriles and carboxylic acid esters remained unaffected. F.e. with LiN(SO$_2$CF$_3$)$_2$ or KBr as supporting electrolyte s. D. Franco, E. Duñach, Tetrahedron Lett. *41*, 7333-6 (2000).

Lithium/N,N′-dimethylethylenediamine/tert-butylamine   ←
N-Debenzylation   NBn → NH
with Na/liq. NH$_3$ cf. *5*, 32; *16*, 24; of tert. N-benzylamides and N-benzyllactams with more manageable Li/MeNHCH$_2$CH$_2$NMe (or ethylenediamine) in low molecular weight amines, also reduction of oximes and nitriles (to prim. amines), Birch-type and further reductions s. M.E. Garst et al., J. Org. Chem. *65*, 7098-104 (2000).

Sodium hydrogen carbonate   NaHCO$_3$
Preferential N-decarbo-*tert*-butoxylation   NCOOBu-*t* → NH
with silica gel cf. *51*, 17; with retention of di-*tert*-butyl imidodicarbonate groups using NaHCO$_3$, also partial cleavage of the latter with ethanolic NaOH, s. S. Dey, P. Garner, J. Org. Chem. *65*, 7697-9 (2000).

Ammonium ceric nitrate   (NH$_4$)$_2$Ce(NO$_3$)$_6$
N-De-*p*-methoxybenzylation   NCH$_2$Ar → NH
s. *49*, 330; cleavage of chiral N-(1-*p*-methoxyphenyl)ethyl as stereo-directing and protective group in the Staudinger synthesis (diazoketone-variant) of chiral 2-azetidinones s. J. Podlech, S. Steurer, Synthesis *1999*, 650-4.

Citric buffer (pH 7)   ←
N-De-*tert*-alkylation of N,N-disubst. carboxylic acid amides   NR → NH
in 98% H$_2$SO$_4$ cf. *16*, 28; in pH 7 citric buffer/methanol **under neutral conditions** s. N. Auzeil, M. Largeron, Tetrahedron Lett. *41*, 8781-5 (2000).

Trifluoroacetic acid/dimethyl sulfide   CF$_3$COOH/Me$_2$S
Cleavage of polymer supports by N-decarbalkoxylation   NCOOR → NH
s. *27*, 110s*53*; cleavage of the carbamate linker of hydroxymethyl polystyrene with CF$_3$COOH/Me$_2$S, and application to Fischer indole and Pictet-Spengler cyclization s. L. Yang, Tetrahedron Lett. *41*, 6981-4 (2000).

Methyltrichlorosilane/triethylamine
Selective N-decarbomethoxylation

MeSiCl₃/Et₃N
NCOOMe → NH

2-Amino-2-deoxyglycosides. 5 eqs. MeSiCl₃ added to a soln. of startg. protected glycoside and 5 eqs. Et₃N in dry THF, the reaction flask sealed under N₂, heated to 60° for 48 h, diluted with water or 1:1 THF/water, and stirred until TLC indicated complete hydrolysis of the intermediate isocyanate → product. Y 93%. Deprotection by the action of MeSiCl₃ is compatible with a variety of functional groups, including acid-labile acetals and bis-ketals, ethers, acoxy compds., and silyl ethers, as well as other N-protecting groups such as azides and carbo-2,2,2-trichloroethoxyamines. F.e.s. B.K.S. Yeung, P.A. Petillo et al., Org. Lett. 2, 3135-8 (2000).

Methanesulfonic or Sulfuric acid
Selective N-decarbo-*tert*-butoxylation

MeSO₃H or H₂SO₄
NCOOBu-*t* → NH

with retention of *tert*-butyl esters using cation exchangers in aq. methanol cf. 25, 21; cleavage of peptide and amino acid derivs. with MeSO₃H in 4:1 *t*-BuOAc/CH₂Cl₂ or concd. H₂SO₄ in *t*-BuOAc s. L.S. Lin et al., Tetrahedron Lett. 41, 7013-6 (2000).

Tetra-n-butylammonium fluoride
**Protection of amino groups
as carbo-2-trimethylsilylethoxyamines and removal of the protective group**

Bu₄NF
NCOOC-C-Si≤ → NH

with Et₄NF cf. 34, 23; more facile cleavage of carbo-2-phenyl-2-trimethylsilylethoxyamines under milder conditions with Bu₄NF s. M. Wagner, H. Kunz, Synlett 2000, 1753-6.

Tris(2,2'-bipyridyl)nickel(II) fluoroborate s. under ⁷ *Ni(bipy)₃(BF₄)₂*

Formation of H-S Bond

Exchange   ⇅

Sulfur ↑   HS ⇅ S

Sodium tetrahydridoborate/zirconium tetrachloride
Mercaptans from disulfides

NaBH₄/ZrCl₄
(RS)₂ → 2 RSH

with NaBH₄ cf. 17, 45; rapid, general procedure with 4 eqs. NaBH₄ and 1 eq. ZrCl₄, selectivity, s. K. Purushothama Chary, D.S. Iyengar et al., Synth. Commun. 30, 3905-11 (2000).

Carbon ↑   HS ⇅ C

Tetraalkylammonium fluoride
**Selenothiocarboxylic acid ammonium salts
from selenothiolic acid 2-(trimethylsilyl)ethyl esters**

R₄NF
C(Se)S⁻·R₄N⁺

The first isolation and characterization of a selenothiocarboxylic acid salt is reported. **E:** Startg. selenothiolic acid 2-(trimethylsilyl)ethyl ester treated with a soln. of Bu₄NF in THF at 0° for 3 h, the mixture concentrated, and the residue washed with hexane → product. Y ca. 100% (>90% purity). The products are deep blue oils or green-purple solids (as their tetramethylammonum salts). Reaction with electrophiles (e.g. alkyl or acyl halides) took place at the Se-atom. F.e.s. T. Murai, S. Kato et al., J. Am. Chem. Soc. *122*, 9850-1 (2000).

Formation of H-C Bond

Uptake ⇓

Addition to Oxygen and Carbon   HC ⇓ OC

Cuprous acetate s. under PhMe$_2$SiH   *CuOAc*
Magnesium bromide s. under Bu$_3$SnH   *MgBr$_2$*

Borane-tetrahydrofuran/(S)-2-[p-(trifluoromethyl)anilinomethyl]indoline   ←
Asym. reduction of ketones   CO → CHOH
with borane-Me$_2$S and a little chiral 2-aminoalcohol or 2-sulfonylaminoalcohol cf. *33*, 43s*59*; with borane-THF and a little chiral diamine, e.g. (S)-2-[*p*-(trifluoromethyl)anilinomethyl]indoline, s. M. Asami, H. Watanabe et al., Chem. Lett. *2000*, 990-1; f. chiral 1,2-diamines s. Tetrahedron: Asym. *11*, 4329-40 (2000).

Borane-dimethyl sulfide/chiral 2-aminoalcohols/trimethyl borate   ←
Asym. reduction of ketones
s. *58*, 17; with camphor-derived 2-aminoalcohols s.a. V. Santhi, J.M. Rao, Tetrahedron:Asym. *11*, 3553-60 (2000); C$_2$-symmetric chiral diols from diketones s. D.J. Aldous, P.G. Steel et al., ibid. 2455-62.

Borane-dimethyl sulfide/titanium tetrachloride   *BH$_3$-Me$_2$S/TiCl$_4$*
***anti*-2-Nitroalcohols from α-nitroketones**

19.

Ph–CO–C(NO$_2$)–Bu ⟶ Ph–CH(OH)–C(NO$_2$)–Bu

1.5 eqs. TiCl$_4$ added to a soln. of startg. α-nitroketone in dry dichloromethane at -78°, stirred for 15 min, a soln. of 1.5 eqs. BH$_3$-Me$_2$S in the same solvent added, after 15 min the mixture was quenched with 1 *N* HCl, and warmed to room temp. → 2-nitro-1-phenylheptan-1-ol. Y 95% (>99% *anti*). Chelation by the Lewis acid is responsible for the *anti*-stereospecificity while the familiar Henry reaction is normally *syn*-specific. F.e. and with retention of ar. nitro groups s. R. Ballini et al., J. Org. Chem. *65*, 5854-7 (2000).

Sodium tetrahydridoborate/D-tartaric acid   ←
Asym. reduction of functionalized ketones
s. *38*, 24s*46*; chiral β-hydroxy esters s. D.V. Johnson, H. Griengl et al., Tetrahedron *56*, 9289-95 (2000).

Lithium tetrahydroaluminate/chiral 2-aminoalcohols   ←
Asym. reduction of ketones
s. *33*, 43s*57*; of fluorenones with (2R)-2-isoindolin-2-ylbutanol as ligand, selectivity, s. Z.R. Yu, D. Velasco et al., Tetrahedron:Asym *11*, 3227-30 (2000).

Sodium tetrahydridoaluminate/chiral 2-aminoalcohols/4-dimethylaminopyridine ←
1,1-Alkoxyacoxy compds. from carboxylic acid esters　　　　COOR → CH(OAc)OR
Asym. reduction

20.

$$\text{BnO}\underset{\text{OEt}}{\overset{\text{O}}{\diagdown}} \xrightarrow[\underset{\text{NHMe}}{\overset{\text{Ph}\diagdown}{\text{OH}}}]{\text{NaAlH}_4} \left[\text{BnO}\underset{\text{OEt}}{\diagdown}^{\text{OAlLn}} \right] \xrightarrow{\text{Ac}_2\text{O}} \text{BnO}\underset{\text{OEt}}{\diagdown}^{\text{OAc}}$$

A soln. of 3 eqs. (1R,2S)-ephedrine in THF added over 4 min to 3 eqs. 2 M ethereal NaAlH₄ in the same solvent, stirred for 15 min, cooled to -78°, ethyl O-benzylglycolate added, stirred for 2 h at the same temp., 6 eqs. acetic anhydride and 4 eqs. DMAP added, warmed to room temp. over 16 h, and diluted with ether and 1 N HCl → product. Y 55% (e.e. 83%). This is the first instance of an asym. reduction of a carbalkoxy group. However, the method is not applicable to aromatic and α,β-unsatd. esters. F.e.s. S.D. Rychnovsky, B.M. Bax, Tetrahedron Lett. *41*, 3593-6 (2000).

Trialkylboranes/1-ethyl-3-methylimidazolium hexafluorophosphate ←
Prim. alcohols from aldehydes　　　　　　　　　　　　　　　CHO → CH₂OH
Reduction in ionic liquids

21.

$$\text{Br}\diagdown\text{C}_6\text{H}_4\diagdown\text{CHO} \longrightarrow \text{Br}\diagdown\text{C}_6\text{H}_4\diagdown\text{CH}_2\text{OH}$$

A mixture of 106 mg. startg. aldehyde, 250 mg. 1-ethyl-3-methylimidazolium hexafluorophosphate, and 182 mg. tributylborane stirred at 100° for 16 h → product. Y 100% (NMR yield). Significantly, reduction requires a much higher temp. in traditional organic media, while in ionic liquids reaction can even be carried out at room temp. (over 48 h) with isolated yields >94%. F.e.s. G.W. Kabalka, R.R. Malladi, Chem. Commun. *2000*, 2191-1.

Lithium hydridotri-tert-butoxoaluminate　　　　　　　　　　　　　LiAlH(OBu-t)₃
Sec. alcohols from ketones　　　　　　　　　　　　　　　　　　CO → CHOH
s. *39*, 32; *32*, 26s33; 2-phthalimidoalcohols with asym. induction s. S. Sengupta, D.S. Sarma, Tetrahedron:Asym. *10*, 4633-7 (1999).

Borane-amine complexes　　　　　　　　　　　　　　　　　　　　BH₃·R₃N
Alcohols from oxo compds.
s. *15*, 39s41; prepn. of tert. benzylamine-borane complexes s. C.J. Collins, B. Singaram et al., J. Org. Chem. *64*, 2574-6 (1999).

Diisobutyl(pyrrol-1-yl)alane　　　　　　　　　　　　　　　　　　i-Bu₂AlNR₂
Sec. alcohols from ketones
with i-Bu₂AlCl cf. *51*, 23; reduction of cyclic ketones more efficiently with 0.5 eq. diisobutyl(pyrrol-1-yl)alane (utilizing *both* isobutyl groups), stereoselectivity, s. O.O. Kwon, J.S. Cha, Bull. Korean Chem. Soc. *21*, 659-61 (2000).

Chiral 1,3,2-oxazaborolidines/borane-dimethyl sulfide ←
Asym. reduction of ketones
update s. *43*, 45s58; of 2-alkylidenecycloalkanols, effect of alkene geometry on face selectivity, s. A.F. Simpson, T. Gallagher et al., J. Chem. Soc. Perkin Trans. 1 *2000*, 3047-54; of N-protected silylethynyl α-aminoketones s. C. Alemany, J. Bach et al., Org. Lett. *1*, 1831-4 (1999); of silylethynyl α,β-ethyleneketones s. J. Garcia et al., Tetrahedron:Asym. *10*, 2617-26 (1999); with chiral 1,3,2-oxazaborolidines based on (1R)-camphor s. V. Santhi, J.M. Rao, Synth. Commun. *30*, 4329-41 (2000); chiral spiro[4.4]nonane-1,6-diol from -1,6-dione s. C.-W. Lin, A.S.C. Chan et al., Tetrahedron Lett. *41*, 4425-9 (2000); with chiral 1,3,2-oxazaborolidin-5-ones based on L-α-amino acids cf. S. Huang, D. Zhao et al., Synth. Commun. *30*, 2423-9 (2000).

Chiral 1,3,2-oxazaborolidines/borane-aniline complexes ←
Asym. reduction of α-oxyketones CO → CHOH
of α-tetrahydropyran-2-yloxyketones cf. *57*, 20s*58*; of α-sulfonyloxyketones with borane-N-ethyl-N-isopropylaniline/chiral 3,3-diphenyl-1-methylpyrrolidino[1,2-*c*]-1,3,2-oxazaborolidine, also conversion to chiral epoxides, s. O.K. Choi, B.T. Cho, Org. Prep. Proced. Int. *32*, 493-7 (2000).

(Tricyclohexylphosphine)indium trihydride $[InH_3P(C_6H_{11})_3]$
Sec. alcohols from ketones
meso-glycols from α-diketones or α-hydroxyketones with LiInH$_4$ or LiInR$_2$H$_2$ cf. *54*, 27; with stable, more manageable $[InH_3P(C_6H_{11})_3]$ s. C.D. Abernethy, C. Jones et al., Tetrahedron Lett. *41*, 7567-70 (2000).

Diisopinocampheylchloroborane R_2BCl
Asym. reduction of ketones
s. *41*, 37s*47*, *48*; chiral β-hydroxycarboxylic acids s. Z. Wang et al., Tetrahedron:Asym. *10*, 225-8 (1999).

Boron fluoride s. under Et$_3$SiH and Bu$_3$SnH BF_3
Gallium trichloride s. under Bu$_3$SnH $GaCl_3$

Samarium diiodide/samarium/methanol $SmI_2/Sm/MeOH$
Stereospecific reduction of β-oxyketones under chelation control
anti-1,3-diols s. *57*, 22; *anti*-1,3-diol monoethers with added Sm (1 eq.) s. G.E. Keck, C.A. Wagner, Org. Lett. *2*, 2307-9 (2000).

1-Ethyl-3-methylimidazolium hexafluorophosphate s. under R$_3$B ←

Dehydrogenase ←
**δ-Hydroxy-β-ketocarboxylic from β,δ-diketocarboxylic acid esters
via asym. enzymatic reduction**

22.

9.9 mmol Startg. diketoester in isopropanol added to triethanolamine/HCl buffer (pH 7.0) containing MgCl$_2$ (1 mmol), the mixture stirred vigorously for 10 min, stirring rate adjusted to 60 rpm, 33 mmol NADP and 660 units *Lactobacillus brevis* alcohol dehydrogenase (from recombinant *E. coli*) added, and stirred at room temp. for 24 h → product. Y 77% (e.e. >99.4%). The procedure is highly efficient and can be scaled up to the 75 g level. The excess of isopropanol is the driving force of the reaction. F.e. and with retention of ε-chlorine s. M. Wolberg, M. Müller, Angew. Chem. Int. Ed. Engl. *39*, 4306-8 (2000).

Immobilized lipase/ruthenium hydride complex ←
Acoxy compds. from ketones CO → CHOH
Asym. reductive O-acylation

23.

Startg. ketone, 2 mol% ruthenium hydride complex, and Novozym-435 in ethyl acetate hydrogenated under 1 atm. H$_2$ at 70° for 76 h → product. Y 89% (99% e.e.). Hydrogenation initially affords a racemic mixture of the corresponding alcohols, of which the fast reacting (R)-enantiomer is preferentially O-acylated in the presence of the enzyme, while the (S)-enantiomer reverts to the ketone under the same catalytic control. F.e., **also from enol acetates** (without an acyl donor), and with formic acid as H-donor s. H.M. Jung, M.-J. Kim, J. Park et al., Org. Lett. *2*, 2487-90 (2000).

Fungi or Yeast
Asym. reduction of ketones ← CO → CHOH
with yeast s. *29*, 36s*58*; of 2-keto-4-arylbutyrates s. D.H. Dao, A. Ohno, Tetrahedron:Asym. *9*, 2725-37 (1998); of γ-chloro-β-diketones to γ-chloro-β-hydroxyketones s. J.-N. Cui, M. Utaka et al., ibid. 2681-92; asym. reduction of α-ketoalkoximes s. O.C. Kreutz, P.J.S. Moran et al., ibid. *11*, 2107-15 (2000); with more manageable *lyophilized* yeast s. F. Molinari et al., ibid. *10*, 3515-20 (1999); asym. reduction **of α,β-ethylenealdehydes** (adsorbed on resin XAD 1180) (cf. *35*, 29s*58*) s. C. Fuganti, S. Serra, J. Chem. Soc. Perkin Trans. 1 *2000*, 3758-64; of α-cyanoketones with the fungus *Curvularia lunata* s. J.R. Dehli, V. Gotor, Tetrahedron:Asym. *11*, 3693-700 (2000).

Fungus ←
Sec. alcohols from ketones by asym. enzymatic reduction in water with a hydrophobic polymer as solid organic co-solvent

24. Ph-CO-CH₂CH₂CH₃ → Ph-CH(OH)-CH₂CH₂CH₃

15 g (wet weight) of *Geotrichum candidum* IFO 4597 cells added to a mixture of 2.03 mmol 1-phenylbutan-3-one and 18 g Amberlite XAD-7 (a hydrophobic polymer based on acrylic ester) in water, and the mixture shaken at 130 rpm for 24 h at 30° → (S)-1-phenylbutan-3-ol. Y 90% (e.e. >99%). The hydrophobic resin does not damage cell structure as much as organic solvents; it also reduces the concentration of substrates and products in the aqueous phase where reduction takes place, thus increasing stereoselectivity. The procedure is especially effective for reducing simple ketones. F.e.s. K. Nakamura, Y. Ida, J. Chem. Soc. Perkin Trans. 1 *2000*, 3205-11; asym. reduction of α,β-ethylenealdehydes s. *29*, 36s*60*.

Triethylsilane/boron fluoride Et_3SiH/BF_3
Diol monoethers from cyclic acetals
cleavage of carbohydrate 4,6-O,O-benzylidene derivs. with Et₃SiH/TfOH cf. *39*, 33s*59*; prepn. of 6-O-benzyl derivs. with Et₃SiH/BF₃-etherate s. S.D. Debenham, E.J. Toone, Tetrahedron:Asym. *11*, 385-7 (2000).

Dimethylphenylsilane/cuprous acetate $PhMe_2SiH/CuOAc$
Reduction of aryl ketones CO → CHOH
with PhMe₂SiH/CF₃COOH/Bu₄NF cf. *40*, 22; with a copper hydride complex generated *in situ* with >5:1 PhMe₂SiH/CuOAc, also alkylarenes from styrenes (cf. *45*, 23) with CuCl, s. H. Ito, A. Hosomi et al., Synlett *2000*, 479-82.

Organosilicon hydrides/bis(cyclooctadiene)rhodium(I) fluoroborate/ ←
 chiral ferrocenyldiphosphines
Sec. alcohols via asym. hydrosilylation of ketones
s. *49*, 42s*57*; with *trans*-chelating chiral peralkylbis(phosphine) ligands possessing *prim*-alkyl groups on phosphorus s. R. Kuwano, Y. Ito et al., Bull. Chem. Soc. Jpn. *73*, 485-96 (2000); with 1,1'-bis[(*tert*-butyl)methylphosphino]ferrocene as ligand s. H. Tsuruta, T. Imamoto, Tetrahedron:Asym. *10*, 477-82 (1999).

Titanium tetrachloride s. under BH₃-Me₂S $TiCl_4$

Titanium tetraiodide TiI_4
α-Hydroxyketones from α-diketones

25. Ph-CO-CO-Ph --TiI₄→ Ph-CO-CH(OH)-Ph | Ph-CO-CO-Me --TiI₄→ Ph-CO-CH(OH)-Me + Ph-CH(OH)-CO-Me
                                                                            Y 54%   Y 13% (by NMR)

Acetonitrile added under argon to 1.5 eqs. TiI₄ at room temp., stirred for 10 min, a soln. of benzil in acetonitrile added at 0°, stirring continued at 0° to room temp. for 5.3 h, and the mixture quenched with satd. aq. NaHCO₃ followed by 10% aq. NaHSO₃ → benzoin. Y 92%. The method is convenient, mild, efficient, experimentally simple, high-yielding, and is based on a commercially available and inexpensive reagent. It is also applicable to both sym. and unsym. α-diketones (the latter regioselectively). F.e.s. R. Hayakawa, M. Shimizu et al., Tetrahedron Lett. *41*, 7939-42 (2000).

Tri-n-butyltin hydride/magnesium bromide $Bu_3SnH/MgBr_2$
Diol monoethers from cyclic acetals
Regiospecific reductive ring opening

26.

Startg. 1,3-dioxolane, 1.2 eqs. Bu$_3$SnH, and 2 eqs. MgBr$_2$-etherate in methylene chloride allowed to react at room temp. for 5 min → product. Y 99%. The unusual regioselectivity (the reverse of that reported with DIBAH) is ascribed to the formation of an intermediate five-membered Mg-chelate, and is useful for the preparation of differentially O^1,O^3-diprotected glycerol derivs. F.e. incl. cleavage of 1,3-dioxane analogs s. B.-Z. Zheng, O. Yonemitsu et al., Tetrahedron Lett. *41*, 6441-5 (2000).

Tri-n-butyltin hydride/boron fluoride Bu_3SnH/BF_3
Chelation-controlled radical reduction of oxyketones CO → CHOH
glycol monoethers with Me$_3$Al as Lewis acid cf. 56, 30; glycols, 1,3-diols and *syn*-1,4-diols (from the corresponding hydroxyketones) with BF$_3$ as Lewis acid s. T. Ooi, K. Maruoka et al., Org. Lett. *2*, 2015-7 (2000).

Tri-n-butyltin hydride/gallium trichloride $Bu_3SnH/GaCl_3$
Lewis acid-catalyzed reactions of *o*-acetylenealdehydes CHO → CH$_2$OH
under bidentate chelation

27.

An equimolar mixture of startg. *o*- and *p*-acetylenealdehyde in methylene chloride treated with 1 eq. GaCl$_3$ at -78° for 30 min, followed by addition of 1 eq. Bu$_3$SnH → product. Y 85% (with 2% of the *p*-alkynylbenzyl alcohol). Failure of the *p*-alkynyl deriv. to react is evidence for the unprecedented **coordination of Lewis acids to π-electrons of the triple bond.** The same pattern was observed for Lewis acid-catalyzed **nucleophilic allylation** with allyltributylstannane. 2-Ethynylcyclohexanecarboxaldehyde reacted similarly in preference to cyclohexanecarboxaldehyde or its 2-ethyl deriv. F.e. and comparison of Lewis acids s. N. Asao, Y. Yamamoto et al., J. Am. Chem. Soc. *122*, 4817-8 (2000).

Nickel Ni
Prim. alcohols from aldehydes
with Raney nickel s. 2, 75; of aryl or aliphatic aldehydes with retention of aryl, alicyclic or aliphatic ketones s. A.F. Barrero et al., Synlett *2000*, 197-200.

Ruthenium hydride complex s. under Lipase ←

Dichloro(p-cymene)ruthenium(II) dimer/chiral 2-aminoalcohols/isopropanol/ ←
 potassium hydroxide
Asym. transfer-hydrogenation of ketones CO → CHOH
s. *51*, 26s59; with 2-azanorbornylalcohols as ligand s. D.A. Alonso, P.G. Andersson et al., J. Org. Chem. *65*, 3116-22 (2000); of ketoisophorone with various chiral 2-aminoalcohols, also isolation of the catalyst precursors, s. M. Hennig, K. Püner et al., Tetrahedron:Asym. *11*, 1849-58 (2000); of aroylacetic acid esters s. K. Everaeve, J.-F. Carpentier et al., ibid. *10*, 4663-6 (1999); with a chiral N-cyclohexylmethyl-2-aminoalcohol as ligand s. C.G. Frost, P. Mendonça, ibid. *11*, 1845-8 (2000).

Dichloro(p-cymene)ruthenium(II) dimer/(1R,2R)-N-(p-tosyl)-1,2-diphenylethylene- ←
   diamine/formic acid/triethylamine
Asym. transfer-hydrogenation of functionalized ketones   CO → CHOH
chiral glycols from α-diketones via α-hydroxyketones cf. *59*, 33; regioselective formation of chiral α-hydroxyketones s. T. Koike, K. Murata et al., Org. Lett. *2*, 3833-6 (2000); chiral 2-aminoalcohols from α-aminoketones s. A. Kawamoto, M. Wills, Tetrahedron:Asym. *11*, 3257-61 (2000).

(η⁶-p-Cymene)((1R,2R)-N-tosyl-1,2-diphenylethylenediamine)ruthenium(II)   ←
   chloride/dideuterioformic acid/triethylamine
Prim. 1-deuterioalcohols from aldehydes   CHO → CH(OH)D
Asym. transfer-hydrogenation

28.

MeO-C₆H₄-CHO → MeO-C₆H₄-CH(OH)D

A mixture of *p*-anisaldehyde, 0.5 mol% (R,R)-Ru catalyst, 1 eq. DCO$_2$D, and 1 eq. Et$_3$N in acetonitrile stirred at 28° for 14 h → (S)-product. Y 96% (e.e. 99% with >99% isotopic purity). Phenolethers, ar. bromides and trifluoromethyl groups remained unaffected. F.e., also by asym. reduction of 1-deuterioaldehydes with isopropanol as H-donor, s. I. Yamada, R. Noyori, Org. Lett. *2*, 3425-7 (2000).

Benzene(dichloro)ruthenium(II) dimer/chiral 3,3'-bis(diphenylphosphino)-2,2'-biindoles   ←
Asym. homogeneous hydrogenation   CO → CHOH
with the Noyori complex cf. *43*, 51; of α- and β-keto esters with chiral 3,3'-bis(diphenylphosphino)-2,2'-biindoles as ligand s. T. Benincori et al., J. Org. Chem. *65*, 8340-7 (2000).

Chiral dichlororuthenium(II) phosphine complex   ←
1,3-Diols from β-diketones
Asym. homogeneous hydrogenation
with Ru$_2$Cl$_4$((R)-BINAP)$_2$(NEt$_3$) cf. *49*, 47; of β,δ-diketo-ε-dicarboxylic acid esters with RuCl$_2$[(S)-BINAP](*p*-cymene) s. J. Kiegel et al., Tetrahedron Lett. *41*, 4959-63 (2000).

Chiral dichlororuthenium(II) 2,2'-di(phosphino)biphenyl complexes   ←
Asym. homogeneous hydrogenation of ketones
s. *55*, 26s57; of β-keto esters, effect of bite angle on enantioselectivity, s. Z. Zhang, X. Zhang et al., J. Org. Chem. *65*, 6223-6 (2000).

Dichloro[2,2',6,6'-tetramethoxy-4,4'-bis(diphenylphosphino)-3,3'-bipyridyl]-   ←
   ruthenium(II)
β-Hydroxy- from β-keto-carboxylic acid esters
Asym. homogeneous hydrogenation

29.

Cl-CH$_2$-CO-CH$_2$-COOEt →[H$_2$] Cl-CH$_2$-CH(OH)-CH$_2$-COOEt

A stirred mixture of startg. β-keto ester (0.1-0.17 *M* in 10:1 methanol/dichloromethane) and 0.0036 mol% chiral ruthenium(II) complex (DMF solvate) hydrogenated in methanol in a glass-lined stainless steel autoclave under a pressure of 50 psi at 80° for 12-36 h → product. 100% conversion; e.e. 98%. The novel, atropisomeric bipyridyl ligand is easy to prepare, forms well-defined ruthenium complexes, and affords high levels of asym. induction in the hydrogenation of ketones and electron-deficient olefins (the latter with the bis(acetoacetonato) complex, optionally with added H_3PO_4). A significant advantage over Ru-BINAP complexes is that the *basic* catalyst can be easily separated in an aqueous acidic extract (HCl). F.e.s. C.-C. Pai, A.S.C. Chan, W.T. Wong et al., J. Am. Chem. Soc. *122*, 11513-4 (2000).

Dichloro[2,2'-bis[bis(3,5-dimethylphenyl)phosphino]biphenyl]ruthenium(II)/ ←
(S,S)-1,2-diphenylethylenediamine
Asym. homogeneous hydrogenation of ketones CO → CHOH
Asym. activation of racemic ruthenium(II) di(phosphine) complexes
with a racemic ruthenium di(phosphino)binaphthyl complex cf. *54*, 30; improved enantioselectivity with racemic conformationally flexible 2,2'-bis[bis(3,5-dimethylphenyl)phosphino]biphenyl as ligand s. K. Mikami, R. Noyori et al., Angew. Chem. Int. Ed. Engl. *38*, 495-7 (1999).

Dichloro[2,2'-bis(di-3,5-xylylphosphino)-1,1'-binaphthyl]ruthenium(II)/(R)-2,2'- ←
diamino-3,3'-dimethyl-1,1'-binaphthyl/(S,S)-1,2-diamino-1,2-diphenylethane
Asym. activation/deactivation of a racemic catalyst
by sequential complexation with two different chiral auxiliaries
Asym. homogeneous hydrogenation of ketones

30.

The concept of asym. activation of a racemic catalyst with a chiral auxiliary (cf. *54*, 30) has been extended to an asym. activation/deactivation protocol involving *two* different chiral [diamine-type] auxiliaries: one serving to complex and *activate* one of the catalyst enantiomers, and the other to complex and *deactivate* the other. The result is a maximization of the difference in the catalytic activity between the two enantiomers, thereby achieving higher enantioselectivity than in the simple activation protocol irrespective of the nature of the substrate. E: 10 mmol racemic [RuCl$_2$(dm-binap)(DMF)n] and 6 mmol (R)-dm-DABN (for complexation with and *activation* of [RuCl$_2$((R)-dm-binap)]) in methylene chloride, stirred under argon at room temp. in an autoclave, the solvent removed under reduced pressure, the autoclave again purged with argon after addition of 4.5 mmol (S,S)-DPEN (for complexation with and *deactivation* of [RuCl$_2$((S)-dm-binap]), diluted with isopropanol, followed by 20 mmol KOH in isopropanol with stirring at room temp. for 30 min, 2.5 mmol 1'-acetonaphthone added at the same temp. under argon, H$_2$ introduced to 8 atm., and vigorously stirred for 4 h at room temp. → product. Y >99% (e.e. (R)-enantiomer 96%). F.e.s. K. Mikami, R. Noyori et al., Angew. Chem. Int. Ed. Engl. *39*, 3707-10 (2000).

Palladous chloride/formic acid/sodium hydroxide ←
α-Hydroxy- from α-keto-carboxylic acids s. *60*, 37 CO → CHOH

Chloro(cyclooctadiene)iridium dimer/chiral 2-aminothioethers/formic acid/triethylamine ←
Asym. transfer-hydrogenation of aryl ketones
with chiral bis(Δ^2-oxazolines) as ligand and *i*-PrOH as H-donor cf. *46*, 42; with chiral 2-aminothioethers as ligand, and HCOOH as H-donor s. D.G.I. Petra, P.W.N.M. van Leeuwen et al., J. Org. Chem. *65*, 3010-7 (2000).

Addition to Nitrogen and Carbon HC ⇓ NC

Lithium/ethylenediamine/n-propylamine ←
Prim. amines from nitriles s. 5, 32s60 $CN \rightarrow CH_2NH_2$

1-Benzyl-4-aza-1-azoniabicyclo[2.2.2]octane tetrahydridoborate ←
Sec. amines from azomethines s. 60, 32 $C=NR \rightarrow CHNHR$

Zinc bis(tetrahydridoborate) $Zn(BH_4)_2$
Sec. amines from azomethines
s. 46, 44; from chiral ketimines **with asym. induction** s. C. Cimarelli, G. Palmieri, Tetrahedron:Asym. 11, 2555-63 (2000).

Zirconium(IV) tetrahydridoborate $Zr(BH_4)_4$
Sec. amines from azomethines s. 60, 39

Dimethylphenylsilane/tris(pentafluorophenyl)borane $Me_2PhSiH/(C_6F_5)_3B$
Sec. amines from azomethines
with Me_2PhSiH/CF_3COOH cf. 35, 23s44; with Me_2PhSiH and a little $(C_6F_5)_3B$ s. J.M. Blackwell et al., Org. Lett. 2, 3921-3 (2000).

Cationic ruthenium(II) formyl complex ←
1,4-Dihydropyridines from pyridinium salts Ⓗ
Biomimetic reduction by regiospecific hydride transfer

31.

A soln. of startg. pyridinium salt and 1.5 eqs. cis-[Ru(2,2'-bipyridyl)$_2$(CO)(CHO)]PF$_6$ in acetonitrile-d_3 allowed to react at 0° under argon in dim light for 15 min (with ^{1}H and ^{13}C NMR monitoring) → 1-benzyl-1,4-dihydronicotinamide. Y 100% (100% conversion). This serves as a model for NAD$^+$-mediated oxidation of formaldehyde to formic acid by aldehyde dehydrogenases, and presents a system of potential in the synthesis of various 1,4-NADH model compds. In this respect, metal hydride reduction is limited in scope, and the conventional route with Na-dithionite (cf. 11, 73) normally affords a mixture of dihydro isomers. F.e.s. H. Konno, O. Ishitani et al., Angew. Chem. Int. Ed. Engl. 39, 4061-3 (2000).

Chiral rhodium di(phosphine) or di(phosphinite) complexes ←
Sec. amines from azomethines - Asym. homogeneous hydrogenation $C=NR \rightarrow CHNHR$
with chiral rhodium phosphine complexes cf. 29, 42s44; comparison of chiral di(phosphines) and di(phosphinites) as ligand s. V.I. Tararov, A. Börner et al., Tetrahedron:Asym. 10, 4009-15 (1999).

Chiral iridium(I) amidophosphine-phosphinite complexes ←
Chiral iridium(I) di(phosphine) complex/montmorillonite ←
Sec. amines from azomethines - Asym. homogeneous hydrogenation
s. 46, 47s52,53; with recyclable [Ir(cod)[(S,S)-2,4-bis(diphenylphosphino)pentane]iridium(I) hexafluorophosphate adsorbed on montmorillonite K10 s. R. Margalef-Català, P. Salagre et al., Tetrahedron:Asym. 11, 1469-76 (2000); with chiral iridium(I) amidophosphine-phosphinite complexes s. E.A. Broger, M. Scalone et al., ibid. 9, 4043-54 (1998).

Addition to Carbon-Carbon Bonds HC ⇓ CC

Electrolysis ↯
Electrocatalytic hydrogenation $C=C \rightarrow CHCH$
at an activated Pt-cathode cf. 30, 22; with atomic hydrogen permeating through a Pd sheet cathode for hydrogenation of allyl- and homoallyl-arenes with retention of allyl and benzyl ethers s. S. Maki et al., Synth. Commun. 30, 3575-83 (2000); using poly[N-(5-carboxyhexyl)-

pyrrole] film-coated electrodes incorporating Pd cf. N. Takano et al., Bull. Chem. Soc. Jpn. *73*, 745-6 (2000).

Lithium/ethylenediamine/n-propylamine/tert-butanol ←
Birch-type reduction s. *5*, 32s60 ⊕

Cuprous chloride s. under Me₂PhSiH *CuCl*

Borane-dimethyl sulfide *BH₃-Me₂S*
Reduction of heterofunctionalized ethylene derivs. C=C(X) → CHCH(X)
of enamines with diborane cf. *19*, *60*; of α,β-ethylenephosphine oxides with BH₃-Me₂S s. G. Keglevich et al., Synth. Commun. *30*, 4221-31 (2000).

Sodium tetrahydridoborate/tetrakis(triphenylphosphine)palladium(0)/sodium hydroxide ←
Ethylene from acetylene derivs. C≡C → CH=CH
cis-ethylene derivs. with NaBH₄/PdCl₂/PEG cf. *7*, 91s38; partial hydrogenation of terminal alkynes with NaBH₄/Pd(PPh₃)₄ and 0.1 *M* NaOH in 1:1 *aq*. methanol, also with retention of 1,3-enyne groups, s. W.X. Gu, A.X. Wu, X.F. Pan, Chin. Chem. Lett. *11*, 847-8 (2000).

1-Benzyl-4-aza-1-azoniabicyclo[2.2.2]octane tetrahydridoborate ←
Selective reductions ←
with 1-benzyl-4-aza-1-azoniabicyclo[2.2.2]octane tetrahydridoborate

32.

of enamines. Startg. enamine added to a stirred soln. of 1 eq. 1-benzyl-1-azonia-4-aza-bicyclo[2.2.2]octane tetrahydridoborate *in* tert-*butanol*, and stirred at room temp. until TLC indicated completion of reaction (0.5 h) → product. Y 84%. The procedure is rapid, inexpensive and high-yielding, and does not require pH control. F.e., also hydroxylamines from oximes, sec. amines from azomethines, and reductive N-alkylation with oxo compds., s. A.R. Hajipour, I. Mohammadpoor-Baltork et al., Indian J. Chem. *39B*, 239-42 (2000).

Lithium tetrahydridoaluminate/bis(η⁵-pentamethylcyclopentadienyl)titanium dichloride ←
Preferential hydrogenation of ethylene derivs. C=C → CHCH
1,3-dienes from 1,3,5-trienes with ca 1:1 LiAlH₄/Cp₂TiCl₂ cf. *44*, 48; ethylene derivs. from terminal 1,3-dienes with retention of 1,1- and 1,2-disubst. olefin groups [and without double bond shift] using ca. 2:1 LiAlH₄/[C₅Me₅]₂TiCl₂ s. H.S. Lee, H.Y. Lee, Bull. Korean Chem. Soc. *21*, 451-2 (2000).

Imidazolium hexafluorophosphates s. under Ruthenium complexes ←

Immobilized lipase/ruthenium hydride complex ←
Enzyme-mediated asym. hydrogenation of enolesters s. *60*, 22   C(OAc)=C → CH(OAc)CH

Reductase (plant cells) ←
Ketones from α,β-ethyleneketones by asym. reduction C=C → CHCH
with yeast reductase cf. *35*, 39s56; asym. reduction of cyclic α,β-ethyleneketones with the reductase from a *Nicotiana tabacum* cell culture s. T. Hirata et al., Chem. Lett. *2000*, 850-1.

Yeast ←
Alcohols from α,β-ethylenealdehydes
Asym. reduction s. *29*, 36s60

Dimethylphenylsilane/cuprous chloride *Me₂PhSiH/CuCl*
Alkylarenes from styrenes s. *45*, 23s60

Bis(η⁵-pentamethylcyclopentadienyl)titanium dichloride s. under LiAlH₄ *Cp*₂TiCl₂*

Chlorobis(cyclopentadienyl)hydridozirconium   *$Cp_2Zr(Cl)H$*
Stereospecific hydrozirconation of functionalized acetylenes $C\equiv CX \rightarrow CH=CHX$
(E)-α,β-ethylenesulfones cf. *48*, 48s*58*; (Z)-α,β-ethylenesulfoxides from α,β-acetylenesulfoxides, also quenching with I_2, NBS or α,β-acetyleneiodonium salts s. P. Zhong, M. Ping-Guo et al., Tetrahedron *56*, 8921-5 (2000).

Ruthenium-carbon   *Ru-C*
Cyclohexylcarbinols from acylophenones   Ⓗ
with $Rh-Al_2O_3$ cf. *26*, 52; with Ru-C s. Japanese patent JP-2000103752 (Dainippon Ink & Chem. Inc.).

Ruthenium hydride complex s. under Lipase   ←

Chiral ruthenium(II) phosphine complexes   ←
Noyori-type asym. homogeneous hydrogenation $C=C \rightarrow CHCH$
of eneurethans cf. *42*, 45s*55*; of exocyclic eneurethans s. P. Dupau, C. Bruneau et al., Tetrahedron: Asym. *10*, 3467-71 (1999); of α,β-ethylenecarboxylic acids with steroidal BINAPs as ligand s. V. Enev, J.T. Mohr et al., ibid. *11*, 1767-79 (2000); of β-hydroxy-α-methylene carboxylic acid esters with added *p*-methoxyphenol/didodecyl 3,3'-thiodipropionate in methanol s. Japanese patent JP-2000128832 (Nippon Shokubai Co. Ltd.).

Diacetato[(R)-2,2'-bis(di-p-tolylphosphino)-1,1'-binaphthalene]ruthenium(II)/   ←
1-n-butyl-3-methylimidazolium hexafluorophosphate
Asym. hydrogenation in an ionic liquid/water 2-phase medium
with product extraction into supercritical carbon dioxide

1.1 mmol Tiglic acid and 2 mol% $Ru(OAc)_2$((R)-tolBINAP) in 3:2 1-*n*-butyl-3-methylimidazolium hexafluorophosphate/water stirred under N_2 in a steel vessel, H_2 introduced to 5 atm. at 25°, warmed to 35° after 18 h, and supercritical CO_2 (at 175 atm. and at a flow rate of 1 ml/min) bubbled through the mixture and vented through a JASCO back-pressure regulator into a cold trap (containing isopropanol) during ca. 18 h → product. Conversion 99% (e.e. 85%). The product is extracted with high enantioselectivity from the ionic liquid by the CO_2 flow, the catalyst remaining in the ionic liquid for repeated use (at least 4 times) without significant loss of activity. Reaction also proceeds without the need for a fluorous or water-soluble catalyst., F.e.s. R.A. Brown, E. McKoon et al., J. Am. Chem. Soc. *123*, 1254-5 (2001).

Rhodium-alumina or Rhodium-carbon   *$Rh-Al_2O_3$ or Rh-C*
Benzene ring hydrogenation   Ⓗ
s. *23*, 66; functional group-directed diastereoselective hydrogenation of indans and tetralins s. V.S. Ranade, R. Prins et al., J. Org. Chem. *64*, 8862-7 (1999); of prim. benzylamines and 2-phenethylamines *at atm. pressure* and room temp. *in water* s. M. Strotmann, H. Butenschön, Synth. Commun. *30*, 4173-6 (2000).

Chiral rhodium(I) di(phosphines), di(phosphinites), diaminodi(phosphines),   ←
   aminophosphine-phosphinites, or amidophosphine-phosphinites
Asym. homogeneous hydrogenation $C=C \rightarrow CHCH$
of enacylamines, updates s. *27*, 57s*51*-9; with 3,4-O-isopropylidene-(3S,4S)-dihydroxy-(2R,5R)-bis(diphenylphosphino)hexane as ligand s. W. Li, X. Zhang, J. Org. Chem. *65*, 5871-4 (2000); with bidentate glucofuranoside-based di(phosphines) s. M. Diéguez et al., Tetrahedron:Asym.

11, 4701-8 (2000); with chiral spirocyclic di(phosphinites) s. X. Li, T.-K. Yang et al., ibid. *10*, 3863-7 (1999); with C$_2$-symmetric ferrocenyl diaminodi(phosphines) s. J.J. Almena Perea, P. Knochel et al., ibid. 375-84; with chiral aminophosphine-phosphinites s. X. Li, C.-H. Yeung et al., ibid. *11*, 2077-82 (2000); s.a. Y. Xie, Y. Jiang et al., ibid. 1487-94; s.a. D. Moulin, S. Jugé et al., ibid. *10*, 4729-43 (1999); with chiral amidophosphine-phosphinites s. E.A. Broger, M. Scalone et al., ibid. *9*, 4043-54 (1998).

Chiral rhodium(I) bis(phospholane) or bis(phosphetane) or 2,2′-biphosphol-3-ene ←
complexes
Asym. homogeneous hydrogenation C=C → CHCH
s. *47*, 48s*57*; with chiral mannitol-derived 1,2-bis(phospholan-1-yl)benzenes as ligand s. W. Li, X. Zhang et al., J. Org. Chem. *65*, 3489-96 (2000); with 2,2′-biphosphol-3-ene as ligand cf. F. Bienewald et al., Tetrahedron:Asym. *10*, 4701-7 (1999); with further 1,1′-bis(phosphetano)-ferrocenes as ligand (cf. *47*, 48s*58*; *59*, 39) s. A. Marinetti, J.-P. Genêt et al., Synlett *1999*, 1975-7.

Chiral rhodium(I) 1,1′-binaphthalene-2,2′-diyl phosphites, phosphonites or ←
phosphoromonoamidites
Asym. homogeneous hydrogenation with chiral monodentate ligands

34.

A soln. of methyl 2-acetamidoacrylate in methylene chloride hydrogenated at room temp. for 20 h under *1.3 atm.* H$_2$ in the presence of 0.1 mol% chiral Rh(I)-phosphonite complex (prepared as precatalyst by mixing Rh(cod)$_2$BF$_4$ with 2 eqs. of the phosphonite ligand) → product. Conversion 100%; e.e. 93%. Although enantioselectivities are generally lower than those reported for catalysts based on chiral cyclic bis(phosphonites), the advantages of using *monocyclic* chiral phosphonites is that they are more readily prepared and can be more easily modified by adjusting the organyl group on phosphorus. Most reactions were complete within 3-4 h. F.e. and comparison of ligands s. M.T. Reetz, T. Sell, Tetrahedron Lett. *41*, 6333-6 (2000); with 1,1′-binaphthalene-2,2′-diyl 1(R)-phenylethyl phosphite as ligand cf. Angew. Chem. Int. Ed. Engl. *39*, 3889-90 (2000); rapid procedure with 1,1′-binaphthalene-2,2′-diyl N,N-dimethylphosphoramidite as ligand **at high pressure** (60 atm.) cf. M. van den Berg, B.L. Feringa et al., J. Am. Chem. Soc. *122*, 11539-40 (2000).

Polymeric rhodium(I) phosphine complex ←
Heterogeneous hydrogenation in a protic medium

35.

A reusable, highly cross-linked, macroporous polymeric rhodium(I) complex is now available for hydrogenation in a broad range of solvents, incl. polar aprotic solvents, such as methanol (cf.

42, 48). **E:** A soln. of startg. alkene *in methanol* hydrogenated for 12 h under 100 psi H_2 with 3% polymeric Rh(I) complex (prepared by co-polymerization of a *p*-(isopropenylphenyl)-substituted cationic Rh(I) 1,2-di(phosphine) complex with ethylene glycol dimethacrylate) → product. Y 93% (100% conversion). THF and methylene chloride were also suitable solvents. Conversions were high, yields excellent, and work-up (filtration) simple. Furthermore, the catalyst can be reused up to 6 times with only a slight decrease in its activity. F.e., selectivity and diastereoselectivity, also hydroboration of styrene with the same complex, s. R.A. Taylor, M.R. Gagné et al., Org. Lett. *2*, 1781-3 (2000).

Chiral rhodium(I) phosphine-phosphite complexes ←
Asym. homogeneous hydrogenation $C{=}C \rightarrow CHCH$

36.

Inexpensive carbohydrate-based phosphine-phosphite ligands are more efficient in terms of enantioselectivity and reaction rate than the corresponding di(phosphines) or di(phosphites) (cf. 27, 57s57,58) in asym. homogeneous hydrogenation of α,β-ethylenecarboxylic acid derivs. **E:** A soln. of methyl N-acetamidoacrylate, 1 mol% [Rh(cod)$_2$]BF$_4$, and 1.1 mol% xylofuranoside-based phosphine-phosphite ligand in methylene chloride purged three times with H_2 and vacuum in a Schlenk tube, and shaken under 1 atm. H_2 at 20° for 20 min → (S)-product. Conversion 100% (e.e. 98.3%; TOF 330 after 5 min). (R)-BINOL derivs. gave the (R)-product. F.e. and comparison of phosphite ligands s. O. Pàmies, A. Ruiz et al., Chem. Commun. *2000*, 2383-4.

Palladium-carbon *Pd-C*
Hydrogenation of electron-deficient ethylene derivs.
of enones cf. *12*, 103s*15*; of 1-cyanomethylenesugars s. Y. Kakhrissi, Y. Chapleur et al., Tetrahedron: Asym. *11*, 417-21 (2000).

Palladium-carbon/ethylenediamine or squaric acid derivs. ←
Selective hydrogenation of carbon-carbon double bonds
with retention of N-Cbz groups using ethylenediamine as additive cf. *51*, 32s*56*; also with retention of N-Fmoc and BnO groups using squaric acid derivs. as additive, and cleavage of benzyl esters, s. T. Shinada, Y. Ohfune et al., Synthesis *2000*, 1506-8; with retention of *tert*-butyl-dimethylsilyl ethers (using ethylenediamine), also hydrogenolytic O-debenzylation and reduction of [ar.] nitro compds., s. K. Hattori et al., Tetrahedron Lett. *41*, 5711-4 (2000).

Palladous chloride/formic acid/sodium hydroxide $PdCl_2/HCOOH/NaOH$
Transfer-hydrogenation in aq. alkaline medium

37.

Carboxylic acids from α,β-ethylenecarboxylic acids. 7.5 Mol% PdCl$_2$ added to a stirred soln. of cinnamic acid in 2.5 *M* aq. NaOH, ca. 4 eqs. formic acid added dropwise, heated to 65° for 16 h, then the mixture neutralized with 2 *M* HCl → hydrocinnamic acid. Y 98%. This method does not require an organic solvent, is economical, safe, and environmentally friendly. It utilizes inexpensive reagents, and the catalyst is non-pyrophoric. F.e., also α-acylaminocarboxylic acids from 4-alkylidene-Δ^2-5-oxazolones, and reduction of α-keto acids, s. J.B. Arterburn et al., Tetrahedron Lett. *41*, 7847-9 (2000).

Tetrakis(triphenylphosphine)palladium(0) s. under NaBH$_4$ $Pd(PPh_3)_4$

Exchange ↕

Oxygen ↑   HC ↕ O

Irradiation s. under Bu₃SnH   ⇜
Indium/ammonium chloride   In/NH₄Cl
Oximes from 1,1-nitroethylene derivs.   C=C(NO₂) → CHC=NOH
ketoximes with Zn/MeNH₂ cf. *43*, 60; also aldoximes [from β-nitrostyryl derivs.] with In/satd. aq. NH₄Cl in methanol, and α-alkoxyketoximes with In/Me₃SiCl in anhydrous MeOH, s. J.S. Yadav et al., Synlett *2000*, 1447-9.

Zirconium(IV) tetrahydridoborate   Zr(BH₄)₄
Reductions using zirconium(IV) tetrahydridoborate   ←

38.   n-C₁₅H₃₁COOH   ⟶   n-C₁₅H₃₁CH₂OH

Prim. alcohols from carboxylic acids. A soln. of Zr(BH₄)₄ (3 *M* in hydride ion, and prepared from 1:4 ZrCl₄ and NaBH₄ in dry THF under N₂ at room temp. over 2 d) in THF added to a soln. of palmitic acid in THF, and the mixture stirred at room temp. for 1 h → 1-hexadecanol. Y 95%. The method is simple, smooth, effective, economical, safe, high-yielding and rapid at room temp. F.e., incl. diols from ethylenecarboxylic acids, also sec. ar. amines from ar. acylamines, alcohols from carboxylic acid esters, and amines from azomethines, s. S. Narasimhan, R. Balakumar, Synth. Commun. *30*, 4387-95 (2000).

Triethylsilane s. under Os₃(CO)₁₂   Et₃SiH

Diphenylsilane/di-tert-butyl peroxide   Ph₂SiH₂/t-BuOOBu-t
Radical deoxygenation of alcohols via trifluoroacetic acid esters   OCOCF₃ → H

39.

A mixture of O⁶-trifluoroacetyl-1,2:3,4-di-O-isopropylidene-α-D-galactopyranose, 5 eqs. diphenylsilane and 1 eq. di-*tert*-butyl peroxide sealed in an ampoule under argon, and heated at 140° for 15 h → 6-deoxy-deriv. Y 82%. This provides an efficient method for deoxygenation of prim. or sec. alcohols under neutral conditions, avoiding the use of toxic, difficult to remove tin hydrides, and expensive and moisture-sensitive derivatizing agents. Tert. alcohols afforded somewhat lower yields, however. F.e.s. D.O. Jang et al., Tetrahedron Lett. *42*, 1073-5 (2001).

1,1,2,2-Tetraaryldisilanes/azodiisobutyronitrile   Ar₂Si(H)Si(H)Ar₂/AIBN
Deoxygenation of alcohols via xanthates   OC(S)- → H
or thionocarbamic acid esters with Et₃SiH/*tert*-dodecyl mercaptan/1,1-bis(*tert*-butylperoxy)-cyclohexane cf. *46*, 65s48; with tetraphenyldisilane/AIBN, also Giese-type 1,4-addition to α,β-ethylenesulfones, and radical C-alkylation of N-heteroarene salts s. H. Togo et al., J. Org. Chem. *65*, 2816-9 (2000).

Tri-n-butyltin hydride/azodiisobutyronitrile/irradiation   Bu₃SnH/AIBN/⇜
Deoxygenation of sec. alcohols via thionocarbamic acid esters   OC(S)N< → H
s. *31*, 49s38; as part of a multistep preparation of nitriles from hindered aldehydes via cyanohydrins with retention of keto groups s. K.C. Nicolaou et al., Org. Lett. *2*, 1895-8 (2000).

Diphenylphosphine oxide/di-tert-butyl peroxide   Ph₂P(O)H/t-BuOOBu-t
Deoxygenation of sec. alcohols via xanthates   OC(S)SR → H
s. *54*, 44s56; α,α-difluoro- from β-(dithiocarbalkoxyoxy)-α,α-difluoro-carboxylic acid esters s. O. Jiménez, M.P. Bosch et al., Synthesis *2000*, 1917-24.

Hypophosphorous acid/iodine/acetic acid
Diarylmethanes from diaryl ketones ←
with HI/red phosphorus cf. 5, 63; with HI generated *in situ* from 1:2:2 $H_3PO_2/I_2/AcOH$ (cf. 60, 45) s. L.D. Hicks et al., Tetrahedron Lett. *41*, 7817-20 (2000). $CO \rightarrow CH_2$

Triosmium dodecacarbonyl/triethylsilane/diethylamine $Os_3(CO)_{12}/Et_3SiH/Et_2NH$
Amines from carboxylic acid amides $CON< \rightarrow CH_2N<$

40.

A mixture of N-acetylpiperidine, 3-3.5 eqs. triethylsilane, 1 mol% $Os_3(CO)_{12}$, and 5 mol% diethylamine in toluene heated at 100° under argon for 16 h → N-ethylpiperidine. Y 99.8% (by GLC). The monohydrosilane is more stable and less expensive than previously used polyhydrosilanes (under Rh-catalysis). Other hydrosilanes (incl. phenyldimethyl-, *tert*-butyldimethyl-, triisopropyl-, chlorodimethyl-, ethoxydimethyl-, diethoxymethyl-, and triethoxy-silane) gave satisfactory results. Use of pyridine as co-catalyst gave similar results to diethylamine, but triethylamine and *t*-butylamine were ineffective. F.e. and using Ru-complexes with EtI (or MeI or I_2) as additional co-catalyst s. M. Igarashi, T. Fuchigami, Tetrahedron Lett. *42*, 1945-7 (2001).

Nitrogen ↑ HC ↓↑ N

Chlorobis(cyclopentadienyl)hydridozirconium $Cp_2Zr(H)Cl$
Aldehydes from N,N-disubst. carboxylic acid amides $CON< \rightarrow CHO$
Selective reduction under mild conditions

41. $MeOCO(CH_2)_8CONEt_2 \longrightarrow MeOCO(CH_2)_8CHO$

A soln. of startg. amide in dry THF added under argon to 1.5-2 eqs. $Cp_2Zr(H)Cl$ at room temp., allowed to react for 15 min, concentrated, and worked up by short path silica gel chromatography (hexanes/ethyl acetate) → product. Y 74%. Such a reduction of a tert. amide with retention of an ester group is apparently unprecedented. The method is generally applicable, uses a commercial reagent, and proceeds with short reaction times under mild conditions. Hydroxamic acid esters (Weinreb amides), N-acyl-2-oxazolidones and hindered amides were all reduced. Furthermore, the presence of electron-withdrawing groups such as *p*-nitro or electron-donating groups such as *p*-methoxy was not detrimental, and cyano groups remained intact. However, in the case of N-methoxy-N-methylcinnamide, the aldehyde was obtained in 87% yield together with 12% of the allylic alcohol, and *p*-acetyl-N,N-diethylbenzamide afforded the hydroxyaldehyde. F.e. incl. a deuterio-deriv. s. J.M. White, A.R. Tunoori, G.I. Georg, J. Am. Chem. Soc. *122*, 11995-6 (2001).

Halogen ↑ HC ↓↑ Hal

Microwaves s. under PdCl$_2$ ←

Aluminum amalgam/triethylamine $Al,Hg/Et_3N$
β-Hydroxycarboxylic from α-bromo-β-hydroxycarboxylic acid derivs. $Br \rightarrow H$
β-hydroxy esters cf. 59, 45; chiral β-hydroxy acids from 3-(α-bromo-β-hydroxyacyl)-2-oxazolidinethiones (with retention of chirality) in the presence of Et_3N or DMAP s. Y.-C. Wang, T.-H. Yan, J. Org. Chem. *65*, 6752-5 (2000); chiral β-hydroxy esters (in one pot) s. Tetrahedron:Asym. *11*, 1797-800 (2000).

Indium/sodium dodecyl sulfate $In/NaOSO_2OC_{12}H_{25}$
Replacement of halogen by hydrogen $Hal \rightarrow H$
carbonyl compds. from α-iodo- or α-bromo-carbonyl compds. cf. 57, 44; from α-chloro- or α-bromo-carbonyl compds. with sodium dodecyl sulfate as surfactant s. L. Park, Y. Kim et al., J. Chem. Soc. Perkin Trans. 1 *2000*, 4462-3.

Lithium tetrahydridoaluminate s. under CrCl$_3$ $LiAlH_4$
Triethylsilane s. under PdCl$_2$ Et_3SiH

Tri-n-butylphosphine $\qquad Bu_3P$
Replacement of bromine by hydrogen $\qquad Br \rightarrow H$
with Ph_3P cf. *17*, 118; 6-subst. penicillanates from 6-subst. 6-bromopenicillanates with Bu_3P, inversion of configuration, s. A. Ishiwata, S. Mobashery et al., Org. Lett. *2*, 2889-92 (2000).

Chromium(III) chloride/lithium tetrahydridoaluminate $\qquad CrCl_3/LiAlH_4$
Ethylene derivs. from β,γ-ethylenechlorides $\qquad C{=}C\text{-}C(Cl) \rightarrow CHC{=}C$
Regiospecific reduction under nearly neutral conditions

42.

5 eqs. Water and a soln. of geranyl chloride in DMF (with decane as internal standard) added to a suspension of 2.5 eqs. $CrCl_3$ and 1.25 eqs. LAH in 2:1 DMF/THF, and stirred under an inert atmosphere for 15-24 h at room temp. → 3,7-dimethylocta-1,6-diene. The less-substituted olefin is formed. However, the procedure is incompatible with aldehyde groups. F.e. and stereoselectivity, also with methanol or ethanol as proton donor, s. M. Omoto, N. Kato et al., Tetrahedron Lett. *42*, 939-41 (2001).

Hydrogen iodide $\qquad HI$
Replacement of chlorine by hydrogen $\qquad Cl \rightarrow H$
Vinylogous formamidinium salts

43.

1.1 eqs. HI (57 wt% aq. soln.) added to a soln. of startg. chloroamidinium salt in dioxane, and stirred at 25° for 2.5 h → product. Y 83%. Attempted reduction of bromo- and iodo-analogs led to decomposition. F.e.s. I.W. Davies et al., Org. Lett. *2*, 3385-7 (2000).

Iron s. under Pd $\qquad Fe$

Palladium-carbon/triethylamine $\qquad Pd\text{-}C/Et_3N$
Replacement of ar. iodine by hydrogen $\qquad I \rightarrow H$
by hydrogenolysis with NaOAc cf. *11*, 633; with Et_3N, retention of ethylene, nitro and aldehyde groups, s. Y. Ambroise, B. Rousseau et al., J. Org. Chem. *65*, 7183-6 (2000).

Palladium/iron $\qquad Pd/Fe$
Replacement of chlorine by hydrogen $\qquad Cl \rightarrow H$
in polychloroarenes under Pd-catalysis cf. *33*, 76s59; eco-friendly dechlorination of PCBs with zero-valent metal or bimetallic mixtures, e.g. Pd/Fe, *in supercritical carbon dioxide* s. Q. Wu, W.D. Marshall et al., Green. Chem. *2*, 127-32 (2000).

Palladous chloride/triethylsilane/microwaves
Replacement of halogen by hydrogen $\qquad Hal \rightarrow H$
with $PdCl_2/Et_3SiH$ cf. *52*, 38; rapid procedure in dry acetonitrile under microwave irradiation, also aldehydes from carboxylic acid halides, and selectivity, s. D. Villemin, B. Nechab, J. Chem. Res., Synop *2000*, 432-4.

Sulfur ↑ $\qquad$ HC ↓↑ S

Triphenyltin hydride/diphenylsilane/triethylborane $\qquad \leftarrow$
Ethers or cyclic ethers from thiono-carboxylic acid esters or -lactones $\qquad C(S) \rightarrow CH_2$
with 3 eqs. $Bu_3SnH/AIBN/Et_3B$ cf. *51*, 42s56; with 3 eqs. Ph_2SiH_2 and 0.5 eq. Ph_3SnH s. D.O. Jang, S.H. Song, Synlett *2000*, 811-2.

Nickel $\qquad Ni$
Hydrocarbons from dithiocarbonic acid derivs. $\qquad SC(S)\text{-} \rightarrow H$
from xanthates cf. *3*, 73; arenes from thiolcarbamic acid aryl esters s. M. Allegretti et al., World Intellectual Property Organization patent WO-200026176 (Dompe Spa).

Nickel boride Ni_2B
Desulfuration   ←

of mercaptals, thioamides and benzothiophenes s. *45*, 43; cf. *20*, 81; chiral ketones and esters from the corresponding γ-(hetarylsulfinyl) derivs., and 2-step procedure via the corresponding γ-(hetarylthio) derivs. s. M. Casey et al., Synlett *2000*, 1725-8.

Remaining Elements ↑   HC ↓↑ Rem

Potassium tert-*butoxide/tetra-*n-*butylammonium fluoride/18-crown-6 polyether/methanol*   ←
Ethylene derivs. from enesilanes $C=C(Si\leqslant) \rightarrow C=CH$

44.

Terminal ethylene derivs. 5 eqs. 1 M n-Bu₄NF in THF added dropwise to a stirred soln. of startg. vinylsilane in DMSO, the mixture cooled to 0°, treated with 1.1 eqs. *t*-BuOK and 0.2 eq. 18-crown-6 polyether, warmed to room temp. after 1 h, stirred for 36 h, then treated with 2 eqs. methanol → product. Y 82%. The procedure is high yielding, mild, and notably applicable to terminal enesilanes containing other sensitive functional groups (cf. *31*, 65). Boc-amines, triisopropylsilyl ethers, and carboxylic acid amides were unaffected. An alternative hydroboration/ Peterson olefination protocol proved unsuitable for these substrates. F.e., and without Bu₄NF s. J.C. Anderson, A. Flaherty, J. Chem. Soc. Perkin Trans. 1 *2000*, 3025-7.

Carbon ↑   HC ↓↑ C

Sodium or Magnesium chloride $NaCl$ or $MgCl_2$
Methyl groups from malonic acid esters $CH(COOR)_2 \rightarrow CH_3$
with aq. HCl cf. *13*, 799; *o*-nitrotoluenes with MgCl₂·6H₂O in dimethylacetamide or NaCl in DMSO s. M. Gurjar et al., Synthesis *2000*, 1659-61.

Elimination   ↑

Oxygen ↑   HC ↑ O

Hypophosphorous acid/iodine/acetic acid $H_3PO_2/I_2/AcOH$
Diarylmethanes from benzhydrols $OH \rightarrow H$

45.

ca. 5 eqs. 50% aq. H₃PO₂ (1 part) added to a stirred mixture of ca. 1:1 I₂ and 4-methylbenzhydrol in acetic acid (12.5 parts), and the mixture heated at 60° for 24 h → 4-methyldiphenylmethane. Y 100%. This straightforward method is applicable to a wide variety of benzhydrols bearing electron-donating or -withdrawing groups, and proceeds cleanly and in very high yield after a simple work-up. The active reducing agent is believed to be HI, with reaction taking place via carbocation formation. Substrates bearing nitro or ester groups were not investigated since the former are known to be reduced by HI and the latter hydrolyzed under these conditions. Furthermore, in 4-methoxybenzhydrol O-demethylation and partial esterification was observed, and hydrogenolysis of halogen atoms occurred during reduction of 4,4'-diiodobenzhydrol (but not 4-bromo-, 4-chloro- or 4,4'-dichloro-benzhydrol). F.e.s. P.E. Gordon, A.J. Fry, Tetrahedron Lett. *42*, 831-3 (2001).

Via intermediates   v.i.
Nitriles from cyanohydrins via O-thionocarbamylation s. *31*, 49s60

Carbon ↑   HC ⇑ C

Sodium hydroxide/irradiation   NaOH/⫽
Photodecarboxylation   COOH → H
with Ph$_2$CO as sensitizer cf. *27*, 93; aryl ketones from α-(ketoaryl)carboxylic acids with NaOH in 1:1 water/acetonitrile at pH 7, also **deformylation**, s. M. Xu, P. Wan, *Chem. Commun. 2000*, 2147-8.

Boron fluoride   BF$_3$
4-Acetylenealcohols from 4-acetylene-1,3-diol 1-(monobenzyl ethers)   ←
via dicobalt hexacarbonyl complexation
Asym. induction via 1,5-hydride shift

46.

1 eq. BF$_3$·OEt$_2$ added to a soln. of startg. chiral Co$_2$(CO)$_6$-complexed alkyne (prepared from the parent alkyne and Co$_2$(CO)$_8$ in methylene chloride at room temp.) in methylene chloride at -20°, and the resulting complexed bishomopropargyl alcohol demetalated with (NH$_4$)$_2$Ce(NO$_3$)$_6$ in acetone at 0° → product. Y 79% (from the uncomplexed benzyl ether). Significantly, intramolecular reduction of such tertiary propargyl alcohols, where the α-alkyl group is not very bulky, affords the corresponding *sec*-alkylacetylene as a unique diastereoisomer regardless of the stereochemistry of the propargylic carbinol centre. However, stereoselectivity was low for substrates with the bulky *tert*-butyl group at the carbinol site. Reaction is thought to involve stereospecific hydride transfer to the incipient propargylic (Nicholas-type) carbonium ion. F.e., stereospecific deuterium-labelling, and conversion to **α,γ-disubst. γ-lactones** with retention of configuration, s. D. Díaz, V.S. Martín, *Org. Lett. 2*, 335-7 (2000).

Formation of O-N Bond

Uptake   ⇓

Addition to Hydrogen and Nitrogen   ON ⇓ HN

Potassium peroxymonosulfate/silica gel/microwaves   ←
Hydroxylamines from amines under microwave irradiation   ⩾NH → ⩾NOH
in the absence of solvent s. *49*, 85s60

Addition to Nitrogen   ON ⇓ N

Microwaves s. under KHSO$_5$ ←
Urea-hydrogen peroxide/trifluoroacetic anhydride   (H$_2$N)$_2$CO·H$_2$O$_2$/(CF$_3$CO)$_2$O
N-Oxidation   ⩾N → ⩾NO
with added phthalic anhydride cf. *46*, 123s48; pyridine N-oxides from electron-poor pyridines with added (CF$_3$CO)$_2$O s. S. Caron, J.E. Sieser et al., *Tetrahedron Lett. 41*, 2299-302 (2000).

Potassium peroxymonosulfate/silica gel/microwaves   ←
N-Oxidation
of aminopyridines with KHSO$_5$/KOH cf. *49*, 85; N-oxidation of tert. amines and pyridines with Oxone/silica gel [or alumina], also under microwave irradiation without solvent, and hydroxylamines from prim. or sec. amines, s. J.D. Fields, P.J. Kropp, *J. Org. Chem. 65*, 5937-41 (2000).

Exchange ↕

Hydrogen ↑ ON ↓↑ H

Titanium(IV)-supported superoxide/hydrogen peroxide ←
Ar. nitro compds. from amines $NH_2 \rightarrow NO_2$
Heterogeneous catalytic oxidation with a supported superoxide

47.

PhNH$_2$ $\longrightarrow$ PhNO$_2$

6 eqs. 50% aq. H$_2$O$_2$ added slowly to a mixture of aniline, 25 wt% titanium(IV)-supported superoxide (prepared as a solid matrix of Ti content 41.7% by action of 50% H$_2$O$_2$ on Ti(OPr-i)$_4$ in methanol), and anhydrous methanol under N$_2$ with stirring over 10 min (exothermic), and worked up after 30 min → nitrobenzene. Y 98.18% (100% conversion). The supported superoxide is exceptionally stable, easy to prepare, and readily recoverable for reuse. Selectivity is high with a H$_2$O$_2$/amine ratio of 6:1 (even for ar. amines bearing electron-withdrawing groups, such as COOH and NO$_2$). Aliphatic prim. amines, however, gave **oximes** (together with 5-12% of the corresponding ketones). A catalytic cycle is invoked. F.e.s. G.K. Dewkar, A. Sudalai et al., Angew. Chem. Int. Ed. Engl. *40,* 405-8 (2001).

Hypofluorous acid HOF
Ar. nitro compds. from amines
s. *46,* 90; from electron-deficient ar. amines with HOF generated *in situ* from 20% F$_2$/He in 3% water/acetonitrile/ethyl acetate, selectivity, s. S.M. Dirk, J.M. Tour et al., Org. Lett. *2,* 3405-6 (2000).

Elimination ↑

Oxygen ↑ ON ↑ O

Alumina Al_2O_3
Furoxans from α-nitrooximes under mild conditions O

48.

A soln. of startg. α-nitroketoxime in acetonitrile added to a well-stirred suspension of *acidic* Al$_2$O$_3$ (Brockmann I; 1 g per mmol substrate) in the same solvent, and stirring continued *at 60°* for 1 h → 4,5,6,7-tetrahydro-2,1,3-benzoxadiazole N-oxide. Y 93%. Reaction is relatively fast and high-yielding, the startg. m. are readily available, and work-up is simple. F.e.s. M. Curini, R. Ballini et al., Tetrahedron Lett. *41,* 8817-20 (2000); 1,2-benzisoxazol-3(2*H*)-ones from *o*-hydroxyhydroxamic acids with Ph$_3$P/EtOOCN=NCOOEt s. G. Shi, Tetrahedron Lett. *41,* 2295-8 (2000).

Formation of O-S Bond

Uptake ⇓

Addition to Sulfur   OS ⇓ S

Myoglobin mutants or Enzymes   ←
Sulfoxides from thioethers by asym. oxidation   >S → >SO
with a peroxidase cf. *24*, 100s58; chiral β-ketosulfoxides with chloroperoxidase s. R.R. Vargas et al., Tetrahedron:Asym. *10*, 3219-27 (1999); with the vanadium bromoperoxidase from *Ascophyllum nodosum* s. H.B. ten Brink et al., ibid. 4563-72; with sperm whale myoglobin active site mutants cf. S. Ozaki, Y. Watanabe et al., ibid. 183-92.

Hetaryl iodosoacetates   *ArI(OAc)$_2$*
S-Oxidation s. *43*, 213s60   >S → >SO or >SO$_2$

N-Condensed 1,2-benziodazol-3-one 1-oxides/trifluoroacetic acid   ←
Sulfoxides from thioethers s. *50*, 123s60   >S → >SO

Titanium tetraisopropoxide/L-diethyl tartrate/hydroperoxides   ←
Sulfoxides from thioethers by catalytic asym. oxidation
s. *39*, 83; chiral β-ketosulfoxides with 2-α-hydroperoxyfurans as reoxidant s. A. Lattanzi et al., Tetrahedron:Asym. *9*, 2619-25 (1998).

Vanadyl acetoacetonate/chiral 2-(salicylideneamino)alcohols/hydrogen peroxide   ←
Sulfoxides from thioethers by catalytic asym. oxidation
s. *51*, 46s56; C$_2$-symmetric di(sulfoxides) and chiral (arylthio)sulfoxides from di(thioethers) s. J. Skarzewski et al., Tetrahedron:Asym. *10*, 3457-61 (1999).

N-Sulfonyloxaziridines   ←
Sulfoxides from thioethers
s. *35*, 57; **asym. oxidation** with chiral 2,3-oxido-1,2-benzisothiazoline 1,1-dioxides s. D. Bethell, P.C. Bulman Page et al., J. Org. Chem. *65*, 6756-60 (2000).

Exchange ⇕

Nitrogen ↑   OS ⇕ N

Ethyldiisopropylamine   *i-Pr$_2$NEt*
Polymer-based O-triflylation   OH → OTf

49. PhOH + Tf$_2$N–C$_6$H$_4$–[O–CH$_2$CH$_2$]$_n$–O–C$_6$H$_4$–NTf$_2$ ⟶ PhOTf + [TfHN–C$_6$H$_4$–[O–CH$_2$CH$_2$]$_n$–O–C$_6$H$_4$–NHTf]

The first polymer-supported triflylating agent is reported. **E: Aryl triflates.** Phenol, 2 eqs. soluble, PEG-based triflylating agent, and *i*-Pr$_2$NEt added to methylene chloride, and stirred for

8 h at room temp. → phenyl triflate. Y 97% (88% with N-phenyltriflimide, cf. *38*, 73). Reaction is generally applicable to a range of phenols (such as those bearing methoxy, nitro, nitrile or formyl groups) and the procedure is well suited to high-throughput synthesis. Work-up is simple, and the reagent can be readily regenerated by precipitation of the by-product, filtration, and re-triflylation (with Tf$_2$O/trioctylamine). The reagent is also very stable, there being <10% hydrolysis on standing in air for 2 days. F.e. and **regiospecific formation of enol triflates from ketones** under *kinetic* control s. A.D. Wentworth, K.D. Janda et al., Org. Lett. *2*, 477-80 (2000).

m-*Chloroperoxybenzoic acid/hydrogen chloride*   ArCOO$_2$H/HCl
Sulfones from sulfoximines   >S(O)=NR → >SO$_2$

50.

Chiral sulfones. 0.75 Mol% 0.1 *M* aq. HCl added to a soln. of startg. (S$_S$)-sulfoximine and 2 eqs. *m*-chloroperoxybenzoic acid (15% water) in THF, and heated for 8 h to reflux → (E,R)-4-phenyl-sulfonylhept-4-en-3-ol. Y 97%. The N-atom may be unsubstituted, or substituted by alkyl or aryl groups. There was no epoxidation, elimination or isomerization under these conditions, and sec. alcohols and siloxy groups were unaffected. F.e. and from polymer-based sulfoximines s. J. Hachtel, H.-J. Gais, Eur. J. Org. Chem. *2000*, 1457-65 (2000).

Halogen ↑   OS ↓↑ Hal

Without additional reagents   w.a.r.
Sulfuric acid amide esters from alcohols in the absence of base   OH → OSO$_2$NH$_2$

51.

2 eqs. Sulfamoyl chloride added to a mixture of startg. hydroxy compd. in dimethylacetamide (DMA) with ice-cooling, and stirred at room temp. for 6 h → 17-oxo-17a-homo-17a-oxaestra-1,3,5(10)-trien-3-yl sulfamate. Y 89%. The method is high-yeielding, cheap, easily scaled-up (to the 20 g level), and applicable to phenols, aliphatic prim. and cyclic sec. alcohols (not tert. alcohols). The rate of reaction was slow in DME, dichloromethane or DMF, but fast in NMP or DMA with minimal by-product formation. The latter solvents were assumed to work as moderate bases without wasting sulfamoyl chloride. Addition of excessive Et$_3$N as base decreased the yield as a result of further sulfamoylation and decomposition of the sulfamoyl chloride. Interestingly, elimination of a base led to the highest yield. Functional groups such as ketones, phenolethers, ketals and ethylene derivs. were tolerated. F.e. and with NMP as solvent s. M. Okada, N. Koizumi et al., Tetrahedron Lett. *41*, 7047-51 (2000).

Formation of O-Rem Bond

Uptake ⇓

Addition to Hydrogen and Oxygen ORem ⇓ HO

N-2,4,6-Triisopropylbenzenesulfonyltetrazolide/pyridine
O-Phosphorylation

$\quad\quad\quad\quad\quad\quad\quad\quad\quad\quad\quad\quad\quad\quad\quad\quad\quad\quad\quad\quad\quad\quad\quad\quad\quad$ ←
$\quad\quad\quad\quad\quad\quad\quad\quad\quad\quad\quad\quad\quad\quad\quad\quad\quad\quad\quad\quad\quad\quad$ OH → OPO(OH)$_2$

52.

of nucleosides. 5 eqs. Disodium 2-O-(4,4'-dimethoxytrityloxy)ethylsulfonylethan-2'-yl phosphate, 10 eqs. N-2,4,6-triisopropylbenzenesulfonyltetrazolide, and pyridine (excess) added under N$_2$ to startg. anhydrous nucleoside (0.2 mmol scale), refrigerated for 1 h, kept at 25° for 2 h, partitioned between methylene chloride and satd. aq. NaHCO$_3$, the organic layer dried and evaporated, the residue dissolved in pyridine, treated with 2 M ethanolic NaOH, and stirred at 25° for 20 min → product. Y 91%. The procedure is high-yielding, applicable to the phosphorylation of prim. or sec. hydroxyl groups, and of potential in the synthesis of oligonucleotides, carbohydrates and other biomolecules. F.e.s. M. Taktakishvili, V. Nair, Tetrahedron Lett. *41*, 7173-6 (2000); regiospecific O^2-phosphorylation of aldoses with diamidophosphate, (NH$_2$)$_2$P(O)O, also isolation of the intermediate 1,2-cyclophosphates (with amidotriphosphate, H$_2$NPO(O$^-$)OPO(O$^-$)OPO(O$^-$)$_2$ or diamidophosphate), s. R. Krishnamurthy, A. Eschenmoser et al., Angew. Chem. Int. Ed. Engl. *39*, 2281-5 (2000).

Via intermediates $\quad\quad\quad\quad\quad\quad\quad\quad\quad\quad\quad\quad\quad\quad\quad\quad\quad\quad\quad\quad\quad\quad\quad\quad\quad\quad\quad\quad\quad$ v.i.
Glycosyl dihydrogen phosphates from aldoses
with H$_3$PO$_4$/propylene oxide/pyridine cf. *50*, 43; with tris(1-imidazolyl)phosphine via glycosyl H-phosphonates and phosphates (as the triethylammonium salt) s. D.V. Yashunsky, A.V. Nikolaev, J. Chem. Soc. Perkin Trans. 1 *2000*, 1195-8.

Addition to Hydrogen and ORem ⇓ HRem
Remaining Elements

Dichloro(p-cymene)ruthenium(II) dimer $\quad\quad\quad\quad\quad\quad\quad\quad\quad\quad\quad\quad\quad\quad\quad\quad\quad\quad\quad\quad$ ←
Silanols from organosilicon hydrides $\quad\quad\quad\quad\quad\quad\quad\quad\quad\quad\quad\quad\quad\quad\quad\quad$ ⩾SiH → ⩾SiOH
Ruthenium-catalyzed oxidation under mild conditions

53.

2 Mol% dichloro(*p*-cymene)ruthenium(II) dimer and 2 eqs. deionized water added to a soln. of dimethylphenylsilane in acetonitrile, and the mixture stirred under air for 10 min at room temp. → dimethylphenylsilanol. Y 95%. These mild conditions afforded excellent yields with the highest selectivity for silanol over disiloxane formation, even for unhindered silanes, and regardless of the substituents on silicon. The method is notably applicable to ene- and 1-acetylene-silanol

formation without simultaneous oxidation of the unsaturated groups; halides, thienyl sulfur and other silane groups were also unaffected. Oxidation of optically active silanes proceeded **with inversion of configuration**, the first such example using a homogeneous catalyst. Disubstituted silanes gave **silanediols** but with more steric sensitivity (resulting in higher reaction temp.). F.e.s. M. Lee, S. Ko, S. Chang, J. Am. Chem. Soc. *122*, 12011-2 (2000).

Exchange ⇅

Hydrogen ↑ ORem ⇵ H

Microwaves s. under NaOH ←

Sodium hydroxide/microwaves ←
Sym. triaryl phosphates from phenols   3 ArOH → (ArO)$_3$P(O)
from the Na$^+$ salt in ether cf. *13*, 156; with NaOH under microwave irradiation with a few drops of water s. A.D. Sagar, B.P. Bandgar et al., Org. Prep. Proced. Int. *32*, 269-71 (2000).

Pyridine C_5H_5N
Cyclic phosphoric acid diesters from diols   ○
s. *38*, 86s53; nucleoside 2',3'-cyclic thiolphosphates from 5'-protected nucleosides with diphenyl phosphite/S$_8$ s. J. Jankowska, A. Kraszewski et al., Tetrahedron Lett. *41*, 2227-9 (2000).

Dibenzo[b,d]thiophene S-oxide/trifluoromethanesulfonic anhydride/   ←
  2,4,6-tri-tert-butylpyridine
Glycosyl phosphates from aldoses   OH → OPO(OR)$_2$

54.

β-Glycosyl phosphates. 1.4 eqs. Tf$_2$O added to a soln. of startg. aldose, 2 eqs. dibenzo[*b,d*]-thiophene S-oxide [DBTO], and 5 eqs. 2,4,6-tri-*tert*-butylpyridine (as acid scavenger) in methylene chloride at -78°, stirred at -45° for 1 h, 3 eqs. dibenzyl hydrogen phosphate added, stirred at -45° for 1 h, at 0° for 30 min, and at 23° for 1 h, and neutralized with 10 eqs. Et$_3$N prior to immediate concentration and purification by flash column chromatography → product. Y 93%. Reaction is thought to proceed through an anomeric oxosulfonium triflate (or a sulfurane species). 1,2-*trans*-Derivs. were obtained from O^2-acylaldoses as a result of neighbouring group participation. F.e. and **α-glycosyl phosphates** by controlled anomerization with BF$_3$-etherate/2-chloropyridine s. B.A. Garcia, D.Y. Gin, Org. Lett. *2*, 2135-8 (2000).

Oxygen ↑ ORem ⇵ O

Without additional reagents   *w.a.r*
Phosphonous acid monoesters from phosphonous acids   P-OH → P-OR
and tetraalkoxysilanes s. *60*, 56

Nitrogen ↑ ORem ⇵ N

2,4-Dinitrophenol   *ArOH*
Phosphorous acid esters from phosphoromonoamidites   P-N< → P-OR
with 1*H*-tetrazole cf. *43*, 83; improved procedure with 2,4-dinitrophenol for coupling phosphorus(III) amides bearing strongly electron-attracting groups at the phosphorus centre, also **dinucleoside phosphoromonoamidites** by coupling the free nucleoside with bis(2,4-dinitrophenyl) N,N-dialkylphosphoromonoamidites, s. W. Dabkowski, J. Michalski et al., Tetrahedron Lett. *41*, 7535-9 (2000).

Iodine I_2
Catalytic O-silylation under mild conditions OH → OSi≤

55.

$$Ph_2C(Me)OH + [(Me_3Si)_2\overset{+}{N}HI] \xrightarrow{\underset{}{\downarrow I_2} \; (Me_3Si)_2NH} Ph_2C(Me)OSiMe_3 \; [+ I_2 + NH_3]$$

of alcohols. 0.8 eq. Hexamethyldisilazane in methylene chloride added dropwise during 5 min to a stirred soln. of startg. alcohol and *10 mol%* I_2 in the same solvent, treated portionwise after 30 min with powdered $Na_2S_2O_3$, and stirring continued for 30 min → product. Y 98%. The procedure is generally applicable to prim., sec., and tert. alcohols, incl. highly *acid-sensitive,* hindered alkyl(diaryl)carbinols. Reaction is presumed to take place in a catalytic cycle wherein I_2 initially polarizes the Si-N bond of the disilazane to give a reactive silylating agent, the catalyst being liberated to complete the cycle with evolution of NH_3. F.e.s. B. Karimi, B. Golshani, J. Org. Chem. *65,* 7228-30 (2000).

Halogen ↑ ORem ↓↑ Hal

Pyridine/4-dimethylaminopyridine/imidazole ←
O-Silylation OH → OSi≤
with simultaneous O-tritylation and O-acylation of carbohydrates s. *60,* 90

Silver trifluoromethanesulfonate AgOTf
Protection of carboxyl groups as tris(2,6-diphenylbenzyl)silyl esters s. *60,* 4 COOSi≤

Remaining Elements ↑ ORem ↓↑ Rem

Without additional reagents w.a.r
Phosphonous acid monoesters P-OH → P-OR
from phosphonous acids and tetraalkoxysilanes

56.

$$Ph-\underset{OH}{\overset{O}{\overset{\|}{P}}}{}^{H} + Si(OCH_2CH=CH_2)_4 \longrightarrow Ph-\underset{OCH_2CH=CH_2}{\overset{O}{\overset{\|}{P}}}{}^{H}$$

An equimolar mixture of phenylphosphinic acid and tetraallyloxysilane in toluene refluxed for 24 hrs. under N_2 → product. Y 94%. Phosphinic and phosphonic acids did not react under these conditions. For methyl esters, alkyltrimethoxysilanes can be used to avoid handling the toxic tetramethoxysilane. F.e. and one-pot esterification with alcohols and commercially available silanes s. Y.R. Dumond, J.-L. Montchamp et al., Org. Lett. *2,* 3341-4 (2000).

***p*-Quinol silyl ethers from *p*-quinones** ←
and hexamethyldisilane/I_2 cf. *38,* 91; and trimethylsilyl phenyl telluride without additional catalyst or promoter s. S. Yamago, J. Yoshida et al., Org. Lett. *2,* 3671-3 (2000).

Carbon ↑ ORem ↓↑ C

Microencapsulated scandium(III) triflate $Sc(OTf)_3$
O-Silylation with 2-ethylenesilanes OH → OSi≤
using a little TsOH cf. *36,* 101; methallylation of alcohols or phenols using readily recoverable microencapsulated $Sc(OTf)_3$ s. T. Suzuki, T. Oriyama et al., Tetrahedron Lett *41,* 8903-6 (2000).

Formation of O-C Bond

Uptake ⇓

Addition to Hydrogen and Carbon OC ⇓ HC

Hydrogen peroxide/selenium dioxide H_2O_2/SeO_2
Carboxylic acids from aldehydes CHO → COOH

57. n-C_9H_{19}CHO ⟶ n-C_9H_{19}COOH

A mixture of decanal, 2 eqs. 30% H_2O_2, and 5 mol% SeO_2 in THF (freshly distilled from LiAlH$_4$) stirred under reflux for 4 h, then a little Pd/asbestos (10%) added to decompose the peroxides → decanoic acid. Y 100%. Aliphatic aldehydes undergo oxidation to carboxylic acids substantially faster than aromatic and heteroaromatic aldehydes, reaction being more efficient than that with benzeneseleninic acid as catalyst (cf. *44*, 97). Nonetheless, yields were high with aromatic aldehydes (notably bearing methyl groups or one or two electron-withdrawing groups). However, aromatic aldehydes with two electron-donating methoxy groups produced phenols as the major product, while 3-indolecarboxaldehyde gave a mixture of several products. The oxidation was less selective in other solvents, such as DMF, MeOH, acetone, MeCN, dioxane or *t*-BuOH. F.e. incl aromatic, dicarboxylic, and heteroaromatic acids s. M. Brzaszcz, J. Mlochowski, Synth. Commun. *30*, 4425-34 (2000).

Cationic (6,6'-dichloro-2,2'-bipyridyl)ruthenium(VI) oxo complexes/ ←
tert-butyl hydroperoxide
Catalytic oxidation of unactivated hydrocarbons H → OH
with carbonyl[tetrakis(pentafluorophenyl)porphyrinato]ruthenium(II) cf. *50*, 53s*52*; with cationic (6,6'-dichloro-2,2'-bipyridyl)ruthenium(VI) oxo complexes, and further oxidations, s. C.-M. Che et al., J. Org. Chem. *65*, 7996-8000 (2000).

Addition to Oxygen-Oxygen Bonds OC ⇓ OO

Azafullerenes/irradiation ←
Ene reaction with singlet oxygen ←
with Rose Bengal as sensitizer cf. *48*, 106s*54*; with azafullerenes, e.g. C_{59}NH, also [4+2]-cycloaddition, s. N. Tagmatarchis, H. Shinohara, Org. Lett. *2*, 3551-4 (2000).

Addition to Oxygen and Carbon OC ⇓ OC

Lithium bromide LiBr
1,3-Dioxolan-2-ones from epoxides and carbon dioxide s. *42*, 108s*60* ○

1,4-Bis(9-O-dihydroquin[id]ine)anthraquinone ←
Dicarboxylic acid monomethyl esters from their anhydrides C
Desymmetrization
with quin[id]ine cf. *41*, 118s*56*; details s. C. Bolm et al., J. Org. Chem. *65*, 6984-91 (2000); application to axinellamine synthesis s. J.T. Starr, E.M. Carreira et al., J. Am. Chem. Soc. *122*, 8793-4 (2000); with recoverable 1,4-bis(9-O-dihydroquin[id]ine)anthraquinone (e.e. 90-8%) and comparison of ligands s. Y. Chen, L. Deng et al., J. Am. Chem. Soc. *122*, 9542-3 (2000).

Dimethylformamide HCONMe$_2$
1,3-Dioxolan-2-ones from epoxides and carbon dioxide ○
with Et$_4$NBr under ca. 80 atm. CO_2 cf. *23*, 139; at atm. pressure with LiBr (cf. *42*, 108s*49*) s. T. Iwasaki, T. Endo et al., Bull. Chem. Soc. Jpn. *73*, 713-9 (2000); in supercritical carbon dioxide/dimethylformamide in the absence of additives s. H. Kawanami, Y. Ikushima, Chem. Commun. *2000*, 2089-90.

Chiral cyclic dipeptide/L-diisopropyl tartrate ←
Immobilized lipase ←
α-Acylaminocarboxylic acid esters from Δ²-5-oxazolones C
with dynamic kinetic resolution using lipozyme cf. *50*, 54; improved procedure with Novozyme s. S.A. Brown, N.J. Turner et al., Tetrahedron:Asym. *11*, 1687-90 (2000); **asym. alcoholysis** with cyclo[(S)-His-(S)-Phe]/L-diisopropyl tartrate (e.e. 20-39%) s. L. Xie, W. Hua et al., Tetrahedron:Asym. *10*, 4715-28 (1999).

Immobilized esterase ←
Dicarboxylic acid monoesters from anhydrides
Desymmetrization
with lipase cf. *41*, 118s*44*; with pig liver esterase immobilized in a hollow fibre ultrafiltration membrane s. Tetrahedron:Asym. *11*, 929-34 (2000).

Hydrolase or Yeast ←
Glycols from epoxides with kinetic resolution   △O → C(OH)C(OH)
with pig or rabbit microsomal hydrolase cf. *48*, 108s*54*; kinetic resolution of phenyloxiranes with soluble hydrolases from a variety of sources, regio- and face-selectivity, s. K.C. Williamson, B.D. Hammock et al., Tetrahedron:Asym. *11*, 4451-62 (2000); isolation of pure (S)-2-pyridyloxirane s. Y. Genzel et al., ibid. 3041-4; kinetic resolution with yeast (*Rhodotorula, Rhodosporidium or Trichosporon* sp.) s. A.L. Botes, M.S. van Dyk et al., Tetrahedron:Asym. *10*, 3327-36 (1999).

Hydrogen peroxide s. under $(NH_4)_6Mo_7O_{24}$ H_2O_2

Ammonium molybdate/hydrogen peroxide $(NH_4)_6Mo_7O_{24}/H_2O_2$
α-Hydroxyketones from epoxides   △O → COC(OH)
Regiospecific oxidative ring opening under mild conditions

58.   PhO–△O → PhO–C(O)–OH

A stirred mixture of startg. epoxide and 1 eq. $(NH_4)_6Mo_7O_{24}\cdot 4H_2O$ in THF treated with 30% H_2O_2 at room temp., and worked up after 60 min → product. Y 96%. Progress of the reaction could be followed by measuring the colour change from white to yellow. N-Functionalized or styrene epoxides reacted faster than O-functionalized derivs. Phenolethers, acylamines and cyclic imides remained unaffected. Reaction presumably involves selective oxidation of the intermediate glycol. F.e.s. N. Ismail, R.N. Rao, Chem. Lett. *2000*, 844-5.

Silicotungstic acid $H_4SiW_{12}O_{24}$
Thiele-Winter reaction ←
with CF_3SO_3H cf. *8*, 165s*53*; with $HClO_4$, $ClSO_3H$ or TfOH, or with silicotungstic acid under heterogeneous catalysis s. D. Villemin et al., J. Chem. Res., Synop *2000*, 356-8.

Ferric trifluoroacetate $Fe(OCOCF_3)_3$
Solvolytic ring opening of epoxides   △O → C(OH)C(O-)
with $Fe(ClO_4)_3$ cf. *50*, 55s*59*; alcoholysis, hydrolysis and acetolysis with non-hygroscopic $Fe(OCOCF_3)_3$, **also 1,3-dioxolanes** (acetonide derivs.) with acetone, s. N. Iranpoor, H. Adibi, Bull. Chem. Soc. Jpn. *73*, 675-80 (2000).

Polymer-based ferric chloride ←
Acylals from aldehydes   CHO → $CH(OAc)_2$
with $FeCl_3$ cf. *38*, 100; general procedure with readily retrievable and reusable poly(vinyl chloride)-supported $FeCl_3$ (0.03 eq. Fe^{3+}) s. Y.-Q. Li, Synth. Commun. *30*, 3913-7 (2000).

Chiral dendritic cobalt(III) salen complex
Glycols from epoxides with kinetic resolution $\quad\overset{\triangle}{\text{O}} \to \text{C(OH)C(OH)}$
with chiral Co(III) salen complexes s. *53*, 59s56; chiral 3-amino-1,2-diols s. X.-L. Hou et al., Tetrahedron:Asym. *10*, 2319-26 (1999); asym. hydrolysis of β,γ-epoxyphosphonic acid esters s. A.E. Wroblewski, A. Halajewska-Wosik, ibid. *11*, 2053-5 (2000); with dendritic complexes **under cooperative catalysis** s. R. Breinbauer, E.N. Jacobsen, Angew. Chem. Int. Ed. Engl. *39*, 3604-7 (2000).

Dicarbonyl(chloro)rhodium(I) dimer $\qquad\qquad\qquad\qquad\qquad\qquad$ *[Rh(CO)$_2$Cl]$_2$*
Regio- and stereo-specific nucleophilic ring opening of 3-ethyleneepoxides $\quad$ C
under mild, neutral conditions

trans-3-Ene-1,2-diol 2-monoethers. A soln. of startg. epoxide in THF treated with 1-2 mol% [Rh(CO)$_2$Cl]$_2$ and 10 eqs. methanol at room temp. until reaction complete → product. Y 94% (regio- and diastereo-selectivity 20:1). The procedure is simple (no special precautions being required), and a useful alternative to traditional Brønsted- and Lewis acid-mediated routes. The regio- and stereo-selectivity are complementary with that obtained under Pd-catalysis (cf. *38*, 97s*44*; *44*, 268), the configuration at the attacked C-atom being inverted. The procedure is generally applicable to the addition of prim. and sec. alcohols as well as ar. amines [5 eqs.], but yields were lower with phenol while benzyl alcohol and aliphatic amines were unreactive. F.e.s. K. Fagnou, M. Lautens, Org. Lett. *2*, 2319-21 (2000).

Addition to Nitrogen and Carbon $\qquad\qquad\qquad\qquad\qquad$ OC ⇓ NC

Microwaves s. under HClO$_4$ $\qquad\qquad\qquad\qquad\qquad\qquad\qquad\qquad\qquad\qquad\qquad\qquad$ ←

m-Chloroperoxybenzoic acid $\qquad\qquad\qquad\qquad\qquad\qquad\qquad\qquad$ m-ClC$_6$H$_4$COO$_2$H
Carboxylic acid amides from aldimines $\qquad\qquad\qquad\qquad\qquad\qquad$ CH=NR → CONHR
with KMnO$_4$ cf. *18*, 180; with *m*-CPBA without solvent s. K. Mogilaiah, R.B. Rao, Indian J. Chem. *39B*, 145-6 (2000).

Perchloric acid/microwaves $\qquad\qquad\qquad\qquad\qquad\qquad\qquad\qquad\qquad\qquad\qquad$ ←
2-Amino-3-ethylenealcohols from 2-vinylaziridines s. *13*, 361s*60* $\qquad\qquad\qquad$ C

Hydroxyapatite-supported ruthenium(III) $\qquad\qquad\qquad\qquad\qquad\qquad\qquad\qquad\qquad$ ←
Carboxylic acid amides from nitriles s. *60*, 167 $\qquad\qquad\qquad\qquad\qquad$ CN → CONH$_2$

Addition to Carbon-Carbon Bonds $\qquad\qquad\qquad\qquad\qquad\qquad$ OC ⇓ CC

Microwaves s. under I$_2$ $\qquad\qquad\qquad\qquad\qquad\qquad\qquad\qquad\qquad\qquad\qquad\qquad\qquad$ ←

Potassium hydroxide/cesium fluoride $\qquad\qquad\qquad\qquad\qquad\qquad\qquad\qquad\qquad$ KOH/CsF
O-Vinylation of ketoximes $\qquad\qquad\qquad\qquad\qquad\qquad$ C=NOH → C=NOCH=CH$_2$
with CaC$_2$/KOH/water under forcing conditions cf. *27*, 122; improved method with KOH/CsF in a 2-phase medium (DMSO/pentane) s. B.A. Trofimov et al., Synthesis *2000*, 1125-32.

Sodium percarbonate s. under MeReO$_3$ $\qquad\qquad\qquad\qquad\qquad\qquad\qquad\qquad\qquad$ ←

Lithium bromide $\qquad\qquad\qquad\qquad\qquad\qquad\qquad\qquad\qquad\qquad\qquad\qquad\qquad$ *LiBr*
Protection of hydroxyl groups as tetrahydropyran-2-yl ethers $\qquad\qquad$ OH → OThp
with LiBF$_4$ cf. *56*, 73s*57*; with LiBr for protection of prim., sec. or tert. alcohols and phenols, also deprotection using MeOH/LiBr (1.5 eqs.) at reflux for 6 h, s. M.A. Reddy, K. Rama Rao et al., Synth. Commun. *30*, 4323-8 (2000).

4-Dimethylaminopyridine   DMAP
Protection of glycols as 1,3-dioxolanes
s. *17*, 234; *19*, 221; as 2-(carbomethoxymethyl-)- and 2-(carbo-*tert*-butoxymethyl)-1,3-dioxolanes by reaction with the corresponding alkyl propiolate in the presence of DMAP, also removal of the protective group with ca. 2:1 pyrrolidine/BuLi in THF or with neat pyrrolidine s. X. Ariza, J. Vilarrasa et al., Org. Lett. *2*, 2809-11 (2000).

Cupric chloride s. under PdCl$_2$ $CuCl_2$

Diethylzinc/(1R,2R)-N-methylpseudoephedrine
Asym. epoxidation with molecular oxygen   C=C → ᐯₒ̸
of enones cf. *52*, 68; of 1,1-nitroethylene derivs. (e.e. 36-82%) s. D. Enders et al., Tetrahedron: Asym. *9*, 3959-62 (1998).

Diborane B_2H_6
Rapid hydroboration at low temperature in chlorohydrocarbon medium   C(OH)CH

60.

An excess of diborane gas passed slowly into methylene chloride *at -16°* under N_2 to a concentration of 0.5 *M*, α-pinene added slowly, stirred for 10 min, residual active hydride destroyed by careful addition of methanol, H_2 flushed out, 3 *M* aq. NaOH added, followed by 1.2 eqs. H_2O_2, and the aqueous layer worked up → isopinocampheol. Y 88%. ^{11}B NMR examination indicated that dissolved B_2H_6 is in equilibrium with 8% BH_3·methylene chloride complex, the system being unusually active at low temp. (instantaneously at -16°) for the hydroboration of a wide range of olefins. Regioselectivity is similar to that reported for hydroboration with BH_3·THF or BH_3·Me_2S. F.e. and in chloroform, 1,2-dichloroethane or 1,1,2,2-tetrachloroethane s. J.V.B. Kanth, H.C. Brown, Tetrahedron Lett. *41*, 9361-4 (2000).

Borane-tetrahydrofuran BH_3-THF
Hydroboration of enamines   C=C(N<) → C(OH)C(N<)
s. *19*, 188s26; **with asym. induction** using an inexpensive chiral amine as auxiliary s. H. Pennemann, J. Martens et al., Tetrahedron:Asym. *11*, 2133-42 (2000).

Sodium tetrahydridoborate/Aliquat 336 $NaBH_4/Me(C_8H_{17})_3NCl$
Alcohols from ethylene derivs. by modified Brown hydration   C=C → CHC(OH)
with $NaBH_4/(MeO)_2SO_2$-H_2O_2/NaOH cf. *25*, 104; with 30% H_2O_2/NaOH in a 2-phase medium using Aliquat 336 as phase transfer catalyst, regio- and stereo-selectivity, s. D. Albanese, D. Landini et al., Synlett *2000*, 997-8.

Zirconium(IV) tetrahydridoborate $Zr(BH_4)_4$
Diols from ethylenecarboxylic acids s. *60*, 39

Sodium trihydridoacetoxoborate/pyridinium chlorochromate $NaBH_3OAc/C_5H_5NH·CrO_3Cl$
Oxo compds. from ethylene derivs. via hydroboration   C=CH → CHCO
ketones with BH_3-THF/pyridinium chlorochromate cf. *46*, 116; also aldehydes with $NaBH_3OAc$, and preferential formation of ketones (with retention of the *less*-substituted alkene group) s. R.S. Dhillon, G. Kaur et al., J. Indian Chem. Soc. *77*, 453-4 (2000).

Borane-N,N-diethyl-1,1,3,3-tetramethylbutylamine $H_3B·Et_2NCMe_2CH_2Bu$-t
Hydroboration   C=C → C(OH)CH
with borane-N-ethyl-N-isopropylaniline cf. *21*, 174s56; with highly reactive borane-N,N-diethyl-1,1,3,3-tetramethylbutylamine for regiospecific hydroboration of mono-, di-, tri-, or tetra-subst. ethylenes s. H.C. Brown et al., J. Org. Chem. *65*, 4655-61 (2000).

Catecholborane s. under Chiral rhodium phosphine complexes   ←

Sodium perborate/poly-L-leucine ←
Juliá-Colonna asym. epoxidation of α,β-ethyleneketones   C=C →
with Na-percarbonate/immobilized poly-L-leucine cf. *57*, 60; with NaBO$_3$·4H$_2$O/poly-L-leucine under ultrasonication (in high chemical yield but moderate enantioselectivity) s. R.M. Savizky, J.L. Bové et al., Tetrahedron:Asym. *9*, 3967-9 (1998).

Hydrotalcite s. under H$_2$O$_2$ ←

Zeolite/silanized titanium dioxide/hydrogen peroxide ←
Heterogeneous epoxidation s. *36*, 235s60

Dichloroborane or Dichloroborane-dimethyl sulfide   BHCl$_2$ or BHCl$_2$-Me$_2$S
Hydroboration   C=C → C(OH)CH
with BH$_3$-Me$_2$S s. *21*, 174s29; of terminal (perfluoroalkyl)ethylenes and pentafluoroethene with BH$_3$-Me$_2$S or, preferably, BHCl$_2$ or BHBr$_2$ (free or as the Me$_2$S complex) s. H.C. Brown et al., Angew. Chem. Int. Ed. Engl. *38*, 2052-4 (1999); comparison of thioethers as borane (or chloroborane) carriers s. J. Org. Chem. *65*, 6697-702 (2000).

Lanthanum triisopropoxide/(R)-1,1'-bi-2-naphthol/triphenylarsine oxide/ ←
  tert-*butyl hydroperoxide*
Asym. epoxidation of α,β-ethyleneketones   C=C →
s. *53*, 61s56; improved procedure with added Ph$_3$AsO (in place of Ph$_3$PO) s. T. Nemoto, M. Shibasaki et al., Tetrahedron Lett. *41*, 9569-74 (2000).

Poly-L-leucine s. under H$_2$O$_2$ ←
tert-*Butyl hydroperoxide s.a. under Ln(OPr-i)$_3$ and VO(OPr-i)$_3$*   t-BuOOH

tert-*Butyl hydroperoxide/titanium tetraisopropoxide/chiral polymer-based tartrate* ←
Heterogeneous Sharpless epoxidation
of 2-ethylenealcohols s. *55*, 64; of 3-ethylenealcohols s. J.K. Karjalainen, O.E.O. Hormi et al., Tetrahedron:Asym. *9*, 3895-901 (1998).

tert-*Butyl hydroperoxide/zirconium tetraisopropoxide/(-)-diisopropyl tartrate* ←
Zirconium(IV)-mediated Sharpless-type epoxidation of tert. 1,4-dien-3-ols with desymmetrization

3.3 eqs. (-)-Diisopropyl tartrate added to a soln. of 3 eqs. commercial Zr(OPr-*i*)$_4$·*i*-PrOH in methylene chloride (containing 4 Å molecular sieves) at -20°, followed by 3.3 eqs. 3.8 M *t*-BuOOH in toluene, stirred at the same temp. for 30 min, startg. diallyl alcohol in methylene chloride (at -20°) added, stirred for a further 30 min, and kept in a freezer for 3 days at -20° → (+)-product. Y 44% (e.e. >95%). Enantioselectivity for epoxidation of such tert. alcohols was poor (e.e. 10-20%) by the classical Sharpless procedure. The (-)-product was obtained with L-(+)-diisopropyl tartrate. F.e.s. A.C. Spivey et al., Angew. Chem. Int. Ed. Engl. *40*, 769-71 (2001).

Trityl hydroperoxide s. under VO(OPr-i)$_3$ TrOOH

Carbohydrate-based hydroperoxides/1,8-diazabicyclo[5.4.0]undec-7-ene   ROOH/DBU
Asym. Weitz-Scheffer epoxidation
of enones with (S)-1-phenylethyl hydroperoxide/KOH cf. *59*, 73; **of quinones** (acylamino-*p*-benzoquinones and 1,4-naphthoquinones) with carbohydrate-based hydroperoxides/DBU (method s. *51*, 62) s. C.L. Dwyer, R.J.K. Taylor et al., Synlett *2000*, 704-6.

Dimethyldioxirane/1,1,1-triacetoxy-1,2-benziodoxol-3-one/sodium hydrogen carbonate ←
α-Sulfonyloxyketones from ethylene derivs. $C=C \rightarrow C(OSO_3R)CO$

62.

Polymer-based α-sulfonyloxyketones. A soln. of 2.5 eqs. startg. olefin and 4 eqs. dimethyldioxirane (ca. 0.1 M in acetone) in methylene chloride allowed to react for 1 h at 25°, polystyrene-supported arylsulfonic acid hydrate added, allowed to react for 4 h at the same temp., and treated with 6 eqs. NaHCO$_3$ and 2 eqs. Dess-Martin periodinane at 25° for 12 h → supported α-sulfonyloxyketone. Y 95% (based on conversion to the corresponding α-hydroxyketone by refluxing with 1 eq. K$_2$CO$_3$ in 1:1 THF/water for 30 min). The geometry of the olefin may be *cis* or *trans*. The products are remarkably stable and suitable for diversification by traceless **nucleophilic substitution of the sulfonyloxy linker,** as in a novel **heterocycle release strategy.** E: A mixture of the startg. supported sulfonyloxyketone and 10 eqs. 2-amino-5-methylphenol in benzene refluxed with a little pyridinium tosylate under a Dean-Stark trap for 24 h → product. Y 58%. F.e. and nucleophilic displacements, also *solution-phase* procedures s. K.C. Nicolaou et al., J. Am. Chem. Soc. *122*, 10246-8 (2000).

m-Chloroperoxybenzoic acid $m\text{-}ClC_6H_4COO_2H$
Baeyer-Villiger oxidation ←
s. *21*, 177; lactones by transannular ring opening of bicyclic lactols via bicyclic alkoxycarbenium ions (for ζ-lactones) cf. K.W. Hunt, P.A. Grieco, Org. Lett. *2*, 1717-9 (2000).

1,1,1-Triacetoxy-1,2-benziodoxol-3-one s. under Dimethyldioxirane ←

Tetra-n-butylammonium 1-oxido-1,2-benziodoxol-3-one ←
α,β-Epoxyketones from α,β-ethyleneketones $C=C \rightarrow \triangle_O$

63.

Startg. α,β-ethyleneketone and 1.2 eqs. tetra-*n*-butylammonium 1-oxido-1,2-benziodoxol-3-one (readily prepared in crystalline form from 1-hydroxy-1,2-benziodoxol-3-one and Bu$_4$NF in THF) in THF stirred at room temp. under N$_2$ for 2 h → product. Y 83% (89% *trans*-selectivity). The (Z)-substrate also produced the *trans*-epoxide predominantly as a result of initial base-catalyzed isomerization of the substrate. Dipolar aprotic solvents such as DMF, DMSO, HMPA, and THF afford the best results. Reaction probably involves initial Michael addition of the reagent, followed by intramolecular nucleophilic displacement of the λ^3-iodanyl group with retention of configuration. F.e.s. M. Ochiai et al., Org. Lett. 2, 2923-6 (2000).

1,1,1,3,3,3-Hexafluoro-2-propanol/hydrogen peroxide ←
1,1,1-Trifluoro-2-propanone/hydrogen peroxide/potassium carbonate ←
Epoxides from ethylene derivs. $C{=}C \rightarrow \overset{\diagdown}{\underset{O}{\diagup}}$
with perfluoroheptadecan-9-one cf. *57, 62*; with trifluoroacetone in acetonitrile at high pH (with K_2CO_3 in aq. EDTA) s. L. Shu, Y. Shi, J. Org. Chem. 65, 8807-10 (2000); with 1,1,1,3,3,3-hexafluoro-2-propanol as medium and activator, also Baeyer-Villiger oxidation, s. K. Neimann, R. Neumann, Org. Lett. 2, 2861-3 (2000).

N-Iodosuccinimide/sodium dithionite $NIS/Na_2S_2O_4$
2-Deoxyaldoses from glycals under mild conditions $C{=}C \rightarrow C(OH)CH$

One-pot procedure. A soln of 3,4,6-tri-O-benzyl-D-galactal in 95:5 acetonitrile/water treated at 0° with 1.1 eqs. NIS, allowed to warm to room temp., stirred for 15 min, solvent removed *in vacuo*, the residue taken up in DMF (1 part), 10 eqs. aq. bicarbonate (1 part) added, followed by 4 eqs. solid $Na_2S_2O_4$, and stirred for 5 h at room temp. → 3,4,6-tri-O-benzyl-2-deoxygalactose (as a mixture of anomers). Y 90%. The reagents are inexpensive, and the procedure is milder (allowing retention of silyl and trityl ethers) and less toxic than hydration by oxymercuration-demercuration (cf. *19, 201s55*). F.e.s. V. Costantino, E. Fattorusso et al., Tetrahedron Lett. 41, 9177-80 (2000).

Silanized titanium dioxide s. under Zeolites ←
Titanium tetraisopropoxide s. under t-BuOOH $Ti(OPr-i)_4$
Zirconium tetraisopropoxide s. under t-BuOOH $Zr(OPr-i)_4$

Lead tetraacetate $Pb(OAc)_4$
Oxocarboxylic acid esters from cyclic α-hydroxyketones C
aldehydo-esters cf. *27, 146s33*; cyclic ethyleneketo-esters s. J.R. Rodríguez, J.L. Mascareñas, Org. Lett. 2, 3209-12 (2000).

Vanadyl triisopropoxide/tert-butyl hydroperoxide/ ←
  chiral N-protected α-aminohydroxamic acids
Asym. epoxidation of 2-ethylenealcohols $C{=}C \rightarrow \overset{\diagdown}{\underset{O}{\diagup}}$

Improved procedure. 1 Mol% $VO(OPr-i)_3$ added to a soln. of 1.5 mol% chiral hydroxamic acid in toluene, stirred for 1 h at room temp., cooled to 0°, 1.5 eqs. 78% *t*-BuOOH and startg. allyl alcohol added, stirred for 6 h at 0°, treated with satd. aq. $Na_2S_2O_3$, stirred for a further 1 h, allowed to warm to room temp., worked up, and the crude residue purified by column chromatography → product. Y 91% (e.e. 62%). The mild conditions, catalyst loading (optimally as low as 0.1 mol%), high yields and e.e. (up to 96%), and the use of a halogen-free solvent offer significant advantages over the prior art (cf. *55, 67s58-60*). The ligand was optimized from three points of molecular diversity. F.e.s. Y. Hoshino, H. Yamamoto, J. Am. Chem. Soc. 122, 10452-3 (2000).

Vanadyl triisopropoxide/trityl hydroperoxide/chiral hydroxamic acids ←
Asym. epoxidation of 2-ethylenealcohols   C=C → \o/
s. 55, 67s58; ligand design and scope s. Y. Hoshino, H. Yamamoto et al., Bull. Chem. Soc. Jpn. 73, 1653-8 (2000).

Hydrogen peroxide s.a. under Zeolites and 1,1,1-Trifluoro-2-propanone H_2O_2

Hydrogen peroxide/hydrotalcite/dodecyltrimethylammonium bromide   ←
Heterogeneous epoxidation
s. 36, 235s51; improved procedure for epoxidation of enones with added dodecyltrimethylammonium bromide in 5:3 n-heptane/water s. K. Yamaguchi, K. Kaneda et al., J. Org. Chem. 65, 6897-903 (2000); epoxidation of alkenes using NaY zeolite loaded with alkylsilane-covered TiO_2 **under phase-boundary catalysis** (cf. 36, 235s57) s. H. Nur, B. Ohtani et al., Chem. Commun. 2000, 2235-6; with delaminated Ti/ITQ-6 (based on a layered titanosilicate precursor) cf. A. Corma et al., ibid. 137-8.

Hydrogen peroxide/poly-L-leucine   ←
Juliá-Colonna asym. epoxidation of α,β-ethyleneketones
s. 37, 127; effect of polyleucine chain length on selectivity s. R. Takagi, K. Ohkata et al., Bull. Chem. Soc. Jpn. 73, 2115-21 (2000).

Sodium dithionite s. under NIS $Na_2S_2O_4$

Trifluoromethanesulfonic acid or Trifluoromethanesulfonimide   *TfOH or Tf$_2$NH*
Ketones from acetylene derivs.   C≡C → COCH$_2$
Regiospecific catalytic hydration in the absence of metal salt

66.   Ph—≡ + H$_2$O → Ph-CO-CH$_3$

3 eqs. Water and 0.1 eq. TfOH added successively to a soln. of phenylacetylene in 1,4-dioxane at room temp., the mixture stirred at 100° for 48 h, then poured into satd. aq. NaHCO$_3$ → acetophenone. Y 93% by GC. This is the first example of alkyne hydration using a catalytic amount of a Brønsted acid, TfOH (or Tf$_2$NH, especially for electron-deficient alkynes for which TfOH was inactive), without a metal salt. Fluorosulfonic acid and sulfuric acid were ineffective. The process provides a wide variety of carbonyl compds. with very high regiospecificity. F.e. incl. enones from 1,3-enynes, **also Rupe rearrangement**, s. T. Tsuchimoto, E. Shirakawa et al., Synlett 2000, 1777-8.

Potassium peroxymonosulfate/chiral ketones or azomethinium salts/potassium carbonate   ←
Asym. epoxidation   C=C → \o/
with *in situ*-generated chiral dioxiranes s. 51, 65s58; of α,β-ethylenecarboxylic acid esters with (2S,5R)-2-fluoro-2-methyl-5-isopropylcyclohexanone s. A. Solladié-Cavallo, L. Bouérat, Org. Lett. 2, 3531-4 (2000); of (E)-stilbene with chiral 8-oxa- or 8-aza-bicyclo[3.2.1]octan-3-ones s. A. Armstrong et al., Tetrahedron:Asym. 11, 2057-61 (2000); of various olefins with chiral aliphatic ketiminium salts (cf. 51, 65s55) s. S. Minakata, M. Komatsu et al., Synlett 2000, 1810-2; with chiral 3,4-dihydroisoquinolinium salts as catalyst cf. P.C. Bulman Page et al., J. Chem. Soc. Perkin Trans. 1 2000, 3325-34.

Potassium peroxymonosulfate/silica-supported trifluoromethyl ketone/potassium carbonate   ←
Heterogeneous dioxirane-mediated epoxidation

67.   Ph-CH=CH-Ph —[silica-O-Si(OMe)-S(CH$_2$)$_{10}$COCF$_3$ / Oxone]→ Ph-CH(-O-)CH-Ph

Immobilized ketones are more resistant to adventitious oxidation with Oxone (e.g. by Baeyer-Villiger oxidation or 1,2,4,5-tetroxane formation) than the homogeneous counterparts, so that dioxirane-mediated epoxidation of alkenes is more efficient with catalytic activity being maintained over 10 cycles. **E:** Stilbene, 0.5 eq. silica-supported trifluoromethyl ketone, 2.07 eq. Oxone, and 8.7 eq. K_2CO_3 in 3:2 acetonitrile/aq. EDTA mixed at room temp. until reaction complete → product. Y 94% (unoptimized). A further advantage is that the supported ketone is easily retrieved (by filtration) for recycling, while the product is readily isolable from the filtrate. F.e. and comparison with a soluble ketone s. C.E. Song et al., Chem. Commun. *2000*, 2415-6.

Potassium peroxymonosulfate/(5S,8R,9R)-8,9-O-isopropylidene-N-tert-butoxycarbonyl- ←
8,9-dihydroxy-1,6-dioxa-3-azaspiro[4.5]decane-2,10-dione/potassium carbonate/
tetra-n-butylammonium hydrogen sulfate
Asym. epoxidation of *cis*-ethylene derivs.   C=C → \o/

68.

A mixture of startg. *cis*-ethylene deriv., a little chiral auxiliary, 1.8 eqs. Oxone (0.212 *M* in aq. EDTA), 0.04 eq. Bu$_4$NHSO$_4$, and 4 eqs. K$_2$CO$_3$ (0.479 *M* in aq. EDTA) in 3:1 dimethoxyethane/dimethoxymethane and pH 8 buffer (0.2 *M* K$_2$CO$_3$-AcOH) stirred at 0° for 3.5 h → product. Y 82% (e.e. 91%). The reagent is readily recoverable for reuse, and is complementary with the all-oxygen analog (cf. *51*, 65s*56*, *60*), which is only efficient with *trans*-olefins. F.e. incl. asym. epoxidation of cycloalkenes s. H. Tian et al., J. Am. Chem. Soc. *122*, 11551-2 (2000).

Pyridinium chlorochromate s. under NaBH$_3$OAc $C_5H_5NH·CrO_3Cl$

Iodine/microwaves   ←
Protection of alcohols as tetrahydropyran-2-yl ethers   OH → OThp
with a little I$_2$ cf. *56*, 73s*58*; rapid procedure under microwave irradiation without solvent, also deprotection in methanol, s. N. Deka, J.C. Sarma, Synth. Commun. *30*, 4435-41 (2000).

Hypofluorous acid-acetonitrile   *HOF·MeCN*
α-Diketones from acetylene derivs.   C≡C → COCO
with I$_2$/DMSO cf. *27*, 145s*46*; with HOF·MeCN s. S. Dayan, S. Rozen et al., J. Org. Chem. *65*, 8816-8 (2000).

Chiral manganese(III) salen complexes/iodosobenzene or sodium hypochlorite/   ←
pyridine N-oxides
Asym. epoxidation   C=C → \o/
s. *46*, 106s*55*,*56*; with 1,1'-binaphthyl-based complexes s. T. Nishida, T. Katsuki et al., Tetrahedron Lett. *41*, 7053-8 (2000); asym. epoxidation of conjugated *trans*-olefins s. Synlett *2000*, 1557-60; regiospecific asym. epoxidation of cyclic 2-sulfonyl-1,3-dienes with (S,S)-N,N'-[bis(3,5-di-*tert*-butylsalicylidene)-1,2-cyclohexanediamine]manganese(III) chloride/NaOCl/4-(3-phenyl-propyl)pyridine N-oxide s. W.L. Jiang, P.L. Fuchs et al., Org. Lett. *2*, 2181-4 (2000).

Fluorous chiral manganese(III) salen complexes/iodosobenzene/pyridine N-oxide   ←
Asym. epoxidation in a fluorous 2-phase medium

69.

Asym. epoxidation is reported with readily recyclable, second generation fluorous chiral Mn-salen complexes (cf. *48*, 134s*50*) bearing perfluoroalkylaryl groups in the 5,5'-positions (for improved shielding of the metal site) and sterically demanding *tert*-butyl groups in the 3,3'-positions. **E:** A 0.2 *M* soln. of triphenylethylene in acetonitrile (5 parts) containing *o*-dichlorobenzene (internal standard) and a 0.25 M soln. of pyridine N-oxide in acetonitrile (1 part) added under N_2 to a soln. of the Mn-catalyst in *n*-perfluorooctane (5 parts) at 100°, 1.5 eqs. PhIO added quickly, and the 2 phases stirred magnetically at 1250-1350 rpm for 0.5 h → product. Y 98% (e.e. 87%). The fluorous layer separates readily on cooling and may be reused up to 3 times. F.e.s. M. Cavazzini, G. Pozzi et al., Chem. Commun. 2000, 2171-2.

Chiral silica-supported or polymer-based manganese(III) salen complex/ ←
m-chloroperoxybenzoic acid/N-methylmorpholine N-oxide
Heterogeneous asym. epoxidation C=C → \o/
with a JandaJel-supported complex cf. 46, 106s56,59; with a polystyrene- and polymethacrylate-based complex, details, s. L. Canali, D.C. Sherrington et al., J. Chem. Soc. Perkin Trans. 1 2000, 2055-6; with a silica gel-based complex (cf. 46, 106s58) cf. D. Pini, P. Salvadori et al., Tetrahedron:Asym. 10, 3883-6 (1999).

5,10,15,20-Tetraarylporphyrinatomanganese(III) chloride/tert-butyl hypochlorite/ ←
4-tert-butylpyridine/tetra-n-butylammonium chloride
Epoxidation
s. 39, 124s59; with a porphyrin-capped molecular clip *host-guest* complex of enhanced activity and stability s. J.A.A.W. Elemans, A.E. Rowan et al., Chem. Commun. 2000, 2443-4.

Methylrhenium oxide/sodium percarbonate/trifluoroacetic acid/pyrazole ←
Epoxides from ethylene derivs. under mild conditions

70. n-Pr—CH=CH—C_5H_{11} $\xrightarrow[MeReO_3]{[H_2O_2]}$ n-Pr—(epoxide)—C_5H_{11}

Cheap, safe, mild yet powerful, and easy-to-handle Na-percarbonate serves, in combination with trifluoroacetic acid, as a slow release form of H_2O_2 for Re-catalyzed epoxidation (cf. 47, 111s57). E: A soln. of *cis*-4-decene in THF treated with 2.5 eqs. Na-percarbonate, 12 mol% pyrazole, and 0.5 mol% $MeReO_3$, stirred vigorously while treated dropwise with 2.5 eqs. trifluoroacetic acid over 2 min, stirred for an additional 3 h, a further 0.5 mol% $MeReO_3$ added, and stirring continued for 3 h → product. Y 96%. In contrast to the original method with aq. H_2O_2 no external cooling is required. Conversion (of styrene) was lower using AcOH in place of TFA, precipitated NaOAc interfering with the release of H_2O_2 (unlike $NaOCOCF_3$). F.e. incl. reaction in methylene chloride for terminal alkenes s. A.R. Vaino, J. Org. Chem. 65, 4210-2 (2000).

Bis(1,1,1,5,5,5-hexafluoroacetylacetonato)cobalt(II)/isobutyraldehyde ←
Catalyzed aerobic epoxidation
s. 47, 113s52,54,56; with bis(1,1,1,5,5,5-hexafluoroacetylacetonato)cobalt(II) for epoxidation of *less* reactive alkenes (terminal or electron-deficient derivs.) s. R. Hunter, S. Rimmer et al., Synth. Commun. 30, 4461-6 (2000).

Polybenzimidazole-based nickel(II) acetoacetonate/isobutyraldehyde ←
Catalyzed heterogeneous aerobic epoxidation
with polyaniline-based $Co(OAc)_2$ cf. 47, 113s53; with polybenzimidazole-based $Ni(acac)_2$ s. B.B. Wentzel, R.J.M. Nolte et al., J. Chem. Soc. Perkin Trans. 1 2000, 3428-31; with Pd-silica cf. H. Gao, R.J. Angelici, Synth. Commun. 30, 1239-47 (2000).

Tetracarbonyldi(μ-formato)bis(triphenylphosphine)diruthenium $[Ru(OCHO)(CO)_2PPh_3]_2$
Addition of carboxylic acids to acetylene derivs. ←
s. 40, 81s58; addition to 1-ethoxy-2-ethynylcyclopropane with simultaneous formation of **1-alkoxy-1-acoxy-3-allenes** s. I. Emme, A. de Meijere et al., Synlett 2000, 1315-7.

Chiral rhodium phosphine complex/catecholborane ←
Asym. hydroboration C=C → C(OH)CH
s. 44, 129; with chiral tricarbonylchromium-complexed α-(2-pyridyl)-*o*-phosphinobenzyl ethers as ligand s. S.U. Son, Y.K. Chung et al., Tetrahedron:Asym. 10, 347-54 (1999).

Palladium-silica/isobutyraldehyde $Pd-SiO_2/Me_2CHCHO$
Catalytic heterogeneous aerobic epoxidation s. 47, 113s60 C=C → \o/

Palladous acetate/isopropanol/pyridine $Pd(OAc)_2$/i-$PrOH$/C_5H_5N
Palladous chloride/cupric chloride $PdCl_2$/$CuCl_2$
Wacker oxidation CH≡C → COCH
in a fluorous 2-phase medium cf. 56, 75; in supercritical CO_2, or preferably with an alcohol as co-solvent, s. H. Jiang et al., Green Chem. *2000*, 161-4; methyl ketones from terminal olefins with $Pd(OAc)_2$/isopropanol/pyridine cf. T. Nishimura, S. Uemura et al., J. Chem. Soc. Perkin Trans. 1 *2000*, 1915-8.

Potassium osmate/1,4-bis(9-O-dihydroquin[id]ine)phthalazine/potassium hexacyanoferrate/potassium carbonate/tert-butanol/methanesulfonamide ←
Asym. dihydroxylation C=C → C(OH)C(OH)
s. *47*, 114s59; chiral acetylene-α,β-dihydroxycarboxylic acid esters s. L. He, R. Bittman, J. Org. Chem. *65*, 7627-33 (2000); improved procedure at constant pH of 12 (for internal olefins) or 10 (for terminal olefins) with omission of $MeSO_2NH_2$ cf. G.M. Mehltretter, M. Beller et al., Tetrahedron Lett. *41*, 8083-7 (2000); asym. dihydroxylation of cinnamic acid esters s. S. Jew, H. Park et al., ibid. *41*, 7925-8 (2000).

Silica-supported osmate/N-methylmorpholine N-oxide ←
Heterogeneous *cis*-dihydroxylation

71.

In the field of heterogeneous *cis*-dihydroxylation of olefins, the problems associated with immobilization of OsO_4 on a polymeric support (notably leaching of the metal after hydrolysis of the intermediate osmium(VI) diolate complex) have now been solved with the design of a hydrolytically stable, silica-supported osmium diolate which remains intact during the catalytic cycle of olefin osmylation, reoxidation and hydrolytic release of the product. E: 100 mg Silica-supported osmate (prepared by surface functionalization of commercial SiO_2 with 9:1 propyl-trimethoxysilane/3-aminopropyltrimethoxysilane, followed by condensation with 3,4-dimethyl-cyclohex-3-ene-1-carboxylic acid chloride, and subsequent osmylation with OsO_4) added to a mixture of 1.6 mmol *cis*-2-hexene, 1.6 mmol N-methylmorpholine N-oxide, and 200 mL water in 2:1 *t*-BuOH/methylene chloride, and stirred at room temp. for 48 h → product. Conversion 99% (selectivity 99%). Reaction is generally applicable to mono-, *cis*- and *trans*-disubst. aliphatic olefins and cyclic olefins, as well as ar. olefins and α,β-ethylenecarboxylic acid esters, but trisubst. olefins reacted more slowly. F.e.s. A. Severeyns, P.A. Jacobs et al., Angew. Chem. Int. Ed. Engl. *40*, 586-9 (2001).

Via intermediates v.i
Ketones from acetylene derivs. C≡C → $COCH_2$
via enamines cf. *7*, 217; via thioenolethers, α-hydroxyketones, s. M.S. Waters et al., Tetrahedron Lett. *41*, 141-4 (2000).

Rearrangement ∩

Hydrogen/Oxygen Type OC ∩ HO

Bis(tributyltin) oxide/zinc triflate $(Bu_3Sn)_2O$/$Zn(OTf)_2$
Hydroxy-O-heterocyclics from epoxyalcohols ÖC
with $La(OTf)_3$/MeOH cf. *54*, 71s58 and with BF_3 cf. *33*, 145s49; 2-α-hydroxyoxepans from 6,7-epoxyalcohols with $(Bu_3Sn)_2O$/$Zn(OTf)_2$, stereoselectivity, s. R. Matsumura, T. Suzuki et al., Tetrahedron Lett. *41*, 7701-4 (2000); *cis*- and *trans*-2,7-disubst. oxepans s. ibid. 7697-700.

Hydrogen/Carbon Type   OC ∩ HC

sec-*Butyllithium/N,N,N',N'-tetramethylethylenediamine*   s-BuLi/TMEDA
Base-catalyzed isomerization of epoxides   ▽$_o$ → CHCO
aldehydes with Li-tetramethylpiperidide cf. *50*, 73; 2-acyl-Δ^2-oxazolines with s-BuLi/TMEDA s. V. Capriati, S. Florio et al., Tetrahedron Lett. *41*, 8835-8 (2000).

Boron fluoride/2-chloropyridine   ←
Anomerization of glycosyl phosphates s. *60*, 54   ←

Ammonium ceric nitrate/montmorillonite or perchloric acid   ←
α-Arylcarboxylic acid esters from aryl ketones   ArCOCH → ArCCOOR
with Tl(NO$_3$)$_3$/HClO$_4$ cf. *27*, 162; with CAN/montmorillonite K10 or HClO$_4$ s. B. Wang, H.Z. Ma, Chin. Chem. Lett. *11*, 849-50 (2000); β-hydroxyketones from 3-ethylenealcohols with Taylor-McKillop ring contraction using CAN in 1:1 AcOH/water, also **α-spiro-γ-lactones** *in pentane*, s. H.M.C. Ferraz et al., Synth. Commun. *30*, 751-62 (2000).

Yttrium trichloride/triethylamine or piperidine   ←
α-Hydroxycarboxylic acid esters from α-ketoaldehydes   COCHO → CH(OH)COOR
with N-diethylcysteamine cf. *11*, 215; with YCl$_3$/Et$_3$N or piperidine, also α-ketoacetals in the absence of base, s. P.R. Likhar, A.K. Bandyopadhyay, Synlett *2000*, 538-40.

Alkyl dichlorovanadates   VO(OR)Cl$_2$
Regiospecific catalytic isomerization of epoxides   ▽$_o$ → CHCO
with Pd(OAc)$_2$/Bu$_3$P cf. *50*, 75; aldehydes from terminal epoxides, and ketones from 1,2-disubst. and 1,1,2-disubst. epoxides, with VO(OR)Cl$_2$ (R = Et or Pr-*i*) s. F. Martínez, E.F. Llama et al., J. Chem. Soc. Perkin Trans. 1 *2000*, 1749-51.

Chromium(III) perchlorate   Cr(ClO$_4$)$_3$
α-Hydroxycarboxylic acid esters from α-ketoaldehydes   COCHO → CH(OH)COOR
Chromium-catalyzed Cannizzaro reaction

72.

A mixture of startg. glyoxal and 10 mol% Cr(ClO$_4$)$_3$·6H$_2$O in 2:1 isopropanol/1,2-dichloroethane stirred at room temp. for 24 h, then quenched with 3 *N* HCl → product. Y 84%. Other effective catalysts include Cu(II)-triflate and -perchlorate, but not perchlorates of Fe, Mg, Al, Li or Y. A phenolic hydroxyl group was unaffected. F.e. and an asym. variation with Cu(II)-triflate in the presence of (S,S)-2,2'-isopropylidenebis(4-phenyl-2-oxazoline) (e.e. 17-33%) s. A.E. Russell, J.P. Morken et al., J. Org. Chem. *65*, 8381-3 (2000).

Chiral cationic rhodium(I) phosphaferrocene complex   ←
Aldehydes from 2-ethylenealcohols   C≡C-CH$_2$OH → CHCHCHO
Asym. isomerization

73.

A soln. of startg. allyl alcohol in THF added to 5 mol% chiral cationic rhodium(I) phosphaferrocene complex (prepared in a Schlenk tube by mixing equivalent (5%) amounts of [Rh(cod)$_2$]BF$_4$ and the free ligand in THF under H_2 for 1 h at room temp.), and stirred at 70° for 48 h → product. Y 87% (e.e. 82%). Conditions were optimal in THF with BF_4^- as counterion. Enantioselectivity was lower from (E)-substrates. However, electronic effects were small, and the olefin may be substituted by ar. or (cyclo)alkyl groups. F.e.s. K. Tanaka, G.C. Fu et al., J. Am. Chem. Soc. *122*, 9870-1 (2000).

Oxygen/Nitrogen Type   OC ∩ ON

Indium(III) chloride $InCl_3$
Beckmann rearrangement s. *60*, 168   ←

Nitrogen/Carbon Type   OC ∩ NC

Triethylamine Et_3N
3-Acyloxazolidines from 2-(alkylideneamino)alcohols   ○
with Ac_2O cf. *17*, 216; polymer-based synthesis with acyl chlorides and Et_3N, followed by oxidative removal of the Wang resin with DDQ, s. H.S. Oh et al., Tetrahedron Lett. *41*, 5069-72 (2000).

Carbon/Carbon Type   OC ∩ CC

Potassium carbonate K_2CO_3
Isomerization of acoxy-2-ethylenes   ←
Baylis-Hillman adducts s. *60*, 76

Gold trichloride $AuCl_3$
Furans from β,γ-acetyleneketones   ○
Gold(III)-catalyzed cycloisomerization under mild conditions

74.

A soln. of startg. acetyleneketone in acetonitrile allowed to react *at room temp.* with *0.1 mol%* $AuCl_3$ until reaction complete → product. Y ca. 100%. Reaction failed with Ag(I), and heating was required for the corresponding Pd(II)-catalyzed conversion (cf. *43*, 131). F. details **and 2-γ-ketofurans from α-alleneketones** and α,β-acetyleneketones s. A.S.K. Hashmi et al., Angew. Chem. Int. Ed. Engl. *39*, 2285-8 (2000).

Zinc iodide/4-dimethylaminopyridine ZnI_2/DMAP
Δ^4-Isoxazolines from 2-acetylenehydroxylamines   ○
under mild conditions

75.

10 Mol% ZnI_2 and 10 mol% DMAP added to a soln. of startg. 2-acetylenehydroxylamine in methylene chloride at 23°, stirred for 1 h, and quenched by partitioning between methylene chloride and aq. NH_4Cl → product. Y 95%. The Zn(II)/DMAP system appears to be unique in facilitating the transformation by activating both the hydroxylamine and the alkyne residue. Reaction failed with protic acid (e.g. TsOH), and other Lewis acids were ineffective. F.e.s. P. Aschwanden, E.M. Carreira et al., Org. Lett. *2*, 2331-3 (2000).

Tris(6,6,7,7,8,8,8-heptafluoro-2,2-dimethyl-3,5-octanedionato)europium(III) $Eu(fod)_3$
Isomerization of methoxyacetoxy-2-ethylenes   ←
s. *52*, 77s*56*; (3E)-5-acoxy-2-alkoxy-1,3-dienes, also isomerization of *p*-methoxybenzyloxy-acetoxy-2-ethylenes, s. W.-M. Dai et al., Tetrahedron Lett. *41*, 7101-5 (2000).

Trimethylsilyl triflate/acetic anhydride/potassium carbonate $Me_3SiOTf/Ac_2O/K_2CO_3$
(E)-α,β-Ethylene-α-(hydroxymethyl)-   ←
from β-hydroxy-α-methylene-carboxylic acid esters

76.

One-pot conversion. 11 Mol% Me$_3$SiOTf added to a soln. of startg. Baylis-Hillman adduct and 1.2 eqs. acetic anhydride in dichloromethane at room temp., the mixture allowed to react for 2 h, dichloromethane removed, the residue in methanol treated with 3 eqs. K$_2$CO$_3$, and stirred for 1 h at the same temp. → product. Y 77%. The method was unsuccessful with non-benzylic β-hydroxyesters. F.e.s. D. Basavaiah et al., Synthesis *2000*, 1662-4; isomerization of the preformed acetates with the same reagent, also nitrile analogs, s. ibid. 545-8.

Exchange   ↕

Hydrogen ↑   OC ↕ H

Electrolysis/sodium iodide $⌇/NaI$
α-Hydroxyketals from ketones   CHCO → C(OH)C(OR)$_2$
with KI/KOH cf. *41*, 154; with NaI/NaOH s. M.N. Elinson et al., Tetrahedron *56*, 9999-10003 (2000).

Cupric trifluoromethanesulfonate/2,2'-bi[(6R,8R)-7,7-dimethyl-   ←
    5,6,7,8-tetrahydro-6,8-methanoquinolyl]/phenylhydrazine
Asym. Kharasch acoxylation of ethylene derivs.   H → OAc
with chiral tris(Δ^2-oxazolines) cf. *51*, 72s59; with 2,2'-bi[(6R,8R)-7,7-dimethyl-5,6,7,8-tetrahydro-6,8-methanoquinolyl] s. A.V. Malkov, P. Kocovsky et al., Org. Lett. *2*, 2047-9 (2000).

Cupric sulfate s. under KMnO$_4$ $CuSO_4$

Copper(II)- and sodium-exchanged zeolite   ←
Kharasch acoxylation of ethylene derivs.
with CuCl$_2$/LiCl cf. *14*, 211s56; heterogeneous conversion with recyclable Cu-Na-HSZ-320 zeolite s. S. Carloni, G. Sartori et al., Tetrahedron Lett. *41*, 8947-50 (2000).

Ammonium ceric nitrate $(NH_4)_2Ce(NO_3)_6$
α-Hydroxyketones from 3-ethylenealcohols
with Taylor-McKillop ring contraction - Also α-spiro-γ-lactones s. *27*, 162s60

N-Hydroxyphthalimide/cobalt(II) acetate   ←
N-Hydroxyphthalimide/nickel(II) acetate/chromium(III) nitrate/benzaldehyde   ←
Imide-mediated aerobic oxidation
aryloxo compds. with Co(OAc)$_2$ s. *51*, 76; arylglyoxylic from arylacetic or mandelic acid esters s. B.B. Wentzel, R.J.M. Nolte et al., Tetrahedron *56*, 7797-803 (2000); oxidation of aryl-methyl, -methylene or -methine groups with Ni(OAc)$_2$/Cr(NO$_3$)$_2$/PhCHO s. P. Alsters, S. Bouttemy, European patent EP-1004566.

2,2,6,6-Tetramethylpiperidine nitroxyl/sodium chlorite/sodium hypochlorite   ←
Carboxylic acids from prim. alcohols   CH$_2$OH → COOH
s. *45*, 120s57; chiral α-hydroxycarboxylic acids s. F.J. Aladro, G.M. Massanet et al., Tetrahedron Lett. *41*, 3209-13 (2000).

1,1,1-Triacetoxy-1,2-benziodoxol-3-one
Oxidation of ar. acylamines with Dess-Martin periodinane

77.

Acylamino-*p*-quinones. A soln. of acetanilide in methylene chloride treated with 4 eqs. Dess-Martin periodinane and 2 eqs. water at 25° for 4 h → product. Y 52%. A range of substituents on the aromatic ring (except NO_2) and on the amide side chain is tolerated. Interestingly, *o*-subst. acetanilides afforded the corresponding *o*-**quinone mono-N-acylimines** which underwent hetero-Diels-Alder reaction with electron-rich olefins. F.e. and Diels-Alder reaction with the acylamino-*p*-quinones s. K.C. Nicolaou et al., Angew. Chem. Int. Ed. Engl. *40*, 207-10 (2001).

**Lactam-condensed 3,4-dihydro-2*H*-1,4-benzoxazines
from ethylenecarboxylic acid anilides
Stereospecific oxidative double ring closure
via intramolecular hetero-Diels-Alder reaction with *o*-quinone mono(N-acylimines)**

78.

2 eqs. Dess-Martin periodinane added in one portion to a soln. of startg. anilide in benzene [or benzotrifluoride], and heated at reflux [or 80-85° for reactions in BTF] for ca. 30 min → product. Y 40%. Reaction likely involves initial displacement of an acetoxy group of the reagent by anilide nitrogen, followed by *ortho*-oxidation to, and subsequent cyclization of, an *o*-quinone monoimine with the creation of three contiguous carbon centres. The method is also generally applicable to allylic N-aryl-carbamates, -thionocarbamates and -ureas. However, ring closure of simple anilides bearing a terminal ethylene group gave low yields. F.e. incl. a bis(cascade) process s. K.C. Nicolaou et al., Angew. Chem. Int. Ed. Engl. *39*, 622-5 (2000).

Bromotrichloromethane/2,4,6-collidine
Protection of hydroxyl groups as tetrahydrofuran-2-yl ethers   OH → OThf
via H-atom abstraction with trichloromethyl radical using $CCl_4/CrCl_2$ cf. *59*, 86; using $BrCCl_3/$ 2,4,6-collidine for protection of prim. and sec. alcohols s. J.M. Barks, A.F. Parsons et al., Tetrahedron Lett. *41*, 6249-52 (2000).

Sodium [hypo]chlorite s. under 2,2,6,6-Tetramethylpiperidine nitroxyl $NaClO_2$ or $NaOCl$

Potassium permanganate/cupric sulfate $KMnO_4/CuSO_4$
Keto from methylene groups CH_2 → CO
under ultrasonic acceleration s. *60*, 109

Cobalt(II) and Nickel(II) acetate s. under N-Hydroxyphthalimide $Co(OAc)_2$ and $Ni(OAc)_2$

Oxygen↑   OC ↓↑ O

Microwaves s. under TsOH

Potassium hydride   KH
Cyclic α,β-ethylene-α-(2-pyridyloxy)ketones
from cyclic α,β-ethyleneepoxides s. *60*, 2

Cesium salt/4-dimethylaminopyridine $Cs^+/DMAP$
Acoxy compds. from sulfonic acid esters $OSO_2R \rightarrow OAc$
with inversion of configuration using $CsOCOCF_3$ or CsF cf. *37*, 148s*54*; using Cs-carboxylates (preferably Cs-benzoate) *in toluene* with DMAP as solubilizer s. N.A. Hawryluk, B.B. Snider, J. Org. Chem. *65*, 8379-80 (2000).

4-Dimethylaminopyridine or N-Methylimidazole   ←
O-Carbo-*tert*-butoxylation of alcohols s. *60*, 135 $OH \rightarrow OCOOBu\text{-}t$

Indium/trimethylsilyl chloride In/Me_3SiCl
α-Alkoxyketoximes from 1,1-nitroethylene derivs. s. *43*, 60s*60* $C(OR)C=NOH$

Sodium tetrahydridoaluminate/chiral 2-aminoalcohols/4-dimethylaminopyridine   ←
1,1-Alkoxyacoxy compds. from carboxylic acid esters $COOR \rightarrow CH(OAc)OR$
Asym. reduction s. *60*, 20

Lithium tetrakis(pentafluorophenyl)borate/lithium trifluoromethanesulfonate/ magnesium oxide   ←
O-Benzylation of alcohols under mild conditions $OH \rightarrow OBn$

79.   Cl~~~OH + $MeSO_2OBn$ ⟶ Cl~~~OBn

A soln. of 4-chloro-1-butanol and 1.2 eqs. benzyl mesylate in 13:7 cyclohexane/dichloromethane added to a mixture of 0.1 eq. $LiB(C_6F_5)_4$, 1 eq. LiOTf and 1.6 eqs. MgO under argon, and the reaction mixture stirred for 24 h at room temp. → benzyl 4-chlorobutyl ether. Y 96%. This convenient method is suitable for the benzylation of alcohols possessing alkali-labile substituents, incl. halogen (chlorine, bromine), ester and ketone. Thus, inter- or intra-molecular etherification or olefin formation was minimal for the example described. LiOTf serves to regenerate the catalyst by trapping the formed benzyl ether, thereby improving the efficiency of the reaction. F.e. incl. *p*-chlorobenzyl ethers s. M. Nakano, T. Mukaiyama et al., Chem. Lett. *2000*, 1352-3.

Montmorillonite   ←
Carboxylic acid esters from acids $COOH \rightarrow COOR$
with zeolites cf. *13*, 239s*53*; with natural or cation-tunable montmorillonite (e.g. Na^+-exchanged montmorillonite), also O-acetylation with acetic acid, s. B.M. Choudary et al., Green. Chem. *2*, 67-70 (2000).

Scandium(III) triflate $Sc(OTf)_3$
Ferrier reaction   ←
with $InCl_3$ cf. *58*, 76; with $Sc(OTf)_3$ in dichloromethane or 9:1 acetonitrile/water for reaction with a variety of prim., sec., benzyl, allyl and propargyl alcohols and phenols, α-selectivity, s. J.S. Yadav et al., Synlett *2000*, 1450-1.

Cerium(III) chloride $CeCl_3$
O-Acylation $OH \rightarrow OAc$
s. *55*, 81; partial O-acylation of glycols, also with $YbCl_3$ or $DyCl_3$, and with desymmetrization using a chiral anhydride s. P.A. Clarke, R.A. Holton et al., Tetrahedron Lett. *41*, 2687-90 (2000).

Amberlyst 15W (erbium(III)- or ytterbium(III)-form)   ←
Acetals from oxo compds. $CO \rightarrow C(OR)_2$
with Ce(III)-exchanged montmorillonite cf. *28*, 141s*51*; carbohydrate O,O-alkylidene derivs. with Amberlyst 15W (erbium(III)- or ytterbium(III)-form), also by transacetalation, s. S. Porwanski, Y. Queneau et al., Synthesis *2000*, 525-8.

Immobilized enzyme/ruthenium hydride complex   ←
Sec. acoxy compds. from ketones $CO \rightarrow CH(OAc)$
Asym. hydrogenation with O-acylation s. *60*, 22

Diphenyl sulfoxide/trifluoromethanesulfonic anhydride/2,4,6-tri-tert-butylpyridine ←
Iterative dehydrative glycosidation   OH → OR

80.

Synth. Meth. *53*, 76 has been adapted iteratively. **E:** 1.4 eqs. Tf$_2$O added to a soln. of 1 eq. 4-O-acetyl-2,3-O-isopropylidene-L-rhamnopyranose, 2.8 eqs. 2,4,6-tri-*tert*-butylpyridine, and 2.8 eqs. diphenyl sulfoxide in 3:1 toluene/methylene chloride at -78°, stirred for 10 min at the same temp., then at -40° for 85 min, a soln. of 2 eqs. 2,3-O-isopropylidene-L-rhamnopyranose in 1:1 toluene/methylene chloride added via cannula at -40°, stirred for 1 h at this temp. then at 0° for 30 min, and treated with 9.5 eqs. Et$_3$N → product. Y 97% (α:β 4.1:1). The chain was simply extended (to the tetrasaccharide) by repeated coupling with the acceptor aldose. Reaction capitalizes on the inherent difference in nucleophilicity between the alkyl and anomeric hydroxyl group of the acceptor. No protective group strategy is required. F.e.s. H.M. Nguyen, D.Y. Gin et al., Angew. Chem. Int. Ed. Engl. *40*, 414-7 (2001).

Carbon tetrabromide/irradiation   CBr$_4$/*hν*
Carboxylic acid esters from acids   COOH → COOMe
Preferential esterification under mild conditions

81. Br(CH$_2$)$_7$COOH   →(MeOH)   Br(CH$_2$)$_7$COOMe
     +   +
    COOH   COOH

A mixture of 8-bromo-1-octanoic acid and 2,4-hexadienoic acid in carbon tetrabromide/anhydrous methanol irradiated through Pyrex with a TLC lamp at room temp. for 30 min, then stirred without irradiation for 24 h (monitored by TLC) → methyl 8-bromo-1-octanoate. Y 98% (and 96% recovery of 2,4-hexadienoic acid). Benzoic acid, sp^2-C tethered and α-amino acids undergo esterification slowly; thus, the method also distinguishes between phenylacetic and benzoic acid, 2-phenylpropanoic and 2,4-hexadienoic acid, and phenylacetic and phenylpropiolic acid. Furthermore, the rate of esterification is much slower with ethanol, and reaction hardly proceeded with isopropanol. F.e.s. A.S-Y. Lee et al., Tetrahedron Lett. *42*, 301-3 (2001).

2-Chloro-1,3-dimethyl-Δ^2-imidazolinium chloride/triethylamine ←
2-Chloro-1,3-dimethyl-Δ^2-imidazolinium chloride as mild dehydration agent ←
s. *58*, 80; urethans and ureas via Lossen-type rearrangement s. T. Isobe, T. Ishikawa, J. Org. Chem. *64*, 5832-5 (1999).

N-Benzyl-N-(6-methylpyrid-2-yl)carbamyl chloride/triethylamine/4-dimethylaminopyridine ←
Carboxylic acid esters from acids   COOH → COOR
with di-2-pyridyl carbonate/DMAP cf. *40*, 99; with N-benzyl-N-(6-methylpyrid-2-yl)carbamyl chloride/Et$_3$N/DMAP, incl. partial and preferential O-benzoylation of diols, and limitations, s. J.I. Lee, S.J. Park, Bull. Korean Chem. Soc. *21*, 141-4 (2000).

Trimethylsilyl chloride s. under In   Me$_3$SiCl

Tetraphenylphosphonium chloride   Ph$_4$PCl
Carboxylic acid aryl esters   COOH → COOAr
from carboxylic acids and oxalic acid aryl esters

82.   PhCO$_2$H + PhO$_2$C-CO$_2$Ph   →   PhCO$_2$Ph [+ PhOH + CO + CO$_2$ + H$_2$O]

A mixture of benzoic acid (5 mmol scale) and 1 eq. diphenyl oxalate heated with stirring at 200°, 5 mol% tetraphenylphosphonium chloride added, and heating continued with liberation of CO and CO_2 until reaction complete → phenyl benzoate. Y 97%. Corrosion was minimal. F.e.s. Japanese patent JP-11217355 (Ube Ind. Ltd.).

Thionyl chloride/Amberlyst A-21 ←
Carboxylic acid esters from acids via carboxylic acid chlorides   COOH → COOR
with $SOCl_2$ cf. *4*, 214s52; rapid, efficient procedure, notably for saturated fatty acids from C_{12} to C_{18} (such as lauric, myristic, palmitic and stearic acids), and unsatd. acids (such as oleic and arachidonic), also carboxylic acid amides, s. C. Girard, I. Tranchant et al., Synlett *2000*, 1577-80.

Trifluoromethanesulfonic anhydride s. under Ph_2SO Tf_2O
Lithium triflate s. under $LiB(C_6H_5)_4$ LiOTf

Nafion-H
Carboxylic acid esters from acids   ←
s. *35*, 105; O-acetylation of prim. and sec. alcohols (incl. allyl alcohols without rearrangement), and per-O-acetylation of monosaccharides, selectivity, also N- and S-acetylation, s. R. Kumareswaran, Y.D. Vankar et al., Synlett *2000*, 1652-4.

p-Toluenesulfonic acid/activated carbon/microwaves   ←
Carboxylic acid esters from acids
under microwave irradiation on a solid support in the absence of solvent
with Al_2O_3 as support cf. *48*, 169s58; more rapidly with faster heat exchange using activated carbon as support, also phthalic acid esters from phthalic anhydride, s. X. Fan, G. Tan et al., Org. Prep. Proced. Int. *32*, 287-90 (2000).

Ruthenium hydride complex s. under Enzymes   ←

Thiolate-bridged diruthenium complex/ammonium fluoroborate   ←
Regiospecific Nicholas-type substitution of propargyl alcohols   ←
via ruthenium allenylidene complexes

83.

A useful *catalytic* alternative to the stoichiometric Co-mediated Nicholas reaction is reported. E: A soln. of startg. propargyl alcohol (0.6 mmol), 5 mol% thiolate-bridged diruthenium complex, and 10 mol% NH_4BF_4 in ethanol heated at 60° for 15 min → product. Y 88%. The unique reactivity and selectivity at the propargylic *ipso*-carbon is due to the dimetallic nature of the catalyst, there being no reaction with the monoruthenium complex. The procedure is applicable to the reaction of 1-monosubst. and 1,1-disubst. propargyl alcohols with O-, N-, S- and P-nucleophiles (the conversion of 1,1-diaryl derivs. being sluggish). Mechanistically, formation of the intermediate ruthenium allenylidene species is preceded, and its reaction with the alcohol followed by, generation of an oxy-functionalized ruthenium vinylidene complex. F.e.s. Y. Nishibayashi, M. Hidai et al., J. Am. Chem. Soc. *122*, 11019-20 (2000).

Tris(dibenzylideneacetone)dipalladium/1,2-bis(diphenylphosphino)ethane $Pd_2(dba)_3$/dppe
4-(1-Aryloxyvinyl)-1,3-dioxolan-2-ones from 2-acetylene-1,4-diol monocarbonates   ○

84.

and phenols. 1.1 eqs. *p*-Methoxyphenol, 5 mol% $Pd_2(dba)_3.CHCl_3$, and 20 mol% dppe added at room temp. to a stirred soln. of startg. diol monocarbonate in dioxane, and stirred for 4 h → product. Y 91%. This convenient and environmentally friendly method appears to be the first involving efficient refixation of the CO_2 molecule after a decarboxylation. It is also applicable to reaction of **O-aryl 2-acetylene-1,4-diol monocarbonates** in the absence of a phenol, crossover experiments indicating that the aroxide ion completely dissociates from the propargyl unit. F.e.s. M. Yoshida, M. Ihara, Angew. Chem. Intern. Ed. Engl. *40*, 616-9 (2001).

Tris(dibenzylideneacetone)dipalladium/1,4-bis(diphenylphosphino)butane $Pd_2(dba)_3/dppb$
O-Allylation of phenols   OH → OC-C≡C
s. *42*, 165; bis(allyloxy)arenes with $Pd_2(dba)_3$/dppb s. R. Kolodziuk, D. Sinou et al., Synth. Commun. *30*, 3955-61 (2000).

Nitrogen ↑   OC ↓↑ N

Without additional reagents   *w.a.r.*
Carboxylic acids from N-acyl-2-oxazolidones   CON< → COOH
with $LiOH/H_2O_2$ cf. *43*, 162; from N-acyl-2-imidazolidones in water at reflux for regenerating chiral α-aminocarboxylic acids s. G. Guillena, C. Nájera, J. Org. Chem. *65*, 7310-22 (2000).

Microwaves s. under $Cu(NO_3)_2$ and $Zn(NO_3)_2$ ←

Potassium hydroxide   *KOH*
Prim. alcohols from amines NH_2 → OH

85.

A mixture of startg. amine and 55 eqs. KOH in diethylene glycol heated at 210° for 3 h → (2R*,6R*,10R*)-10-(benzyloxymethyl)-10-(hydroxymethyl)-6-methylbicyclo[4.4.0]decan-2-ol. Y 65%. This method was also successful with hindered primary amines. F.e.s. S.M.A. Rahman, T. Tanaka et al., Org. Lett. *2*, 2893-5 (2000).

Sodium nitrite/cuprous chloride $NaNO_2/CuCl$
1,1-Alkoxyaroylamines from *o*-aminocarboxylic acid N-*tert*-alkylamides   ←
via 1,5-hydrogen atom migration

86.

5.6 eqs. 1 *M* Methanolic HCl added dropwise at 0° over 10 min to a 0.07 *M* soln. of startg. *o*-aminobenzamide and 1.5 eqs. $NaNO_2$ in dry methanol, stirred for 50 min until formation of the diazonium salt was complete (with TLC monitoring), dry, powdered 3 Å molecular sieves added quickly, the mixture stirred for 10 min, 0.1 eq. anhydrous CuCl added, and stirred at 12-18° for 1.5 h → product. Y 93%. The products serve as storable precursors of N-acylimines. With unsym. *o*-aminobenzamides derived from pyrrolidines and piperidines, there was good regioselectivity for methylene over methine oxidation. F.e. and from N-cumyl derivs., and further reactions s. W. Chao, S.M. Weinreb, Tetrahedron Lett. *41*, 9199-204 (2000).

Disodium hydrogen phosphate Na_2HPO_4
Carboxylic acid esters from amides and oxonium fluoroborates   CON< → COOR
from N-unsubst. or N-monosubst. amides cf. *53*, 82; arylcarboxylic acid esters from N,N-disubst. amides s. G.E. Keck et al., Tetrahedron *56*, 9875-83 (2000).

Cupric nitrate-silica/microwaves   ←
Oxo compds. from oximes   C=NOH → CO
with $Cu(NO_3)_2$-on-bentonite cf. *47*, 146s*51*; rapid cleavage of aldoximes and cyclic or aryl ketoximes with $Cu(NO_3)_2$-on-SiO_2 under microwave irradiation in the absence of solvent s. M. Ghiaci, J. Asghari, Synth. Commun. *30*, 3865-72 (2000); cleavage of aromatic, aliphatic,

polynuclear and heterocyclic oximes with $Zn(NO_3)_2$-on-SiO_2 under microwave irradiation cf. B. Tamami, A.R. Kiasat, ibid. 4129-35.

Cuprous chloride s. under $NaNO_2$ *CuCl*

Zinc nitrate/silica/microwaves   ←
Oxo compds. from oximes   C=NOH → CO
under microwave irradiation on a solid support in the absence of solvent s. 47, 146s60

Inorganic-supported lipase   ←
Preferential O-acylation with acyloximes   OH → OAc
5'-acylation of nucleosides s. 46, 171s48; with diatomite- or ceramic-supported lipase s. R.V. Nair, M.M. Salunkhe, Synth. Commun. 30, 3115-20 (2000).

Phenyl iodosohydroxytosylate   *PhI(OH)OTs*
Carboxylic acids from N,N-disubst. hydrazides s. 60, 87   CONHN< → COOH

Nitric oxide/oxygen   *NO/O_2*
Oxo compds. from oximes   C=NOH → CO
with N_2O_4 cf. 39, 171s42; with 100:1 NO/O_2 in dry acetonitrile s. Y.Z. Mao, L.M. Wu et al., Chin. J. Chem. 18, 789-91 (2000).

Peroxymonosulfuric acid/silica gel   *H_2SO_5/SiO_2*
Oxo compds. from oximes
with $KHSO_5$-on-silica gel under microwaves cf. 48, 175s59; cleavage of aldoximes or ketoximes, incl. acid-sensitive compds., with H_2SO_5-on-silica gel s. B. Movassagh, M.M. Lakouraj et al., Synth. Commun. 30, 4501-6 (2000).

Potassium peroxymonosulfate/sodium hydrogen carbonate/ethylenediaminetetraacetic acid ←
Carboxylic acids from carboxylic acid hydrazides $CONHNH_2$ → CO(OH,OR)

87.

A mixture of startg. hydrazide, water, $NaHCO_3$ and acetone stirred at room temp. for 10 min, a soln. of 1.5 eqs. Oxone in aq. EDTA added dropwise below 2°, then quenched with Na-bisulfite → product. Y 88%. The method is simple, mild, efficient, economical, eco-friendly and high-yielding. Unaffected groups include alkenes, alcohols, ketones, sulfides, ethers, cyclopropyl, and nitriles. F.e. **and carboxylic acid esters** in alcoholic medium s. R. Srinivas et al., J. Chem. Res., Synop. 2000, 376-7; carboxylic acids from N,N-disubst. carboxylic acid hydrazides with PhI(OH)OTs s. P.G.M. Wuts, M.P. Goble, Org. Lett. 2, 2139-40 (2000).

Tris(ethylenediamine)nickel thiosulfate   *[Ni(en)$_3$]S_2O_3*
Oxo compds. from N,N-dimethylhydrazones under mild conditions   C=NNMe$_2$ → CO

88.

0.5 eq. [Ni(en)$_3$]S_2O_3 added (after moistening with a drop of water) to a soln. of startg. N,N-dimethylhydrazone in chloroform, and the mixture stirred at room temp. for 5 min (a change of pink colour to colourless being observed) → 2-bromo-4,5-(methylenedioxy)benzaldehyde. Y 96%. The method is efficient, rapid (5-10 min reaction time), high-yielding, involves simple work-up, and is suitable for regenerating both ketones and aldehydes. It was also successful (although over a longer reaction time) with Hg[Co(SCN)$_4$] and [Mn(acac)$_3$]. It is believed the transition metal ions promote hydrazone hydrolysis via chelation, making the imino bond more electrophilic and stabilizing the leaving group in the transition state. F.e.s. A. Kamal, M.V. Rao et al., Synlett 2000, 1482-4.

Rhodium(II) trifluoroacetate $\quad Rh_2(OCOCF_3)_4$
O-Alkylactims from lactams and diazo compds. $\quad\leftarrow$

89.

A soln. of 2-pyridone in dry 1,2-dichloroethane treated with 2 mol% Rh(II)-trifluoroacetate, heated to reflux, 0.9 eq. *tert*-butyl diazoacetate added over 6 h via syringe pump, and refluxed for a further 2 h → product. Y 84%. Reaction is thought to involve generation of a carbonyl ylid, followed by 1,4-H-shift. N-Methyl-2-pyrrolidone and N-methylbenzamide, however, failed to react, while 5-phenyl-2(3*H*)-furanone gave a low yield. F.e. incl. oxindole and ε-caprolactam derivs., also an S-analog and cyclic enolethers from ketones, s. J. Busch-Petersen, E.J. Corey, Org. Lett. 2, 1641-3 (2000).

Halogen ↑ $\quad$ OC ↓↑ Hal

Irradiation s. under SiO_2 $\quad$ ///

Cesium carbonate/tetra-n-butylammonium iodide $\quad Cs_2CO_3/Bu_4NI$
Carbonic acid esters from alcohols and halides $\quad ROH + HalR' \rightarrow ROCOOR'$
with CO_2 s. 49, 168s58; preparation of polymer-based carbonic acid esters, **also polymer-based urethans from amines**, s. R.N. Salvatore, K.W. Jung et al., Org. Lett. 2, 2797-800 (2000).

N,N,N',N'-Tetramethylethylenediamine $\quad TMEDA$
O-Carbalkoxylation with chloroformic acid esters $\quad OH \rightarrow OCOOR$
with Et_3N cf. 23, 237; O-carballyloxylation and O-carbobenzoxylation of carbohydrate hydroxyl groups *at 0°* with TMEDA, also regiospecific conversion at -78°, s. M. Adinolfi, A. Iadonisi et al., Tetrahedron Lett. 41, 9305-9 (2000).

Pyridine/4-dimethylaminopyridine/imidazole $\quad\leftarrow$
Regiospecific O-tritylation, -acylation and -silylation of carbohydrates in one pot $\quad\leftarrow$

90.

1.25 eqs. Trityl chloride and a little DMAP added to the startg. carbohydrate in pyridine, the mixture stirred at 80°, cooled to 0°, 2 eqs. imidazole and 1.1 eqs. *t*-BuMe$_2$SiCl in DMF added in portions over 2 h, the mixture stirred at room temp. overnight, a soln. of benzoyl chloride in pyridine added, and stirring continued at 50° overnight → product. Y 79%. F.e. and with *tert*-butyldiphenylsilyl in place of trityl chloride s. Y.G. Du, F.Z. Kong et al., Org. Lett. 2, 3797-800 (2000).

Silver trifluoromethanesulfonate s. under Cp_2HfCl_2 $\quad AgOSO_2CF_3$
Mesoporous silica/irradiation $\quad SiO_2$/ ///
Arylcarbonyl compds. from benzyl halides $\quad CH_2Hal \rightarrow COOH$

91. $\quad$ t-Bu—⟨ ⟩—CH$_2$Br $\quad\longrightarrow\quad$ t-Bu—⟨ ⟩—COOH

Heterogeneous photooxidation. A suspension of 4-*tert*-butylbenzyl bromide (50 mg) and mesoporous silica (FSM-16) (100 mg) in dry acetone irradiated in a Pyrex tube at room temp. with a 400 W high-pressure Hg-lamp for 30 hrs. → 4-*tert*-butylbenzoic acid. Y 86%. Substrates possessing an electron-withdrawing group, such as nitro or cyano, give somewhat lower yields with *p*-nitrobenzyl bromide favoring aldehyde formation. Sec. bromides produced the corresponding ketones, while cinnamyl bromide was also oxidized to benzoic acid. The procedure in eco-friendly (no heavy metals being required) and the catalyst is recyclable. However, alkyl bromides were unreactive, and benzylic chlorides gave much lower yields. F.e.s. A. Itoh, Y. Masaki et al., Org. Lett. *2*, 2455-7 (2000).

Hafnocene dichloride/silver trifluoromethanesulfonate $Cp_2HfCl_2/AgOSO_2CF_3$
Trifluoromethanesulfonic acid CF_3SO_3H
Glycosides from glycosyl fluorides   F → OR
s. *44*, 191s52; oligosaccharides (incl. a one-pot synthesis of a linear pentasaccharide) by using an α-di(acetal) O-protective group for reactivity tuning s. D.K. Baeschlin, S.V. Ley et al., Tetrahedron:Asym. *11*, 173-97 (2000); α-glycosides with CF_3SO_3H/5 Å molecular sieves, incl. a **one-pot sequential trisaccharide synthesis** (cf. *50*, 103s60), s. H. Jona, T. Mukaiyama et al., Chem. Lett. 2000, 1278-9.

Ferric chloride $FeCl_3$
Carboxylic acid chlorides from 1,1,1-trichlorides CCl_3 → COCl

92. Cl—C₆H₄—CCl₃ ⟶ Cl—C₆H₄—COCl

A mixture of 4-chlorobenzotrichloride and 2 mol% $FeCl_3$ heated to 60°, 1 eq. water added over 1 h, and the mixture kept at the same temp. for 30 min → 4-chlorobenzoyl chloride. Conversion 100%. The process overcomes the problems of prior art which require high temp. F.e.s. W.L. Nan Den Akker, M. Brink, United States patent US-6069274 (American Cyanamid Co.).

Polymer-based cobalt(II) phosphine complex   ←
Carboxylic acid anhydrides   RCOCl → HOOCR′ → RCO·O·COR′
from carboxylic acid chlorides and carboxylic acids
with $CoCl_2$ cf. *47*, 162; aliphatic or ar. anhydrides with an air-stable, readily recyclable polymer-based cobalt(II) phosphine complex (prepared from commercially available polymer-based PPh_2-polystyrene cross-linked with 2% divinylbenzene and $CoCl_2(PPh_3)_2$) s. N.E. Leadbeater, K.A. Scott, J. Org. Chem. *65*, 4770-2 (2000).

Ruthenium trichloride/sodium periodate $RuCl_3/NaIO_4$
α-Diketones from 1,2-ethylene-1,2-dihalides   C(Cl)=C(Cl) → COCO

93.

A soln. of 7 mol% $RuCl_3·3H_2O$ and 1.5 eqs. $NaIO_4$ in water added to a vigorously stirred soln. of startg. ethylenedihalide in acetonitrile at 0-5° (ice-water bath), and the mixture stirred for 2 min → product. Y 98%. Ethers, cyclic acetals and halides remained unaffected, while alcohol, lactone, aldehyde and carboxylic acid-containing substrates provided mixtures of products. F.e. incl. oxidation of 1,2-ethylene-1,2-dibromides, also subsequent conversion to highly functionalized cyclopentanes (possessing bis(α-chloroester) groups) or bicyclic lactones, s. F.A. Khan et al., J. Am. Chem. Soc. *122*, 9558-9 (2000).

Bis(dibenzylideneacetone)palladium(0)/1,2,3,4,5-pentaphenyl-   ←
1′-(di-tert-butylphosphino)ferrocene/sodium tert-butoxide
Phenolethers from ar. halides   ArHal → ArOR
with $Pd(OAc)_2$/t-Bu_3P/NaOBu-*t* cf. *53*, 86s58; rapid condensation with neutral and electron-poor aryl bromides *at room temp.* with Pd(dba)₂ and 1,2,3,4,5-pentaphenyl-1′-(di-*tert*-butylphosphino)-ferrocene as ligand, also intramolecular variant (cf. *52*, 128), and **aryloxysilanes** from silanolates, s. Q. Shelby, J. Hartwig et al., J. Am. Chem. Soc. *122*, 10718-9 (2000).

Sulfur ↑   OC ↓↑ S

Lithium nitrate s. under NBS $LiNO_3$
Silver trifluoromethanesulfonate s. under NIS $AgOSO_2CF_3$

Triphenylcarbonium tetrakis(pentafluorophenyl)borate $Ph_3C^+B(C_6F_5)_4^-$
s. under *N-(Ethylthio)phthalimide*

Samarium(III) or ytterbium(III) triflate $Sm(OTf)_3$ or $Yb(OTf)_3$
Glycosides from glycosyl 2-pyridyl sulfones $SO_2R \rightarrow OR'$

94.

A mixture of startg. sulfone (α/β-mixture; 0.2 mmol scale) and 20 eqs. methanol in methylene chloride refluxed with 1 eq. Sm(OTf)$_3$ under argon for 18 h → product. Y 92% (α/β 5:1). Formation of α-glycosides was preferred, while benzylated substrates were activated in preference to benzoylated derivs. Furthermore, thioglycosides were not activated under these conditions. However, the pyridyl nitrogen, which is presumed to participate in complexation with the metal, appears essential for high yields. F.e. incl. di- and tri-saccharides containing both furanose and pyranose rings, also with Yb(OTf)$_3$, s. G.X. Chang, T.L. Lowary, *Org. Lett.* **2**, 1505-8 (2000).

N-(Ethylthio)phthalimide/trityl tetrakis(pentafluorophenyl)borate   ←
Glycosides from thioglycosides $SR \rightarrow OR'$
with N-(phenylseleno)phthalimide/Me$_3$SiOTf cf. *39*, 189s*49*; β-glycosides with N-(ethylthio)-phthalimide/TrB(C$_6$F$_5$)$_4$ in PhCF$_3$/*t*-BuCN, also one-pot trisaccharide synthesis (cf. *50*, 103s*60*), s. H. Jona, T. Mukaiyama et al., *Chem. Lett.* **2000**, 1178-9.

Phenyl iodosoacetate $PhI(OAc)_2$
Oxo compds. from mercaptals $C(SR)_2 \rightarrow CO$
with PhI(OCOCF$_3$)$_2$ cf. *44*, 197; from 1,3-dithianes with PhI(OAc)$_2$ s. X.-X. Shi, Q.-Q. Wu, *Synth. Commun.* **30**, 4081-6 (2000).

N-Bromosuccinimide/lithium nitrate or stannous triflate $NBS/LiNO_3$ or $Sn(OTf)_2$
N-Iodosuccinimide/trifluoromethanesulfonic acid NIS/CF_3SO_3H
Glycosides from thioglycosides $SR \rightarrow OR'$
with NIS/CF$_3$SO$_3$H cf. *39*, 189s*59*; with retention of anomeric configuration and with simultaneous cleavage of *p*-methoxybenzyl ethers s. R. Hirschmann, A.B. Smith III et al., *J. Org. Chem.* **65**, 8307-16 (2000); morphine and dihydromorphine 6-β-D-glucuronides, selectivity, s. I. Rukhman, A.L. Gutman et al., *Synthesis* **2000**, 1241-6; polymer-based synthesis of β-(1→6)-linked oligoglycosides with NBS/Sn(OTf)$_2$ via 2-O-benzoyl group participation, also α-anomers with NBS/LiNO$_3$ with 4-azido-3-chlorobenzyl as O-protective group, s. K. Egusa, K. Fukase et al., *Synlett* **2000**, 27-32; 6-phenylthio-6-deoxyglycosides with Me$_3$SiOTf (or MeOTf), also conversion to glucuronides, s. B. Yu, Y. Hui et al., *Org. Lett.* **2**, 2539-41 (2000).

N-Iodosuccinimide/silver trifluoromethanesulfonate $NIS/AgOSO_2CF_3$
Trisaccharides by sequential glycosidation in one-pot   ←
s. *50*, 103; without added CF$_3$SO$_3$H, effect of solvent addition, s. M. Lahmann, S. Oscarson, *Org. Lett.* **2**, 3881-2 (2000)

N-Iodosuccinimide/triethylsilyl triflate NIS/Et_3SiOTf
Glycosides from glycosyl disulfides $SSR \rightarrow OR'$

95.

A soln. of startg. glycosyl disulfide and methanol in methylene chloride treated with NIS and triethylsilyl triflate at 0° for 1 h → product. Y 90% (α:β 9:15). F.e. incl. a disaccharide, also 'traceless' **polymer-based synthesis** with a supported glycosyl disulfide, and preparation of the disulfides **from S-glycosyl methanethiolsulfonates**, s. B.G. Davis et al., *Chem. Commun.* **2001**, 189-90.

Trimethylsilyl iodide $\qquad$ *Me₃SiI*
Oxo compds. from α,β-ethylenesulfoxides under mild conditions $\qquad$ C=CS(O)R → CHCO

96.

ca. 1.5 eqs. Me₃SiI (97%) added at room temp. to a soln. of startg. sulfoxide in anhydrous chloroform at room temp., methanol added after 1.5 h, and stirred for a further 30 min → product (comprising 55% *trans*-fused isomer and 27% *cis*-fused isomer). There was no reaction with the corresponding vinyl sulfones or thioethers. The steric requirement of the alkyl or aryl group directly linked to the sulfinyl vinyl residue only affects the conversion rate. The procedure is general and efficient. F.e.s. M.C. Aversa et al., Tetrahedron Lett. *41*, 4441-5 (2000).

Trifluoromethanesulfonic anhydride/S-(4-methoxyphenyl) benzenethiosulfinate/ ←
*2,6-di-*tert-*butyl-4-methylpyridine*
Glycosides from thioglycosides via glycosyl triflates $\qquad$ SR → OR'
with AgOTf/PhSCl cf. *54*, 92; β-mannosides and α-glucosides with Tf₂O/S-(4-methoxyphenyl) benzenethiosulfinate s. D. Crich, M. Smith, Org. Lett. *2*, 4067-9 (2000).

Iodine mono-chloride or -bromide $\qquad$ *ICl or IBr*
Glycosides from thioglycosides
with I₂/K₂CO₃ cf. *42*, 200s52; methyl glycosides more rapidly with ICl or IBr (notably from *p*-nitrophenyl thioglycosides) s. P. Cura, R.A. Field et al., Synlett *2000*, 1279-80.

Ferric chloride $\qquad$ *FeCl₃*
Oxo compds. from mercaptals $\qquad$ C(SR)₂ → CO
with Fe(NO₃)₃/montmorillonite cf. *24*, 228s53; with FeCl₃·6H₂O under mild conditions, also synthesis of tetrahydropyrrolo[2,1-*c*][1,4]benzodiazepinones, s. A. Kamal, P.S.M.M. Reddy et al., Synlett *2000*, 1476-8.

Remaining Elements ↑ $\qquad$ OC ↓↑ Rem

Montmorillonite ←
Acoxy compds. from alkoxysilanes $\qquad$ OSi≤ → OAc
and carboxylic acids with FeCl₃ cf. *57*, 102; acetoxy compds. with Ac₂O and montmorillonite (with or without solvent) s. B. Movassagh et al., J. Chem. Res., Synop *2000*, 348-9.

Vanadyl acetoacetonate/cumene hydroperoxide/tributylphosphine ←
2-Ethylenealcohols from 2-ethyleneselenides ←
via [2.3]-sigmatropic rearrangement

97.

2-Methylene-*sec*-alcohols. A 0.3 *M* soln. of startg. selenide in methylene chloride treated with 10 mol% VO(acac)₂ in the presence of 4 Å molecular sieves, the green soln. cooled to -10°, treated with 1.8 eqs. cumene hydroperoxide while stirring, and when reaction complete (ca. 30 min) treated with 1.2 eqs. Bu₃P for 5 min → product. Y 84%. This methodology is effective for a broad range of substrates, incl. sterically hindered ones. Use of Bu₃P provides a viable alternative to traditional methods for selenenate cleavage, whereas pyridine and water proved slow and capricious. F.e.s. R.G. Carter, T.C. Bourland, Chem. Commun. *2000*, 2031-2.

Trimethylsilyl triflate Me_3SiOTf
Glycosides from glycosyl phosphates $OPO_3R \rightarrow OR'$
from diphenyl phosphates cf. *45*, 88; α- or β-manno-linked glycosides from 1-O-2,3,4,6-tetra-O-benzyl-D-mannopyranosyl propane-1,3-diyl phosphate s. G. Singh, H. Vankayalapati, Tetrahedron:Asym. *11*, 125-38 (2000).

Methylrhenium oxide/urea-hydrogen peroxide/pyridine/potassium fluoride   ←
α-Hydroxycarbonyl compds. from enoxysilanes $C(OSi\leqslant)=C \rightarrow COC(OH)$
α-hydroxyketones with H₂O₂ as reoxidant cf. *56*, 106; α-hydroxycarboxylic acid esters with urea-hydrogen peroxide as reoxidant (reducing substrate hydrolysis) s. S. Stankovic, J.H. Espenson, J. Org. Chem. *65*, 5528-30 (2000).

Via intermediates   *v.i.*
Alcohols from monoorganomercury chlorides $HgCl \rightarrow OH$
with NaBH₄/O₂ cf. *35*, 137; via 1-alkoxypiperidines with added 2,2,6,6-tetramethylpiperidine nitroxyl, and reduction of the intermediate according to *49*, 175, s. P. Hayes et al., Tetrahedron Lett. *41*, 6175-9 (2000).

Carbon↑   OC ↕ C

Without additional reagents   *w.a.r.*
Alkylation with isoureas $OH \rightarrow OR$
review s. *29*, 547s*35*; protection of carboxylic acids as *p*-methoxybenzyl esters, selectivity, s. M.F. Wang, G.A. Potter, Synth. Commun. *30*, 4197-204 (2000).

Sodium hydrogen sulfate/silica gel $NaHSO_4/SiO_2$
Heterogeneous transesterification $COOR \rightarrow COOR'$
with sulfated SnO₂ cf. *51*, 104; of aliphatic carboxylic acid esters with NaHSO₄/silica gel, retention of arylcarboxylic acid esters (on condensation with ethyl acetate), s. B. Das, B. Venkataiah, Synthesis *2000*, 1671-2.

Zinc bromide $ZnBr_2$
Polymer-based synthesis of carboxylic acid esters $COBr \rightarrow COOR$
from carboxylic acid bromides and carbonic acid esters s. *60*, 159

Boron fluoride BF_3
Carboxylic acid esters from anhydrides and acetals $(RCO)_2O \rightarrow RCOOR'$

98. $(RCO)_2O$ + MeO-CH(Ph)-OMe ⟶ RCOOMe   R = (CH₂)₇CH=CH(CH₂)₇Me

0.2 eq. BF₃-etherate added to a soln. of 0.4 eq. benzaldehyde dimethyl acetal and oleic anhydride in alcohol-free chloroform, refluxed (ca. 60°) with stirring under N₂ for 4 h, cooled to room temp., treated with 0.2 eq. Et₃N, stirred for 30 min, and worked up → methyl oleate. Y 93%. Reaction is mild, highly efficient, based on commonly available reagents, alleviates problems associated with deprotection of the acetal prior to acylation (thereby minimizing side-reactions), and proceeds **with retention of configuration** at the acetal site. A mechanism involving intermediate formation of an oxocarbenium ion is invoked. F.e.s. S.D. Stamatov, J. Stawinski, Tetrahedron *56*, 9697-703 (2000).

99.

Samarium(III) triflate
Glycosides from trichloroacetimidoyl glycosides
Solvent effect on stereoselectivity

$Sm(OTf)_3$
$OC(=NH)CCl_3 \rightarrow OR$

β-Glycosides. A soln. of 2 mol% Sm(OTf)$_3$ *in acetonitrile* added dropwise to a mixture of 1.4 eqs. startg. imidate, 1 eq. startg. glycosyl acceptor, and molecular sieves (4 Å) in the same solvent at -25°, and stirred until TLC indicated completion of reaction (60-90 min) at the same temp. → product. Y 88% (β:α = 10:1). Acetonitrile favored formation of the β-anomer, in contrast to 80% ether in dioxane which yielded mainly the α-anomer. Yields were lower with 2-O-acetylated acceptors. F.e.s. M. Adinolfi, A. Iadonisi et al., Tetrahedron Lett *41*, 9005-8 (2000).

100.

Ammonium ceric nitrate
Oxamic acid esters from acetoacetic acid amides

$(NH_4)_2Ce(NO_3)_6$
$CH_2COMe \rightarrow COOR$

An O$_2$-satd. soln. of *2.5 eqs.* CAN in methanol added to a soln. of startg. acetoacetamide in the same O$_2$-satd. solvent under a continuous O$_2$ purge, and worked up after 15 min → methyl N-phenyloxamate. Y 70%. Reaction is generally applicable to N-aryl- and N-alkyl-derivs., and is thought to involve the intermediate generation of a hydroperoxyl radical (there being no reaction in the absence of O$_2$). F.e.s. V. Nair, V. Sheeba, J. Org. Chem. *64*, 6898-900 (1999).

Amberlyst 15W (erbium(III)- or ytterbium(III)-form)
Carbohydrate O,O-alkylidene derivs.
←
○

by transacetalation with Amberlyst 15W (erbium(III)- or ytterbium(III)-form) s. *28*, 141s60; O,O-isopropylidene derivs. with CAN in anhydrous DMF s. Manzo, M. Parrilli et al., Synthesis *2000*, 887-9.

Lipases or Aminoacylase
Kinetic resolution by asym. O-acylation with enolesters
←
OH → OAc

update s. *44*, 214s59; kinetic resolution of γ-hydroxy-α-methylenecarboxylic acid esters *en route* to chiral α-methylene-γ-lactones s. W. Adam et al., Tetrahedron:Asym. *11*, 2239-43 (2000); of 2-hydroxymethyl-1,2-dihydrobenzofurans s. S. Ramadas, G.L.D. Krupadanam, ibid. 3375-93; of pantolactone s. L. Haughton, J.M.J. Williams et al., ibid. 1697-701; of aryl β-hydroxyketones s. M.S. Nair, S. Joly, ibid. 2049-52; of 3,4-condensed N-hydroxymethyl-2-azetidinones s. ibid. 1593-601; desymmetrization of *meso*-cyclobutene- and *meso*-cyclobutane-1,2-diols at low temperature s. C. Pichon et al., ibid. 2429-34; of *meso*-2,6- and *cis,cis*-2,4,6-subst. piperidines s. R. Chenevert et al., ibid. *10*, 3117-22 (1999); of *meso*-cyclopentene-*cis*-1,4-diol using an immobilised lipase from *Mucor meihei* (Lipozyme/Chirazyme) s. S.R. Ghorpade, T. Ravindranathan et al., ibid. 891-9; kinetic resolution of 2,2'-bis(hydroxymethyl)-1,1'-binaphthyl s. T. Furutani, M. Seki et al., ibid. 4763-8; of malic acid esters with lipase A-on-Celite (from *Candida antarctica*) s. A. Liljeblad, L.T. Kanerva, ibid. 4405-15; of 2-(dialkylaminomethyl)-cyclopentanols and -cycloheptanols s. E. Forro, F. Fulop, ibid. 1985-94; of 2-(N-*tert*-butoxycarbonyl))-N'-methylhydrazino)cycloalkanols s. ibid. 4619-26; of glycerol 1,2-cyclic sulfite s. M. Lemaire, J. Bolte, ibid. 4755-62; of Baylis-Hillman products s. N. Hayashi, S. Tsuboi et al., ibid. *9*, 3820-30 (1998); **double kinetic resolution** on transesterification of (R,S)-α-subst. carboxylic acid vinyl esters with (R,S)-1-phenylethanol s. H. Yang, U.T. Bornscheuer et al., ibid. *10*, 957-60 (1999); improved kinetic resolution of sec. benzyl alcohols with an aminoacylase from *Aspergillus melleus* s. M. Bakker, R.A. Sheldon et al., ibid. *11*, 1801-8 (2000); kinetic resolution **in ionic liquids** for improved isolation of the products (by distillation) and ease of recycling s. S.H. Schofer, U. Kragl et al., Chem. Commun. *2001*, 425-6; enhancement of enantioselectivity by substrate matching with an appropriate enolester (based on an enzyme memory approach) s. D. Lee, M.-J. Kim et al., Org. Lett. *2*, 2553-5 (2000).

Glycosyl transferase ←
Enzymatic transglycosidation OR → OR′
s. *49*, 195s*56*, *58*; **in plasticized glass phases** (the acceptor being at *high* concentration as a solution, as a suspension, a microemulsion or a liquid crystal) s. I. Gill, R. Valivety, Angew. Chem. Int. Ed. Engl. *39*, 3804-8 (2000).

2,3-Dichloro-5,6-dicyanoquinone/molecular sieves ←
O-Tritylation under neutral conditions OH → OTr

101.

A soln. of startg. alcohol and 1.2 eqs. *p*-methoxybenzyl trityl ether in dichloromethane containing activated 4 Å molecular sieves treated with 1.5 eqs. DDQ, the mixture stirred at room temp. for 10 min (monitored by TLC), then quenched with satd. NaHCO₃ soln. → product. Y 92%. The method utilizes a new tritylating reagent which shows improved reactivity compared to benzyl trityl ether. Due to the high oxidation potential of this reagent, reactions proceeded rapidly (10-25 min), irrespective of the nature of the alcohols. High yields and good selectivities were observed for symmetrical and unsymmetrical diols. F.e. and prepn. of the startg. trityl ether s. G.V.M. Sharma et al., Synlett *2000*, 1479-81; with benzyl trityl ether for **preferential O-tritylation** of prim. alcohols (with retention of sec.) s. M. Oikawa et al., ibid. *1998*, 757-60.

Titanium tetrachloride $TiCl_4$
Acoxy compds. from tetrahydropyran-2-yl ethers under mild conditions OThp → OAc

102.

0.2 eqs. 1 *M* TiCl₄ in methylene chloride added dropwise to a soln. of 2-(3-phenylpropoxy)tetrahydropyran and 1.2 eqs. acetic anhydride in the same solvent at 0° under N₂, and the mixture stirred at room temp. for 6 h → product. Y 78%. The mild conditions are compatible with functional groups such as alkenes, acetylenes, bromides and aromatic chlorides and fluorides. Quite remarkably, silyl ethers were also unaffected; however, a 1,3-dioxolane ring was also cleaved to form a 1,2-diacetate. F.e. incl. diacoxy compds. from bis(THP-ethers) s. S. Chandrasekhar et al., J. Org. Chem. *65*, 4729-31 (2000).

Tetraphenylphosphonium chloride Ph_4PCl
Carboxylic acid aryl esters from carboxylic acids COOH → COOAr
and oxalic acid aryl esters s. *60*, 82

Ozone/triphenylphosphine O_3/Ph_3P
Carboxylic acid esters from α,β-ethylenehalides via ozonolysis C(Hal)=C → COOR
from α,β-ethylenechlorides s. *32*, 180; α-fluorocarbocyclic acid esters, also with other nucleophiles (sec. amines, mercaptans, MeONHMe), and from α,β-ethylenefluorides, s. S.J. Brown, J.M. Percy et al., Tetrahedron Lett. *41*, 5269-73 (2000).

Trimethylsilyl triflate
Glycosides from trifluoroacetimidoyl glycosides
Me_3SiOTf
$OC(=NR)CF_3 \rightarrow OR'$

103.

β-Glycosides. A soln. of startg. trifluoroacetimidoyl glycoside (readily prepared from the parent aldose and (N-phenyl)trifluoroacetimidoyl chloride with K_2CO_3) and acceptor in methylene chloride treated with 5 mol% trimethylsilyl triflate and 4 Å molecular sieves at room temp. for 3 h → product. Y 94% (100% β). The new glycosyl donors are accessible, stable, comparable in activity with the more familiar trichloroacetimidoyl glycosides, and potentially tunable by adjusting the N-substituent. Not surprisingly, donors with neighbouring participating groups gave 1,2-*trans*-products exclusively, whilst donors lacking such groups gave mixtures of α and β anomers. F.e. and with steroidal and tert. alcohols, also trisaccharide synthesis, s. B. Yu, H. Tao, Tetrahedron Lett. *42*, 2405-7 (2001).

Iodine/pyridine
Enol phosphates from enol phosphites
I_2/C_5H_5N
$C=COP(OR)_2 \rightarrow C=COPO(OR)OR'$

104.

1.18 eqs. Crystalline I_2 added to 1.35 eqs. O,O-diethyl O-vinyl phosphite in methylene chloride at room temp., stirred for 5 min, treated with startg. hydroxy compd. as a solid followed by 4.2 eqs. neat anhydrous pyridine, and stirred at room temp. for 10 min → O-ethyl O-(2-isopropyl-6-methyl-4-pyrimidyl) O-vinyl phosphate. Y 98%. This is a convenient, practical and highly selective process involving *in situ*-generation of vinyl phosphoromonoiodates. The startg. phosphites may be prepared readily without the need for workup or purification steps by tetrahydrofuran cleavage (cf. *33*, 233) in the presence of diethyl chlorophosphite. Hydroxy compds. such as a pyridinol, 5'-O-tritylthymidine, *p*-nitrophenol, and cholesterol could also be used. F.e. incl. divinyl phosphates s. N. Zhang, J.E. Casida, Synthesis *2000*, 1454-8.

Ferric perchlorate
Catalytic transetherification of alkoxy-2-ethylenes
$Fe(ClO_4)_3$
$OR \rightarrow OR'$
using CAN with inversion of configuration cf. *50*, 115; using $Fe(ClO_4)_3$ with retention of configuration, also transetherification of sec. and tert. benzyl ethers, s. P. Salehi et al., Synth. Commun. *30*, 1743-7 (2000).

Ruthenium tetroxide
Carboxylic acids from benzene ring
RuO_4
C
s. *24*, 245; application to the determination of the absolute configuration of chiral sec. benzyl alcohols s. G. Bringmann et al., Tetrahedron:Asym. *11*, 3167-76 (2000).

Elimination ⇑

Hydrogen ↑   OC ⇑ H

Microwaves s. under Zeolites, HNO_3, SeO_2, CrO_3, and Benzyltriphenylphosphonium chlorochromate ←

Sodium nitrite/acetic anhydride $NaNO_2/Ac_2O$
Aldehydes from prim. alcohols $CH_2OH \rightarrow CHO$
Selective oxidation under mild conditions in the absence of solvent

105. $O_2N-C_6H_4-CH_2OH \longrightarrow O_2N-C_6H_4-CHO$

A mixture of 3 eqs. $NaNO_2$, 4-nitrobenzyl alcohol, and 0.4 eq. acetic anhydride stirred with ice-cooling to control the exotherm, and worked up when TLC indicated completion of reaction (<1 min) → 4-nitrobenzaldehyde. Y 92%. The system is effective for the oxidation of simple, allylic and benzylic prim. alcohols, and proceeds without overoxidation and **with retention of sec. alcohols,** cyano and *tert*-amino groups, and THP and TBDMS ethers. The method is simple, rapid, inexpensive and high-yielding. F.e. incl. oxidation of furyl-, pyridyl- and thienyl-carbinols (without N- or S-oxidation) s. B.P. Bandgar et al., J. Chem. Soc. Perkin Trans. 1 *2000*, 3559-60.

Cupric sulfate s. under $KMnO_4$ $CuSO_4$
Silver(III) periodate complex s. under Periodates ←
Bentonite s. under HNO_3 and MnO_2 ←

Zeolite/microwaves ←
Ketones from sec. alcohols under microwave irradiation without solvent $CHOH \rightarrow CO$
on zeolites with *t*-BuOOH as oxidant cf. *37*, 234s*54*; benzils from benzoins on zeolite A *without* oxidant s. S. Balalaie et al., Green. Chem. *2*, 277-8 (2000).

Aluminum chloride s. under Benzyltriphenylphosphonium peroxymonosulfate $AlCl_3$

2,2,6,6-Tetramethylpiperidine nitroxyl/potassium peroxymonosulfate/ ←
*tetra-*n-*butylammonium bromide*
Oxo compds. from alcohols
ketones with *m*-$ClC_6H_4COO_2H$ as reoxidant cf. *45*, 120s*57*; with Oxone as reoxidant in methylene chloride (aldehydes) or toluene (ketones) s. C. Bolm et al., Org. Lett. *2*, 1173-5 (2000).

2,2,6,6-Tetramethylpiperidine nitroxyl/polymer-based quaternary ammonium ←
diacetoxybromate
Oxo compds. from alcohols
with TEMPO/$PhI(OAc)_2$ cf. *45*, 120s*54*; with TEMPO and recyclable, polymer-based quaternary ammonium diacetoxybromate s. G. Sourkouni-Argirusi, A. Kirschning, Org. Lett. *2*, 3781-4 (2000).

tert-*Butyl hydroperoxide s. under CrO_3 and Cobalt(II) complexes* t-BuOOH

Polymer-based selenocyanates/ammonium persulfate ←
2(5*H*)-Furanones from β,γ-ethylenecarboxylic acids ○
with 10 mol% $PhSeSePh/(NH_4)_2S_2O_8$ cf. *49*, 210; more friendly procedure with a little polymer-based selenocyanate, and further organoselenium(II) reagents s. K. Fujita et al., Synlett *2000*, 1509-11.

Iodosobenzene s. under Cr(III) salen complex PhIO

Phenyl iodosoacetate/sodium acetate $PhI(OAc)_2/NaOAc$
5-Alkoxyoxazoles from α-(alkylideneamino)carboxylic acid esters

106. Me-C_6H_4-CH=N-CH_2-C(O)-OMe → Me-C_6H_4-[oxazole]-OMe

1.2 eqs. Phenyl iodosoacetate and 2 eqs. Na-acetate added in one portion to a soln. of the startg. aldimine in methanol, and stirred for 1 h at room temp. → product. Y 88%. Yields were lower in the absence of Na-acetate. Ar. nitro, halides and phenolethers remain unaffected. F.e.s. M. Xia, J. Chem. Res., Synop *2000*, 382-3.

Phenyl iodosoacetate/iodine $PhI(OAc)_2/I_2$
Ring closures of alcohols via alkoxyl radicals   O
s. *39*, 228s*59*; regiospecific intramolecular oxidative glycosidation s. C.G. Francisco, E. Suárez et al., Tetrahedron Lett. *41*, 7869-73 (2000).

Hetaryl iodosoacetates $ArI(OAc)_2$
Quinones from quinols   ←
with PhI(OAc)$_2$ cf. *43*, 213; with hetaryl iodosoacetates, e.g. 1-trifluoromethanesulfonyl-4-(diacetoxyiodo)pyrazole, also S-oxidation, s. H. Togo et al., J. Org. Chem. *65*, 8391-4 (2000).

N-Condensed 1,2-benziodazol-3-one 1-oxides/trifluoroacetic acid   ←
Oxo compds. from alcohols   CHOH → CO
with o-iodoxybenzoic acid cf. *50*, 123; with non-explosive, readily soluble N-condensed 1,2-benziodazol-3-one 1-oxides/CF$_3$COOH, also S-oxidation, s. V.V. Zhdankin et al., Tetrahedron Lett. *41*, 5299-302 (2000).

Nitric acid/bentonite/microwaves   ←
***p*-Quinones from *p*-quinols**   ←
with concd. HNO$_3$ cf. *18*, 302; rapid, clean procedure with HNO$_3$/bentonite under microwave irradiation in the absence of solvent, also with MnO$_2$/bentonite (cf. *13*, 752s*36*), and under infrared irradiation (with an Osram 250 W lamp), s. J. Gómez-Lara, C. Alvarez-Toledano et al., Synth. Commun. *30*, 2713-20 (2000).

Bismuth s. under Pd   Bi
Hydrogen peroxide s. under Br$_2$, Oxodiiron(III) complexes and H_2O_2
   (6,6'-Dichloro-2,2'-bipyridyl)ruthenium(II) bis(triflate)

Potassium peroxymonosulfate s. under 2,2,6,6-Tetramethylpiperidine nitroxyl $KHSO_5$

Benzyltriphenylphosphonium peroxymonosulfate/aluminum chloride $[BnPh_3P][HSO_5]/AlCl_3$
Oxo compds. from alcohols   CHOH → CO
ketones with KHSO$_5$/Al$_2$O$_3$ cf. *46*, 237; solid-state oxidation of allyl and benzyl alcohols with [BnPh$_3$P][HSO$_5$]/AlCl$_3$ (on grinding in a mortar) s. A.R. Hajipour et al., Chem. Lett. *2000*, 460-1.

Ammonium persulfate s. under Polymer-based selenocyanates $(NH_4)_2S_2O_8$

Selenium dioxide/dimethyl sulfoxide/silica/microwaves   ←
Flavones from *o'*-hydroxychalcones   O
with SeO$_2$ cf. *2*, 288; with SeO$_2$ and 1.2 eqs. DMSO under microwave irradiation on silica gel in the absence of solvent s. M. Gupta, R. Gupta et al., Org. Prep. Proced. Int. *32*, 280-3 (2000).

Chromium trioxide/silica gel/tert-butyl hydroperoxide/microwaves   ←
Aryloxo compds. from benzylalcohols   CHOH → CO
with CrO$_3$/t-BuOOH cf. *42*, 236; preferential oxidation under microwave irradiation on silica gel in the absence of solvent s. J. Singh et al., Synth. Commun. *30*, 3941-5 (2000).

Poly(4-vinylpyridinium fluorochromate)
Oxo compds. from alcohols
s. *42*, 235s*51*; s.a. R. Srinivasan, K. Balasubramanian, Synth. Commun. *30*, 4397-404 (2000).

Benzyltriphenylphosphonium chlorochromate/microwaves   ←
Oxo compds. from alcohols under microwave irradiation
with pyridinium chlorochromate in methylene chloride cf. *26*, 235s*57*; allylic and benzylic oxidation with inexpensive, storable benzyltriphenylphosphonium chlorochromate s. A.R. Hajipour et al., Synth. Commun. *30*, 3855-64 (2000).

Chromium(III) salen complex/iodosobenzene/4-phenylpyridine N-oxide
Oxo compds. from alcohols
Catalytic oxidation under mild conditions
$\leftarrow$
CHOH → CO

107.

Ph⁀⁀OH → Ph⁀⁀=O (via Cr salen complex shown, Cl⁻)

α,β-Ethyleneoxo compds. 0.3 eq. 4-Phenylpyridine N-oxide and 0.15 eq. N,N'-ethylenebis-(salicylideneiminato)chromium(III) chloride added with stirring to a soln. of startg. allyl alcohol in methylene chloride, followed by 1.5 eqs. iodosobenzene under argon, and stirred for 6 h → product. Y 95% (>95% conversion; >95% purity). Reaction is generally applicable to the oxidation of benzylic and allylic alcohols, but the conversion of unactivated alcohols was more sluggish (72 h for 18-21% conversion). The active oxidant is an *in situ*-generated oxochromium(V) salen complex which shows a high preference for allylic oxidation versus epoxidation (except with tetrasubst. derivs.). Furthermore, while (Z)-configured allyl alcohols underwent *cis-trans* isomerization, the stereochemistry of (E)-substrates remained unaffected. The cyclopropane ring was also left intact. F.e.s. W. Adam et al., J. Org. Chem. *65*, 1915-8 (2000); catalyst counter-ion effects, also with PhI(OAc)$_2$, s. Org. Lett. *2*, 2773-6 (2000).

Bromine/hydrogen peroxide
α-Ketocarboxylic from α-hydroxycarboxylic acid esters
Br_2/H_2O_2

108. MeCH(OH)COOEt → MeCOCOOEt

in a 2-phase medium. 6.4 g Br$_2$ and 2.4 g ethyl lactate mixed in 1,2-dichloroethane, a further 21.2 g of the ester and 22 g 35% aq. H$_2$O$_2$ added over 2 h with stirring at 40° in the dark, and heating continued at the same temp. for a further 2.5 h → ethyl pyruvate. Y 86%. The method is high-yielding and applicable on a large scale. F.e. and with HBr in place of Br$_2$ s. Japanese patent JP-11246482 (Toray Ind. Inc.).

Iodine s. under PhI(OAc)$_2$
I_2

Potassium tetrasodium diperiodatoargenate(III)
Oxo compds. from alcohols
$KNa_4[Ag(HIO_6)_2]$
with Bu$_4$NIO$_4$/AlCl$_3$ cf. *52*, 121; aryloxo compds. with potassium tetrasodium diperiodato-argenate(III) dodecahydrate s. Z. Bugaric et al., Monatsh. Chem. *131*, 799-802 (2000).

Bis(collidine)bromine(I) hexafluorophosphate
Oxo compds. from alcohols s. *60*, 114
$\leftarrow$

Polymer-based quaternary ammonium diacetoxybromate
    s. under *2,2,6,6-Tetramethylpiperidine nitroxyl*
$\leftarrow$

Manganese dioxide/bentonite
p-Quinones from p-quinols
under microwave irradiation on a solid support s. *13*, 752s60
$\leftarrow$
$\leftarrow$

Potassium permanganate/cupric sulfate
Heterogeneous permanganate oxidation under ultrasonic acceleration
$KMnO_4/CuSO_4$
$\leftarrow$

109. PhCH$_2$OH → PhCHO

Ultrasonication accelerates heterogeneous permanganate oxidation [with KMnO$_4$/CuSO$_4$] of benzyl alcohols and alkylarenes to the corresponding carbonyl compds. *in methylene chloride* so that reaction can be conducted *at room temp*. **E: Aryloxo compds.** A soln. of 2.8 mmol benzyl alcohol in methylene chloride containing 6.4 g 1:1 KMnO$_4$/CuSO$_4$·5H$_2$O (finely ground as a powder) irradiated in a glass reactor fitted with an ultrasonic JENCONS horn (20 kHz, 400 W) by pulsing at 10 sec intervals *for 30 min* at room temp. under air → benzaldehyde. Y >97% (conversion 74%). At reflux without ultrasonication a 1:1 mixture of benzaldehyde and benzoic acid was obtained after 24 h (100% conversion). Over-oxidation to the carboxylic acid was minimal under ultrasonication. F.e., **also from methyl and methylene groups**, s. M. Meciarova et al., Tetrahedron 56, 8561-6 (2000).

μ-Oxodiiron(III) complex/hydrogen peroxide/trifluoromethanesulfonic acid ←
Oxo compds. from alcohols CHOH → CO
with Fe(III)-β-diketonates/*tert*-butyl hydroperoxide cf. *26, 463s51*; with a non-heme μ-oxo-diiron(III) complex/H_2O_2 and CF_3SO_3H as accelerator s. A.G.J. Ligtenbarg, R. Hage et al., Chem. Commun. *2001*, 385-6.

Polymer-based dichloro[aryl(diphenyl)phosphine](triphenylphosphine)cobalt(II)/ ←
tert-butyl hydroperoxide
Oxo compds. from alcohols
with $CoCl_2(PPh_3)_2$/*t*-BuOOH cf. *26, 463s51*; allylic and benzylic oxidation with air-stable, readily recyclable polymer-based dichloro[aryl(diphenyl)phosphine](triphenylphosphine)cobalt(II) s. N.E. Leadbeater, K.A. Scott, J. Org. Chem. *65*, 4770-2 (2000).

(6,6′-Dichloro-2,2′-bipyridyl)ruthenium(II) bis(triflate)/tert-butyl hydroperoxide ←
Oxo compds. from alcohols
with (N,N′,N″-trimethyl-1,4,7-triazacyclononane/ruthenium(II) bis(trifluoroacetate) cf. *15*, 261s56; ketones from sec. alcohols, and a mixture of aldehydes and carboxylic acids from prim. alcohols, with (6,6′-dichloro-2,2′-bipyridyl)ruthenium(II) bis(triflate) s. C.-M. Che et al., J. Org. Chem. *65*, 7996-8000 (2000).

Dichloro(p-cymene)ruthenium(II) dimer/cesium carbonate/oxygen ←
Oxo compds. from alcohols
Ruthenium-catalyzed aerobic oxidation
with $RuCl_2(Ph_3P)_3$ cf. *24, 261s55*; allylic and benzylic oxidation with $[RuCl_2(p\text{-cymene})]_2$ and 0.1 eq. Cs_2CO_3 s. M. Lee, S. Chang, Tetrahedron Lett. *41*, 7507-10 (2000).

Palladium-platinum-bismuth/carbon/oxygen Pd-Pt-Bi/C/O_2
Oxo compds. from alcohols
Heterogeneous catalytic aerobic oxidation
with Ru/CeO_2 cf. *55, 113*; ketones and aromatic or α,β-unsatd. aldehydes from water-insoluble alcohols *in supercritical CO_2* over Pd-Pt-Bi/C in a continuous fixed-bed reactor (residence time 10-25 sec) s. G. Jenzer, A. Baiker et al., Chem. Commun. *2000*, 2247-8.

Water-soluble iridium(I) complex/acetone/sodium carbonate ←
Ketones from sec. alcohols
Oppenauer-type oxidation in aq. medium

110. Ph-CH(OH)-Me →[Ir(I)]/Me$_2$CO (with dipotassium 2,2′-biquinolyl-4,4′-dicarboxylate ligand)→ Ph-C(O)-Me

The first example of a water-soluble transition metal complex for the Oppenauer oxidation is reported. **E:** A soln. of 1-phenylethyl alcohol in degassed acetone added to a degassed aq. soln. (acetone/water 1:2) of 0.4 mol% [Ir(cod)Cl]$_2$, 6 mol% inexpensive dipotassium 2,2′-biquinolyl-4,4′-dicarboxylate, and 1 eq. Na$_2$CO$_3$ under N$_2$ contained in a glass-lined autoclave, the latter flushed several times with 80 psi N$_2$, and heated at 90° for 4 h → acetophenone. Y 98%. Reaction is applicable to aliphatic sec. and benzyl alcohols at a catalyst loading of 0.4-2.5 mol%. The corresponding water-insoluble complex based on [more expensive] 2,2′-biquinolyl, as well as the corresponding water-soluble Rh analog, were less active. The catalyst is readily retrievable from the aq. phase and reusable up to three times with only a slight decrease in activity. F.e.s. A.N. Ajjou, Tetrahedron Lett. *42*, 13-5 (2001).

Oxygen ↑   OC ⇑ O

Polymer-based carbodiimide/4-dimethylaminopyridine/   ←
4-dimethylaminopyridine hydrochloride
Macrocyclic lactones from hydroxycarboxylic acids   ○
with dicyclohexylcarbodiimide cf. *13*, 317s*41*; with a polymer-based carbodiimide for easy work-up s. G.E. Keck et al., Tetrahedron Lett. *41*, 8673-6 (2000).

Triphenylphosphine/diethyl azodicarboxylate $Ph_3P/EtOOCN=NCOOEt$
Oxo compds. from glycols under neutral conditions $C(OH)C(OH) \rightarrow CHCO$

111.

While mono- and 1,2-di-subst. glycols are normally converted to the epoxide with the Mitsunobu reagent (cf. *34*, 88), *1,1-disubst.* glycols are converted to the corresponding aldehyde or ketone (depending on whether the less subst. carbon is primary or secondary). **E:** A soln. of startg. glycol in benzene allowed to react with Ph_3P and diethyl azodicarboxylate at 0° for 10 h → product. Y 75%. F.e. and stereoselectivity s. A.F. Barrero, R. Chahboun et al., Tetrahedron Lett. *41*, 1959-62 (2000).

N,N'-Sulfonyldiimidazole/sodium hydride   ←
2,3'-Cyclonucleosides by cyclodehydration   ○
with 2-chloro-1,1,2-trifluorotriethylamine cf. *25*, 190; 2,3'-cyclo-2'-deoxynucleosides with N,N'-sulfonyldiimidazole and NaH or K_2CO_3 s. Z. No et al., Synth. Commun. *30*, 3873-82 (2000).

Pyridinium tosylate/p-toluenesulfonic acid $[C_5H_5NH][TsO]/TsOH$
Cyclic ethers from diols with 1,2-arylthio group migration
with TsOH cf. *44*, 238; arylthio-subst. 1,8-dioxa- and 1-oxa-8-thia-spiro[4.5]decanes s. J. Eames, S. Warren et al., J. Chem. Soc. Perkin Trans. 1 *2000*, 1903-14; orthoester-promoted cyclization with pyridinium tosylate/TsOH under kinetic control for preparing chiral 3-acoxy-5-arylthio-tetrahydropyrans, s. D.J. Fox, S. Warren et al., Chem. Commun. *2000*, 1781-2; **4-(arylthio)-tetrahydro-2-furyl -carbinols** with TsOH under thermodynamic control s. ibid. 1779-80; with triols containing three sec. hydroxyl groups s. ibid. 1783-4.

Nitrogen ↑   OC ⇑ N

Hexamethyldisilazane/ammonium sulfate $(Me_3Si)_2NH/(NH_4)_2SO_4$
Release of nucleoside bases on lactonization s. *60*, 121   ○

Halogen ↑   OC ⇑ Hal

Potassium carbonate K_2CO_3
Epoxides from 1,2-bromhydrins $C(OH)C(Br) \rightarrow$ ▽
with inversion of configuration s. *60*, 112

Di-n-butyltin oxide Bu_2SnO
2,2'-Cyclonucleosides   ○
from nucleoside 2'-mesylates with inversion using NaOH cf. *14*, 343; from 2'-iodo-2'-deoxynucleosides using Bu_2SnO s. R. Robles et al., Tetrahedron:Asym. *11*, 3069-77 (2000).

Bis(dibenzylideneacetone)palladium(0)/1,2,3,4,5-pentaphenyl-   ←
  1'-(di-tert-butylphosphino)ferrocene/sodium tert-butoxide
Tris(dibenzylideneacetone)dipalladium/1,1'-bis(diphenylphosphino)ferrocene/   ←
  sodium hydroxide
Palladium-catalyzed intramolecular O-arylation
with $Pd(OAc)_2/(S)$-Tol-BINAP/K_2CO_3 cf. *52*, 128; dibenzoxepino[4,5-*d*]pyrazoles with $Pd_2(dba)_3/$

dppf/NaOH, also with (R)-BINAP as ligand, s. R. Olivera, E. Domínguez et al., Tetrahedron Lett. *41*, 4357-60 (2000); with Pd(dba)$_2$/1,2,3,4,5-pentaphenyl-1'-(di-*tert*-butylphosphino)ferrocene s. *52*, 128s*60*.

Sulfur ↑   OC ↑ S

Sodium hydroxide   *NaOH*
Epoxides from cyclic glycol sulfates

112.

2 eqs. 1 *M* NaOH added to a soln. of startg. cyclic sulfate in 3:1 THF/methanol at room temp., and the mixture stirred for 0.5 h → product. Y 86%. Cyclic sulfates of conformationally rigid rings, such as D-ribose derivs., did not proceed to the epoxide. The method is mild, simple, easy, rapid, and stereospecific **with inversion of configuration**. F.e.s. D.O. Jang et al., Synth. Commun. *30*, 4489-94 (2000); **via 1,2-bromhydrins** (isolated crude on treatment with LiBr, followed by 20% aq. H$_2$SO$_4$) with overall **retention of configuration** (by double inversion), notably for preparing **chiral glycidic acid esters**, s. S.L. He, R. Bittman et al., Tetrahedron Lett. *39*, 2071-4 (1998).

Carbon ↑   OC ↑ C

Microwaves s. under Zn(OTf)$_2$ or BF$_3$ ←
Zinc triflate or boron fluoride/microwaves   ←
2-Oxazolidones from N-carbo-*tert*-butoxyaziridines   ○
with Cu(OTf)$_2$ cf. *59*, 123; cleaner procedure under microwave irradiation (preferably with Zn(OTf)$_2$ or BF$_3$-etherate), notably with N-acyl-2-imidazolidone derivs. (rather than esters), s. G. Cardillo et al., Synlett *2000*, 1309-11.

Boron chloride   *BCl$_3$*
Trifluoroacetic acid   *CF$_3$COOH*
'Traceless' polymer-based synthesis of benzofurans from *o*-hydroxy-β-styryl ethers   ○

113.

Startg. Wang resin-bound phenol swollen in methylene chloride (1.6 parts), treated with 50% trifluoroacetic acid (1 part), stirred for 30 min, and worked up → product. Y 83% (from the Wang ester). The startg. enolether was prepared by the first example of **on-resin Takeda synthesis** (cf. *51*, 423s*53*), avoiding the problematical work-up that resulted in low yields in the solution phase. The ring closure is traceless, and should be applicable to benzofurans bearing substituents in any position. F.e.s. E.J. Guthrie, R.C. Hartley et al., Tetrahedron Lett. *41*, 4987-90 (2000); polymer-based synthesis of **2(3*H*)-benzofuranones** from supported *o*-alkoxythiolacetic acid esters with BCl$_3$, and further syntheses, s. P.J. May, D.C. Harrowven et al., ibid. 1627-30.

m-Chloroperoxybenzoic acid/boron fluoride/1,8-diazabicyclo[5.4.0]undec-7-ene   ←
Carboxylic acid esters from acetals   CH(OR)$_2$ → COOR
with peroxyacetic acid cf. *16*, 342; with *m*-chloroperoxybenzoic acid/BF$_3$-etherate/DBU, also 1,2- and 1,3-diol monobenzoates from cyclic acetals prepared *in situ* from diols and aldehydes or acetals, s. H. Rhee, J.Y. Kim, Bull. Korean Chem. Soc. *21*, 355-7 (2000).

2,3-Dichloro-5,6-dicyanoquinone   *DDQ*
Oxidative cleavage of the Wang resin s. *17*, 216s*60*   ←

Bis(collidine)bromine(I) hexafluorophosphate
Oxidations with bis(collidine)bromine(I) hexafluorophosphate

114.

Oxo compds. from *p*-methoxybenzylalcohols. 1.2 eqs. Bis(collidine)bromine(I) hexafluorophosphate in methylene chloride added over 1 h to a soln. of startg. *p*-methoxybenzyl alcohol in the same solvent, and stirred for 30 min at room temp. → product. Y 91% (and 80-95% *p*-bromoanisole). Reaction is clean and simple, and generally applicable to sec. and tert. *p*-methoxybenzyl alcohols. Furthermore, as the substrates are readily obtained from oxo compds. and *p*-bromoanisole, the procedure can be used for the **protection of carbonyl groups.** F.e., **also oxo compds. from alcohols,** and brominative ring opening of tetralols s. G. Rousseau, S. Robin, Tetrahedron Lett. **41**, 8881-5 (2000).

Diiron nonacarbonyl/fluoroboric acid/cesium carbonate
2-Vinyl-O-heterocyclics from (siloxy)acoxy-2-ethylenes with asym. induction

115.

Startg. (siloxy)acoxy-2-ethylene added to a vigorously stirred soln. of 1.3 eqs. $Fe_2(CO)_9$ in cyclohexane with exclusion of light, air and moisture, satd. with CO (from a CO-balloon), stirred for 24 h at room temp. under CO, solvent removed, the residue taken up in 4:1 ether/methylene chloride, cooled to -23°, 2.5 eqs. HBF_4·etherate added, stirred for 30 min at the same temp., treated with 5.2 eqs. Cs_2CO_3 (to minimize racemization), stirred for 50 min at -10°, and decomplexed by treatment with 9 eqs. CAN in satd. aq. NH_4Cl by vigorously stirring for 12-24 h at room temp. → (S)-(2S)-2-[(E)-2-(phenylsulfonyl)ethenyl]tetrahydrofuran. Y 44% (e.e. 95%). Reaction takes place with chirality transfer via a cationic (η^3-allyl)tetracarbonyliron complex and net **retention of configuration.** However, the ethylenic bond must be substituted by an electron-accepting group (COOMe, SO_2Ph). F.e. incl. tetrahydropyran analogs s. D. Enders, D. Nguyen, Synthesis **2000**, 2092-8.

Formation of N-N Bond

Exchange   ↕

Hydrogen ↑   NN ↕ H

N-Lithio compds.
N-Nitrosation with nitrogen monoxide   NH → N-NO
of acylamines cf. **10**, 251s*53*; of lithium amides s. N.S. Nudelman, A.E. Bonatti, Synlett **2000**, 1825-7.

Sodium nitrite/zinc chloride or aluminum chloride or tungsten hexachloride/silica ←
N-Nitrosation of sec. amines NH → N-NO

Ph$_2$NH [NOCl] → Ph$_2$N-NO

A suspension of diphenylamine, 0.25 eq. WCl$_6$, wet SiO$_2$ (50% w/w,) and 1.5 eqs. NaNO$_2$ in dichloromethane stirred at room temp. for 1 hr. → product. Y 98%. *Caution*: some N-nitrosamines are known carcinogens. This method is mild, simple and chemoselective with no C-nitrosation side-products being observed. F.e. and with AlCl$_3$ or ZnCl$_2$, also with retention of chirality, s. M.A. Zolfigol et al., J. Chem. Res., Synop *2000*, 420-2.

Formation of N-Hal Bond

Exchange ↕

Hydrogen ↑   NHal ↕ H

Sodium hypochlorite NaOCl
N-Chlorination of oxazolidines s. *48*, 392s60   NH → NCl

Formation of N-S Bond

Exchange ↕

Halogen ↑   NS ↕ Hal

Triethylamine/4-dimethylaminopyridine Et$_3$N/DMAP
Sym. disulfimides from sulfonic acid chlorides 2 RSO$_2$Cl → (RSO$_2$)$_2$NH
with NaOH cf. *11*, 362; with Et$_3$N/DMAP s. J. Desmurs et al., World Intellectual Property Organization patent WO-200026161 (Rhodia Chim.).

Carbon ↑   NS ↕ C

Diethyl bromomalonate BrCH(COOEt)$_2$
S-Glycosylsulfenamides from glycosyl thioacetates and sec. amines SAc → SN<

An excess of Et$_2$NH added to a soln. of startg. thioacetate in dry DMF under N$_2$, followed by 2-4 eqs. diethyl bromomalonate, and the mixture stirred at 25° for 14 hrs. → product. Y 81%. Ethers, carboxylic acid esters, thioesters and amides remained unaffected. F.e.s. D.J. Owen, M. von Itzstein, Carbohydr. Res. *328*, 287-92 (2000).

Formation of N-C Bond

Uptake ⇓

Addition to Oxygen and Nitrogen NC ⇓ ON

Without additional reagents w.a.r.
Hydroxyl-directed regio- and stereo-specific ene reaction C=C-C-N(OH)
with nitroso compds.
***threo*-3-Ethylene-2-hydroxylaminoalcohols**

118.

A soln. of startg. allyl alcohol and 1 eq. *p*-nitrosonitrobenzene in chloroform-*d* allowed to react at 0° for 24 h → product. Conversion 51% (>95% purity; mass balance 61% relative to the allyl alcohol; *threo:erythro* >95:5). The *threo*-selectivity (being superior to that obtained by ene reaction with singlet oxygen or triazolinedione enophiles) is interpreted by hydroxyl group directivity, being associated with both 1,3-allylic strain and hydrogen bonding with the incoming nitroso compd. Predictably, diastereoselectivity was much lower in DMSO or methanol, which suppress H-bonding by competitive H-bonding with the solvent molecules. The corresponding (E)-substrate was inactive, demonstrating the importance of a 'twix'-oriented substituent on the olefin residue. F.e.s. W. Adam, N. Bottke, J. Am. Chem. Soc. *122*, 9846-7 (2000).

Addition to Oxygen and Carbon NC ⇓ OC

Without additional reagents w.a.r.
N-Hydroxymethylation NH → NCH$_2$OH
s. *2*, 692; *3*, 657; *6*, 339; *7*, 384; of N-heterocyclics under ultrasonication s. W. Zhong, G. Song et al., Synth. Commun. *30*, 3801-7 (2000).

2-*prim*-Aminoalcohols from epoxides $\diagdown_{\text{O}}\diagup$ → C(OH)C(NH$_2$)
carbohydrate derivs. cf. *14*, 380; β-amino-α-hydroxyphosphonic from α,β-epoxyphosphonic acid esters, stereoselectivity, s. H.-J. Cristau et al., Tetrahedron Lett. *41*, 9781-5 (2000).

2,2′-Dihydroxyamines from epoxides ←
s. *6*, 340; *13*, 359; sym. polyfluoro-2,2′-dihydroxyamines by regiospecific ring opening with aq. NH$_3$, s. B. Charrada, A. Baklouti et al., Tetrahedron Lett. *41*, 7347-9 (2000).

Sodium azide/cupric nitrate NaN$_3$/Cu(NO$_3$)$_2$
2-Azidoalcohols from epoxides $\diagdown_{\text{O}}\diagup$ → C(OH)C(N$_3$)
with NaN$_3$/AcOH in aq. media cf. *10*, 262s59; β-azido-α-hydroxycarboxylic from glycidic acids *in water* with a little Cu(NO$_3$)$_2$, regio- and stereo-selectivity, s. F. Fringuelli, F. Pizzo et al., Synlett *2000*, 311-4.

Sodium azide/aluminum chloride NaN$_3$/AlCl$_3$
Lewis-acid catalysis with aluminum chloride in water
β-Azido-α-hydroxycarboxylic from glycidic acids
Regio- and stereo-specific ring opening

119.

The use of AlCl$_3$ as catalyst in water has been reported for the first time. **E:** 20 μL (*1 mol%*) 0.5 M aq. AlCl$_3$ added to a soln. of 1 mmol startg. epoxy acid and 5 eqs. NaN$_3$ *in water* with stirring at 30°, the pH adjusted to 4 by addition of 50% H_2SO_4, stirred for 3.5 h at pH 4, cooled to 0°, and acidified to pH 2 → β-adduct. Y 93-5%. It is thought that an aluminum aqua ion, $Al(H_2O)_6^{3+}$ is the active catalyst, which coordinates to both the epoxy acid and azide ion to form a reactive complex. The mother liquor after work-up could be re-used up to three times. F.e.s. F. Fringuelli et al., Tetrahedron Lett. *42*, 1131-3 (2001).

Cupric nitrate s. under NaN_3 $Cu(NO_3)_2$
Hexafluoroisopropanol $(CF_3)_2CHOH$
2-(Arylamino)alcohols from epoxides and ar. amines   ◯ → C(OH)C(NHAr)
Stereospecific ring opening in hexafluoroisopropanol

120.

1.1 eqs. *o*-Toluidine added to a soln. of cyclohexene oxide in hexafluoroisopropanol, and refluxed under argon for 2.5 hrs. → product. Y 92% (100% *trans*-selectivity). The solvent (pK_a = 9.3) facilitated ring opening **under neutral conditions** without the need for a catalyst, but more nucleophilic amines, such as alkylamines, were deactivated. F.e.s. U. Das, J.-P. Bégué et al., J. Org. Chem. *65*, 6749-51 (2000).

Hexamethyldisilazane/ammonium sulfate/trimethylsilyl chloride/trimethylsilyl triflate
Nucleoside base exchange in 2'-C-carboxymethyl-2'-deoxynucleosides   ←C
via stereospecific ring opening of anomerically-fused γ-lactones

121.

A mixture of startg. nucleoside, hexamethyldisilazane, and ammonium sulfate heated at reflux for 4 h → intermediate γ-lactone (Y 57%), in dry acetonitrile at 0° treated with 1.2 eqs. hexamethyldisilazane and 1.5 eqs. thymine, followed by 1.2 eqs. trimethylsilyl chloride and 3.3 eqs. trimethylsilyl triflate, warmed to room temp., and stirred for 18 h → product (Y 79%). F.e.s. V. Fehring, R. Cosstick et al., J. Chem. Soc. Perkin Trans. 1 *2000*, 3185-7.

Trimethylsilyl azide/tetra-n-butylammonium chloride/p-toluenesulfonic acid   ←
2-Azidoalcohols from epoxides   ◯ → C(OH)C(N_3)
with Me_3SiN_3/BF_3, cf. *40*, 176s*41*, 42; under *neutral* conditions with Me_3SiN_3 and a little Bu_4NCl/TsOH, regio- and stereo-selectivity, s. C. Schneider, Synlett *2000*, 1840-2.

Tantalum pentachloride/silica $TaCl_5/SiO_2$
2-(Arylamino)alcohols from epoxides and ar. amines   ◯ → C(OH)C(N<)

122.

Startg. epoxide and 1.2 eqs. aniline added to a stirred soln. of 10 mol% $TaCl_5$-SiO_2 in anhydrous methylene chloride, and stirred at room temp. under N_2 until TLC indicated completion of reaction → product. Y 85%. Interestingly, aliphatic amines were unreactive. F.e. incl. ring opening with N-alkylanilines and hindered amines s. S. Chandrasekhar et al., Synthesis *2000*, 1817-8.

Dicarbonyl(chloro)rhodium(I) dimer [Rh(CO)$_2$Cl]$_2$
Regio- and stereo-specific ring opening of 3-ethyleneepoxides ▽$_\text{O}$′ → C(OH)C(N<)
with ar. amines s. *60*, 59

Addition to Nitrogen and Carbon   NC ⇓ NC

Without additional reagents or Triethylamine   *w.a.r or Et$_3$N*
Amidoximes from nitriles   CN → C(=NOH)NH$_2$
with K$_2$CO$_3$ cf. *4*, 311; large-scale procedure with Et$_3$N s. C. Besse et al., World Intellectual Property Organization patent WO-200024740 (Sanofi-Synthelabo); in water without base cf. J.J. Sahbari, W.J. Russell, WO-200032565 (Silicon Valley Chemlabs. Inc.).

Irradiation s. under N-Benzyl-1,4-dihydronicotinamide   ⫼
Microwaves s. under NaN$_3$ ←

Sodium azide/ammonium chloride/microwaves   ←
Tetrazoles from nitriles   ○
s. *13*, 371s55; rapid procedure for 5-vinyl- and 5-aryl-tetrazoles under microwave irradiation, also one-pot prepn. from ar. halides, s. M. Alterman, A. Hallberg, J. Org. Chem. *65*, 7984-9 (2000).

Sodium iodide/trimethylsilyl chloride   *NaI/Me$_3$SiCl*
Regiospecific ring opening of N-protected aziridines with amines   C
using Sn(OTf)$_2$ or Cu(OTf)$_2$ cf. *50*, 151s57; 2,3-diaminoalcohols from aziridinylcarbinols, reversal of regioselectivity using NaI/Me$_3$SiCl, s. S.-H. Shin, W.K. Lee et al., Tetrahedron:Asym. *11*, 3293-301 (2000).

N-Benzyl-1,4-dihydronicotinamide/irradiation   ←
Tert. amines from azomethines and halides   C=N → CHNR
via hydrostannylation cf. *51*, 126; from ar. aldimines or cinnamaldimines with N-benzyl-1,4-dihydronicotinamide under irradiation s. M. Jin, Z. Liu et al., Tetrahedron Lett. *41*, 7357-60 (2000).

Trimethylsilyl azide   *Me$_3$SiN$_3$*
Trimethylsilyl azide/tetra-n-butylammonium fluoride   *Me$_3$SiN$_3$/Bu$_4$NF*
2-Aminoazides from aziridines   C

123.

Ph–△–NBn + Me$_3$SiN$_3$ ⟶ Ph–CH(N$_3$)–CH$_2$–NHBn

Regiospecific ring opening. A soln. of startg. aziridine and Me$_3$SiN$_3$ in acetonitrile allowed to react *at room temp.* for 2 h → product. Y 83%. The procedure is applicable to *non-activated* (N-alkyl and N-aryl)aziridines and does not require a Lewis acid. In fact yields were lower in the presence of Sn(OTf)$_2$. With condensed aziridines the *trans*-products were obtained. F.e. and **with asym. induction** s. M. Chandrasekhar, V.K. Singh et al., Tetrahedron Lett. *41*, 10079-83 (2000); *trans*-2-aminoazides with Me$_3$SiN$_3$ and 5 mol% Bu$_4$NF, also from N-tosyl-, N-acyl-, N-Boc- and N-benzyl-aziridines, and ring opening with Me$_3$SiCN and Me$_3$SiCl, s. J. Wu, X.-L. Hou et al., J. Org. Chem. *65*, 1344-8 (2000).

Trimethylsilyl chloride s. under NaI   *Me$_3$SiCl*

Sulfuric acid   *H$_2$SO$_4$*
Ritter reaction of tert. alcohols with chloroacetonitrile s. *60*, 146   OH → NHCOR

Palladous acetate/triphenylphosphine   *Pd(OAc)$_2$/Ph$_3$P*
Ring expansion of aziridines with heterocumulenes   ○
2-imidazolidones from isocyanates with PdCl$_2$(PhCN)$_2$ cf. *47*, 243s52; 4-vinyl-derivs. with Pd(OAc)$_2$/Ph$_3$P *at room temp.*, also ring expansion with carbodiimides, s. D.C.D. Butler, H. Alper et al., J. Org. Chem. *65*, 5887-90 (2000).

Addition to Carbon-Carbon Bonds NC ⇓ CC

Without additional reagents w.a.r.
1,2-Nitramines from 1,1-nitroethylene derivs. $C(NO_2)=C \rightarrow CH(NO_2)C(N<)$
carbohydrate derivs. s. *16*, 396; *43*, 455; 2-nitro-2-deoxyglycosylamines, regio- and stereoselectivity, s. G.A. Winterfeld, R.R. Schmidt et al., Eur. J. Org. Chem. *2000*, 3047-50.

Michael addition of amines to α,β-ethylenesulfones $C=C \rightarrow CHC(N<)$
s. *17*, 653s54; 2-amino-3-sulfonyl-2,3-dideoxyglycosides, stereoselectivity, s. B. Ravindran, T. Pathak et al., J. Org. Chem. *65*, 2637-41 (2000).

Polymer-based synthesis of 3,4-dihydro-2-pyridones $C\equiv C \rightarrow CH=C(N<)$
from acetylene derivs. via enamines

124.

3,4-Dihydro-2-pyridone-5-carboxylic acid amides. 20 eqs. Isopropylamine added to a suspension of startg. Sieber resin-supported ethylenecarboxylic acid amide in DMSO, and allowed to react for 12 h at room temp. → crude intermediate (Y unspecified), in dichloromethane treated with 10 eqs. acrylic anhydride, stirred at room temp. for 12 h, and the resin cleaved with 3% trifluoroacetic acid in dichloromethane for 2 h at the same temp. → product (Y 95%, overall). α,β-Ethylenecarboxylic acids (with $(PhO)_2P(O)N_3/Et_3N$) could be used in place of the anhydride. F.e., also 5-carboxylic acids from the ester-linked Sasrin resin, s. K. Paulvannan, T. Chen, J. Org. Chem. *65*, 6160-6 (2000).

[3+3]-Cycloaddition with α,β-acetylenenitriles ○
N-condensed 6-imino-1,3-thiazine ring cf. *38*, 496; 4-imino-1,2- and 4-imino-1,4-dihydropyrimidine ring s. J.A. McCauley et al., Org. Lett. *2*, 3389-91 (2000).

3,6-Dihydro-2H-1,2-oxazines from 1,3-dienes
Asym. heterodiene synthesis
with a steroidal 1,1-nitrosochloride cf. *40*, 186; with a carbohydrate-based deriv. s. A. Defoin et al., Synthesis *2000*, 1719-26.

Sodium sulfite Na_2SO_3
N-Protected 1,2-diamines from ethylene derivs. $C=C \rightarrow C(N<)C(N<)$
Regio- and stereo-specific conversion

125.

N-Protected *anti*-α,β-diaminohydrocinnamic acid esters. 1.5 eqs. N,N-Dichloro-*o*-nitrobenzenesulfonamide added to a soln. of methyl *trans*-cinnamate and freshly distilled *acetonitrile* contained in a dry vial, the latter capped (without the need for an inert atmosphere), stirred *at room temp.* for 3 h, and quenched by addition of satd. aq. Na_2SO_3 soln. → product. Y 74%. Reaction is also applicable to more simple olefins (e.g. cyclohexene), but the yield was lower (45%) from stilbene. No catalyst is required. The solvent clearly participates in the reaction, possibly by S_N2 interception of an intermediate N-chloro-N-(*o*-nosyl)aziridinium ion. F.e.s. G. Li et al., Tetrahedron Lett. *41*, 8699-703 (2000).

Mercuric acetate/diisopropylamine   *Hg(OAc)$_2$/i-Pr$_2$NH*
Enamines from acetylene derivs.   C≡C → CH=C(N<)
with Hg(OAc)$_2$/NaBH$_4$ cf. *32*, 287; with Hg(OAc)$_2$/*i*-Pr$_2$NH for prepn. and reactions of polymer-based enamines s. F. Aznar et al., Tetrahedron Lett. *41*, 5683-7 (2000); 2-(5-nitrothien-2-yl)-enamines without reagent (cf. *7*, 217) s. T.J.J. Müller et al., Org. Lett. *2*, 2419-22 (2000).

Ytterbium(III) triflate   *Yb(OTf)$_3$*
Michael addition of N-nucleophiles with asym. induction   C=C → CHC(N<)
of amines to enoates with Ln(OTf)$_3$ cf. *49*, 264; chiral β-hydrazinosulfones with Yb(OTf)$_3$, and reduction to chiral β-amino-derivs. (with BH$_3$·THF), s. D. Enders et al., Eur. J. Org. Chem. *2000*, 879-92.

Ammonia lyase   ←
α-Aminohydrocinnamic from cinnamic acids   C=C → CHC(NH$_2$)
Asym. enzymatic anti-Michael addition of ammonia

126.

Startg. cinnamic acid and recombinant phenylalanine ammonia-lyase (*Petroselinum crispum*) added to an ammonia soln. (freshly prepared by diluting concd. aq. ammonia with water and adjusting the pH to 10 by bubbling CO$_2$), the mixture agitated overnight at 37°, acidified with 5% HCl to pH 1.5, heated to boiling, and filtered → L-*m*-chlorophenylalanine. Y 99% (>99% e.e.). Ar. halides and pyrimidines were unaffected under these conditions. F.e.s. A. Gloge, J. Retey et al., Chem. Eur. J. *6*, 3386-90 (2000).

Trimethylsilyl azide/pivalic acid/β-turn peptides   ←
β-Azido- from α,β-ethylene-carboxylic acid amides   C=C → CHC(N$_3$)
Asym. 1,4-addition under mild conditions

127.

A mixture of 2.5 mol% Boc-His(Bn)-Pro-D-*t*-Leu-(S)-NHCH(Me)(1-naphthyl), 3.8 eqs. trimethylsilyl azide, 1 eq. pivalic acid and startg. amide in toluene allowed to react at 25° for 24 h → product. Y 79% (e.e. 85%). A significant feature of the small β-turn peptide is the τ-benzyl-His residue; significant also is the fact that no metal complex is required. F.e. and comparison of catalysts, also conversion to **chiral N-protected β-aminocarboxylic acids** s. T.E. Horstmann, S.J. Miller et al., Angew. Chem. Int. Ed. Engl. *39*, 3635-8 (2000).

((R)-2,2′-Bis(diphenylphosphino)-1,1′-binaphthyl)palladium(0) bistriflate   ←
Ar. benzylamines from styrenes and ar. amines   C=C → CHC(N<)
Regio- and stereo-specific palladium-catalyzed asym. hydroamination

128.

1.5 eqs. of startg. styrene and aniline added by syringe to a suspension of ((R)-BINAP)Pd(OTf)$_2$ (10 mol%) in toluene, and the mixture stirred at 25° for 72 h → (S)-product. Y 80% (e.e. 81%). This represents the best combination of e.e. and yield for the catalytic hydroamination of ethylene derivs. F.e., also racemic products with Pd(PPh$_3$)$_4$/triflic acid or Pd(OCOCF$_3$)$_2$/dppf/triflic acid, and reaction with sec. ar. amines s. M. Kawatsura, J.F. Hartwig, J. Am. Chem. Soc. *122*, 9546-7 (2000).

Potassium osmate/1,4-bis(9-O-dihydroquinine)phthalazine/lithium hydroxide ←
Regiospecific asym. Sharpless oxyamination C≡C → C(NHAc)C(OH)
chiral *cis*-2-acylaminoalcohols with N-bromoacetamide cf. *53*, 131; f. N-bromacylamines s. Z.P. Demko, K.B. Sharpless et al., Org. Lett. *2*, 2221-3 (2000); scaled-up version with added acetamide, and effect of concentration on chemoselectivity s. P.G.M. Wuts et al., ibid. 2667-9.

Potassium osmate/1,4-bis(9-O-dihydroquinine)phthalazine/sodium hypochlorite/ ←
sodium hydroxide
Regiospecific asym. Sharpless oxyamination C≡C → C(NHCOOR)C(OH)
with *t*-BuOCONH$_2$/*t*-BuOCl/NaOH cf. *51*, 132s*56*; with BnOCONH$_2$/NaOCl/NaOH for asym. oxyamination of α,β-ethylenecarboxylic acid esters s. H. Zhang, W. Zhou et al., Tetrahedron: Asym. *11*, 3439-47 (2000); of β-(N-, S- and O-heteroaryl)acrylates s. D. Raatz, O. Reiser et al., Synlett *1999*, 1907-10.

Potassium osmate/polymer-based 1,4-bis(9-O-dihydroquin[id]ine)phthalazine ←
Heterogeneous asym. Sharpless oxyamination C≡C → C(N<)C(OH)
with a silica-supported 1,4-bis(9-O-dihydroquin[id]ine)phthalazine as ligand and AcNHBr cf. *56*, 131; with a polymer-based analog and MeSO$_2$N(Cl)Na s. A. Mandoli, P. Salvadori et al., Tetrahedron:Asym. *11*, 4039-42 (2000).

Via intermediates *v.i.*
Amines from acetylene derivs. C≡C → CH$_2$CHN<
via titanium(IV)-catalyzed hydroamination
with reduction of the intermediate ketimine using LAH cf. *58*, 126; **prim. amines** via hydrogenation of the intermediate N-(diphenylmethyl)ketimines over Pd/C s. E. Haak, S. Doye et al., Org. Lett. *2*, 1935-7 (2000).

Rearrangement ∩

Hydrogen/Nitrogen Type NC ∩ HN

Hydroxylamine hydrochloride NH$_2$OH·HCl
Dimroth rearrangement ←
review s. *16*, 409s*24*; 5-hydroxylamino-1,2,4-triazines from 5-amino-1,2,4-triazine 4-oxides with NH$_2$OH·HCl as catalyst s. O.N. Chupakhin, V.L. Rusinov et al., Tetrahedron Lett. *41*, 7379-82 (2000).

Palladous chloride/potassium chloride PdCl$_2$/KCl
Pyrroles from (Z)-2-en-4-ynamines ○
Palladium-catalyzed cycloisomerization

129.

A soln. of startg. (Z)-enynamine in anhydrous dimethylacetamide (2 *M*) contained in a Schlenk flask under N$_2$ treated with 1 mol% PdCl$_2$ and 2 mol% KCl, and stirred at 25° until conversion satisfactory according to GLC and/or TLC (6 h) → product. Y 85% (94% by GLC). Contrasting with previously reported cycloisomerizations to furans or thiophenes, PdI$_2$/2KI was no more active as catalyst system than PdCl$_2$/2KCl. C-3 Unsubst. substrates were considerably more reactive than C-3 subst. ones (which required a reaction temp. of 100°). Substrates bearing a terminal triple bond were unstable, undergoing spontaneous pyrrole ring closure. F.e.s. B. Gabriele et al., Tetrahedron Lett. *42*, 1339-41 (2001).

Oxygen/Carbon Type NC ∩ OC

Dimethylamine Me_2NH
O→N-Acyl group migration   ←
1,4-migration cf. *14*, 400s55; N-hydroxypeptides by 1,5-O→N-acyl group migration with Me_2NH s. R. Braslau et al., Org. Lett. *2*, 1399-401 (2000); rearrangement of an N-benzyloxyleucine deriv. s. L. Wang, O. Phanstiel IV, J. Org. Chem. *65*, 1442-7 (2000).

Methanesulfonic acid $MeSO_3H$
Lactams by intramolecular Ritter reaction   O
from ethylenenitriles with PPA cf. *20*, 266; from hydroxynitriles with MsOH, 1,3,4,4a,5,9b-hexahydro-4a*H*-indeno[1,2-*b*]pyrid-2-ones, s. K. Van Emelen, F.A. Compernolle et al., Org. Lett. *2*, 3083-6 (2000).

Pallatous acetate/lithium bromide $Pd(OAc)_2/LiBr$
(E)-2-Ethylenetosylamines from 2-ethylenealcohols   C(OH)C=C → C=C-C(NHTs)
via regio- and stereo-specific intramolecular N-allylation of 2-ethylene-N-tosylurethans

130.

A soln. of startg. allyl alcohol treated with 1.1 eqs. tosyl isocyanate in THF for 20 min, solvent removed, the residue taken up in DMF, treated with 5 mol% $Pd(OAc)_2$ and 4 eqs. LiBr, and stirred at 100° until reaction complete → product. Y 85%. Reaction is presumed to involve Pd(II)-mediated decarboxylative S_N2'-substitution (exclusively at the γ site). The high (E)-selectivity is explained by the favourable conformation of the 6-membered cyclic intermediate and the highly specific *trans*-elimination. F.e.s. A. Lei, X. Lu, Org. Lett. *2*, 2357-60 (2000).

Carbon/Carbon Type NC ∩ CC

Without additional reagents   w.a.r.
Intramolecular hetero-Diels-Alder reaction with acylnitroso compds.   O
s. *41*, 305; effect of bromine substitution of the diene on face selectivity s. H. Abe, C. Kibayashi et al., J. Am. Chem. Soc. *122*, 4583-92 (2000).

Δ²-Pyrroline ring from cyclopropyl ketimines s. *27*, 407s60   O

Hetaryl iodosoacetates/potassium hydroxide $ArI(OAc)_2/KOH$
Urethans from carboxylic acid amides $CONH_2$ → NHCOOR
with $PhI(OAc)_2$ cf. *49*, 274; improved yields with hetaryl iodosoacetates, e.g. 3-(diacetoxyiodo)-thiophene, s. H. Togo et al., J. Org. Chem. *65*, 8391-4 (2000).

p-Toluenesulfonyl chloride/triethylamine $TsCl/Et_3N$
N,N-Disubst. cyanamides from amidoximes   RC(=NOH)NH_2 → RNHCN
via N-alkylative Tiemann rearrangement of O-sulfonylamidoximes
One-pot procedure under mild conditions

131.

A soln. of 1 eq. *p*-toluenesulfonyl chloride in methylene chloride added dropwise during 1 h to a cooled (0°) soln. of startg. amidoxime and 1 eq. Et_3N in the same solvent, the mixture warmed to room temp., stirred for 1 h, 2 eqs. methyl iodide, 30% aq. NaOH and a little benzyltriethylammonium chloride added, the mixture refluxed for 1 h, cooled to room temp., and poured into water → product. Y 85%. Only the most active alkyl halides (MeI, EtBr, allylic and benzylic halides) were suitable, other primary and secondary alkyl chlorides, bromides and iodides leading to poor yields. The method is simple, efficient, and general for aromatic, aliphatic and terpenic amidoximes (incl. those bearing alkenyl, hydroxyl and keto groups as well as chiral chrysanthemyl-derivs.). F.e.s. S.A. Bakunov, A.V. Tkachev et al., Synthesis *2000*, 1148-59.

Exchange ↕

Hydrogen ↑   NC ↕ H

Without additional reagents   w.a.r.
N-Chloroformylation of N-heterocyclics   NH → NCOCl
with $COCl_2$/pyridine cf. *16*, 493; more cleanly with bis(trichloromethyl) carbonate as phosgene substitute for N-chloroformylation of 2-imidazolidones s. W.K. Su, Y.M. Zhang, J. Chem. Res., Synop *2000*, 440-1; 3-chloroformylation of 1-acyl-2-imidazolidones with trichloromethyl chloroformate s. Org. Prep. Proced. Int. *32*, 498-501 (2000).

Sodium azide s.a. under Phenyl iodoso(hydroxy)-p-*nitrobenzenesulfonate* NaN_3

Sodium azide/phenyl iodosoacetate $NaN_3/PhI(OAc)_2$
Carboxylic acid azides from aldehydes   CHO → CON_3
with Me_3SiN_3/CrO_3 cf. *48*, 299; thermally stable and isolable aroyl azides with $NaN_3/PhI(OAc)_2$ s. D.-J. Chen, Z.-C. Chen, Tetrahedron Lett. *41*, 7361-3 (2000).

Chloro(triphenylphosphine)gold(I)/oxygen $Au(PPh_3)Cl/O_2$
N-Formylation by gold-catalyzed oxidative carbonylation   NH → NCHO

132. $H_2N(CH_2)_6NH_2$ + CO $\xrightarrow[O_2]{Au(Ph_3P)Cl}$ $OHCNH(CH_2)_6NHCHO$

Di-N-formylation. A soln. of 1,6-hexanediamine and 0.75-1.6 mol% $Au(PPh_3)Cl$ in methanol charged into a stainless-steel autoclave, pressurized with 4.5 MPa CO and 0.5 MPa O_2 at room temp., and stirred at 175° for 3 h → product. Conversion 100% (selectivity 92%; TOF for product 27.6). Oxygen has a significant effect on selectivity. However, $PdCl_2(PPh_3)_2$ was less effective. F.e.s. F. Shi, Y. Deng et al., Chem. Commun. *2001*, 345-6; **urethans from** [aromatic] **amines and alcohols** (with added Ph_3P) s. ibid. 443-4.

Montmorillonite s. under $Bi(NO_3)_3$ ←
Zeolites s. under Acyl nitrates   ←

Ammonium formate $HCOONH_4$
N-Formylation under mild conditions   NH → NCHO

133. $PhNH_2$ $\xrightarrow{HCOO^- {}^+NH_4}$ $PhNHCHO$

of anilines. 1.5 eqs. Anhydrous ammonium formate added to a soln. of aniline in dry acetonitrile, and heated at 95° for 11 h → formanilide. Y 96%. The reagent is relatively inexpensive and non-toxic, and the procedure is also suitable for the N-formylation **of sec. amines.** Unfortunately, prim. amines (except benzylamine) gave the corresponding alkylammonium formate. F.e. incl. N-formylation of α-aminocarboxylic acid esters with retention of chirality s. P.G. Reddy, S. Baskaran et al., Tetrahedron Lett. *41*, 9149-51 (2000); N-formylation of carbazoles with formic acid s. M. Chakrabarty et al., Synth. Commun. *30*, 187-200 (2000).

N-Hydroxyphthalimide/oxygen/nitrogen dioxide
Catalytic nitration under mild conditions via alkyl radicals $\quad$ H → NO_2

134.

NO_2 added quickly via Hamilton gas-tight syringe to N-hydroxyphthalimide (0.6 mmol) and cyclohexane (5 mL) in a pear-shaped flask, the flask reclosed with a glass stopper, cooled in an ice bath then quickly attached to a condenser, and stirred at 70° for 14 h under air → nitrocyclohexane. Y 70% (with 7% cyclohexyl nitrite and 5% cyclohexanol) based on NO_2 used and averaged over several runs. The [isolated] yield was 53% on a larger scale (50 mL cyclohexane and 33 mmol NO_2). This procedure, the first for catalytic nitration under mild conditions, may be of significance industrially. In the case of substituted alkanes, such as isobutane, 2,5-dimethylhexane, and adamantane, only the tertiary position was nitrated. In conventional alkane nitration by NO_2 the maximum theoretical yield is 66.7% since the proposed mechanism involves conversion of one-third of the NO_2 into NO via HNO_2, the high temperatures required (250-400°) result in poor selectivity (restricting it on the large scale to lower alkanes), and exclusion of air is necessary to prevent formation of complex mixtures. F.e. and method (using 1.1 eqs. NO_2 in benzotrifluoride) s. S. Sakaguchi, Y. Ishii et al., Angew. Chem. Int. Ed. Engl. *40*, 222-4 (2001).

Phenyl iodoso(hydroxy)-p-nitrobenzenesulfonate/sodium azide $\quad$ PhI(OH)ONs/NaN_3
Replacement of hydrogen by azido groups $\quad$ H → N_3
2-ethyleneazides with PhIO/NaN_3 cf. *26*, 325s*42*; α-azidoketones (cf. *30*, 232s*56*) with PhI(OH)ONs/NaN_3 s. J.C. Lee, S. Kim, W.C. Shin, Synth. Commun. *30*, 4271-5 (2000).

Electrochemically-generated tetraethylammonium 2-oxopyrrolidide/tosyl chloride $\quad$ ←
2-Oxazolidones from 2-aminoalcohols $\quad$ O
with electrochemically-generated tetraethylammonium peroxydicarbonate cf. *41*, 352s*57*; improved procedure with electrochemically-generated tetraethylammonium 2-oxopyrrolidide in combination with CO_2 for a mild, efficient carboxylation of 2-aminoalcohols s. M.A. Casadei, A. Inesi et al., J. Org. Chem. *65*, 4759-61 (2000).

tert-*Butoxyformic anhydride/4-dimethylaminopyridine/triethylamine* $\quad$ ←
Cyclocarbonylation with *tert*-butoxyformic anhydride

135.

3-Carbo-*tert*-butoxy-2-oxazolidones from 2-aminoalcohols. 0.5 eq. DMAP and 3 eqs. Et_3N added to 3 eqs. *tert*-butoxyformic anhydride in acetonitrile at 0°, after 5 min serine methyl ester hydrochloride in the same solvent added dropwise during 2 min, and worked up after 1 h → product. Y 93%. **1,3-Dioxolan-2-ones** were obtained similarly **from glycols** using 3 eqs. *tert*-butoxyformic anhydride and 1 eq. DMAP in the same solvent at room temp., while **glycol carbonates** were favoured with N-methylimidazole in toluene. Cyclocarbonylation of 2-aminomercaptans, however, was less efficient. In a broad survey of the reaction of *tert*-butoxyformic anhydride/DMAP with alcohols, **sym. carbonic acid esters** were produced with 0.2-0.5 eq. DMAP in acetonitrile, while **O-carbo-*tert*-butoxylation** took place with <0.1 eq. DMAP in dioxane (or with N-methylimidazole in toluene). F.e. and N-carbo-*tert*-butoxylation of various amines s. Y. Basel, A. Hassner, J. Org. Chem. *65*, 6368-80 (2000).

Phenyl iodosoacetate s. under NaN_3 and Ru(II) complexes $\quad$ PhI(OAc)$_2$
Phenyl iodoso(hydroxy)-p-nitrobenzenesulfonate/sodium azide $\quad$ PhI(OH)ONs/NaN_3
Replacement of hydrogen by azido groups $\quad$ H → N_3

Ammonium nitrate/trifluoroacetic anhydride/1-ethyl-3-methylimidazolium trifluoroacetate ←
Ar. nitration in ionic liquid solvents $H \rightarrow NO_2$

136.

Naphthalene added to a mixture of 0.5 eq. NH_4NO_3 in 2 eqs. 1-ethyl-3-methylimidazolium trifluoroacetate at 0°, 2.5 eqs. trifluoroacetic anhydride added slowly, stirred for 10-30 min (resulting in a one-phase mixture), excess of the anhydride removed under vacuum, ether and *i*-Pr_2NEt added until phase separation was observed, and the nitroarene extracted with ether → product. Y 100% (regioselectivity 92.9%). Nitration also took place with anisoles, alkylbenzenes and halobenzenes in good to excellent yield but there was no reaction with nitrobenzene. The ionic solvent is easily recovered for reuse without need for aq. work-up, and problems associated with large quantities of strong acid are avoided. Yields and isomer distributions were comparable to those from conventional systems. However, other nitrating systems were less efficient. F.e., also use of isoamyl nitrate in 1-ethyl-3-methylimidazolium triflate, s. K.K. Laali, V.J. Gettwert, J. Org. Chem. *66*, 35-40 (2001).

Acyl nitrates/zeolites ←
Heterogeneous ar. nitration
s. *1*, 343s*51*,57; of deactivated arenes with *in situ*-generated trifluoroacetyl nitrate and zeolite Hβ s. K. Smith et al., J. Chem. Soc. Perkin Trans. 1 *2000*, 2753-8.

Bismuth(III) nitrate/montmorillonite ←
Regiospecific heterogeneous ar. nitration
with $Al(NO_3)_3$-clay cf. *30*, 239s*44*; with $Bi(NO_3)_3$/montmorillonite for nitration of a variety of arenes and polyarenes, also non-selective nitration of phenol and estrone, s. S. Samajdar, B.K. Banik et al., Tetrahedron Lett. *41*, 8017-20 (2000).

Hydrogen peroxide/triethylamine H_2O_2/Et_3N
Benzoxazole-2-thiones from *o*-aminophenols O
with CS_2/KOH cf. *6*, 486; with CS_2 and 30% H_2O_2/Et_3N (cat.) s. A. Harizi et al., Tetrahedron Lett. *41*, 5833-5 (2000).

Sulfonyl azides/1,8-diazabicyclo[5.4.0]undec-7-ene or pyridine RSO_2N_3/DBU or C_5H_5N
Diazo group transfer $CH_2 \rightarrow CN_2$
with TsN_3/Cs_2CO_3 cf. *20*, 271s*51*; α-diazo-α-nitrocarbonyl compds. with readily prepared $CF_3SO_2N_3$/py s. A.B. Charette et al., J. Org. Chem. *65*, 9252-4 (2000); aroyl(perfluoroacyl)diazomethanes with 4-methyl-3-nitrobenzenesulfonyl azide and DBU cf. G.P. Kantin, V.A. Nikolaev, Russ. J. Org. Chem. *36*, 486-8 (2000).

Tungsten hexacarbonyl/iodine/potassium carbonate ←
Sym. ureas from amines $2 RNH_2 \rightarrow (RNH)_2CO$
cyclic ureas cf. *58*, 133; sym. ureas from prim. amines s. J.E. McCusker, L. McElwee-White et al., J. Org. Chem. *65*, 5216-22 (2000).

Tetra-n-butylammonium periodate Bu_4NIO_4
N-Acyl-3,6-dihydro-2H-1,2-oxazines from 1,3-dienes O
and *in situ*-generated acylnitroso compds.
with asym. induction using Et_4NIO_4 cf. *30*, 244; chiral 2-acyl-6-sulfinyl-3,6-dihydro-2H-1,2-oxazines from 1,3-dienesulfoxides using Bu_4NIO_4 s. C. Arribas, M.C. Carreño et al., Org. Lett. *2*, 3165-8 (2000).

5,10,15,20-Tetrakis(pentafluorophenyl)porphyrinatomanganese(III) chloride or ←
  (1,4,7-Trimethyl-1,4,7-triazacyclononane)ruthenium(III) tris(trifluoroacetate)/
  phenyl iodosoacetate
Tosylamination $H \rightarrow NHTs$
benzylic tosylamination with [Ru(Me_3tacn)(NHTs)$_2$(OH)] and PhI=NTs cf. *56*, 162; details, also benzylic acylamination and mesylamination with [Ru(Me_3tacn)(CF_3CO_2)·H_2O] s. S.-M. Au,

C.-M. Che et al., J. Org. Chem. 65, 7858-64 (2000); sulfonylamination and trifluoroacetylamination of hydrocarbons with 5,10,15,20-tetrakis(pentafluorophenyl)porphyrinatomanganese(III) chloride s. X.-Q. Yu, C.-M. Che, Org. Lett. 2, 2233-6 (2000).

Oxygen ↑   NC ↓↑ O

Without additional reagents   *w.a.r.*
N-Acylation with polymer-based *o*-acoxynitro compds.   NH → NAc
of weakly nucleophilic N-heteroarenes s. 60, 159

N-Acylation with N-hydroxysuccinimide esters
s. 32, 317; with high-loading ROMPGEL polymer-based N-hydroxysuccinimide esters, also selective N-acylation with retention of alcohols, s. A.G.M. Barrett et al., Org. Lett. 2, 261-4 (2000).

α-Aminonitriles from cyanohydrins   OH → NR$_2$
s. 17, 405; chiral α,β-diaminonitriles in DMF/2,2,2-trifluoroethanol (without racemization) s. A.G. Myers et al., Org. Lett. 2, 3337-40 (2000).

Azomethines from aldehydes and amines   CHO → CH=NR
s. 1, 391; as a suspension **in water** for simplified isolation s. K. Tanaka, R. Shiraishi, Green Chem. 2, 272-3 (2000); scavenging of amines (cf. 41, 286s54) and hydrazines with poly[(*exo*-3,6-epoxy-1,2,3,6-tetrahydrophthalic anhydride)-co-norbornadiene] s. T. Arnauld, A.G.M. Barrett et al., Org. Lett. 2, 2663-6 (2000); overview of covalent scavengers for prim. and sec. amines s. J.C. Hodges, Synlett 1999, 152-8.

β-Amino-α,β-ethylenecarbonyl from β-ketocarbonyl compds.   COC=C(N<)
ketone derivs. cf. 26, 331; also ester derivs. from prim. amines *in water* with a simple product isolation s. H.A. Stefani et al., Synthesis 2000, 1526-8.

Microwaves (s.a. under Al$_2$O$_3$ and NH$_4$OAc)   ←
β-Ketocarboxylic acid amides from esters and amines   COOR → CON<
N-acetoacetylation with pyridine in xylene cf. 12, 453; of ar. and heteroar. amines under microwave irradiation in the absence of solvent s. O.P. Suri et al., Synth. Commun. 30, 3709-18 (2000).

Sodium azide s. under SOCl$_2$ *NaN$_3$*

Sodium sulfate   *Na$_2$SO$_4$*
Pyrazoles from β-diketones   ○
s. 2, 368, 403; 3-silylpyrazoles from β-ketoacylsilanes with a little Na$_2$SO$_4$ s. S. Gérard, C. Portella et al., Tetrahedron Lett. 41, 9791-5 (2000).

Dimethylamine/trimethylamine   *Me$_2$NH/Me$_3$N*
Lactams from lactones and prim. amines   ←
N-subst. 2-pyrrolidones at 280° in a steel bomb cf. 5, 279; at 270°/71 atm. in the presence of Me$_2$NH/Me$_3$N and 1.8 eqs. water s. K. Takahashi et al., European patent EP-1004577 (Mitsubishi Chem. Corp.).

Ethyldiisopropylamine   *i-Pr$_2$NEt*
Polymer-based synthesis of urethans   OCOOR → OCON<
from carbonic acid esters s. 60, 159

4-Dimethylaminopyridine   *DMAP*
N-Trifluoroacetylation   NH → NCOCF$_3$

137.   Ar-NH$_2$ + EtOCOCF$_3$ ⟶ Ar-NHCOCF$_3$

of prim. ar. amines. A mixture of *o*-toluidine, 2 eqs. ethyl trifluoroacetate, and 1 eq. 4-dimethylaminopyridine in THF refluxed for 22 h → product. Y 98%. There was little or no reaction without DMAP. Prim. anilines, incl. hindered ones, containing other functional groups, such as alcohols, phenols, hindered sec. amines and sec. anilines, were selectively trifluoroacetylated in high yield. F.e.s. M. Prashad et al., Tetrahedron Lett. *41*, 9957-61 (2000).

Planar-chiral ferrocenyl-type 4-aminopyridine
Non-enzymatic kinetic resolution of amines ←
via N-acylation with 5-(carbalkoxyoxy)oxazoles NH → NAc

138.

The first effective non-enzymatic catalytic N-acylation with kinetic resolution is reported. **E**: 1-Phenylethylamine (17 mg) and chloroform added to 0.1 eq. (-)-PPY in a Schlenk flask under argon, the resulting purple soln. cooled to -50°, a soln. of 0.3 eq. O-acylated azlactone in chloroform added via syringe, followed after 4 h by an additional 0.3 eq., and after 24 h (total reaction time) the mixture subjected to flash chromatography → urethan (7.3 mg; e.e. 79%) and amine (11.4 mg after N-acylation; e.e. 42%); selectivity factor 13 at 35% conversion. An ortho substituent on the benzylamine increased the selectivity factor, as did a *m*-methoxy group, but electronic effects due to *para* substituents appeared not to affect the stereoselection significantly. F.e.s. S. Arai, S. Bellemin-Laponnaz, G.C. Fu, Angew. Chem. Int. Ed. Engl. *40*, 234-6 (2001).

Sodium triacetoxyhydridoborate/acetic acid $NaBH(OAc)_3/AcOH$
2,3-Piperazinediones via reductive N-alkylation-lactamization ○
from N-(2-aminoethyl)oxamic acid esters cf. *49*, 300s56; from α-(alkoxalylamino)ketones with asym. induction s. D.C. Beshore, C.J. Dinsmore, Tetrahedron Lett. *41*, 8735-9 (2000).

γ-Alumina $γ$-Al_2O_3
Ureas from two different amine molecules via urethans >NH → >NCOOR → >NCON<
with bis(4-nitrophenyl) carbonate cf. *45*, 192; simple, eco-friendly method with dimethyl carbonate using γ-Al_2O_3 (for both steps) s. I. Vauthey, M. Lemaire et al., Tetrahedron Lett. *41*, 6347-50 (2000).

Alumina/microwaves ←
N-Aminomethylation NH → NCH$_2$N<
s. *19*, 419; *26*, 871; of sec. amines, acylamines or imides under microwave irradiation on acidic alumina in the absence of solvent, also condensation with ar. aldehydes, s. A. Sharifi et al., J. Chem. Res., Synop *2000*, 394-6.

N-Subst. carboxylic acid amides from carboxylic acids COOH → CON<
on alumina cf. *26*, 386; under microwave irradiation in the absence of solvent (cf. *45*, 204s57), cephalosporins, s. M. Kidwai, M. Singh et al., Monatsh. Chem. *131*, 937-43 (2000).

Alumina or zeolites or silica gel/microwaves ←
Imidazoles from α-diketones and aldehydes ○
in AcOH cf. *23*, 423; highly functionalized 2,4,5-tri- and 1,2,4,5-tetra-subst. imidazoles under microwave irradiation on acidic Al_2O_3 in the absence of solvent s. A. Ya. Usyatinsky, Y.L. Khmelnitsky, Tetrahedron Lett. *41*, 5031-4 (2000); 2,4,5-triarylimidazoles on zeolites s. S. Balalaie et al., Monatsh. Chem. *131*, 945-8 (2000); tetrasubst. imidazoles (with zeolite HY or silica gel) s. Green Chem. *2*, 274-6 (2000).

Samarium s. under $TiCl_4$ Sm

Scandium(III) triflate $Sc(OTf)_3$
Hydroxamic acids from carboxylic acid esters COOR → CONHOH
with Na in ethanol cf. *8*, 455; with Sc(OTf)$_3$ s. Japanese patent JP-2000136177 (Fuji Photo Film Co. Ltd.).

Ammonium acetate/microwaves
Nitriles from aldehydes under microwave irradiation   CHO → CN
with NH$_2$OH·HCl in N-methyl-2-pyrrolidone cf. *55*, 146s*58*; improved, general procedure with NH$_4$OAc in the absence of solvent s. B. Das, C. Ramesh et al., Synlett *2000*, 1599-600.

4-(4,6-Dimethoxy-1,3,5-triazin-2-yl)-4-methylmorpholinium chloride/   ←
  ethyldiisopropylamine
Peptide synthesis   COOH → CON<
with 2-chloro-4,6-dimethoxy-1,3,5-triazine/N-methylmorpholine cf. *41*, 344; solid-phase peptide synthesis with preformed 4-(4,6-dimethoxy-1,3,5-triazin-2-yl)-4-methylmorpholinium chloride as a useful alternative to PyBOP s. A. Falchi, M. Taddei et al., Synlett *2000*, 275-7.

Polymer-based 2-(3-oxidobenzotriazol-1-yl)-1,1,3,3-tetramethyluronium fluoroborate/   ←
  pyridine
Peptide synthesis
with S-(1-oxido-2-pyridyl)-1,1,3,3-tetramethylthiouronium fluoroborate cf. *28*, 144s*58* and with HATU cf. *28*, 144s*56*; with hydrolytically stable polymer-based 2-(3-oxidobenzotriazol-1-yl)-1,1,3,3-tetramethyluronium fluoroborate for coupling of N-Boc- or N-Cbz-protected amino acids s. R. Chinchilla, C. Nájera et al., Tetrahedron Lett. *41*, 2463-6 (2000); N-unsubst. carboxamides from ar. and aliphatic carboxylic acids, incl. N-protected α-aminocarboxylic acids, with (S)-(1-oxido-2-pyridyl)-1,1,3,3-tetramethylthiouronium hexafluorophosphate s. M.A. Bailén, C. Nájera et al., ibid. 9809-13.

Dicyclohexylcarbodiimide   RN=C=NR
Polymer-based synthesis of N-heterocyclics by N-C bond formation   ○
s. *43*, 316s*59*; of 2-amino-5-hydroxy-1,4,5,6-tetrahydropyrimidines from isothiocyanates and amines s. V. Jammalamadaka, J.G. Berger, Synth. Commun. *30*, 2077-82 (2000); of 2-amino-4(3*H*)-quinazolones from 3,1-benzoxazine-2,4-diones and isothiouronium salts s. R.-Y. Yang, A. Kaplan, Tetrahedron Lett. *41*, 7005-8 (2000); of 2-amino-Δ^2-5-imidazolones from N-protected α-amino esters, isothiocyanates and amines s. D.H. Drewry, C. Ghiron, ibid. 6989-92; of highly subst. pyrazoles and isoxazoles s. D.-M. Shen et al., Org. Lett. *2*, 2789-92 (2000).

1-Ethyl-3-(3-dimethylaminopropyl)carbodiimide/1-hydroxybenzotriazole/triethylamine   ←
N-Subst. carboxylic acid amides from carboxylic acids   COOH → CON<
peptides s. *12*, 455s*43*; 'light fluorous' synthesis of carboxylic acid anilides from 'lightly'-tagged fluorous carboxylic acids for improved separation by *solid-phase* extraction with fluorous reverse-phase silica gel s. D.P. Curran, Z. Luo, J. Am. Chem. Soc. *121*, 9069-72 (1999).

Lipase   ←
Parallel kinetic resolution of sec. alcohols and prim. amines   NH → NAc
by asym. aminolysis s. *60*, 9

Asym. enzymatic N-acylation
in organic media. s. *44*, 314; kinetic resolution of dimethyl aspartate with *Candida antarctica* lipase A-on-Celite s. A. Liljeblad, L.T. Kanerva, Tetrahedron:Asym. *10*, 4405-15 (1999); *regiospecific* γ-amidation of N-protected D-glutamic acid diesters (cf. *52*, 150) s. S. Conde et al., ibid. *11*, 2537-45 (2000); asym. ammonolysis of phenylglycine esters with NH$_3$ in *tert*-butanol (cf. *45*, 210s*53*) s. M.A. Wegman, R.A. Sheldon et al., ibid. *10*, 1739-50 (1999).

Proteases
Peptide synthesis in anhydrous media or aq. alcohol   COOR → CON<
s. *45*, 209s*56*; α-fluoroalkylated peptides with various proteases (incl. chymotrypsin and clostripain) s. F. Bordusa, B. Koksch et al., Tetrahedron:Asym. *10*, 307-13 (1999); peptide isosteres with clostripain s. J. Org. Chem. *65*, 1672-9 (2000).

Modified subtilisin
Peptide synthesis   ←
with unnatural α-amino esters in aq. DMF cf. *53*, 144; incorporation of D-α-amino esters into dipeptides (cf. *43*, 298) with modified mutant enzymes of subtilisin *Bacillus lentus* s. K. Khumtaveeporn, J.B. Jones et al., Tetrahedron:Asym. *10*, 2563-72 (1999).

Amberlyst A-21 s. under SOCl₂ ←

Acetic acid   AcOH
Oxazoles from α-formamidoketones   ○
and HCOONH₄ cf. *12*, 136; from α-formamidoacetals with NH₄OAc in glacial acetic acid s. M. Botta, F. Corelli et al., J. Org. Chem. *65*, 4736-9 (2000).

3,4-Dihydro-2-pyridones from δ-ketocarboxylic acid esters
and 28% aq. NH₃ cf. *13*, 412; polymer-based synthesis with unbranched, aliphatic prim. amines in glacial AcOH s. A.S. Wagman et al., J. Org. Chem. *65*, 9103-13 (2000).

2-Chloro-1,3-dimethyl-1H-benzimidazolium hexafluorophosphate/ethyldiisopropylamine   ←
Peptide synthesis   COOH → CON<
with 2-bromo-1-ethylpyridinium fluoroborate cf. *58*, 141s59; with 2-chloro-1,3-dimethyl-1*H*-benzimidazolium hexafluorophosphate, notably for rapid coupling of hindered N-methyl and hindered amino acids with low racemization, s. P. Li, J.C. Xu, Tetrahedron *56*, 9949-55 (2000).

Ethyl chloroformate/N-methylmorpholine   ClCOOEt/NMM
Hydroxamic acids from carboxylic acids   COOH → CONHOH
with Et₃N as base cf. *20*, 312; general procedure with N-methylmorpholine as base for the conversion of aromatic or aliphatic carboxylic acids containing phenolic, ar. methoxy, halo, or ester groups as additional substituents s. A. Sekar Reddy, G. Ravindra Reddy et al., Tetrahedron Lett. *41*, 6285-8 (2000).

Isobutyl chloroformate/N-methylmorpholine   ClCOOBu-i/NMM
Carboxylic acid azides from carboxylic acids s. *32*, 324s60   COOH → CON₃

Polymethylhydrosiloxane/titanium tetraisopropoxide   ←
Reductive N-alkylation   NH → NR
of amides with Et₃SiH/CF₃COOH cf. *56*, 146; sec. and tert. amines from prim. and sec. amines, respectively, and oxo compds. with polymethylhydrosiloxane/Ti(OPr-*i*)₄ s. S. Chandrasekhar et al., Synlett *2000*, 1655-7.

Titianium tetraisopropoxide s. under Polymethylhydrosiloxane   Ti(OPr-i)₄

Titanium tetrachloride/samarium   TiCl₄/Sm
3,4-Dihydro-2H-1,2,4-benzothiadiazine 1,1-dioxides   ○
from *o*-nitrosulfonamides and oxo compds.
from aldehydes with SmI₂ cf. *59*, 149; from aliphatic and cycloaliphatic ketones with low-valent titanium (from TiCl₄/Sm) s. W.H. Zhong, Y.M. Zhang et al., Chinese J. Chem. *18*, 786-8 (2000).

Stannous chloride   SnCl₂
Polymer-based synthesis of benzimidazoles from *o*-nitramines and aldehydes
Reductive ring closure

139.

Startg. Wang-linked *o*-nitramine SynPhase Lantern treated with a 0.15 *M* soln. of 2 eqs. benzaldehyde and 10 eqs. 0.75 *M* SnCl₂·2H₂O in DMF at 60° for 3 h, allowed to cool briefly, decanted, the resin washed with DMF and methylene chloride, air-dried, and cleaved in a polypropylene tube with 20% trifluoroacetic acid in methylene chloride for 1 h → product. Crude Y 85% (with ca. 4% benzimidazolium impurity). A library of 25 compds. was prepared. However, the proportion of impurity was greater (up to 25%) with aliphatic aldehydes. F.e.s. Z. Wu et al., Tetrahedron Lett. *41*, 9871-4 (2000).

Stannous chloride/triethylamine $SnCl_2/Et_3N$
α-*tert*-Aminooximes $C=CNO_2 \rightarrow C(N<)C=NOH$
from 1,1-nitroethylene derivs. and sec. amines

140.

10 eqs. Et$_3$N and 2 eqs. morpholine added to a soln. of startg. nitroalkene in acetonitrile with water-cooling, 2 eqs. SnCl$_2$·2H$_2$O added, and the heterogeneous mixture stirred overnight → N-[2-[2-(3,4-dimethoxyphenyl)]acetaldoxime]morpholine. Y 76% (E/Z 1.6:1). Although yields may be only moderate, they are comparable with those reported for existing multistep routes. The reaction is thought to involve an electron-deficient carbene-like bis(dialkylamino)tin(II) species, which participates in an intramolecular deoxygenative Michael addition. F.e.s. L. Gottlieb, H.E. Gottlieb et al., Synth. Commun. *30*, 2445-64 (2000).

Triphenylphosphine/iodine/ethyldiisopropylamine   ←
Adenines from hypoxanthines under mild conditions   OH → N<

141.

Startg. nucleoside added to a stirred suspension of 3.1 eqs. imidazole, 2.4 eqs. triphenylphosphine, 2.4 eqs. iodine and 5.1 eqs. ethyldiisopropylamine in freshly distilled toluene, and the mixture stirred at 95° for 50 min → product. Y 94% (after deprotection). Displacement by sec. amines is also quite effective at ambient temp. F.e. incl. ribonucleoside derivs. s. X.Y. Lin, M.J. Robins, Org. Lett. *2*, 3497-9 (2000).

N,N',N"-Tris[tris(dimethylamino)phosphoranylidene]phosphoric acid triamide   ←
2-Oxazolidones from epoxides and urethans   ◯
5-α-aryloxy-3-aryl-2-oxazolidones with Et$_3$N cf. *19*, 442; with [(Me$_2$N)$_3$P=N]$_3$PO s. Japanese patent JP-2000136186 (Mitsui Chem. Inc.).

Diphenyl phosphorazidate/triethylamine $(PhO)_2P(O)N_3/Et_3N$
Ureas from carboxylic acids and amines   RCOOH → RNHCON<
with PhOP(O)(N$_3$)NHPh/Et$_3$N cf. *38*, 353; polymer-based synthesis of N,N'-disubst. ureas with (PhO)$_2$P(O)N$_3$/Et$_3$N and adaption for scavenging isocyanates, also with collidine as base, s. M.T. Migawa, E.E. Swayze, Org. Lett. *2*, 3309-11 (2000).

Bis(2,4-dichlorophenyl) phosphoromonochloridate/sodium azide/4-dimethylaminopyridine   ←
Azides from alcohols   OH → N$_3$
with (PhO)$_2$P(O)N$_3$/DBU cf. *49*, 312; complementary procedure for the preparation of unactivated prim. and sec. alcohols, as well as relatively unreactive benzyl alcohols, with bis(2,4-dichlorophenyl) phosphoromonochloridate/NaN$_3$/DMAP s. C. Yu, B. Liu, L. Hu, Org. Lett. *2*, 1959-61 (2000).

Thionyl chloride/sodium azide $SOCl_2/NaN_3$
Carboxylic acid azides from carboxylic acids   COOH → CON$_3$
in one pot via crude acyl chlorides cf. *25*, 669; Fmoc-protected α-amino acid azides without intermediate removal of solvent, also with isobutyl chloroformate/N-methylmorpholine (cf. *32*,

324) (where the acyl chloride is unstable), s. V.V.S. Babu et al., J. Chem. Soc. Perkin Trans. 1 *2000*, 4328-31.

Thionyl chloride/Amberlyst A-21 ←
Carboxylic acid amides from acids via acid chlorides s. *4*, 214s*60* COOH → CON<

2-Pyridinesulfonyl chloride RSO_2Cl
N,N-Dimethylformamidines from prim. amines $NH_2 \rightarrow NHCH=NR$
s. *5*, 301s*53*; chiral α-formamidinonitriles from α-aminocarboxylic acid amides s. L. Cai et al., Tetrahedron *56*, 8253-62 (2000).

Nafion-H ←
N-Acetylation s. *35*, 105s*60* NH → NAc

[Bis(2-methoxyethyl)amino]sulfur trifluoride/diisopropylamine $R_2NSF_3/i\text{-}Pr_2NH$
Hydroxamic acid esters from carboxylic acids COOH → CONHOR
via carboxylic acid fluorides

142.

3,4-Dichlorobenzoic acid in DMF treated under argon at 0° with 1.5 eqs. diisopropylamine and 1.2 eqs. [bis(2-methoxyethyl)amino]sulfur trifluoride, the mixture stirred for 15 min, a soln. of 1.5 eqs. N,O-dimethylhydroxylamine in the same solvent added, stirred for 15 min at 0°, warmed to room temp., and stirring continued for 4 h → product. Y 92%. This is a simple, high-yielding, one-pot procedure based on a safe and commercially available reagent (Deoxo-Fluor), and proceeding with retention of chirality. F.e. and in dichloromethane s. A.R. Tunoori, G.I. Georg et al., Org. Lett. *2*, 4091-3 (2000).

Iodine I_2
Nitriles from aldehydes CHO → CN

143. PhCHO + NH$_3$ →[I_2] PhCN

A stirred soln. of benzaldehyde in 10:1 28% aq. ammonia/THF treated at room temp. with *1.1* eqs. I$_2$ (CAUTION: excess reagent must be avoided to prevent formation of NI$_3$·NH$_3$, which is explosive when dry), and stirred until the dark solution became colorless or light grey (30 min) → benzonitrile. Y 96%. This method shows distinct advantages over the prior art. It is economical, environmentally-friendly, quick, operationally simple and efficient, based on readily available aq. ammonia, and is generally applicable to aryl, hetaryl, α,β-unsatd. and aliphatic aldehydes and to aldoses. The absence of a base such as MeONa avoids side reactions. F.e.s. S. Talukdar, J-M. Fang et al., Tetrahedron Lett. *42*, 1103-5 (2001).

Hydrogen bromide/acetic acid HBr/AcOH
Solid-phase peptide synthesis COOH → CON<
update s. *19*, 33s*59*; of C-terminal peptidyl aldehydes from aminoacetals anchored to a backbone amide linker (BAL) handle s. F. Guillaumie, K.J. Jensen et al., Tetrahedron Lett. *41*, 6131-5 (2000); of peptides containing prolyl, N-alkylamino acyl, or histidyl derivs. at the C-terminus s. J. Alsina, G. Barany et al., ibid. 7277-80; of peptidyl α,β-ethylenesulfones and α,β-epoxyketones based on Kenner's safety-catch strategy s. H.S. Overkleeft, H.L. Plegh et al., ibid. 6005-9; of peptidyl sulfonamides s. J. van Ameijde, R.M.J. Liskamp, ibid. 1103-6; of partially modified retro and retro-inverso [NHCN(CF$_3$)]-peptides s. ibid. 6517-21; design of a *fluorescent* α-amino acid for solid-phase peptide synthesis s. I. Dufau, H. Mazarguil, ibid. 6063-6; cobalt- and nickel-mediated *solution-phase* **polypeptide synthesis** by 'living polymerization' cf. T.J. Deming, S.A. Curtin, J. Am. Chem. Soc. *122*, 5710-7 (2000).

Update on the use of polymer supports ←
s. *19*, 33s*59*; preparation of reactive poly(benzopinacolones) s. H. Otsuka, T. Endo et al., Tetrahedron Lett. *41*, 1433-7 (2000); novel polymeric supports with dimensional stability s. T.R.

Webb, World Intellectual Property Organization patent WO-9941216 (Chembridge Corp); novel safety-catch linkers s. Y. Xiao, A.W. Czarnik, European patent EP-937696 (Irori); sensitive visual test for detecting OH groups on resin s. M.E. Attardi, M. Taddei, Tetrahedron Lett. *41*, 7395-9 (2000); s.a. O. Kuisle, R. Riguera et al., Tetrahedron *55*, 14807-12 (1999); monitoring of solid-phase reactions by high-resolution magic angle spinning NMR spectroscopy s. R. Warrass, G. Lippens, J. Org. Chem. *65*, 2946-50 (2000).

Update on screening catalyst libraries ←
s. *19*, 33s*58*; high-throughput screening of heterogeneous catalysts by laser-induced fluorescence imaging s. H. Su, E.S. Yeung, J. Am. Chem. Soc. *122*, 7422-3 (2000); of both heterogeneous and homogeneous catalyst libraries by laser imaging s. M.P. Atkins et al., World Intellectual Property Organization patent WO-9919724 (Laser Catalyst Systems Inc); with a CO_2 heating laser s. P. Cong, W.H. Weinberg et al., Angew. Chem. Int. Ed. Engl. *38*, 484-8 (1999).

Ferric chloride/oxygen $FeCl_3/O_2$
Benzimidazoles from o-diamines and aldehydes
with $PhNO_2$ cf. *46*, 321; with 1 mol% $FeCl_3·6H_2O$ under O_2 (for electron-rich aldehydes) or 3 mol% $FeCl_3$ in DMF (for electron-deficient aldehydes), also imidazopyridines, s. M.P. Singh et al., Synthesis *2000*, 1380-90.

Nickel(II) copper(II) formate/tetra-n-butylammonium persulfate/potassium hydroxide ←
Nitriles from aldehydes via *in situ*-generated aldimines $CHO \rightarrow CN$

144.
$$\text{p-MeOC}_6\text{H}_4\text{CHO} \xrightarrow{\text{NH}_4\text{HCO}_3} [\text{p-MeOC}_6\text{H}_4\text{CH=NH}] \xrightarrow{[O]} \text{p-MeOC}_6\text{H}_4\text{CN}$$

A mixture of *p*-methoxybenzaldehyde in acetonitrile, 1 eq. $(Bu_4N)_2S_2O_8$ in acetonitrile, 1.1 eqs. NH_4HCO_3 in water and 1 eq. KOH in water added to a soln. of 3 mol% $Cu(HCO_2)_2·Ni(HCO_2)_2$ (Ni(II) Cu(II) formate) in water, and the mixture stirred vigorously at room temp. for 1.5 h → *p*-methoxybenzonitrile. Y 92%. The reaction failed to proceed in the absence of the Ni(II)/Cu(II) formate, and failed to reach completion with lesser amount of the catalyst. The method is simple, efficient, mild and general for aliphatic, conjugated, aromatic and heteroaromatic (2-furyl-, 2-pyridyl-, and piperonyl) aldehydes. It is also compatible with phenolethers, aromatic chlorides and nitro compds., and styrenes. F.e.s. F.-E. Chen et al., Synthesis *2000*, 1519-20.

Cationic rhodium(I) complexes ←
Reductive N-alkylation $NH \rightarrow NR$
by heterogeneous hydrogenation over rhodium sulfide-on-carbon cf. *22*, 421; under homogeneous conditions with $[Rh(dppb)(cod)]BF_4$ or $[Rh(dpoe)(cod)]BF_4$ (dpoe = 1,2-bis(diphenylphosphinito)ethane, also α-amino- from α-keto-carboxylic acids, s. V.I. Tararov, A. Börner et al., Chem. Commun. *2000*, 1867-8.

Palladous acetate/tris(m-sulfophenyl)phosphine trisodium salt $Pd(OAc)_2/Ar_3P$
Palladium-catalyzed N-allylation $NH \rightarrow NC\text{-}C{=}C$
s. *42*, 354s*55*, *56*; in benzonitrile *under aq. phase catalysis* with the hydrophilic bio-polymer, cellulose, as support for the water-soluble $Pd(OAc)_2/TPPTS$ system with minimal leaching of the catalyst into the organic phase and with the prospect of good retrievability s. F. Quignard, A. Choplin, Chem. Commun. *2001*, 21-2.

Bis(allylpalladium chloride)/cis,cis,cis-1,2,3,4-tetrakis(diphenylphosphinomethyl)- ←
cyclopentane
Palladium-catalyzed N-allylation with a dramatic rate enhancement

145.
$$\text{Ph}\diagup\!\!\!\diagdown\text{OAc} + \text{HNPr}_2 \xrightarrow{\text{Ph}_2\text{P, PPh}_2, \text{Ph}_2\text{P, PPh}_2} \text{Ph}\diagup\!\!\!\diagdown\text{NPr}_2$$

Palladium-catalyzed N-allylation is dramatically increased with *cis,cis,cis*-1,2,3,4-tetrakis-(diphenylphosphinomethyl)cyclopentane (Tedicyp) as ligand at a very high substrate/catalyst ratio. E: A soln. of cinnamyl acetate and 2 eqs. dipropylamine in distilled THF heated at 50° for 90 h in the presence of *0.1 mol%* Pd-complex (prepared from 2:1 Tedicyp/[PdCl(C$_3$H$_5$)]$_2$) under argon → product. Y 95% (94% regio purity; TOF 45/h). Turnover frequencies as high as 8125/h and turnover numbers of ca. 680,000 were recorded with catalyst loadings as low as 0.001 mol%. The catalyst is more stable and less sensitive to poisoning than complexes based on mono- and di-phosphines, for which conversions were very low (1-3%). F.e.s. M. Feuerstein, H. Doucet et al., Chem. Commun. *2001*, 43-4.

Via intermediates   *v.i.*
Replacement of *tert*-hydroxyl by *prim*-amino groups   OH → NH$_2$
via Ritter reaction with chloroacetonitrile

146.

Acetic acid added to a mixture of the startg. tert. alcohol and 2 eqs. chloroacetonitrile, cooled to 0-3°, 3 eqs. sulfuric acid added dropwise below 10°, allowed to warm to room temp., and stirred for 5 h → intermediate chloroacetamide (Y 95%), refluxed with 1.2 eqs. thiourea in 5:1 ethanol/acetic acid for 10 h → product (Y 85%). 1-Phenylcyclohexanol yielded 1-phenylcyclohexene instead. F.e.s. A. Jirgensons et al., Synthesis *2000*, 1709-12.

Nitrogen ↑   NC ↓↑ N

Without additional reagents   *w.a.r.*
N-Acylation with 1-acylbenzotriazoles   NH → NAc
N-trifluoroacetylation s. *53*, 152; N-acylation of prim. and sec. amines, and prepn. of N-unsubst. amides with NH$_4$OH s. A.R. Katritzky et al., J. Org. Chem. *65*, 8210-3 (2000).

Guanidines from two different amine molecules   ⊃NC(=NH)N⊂
via 1-amidinobenzotriazoles

147.

1 Eq. Pyrrolidine added dropwise at room temp. to a soln. of di(benzotriazol-1-yl)methanimine in dry THF under N$_2$, stirred until TLC indicated completion of reaction, concentrated *in vacuo*, and the residue in dichloromethane washed with aq. 10% Na$_2$CO$_3$ → intermediate 1-amidinobenzotriazole (Y 71%), in THF treated with 1 eq. aniline under argon, and refluxed for 12 h → product (Y 68%). The procedure is mild and generally applicable to the preparation of tri- and tetra-subst. guanidines from sec. (or prim.) amines (incl. hindered ones). The by-product, benzotriazole, can be recovered easily for reuse. F.e.s. A.R. Katritzky et al., J. Org. Chem. *65*, 8080-2 (2000); **polymer-based guanidines** from prim. amines (with *i*-Pr$_2$NEt) cf. M. del Fresno, M. Royo et al., Org. Lett. *2*, 3539-42 (2000).

Triethylamine   *Et$_3$N*
N,N''-Di(carbo-*tert*-butoxy)guanidines from amines   NH → NC(=NBoc)NHBoc
and N,N'-di(carbo-*tert*-butoxy)-N''-triflylguanidine s. *50*, 198s56; aminoglycoside derivs. in water s. T.J. Baker, Y. Tor et al., J. Org. Chem. *65*, 9054-8 (2000).

Pyridinium salts by interchange   ←
s. *20*, 369; polymer-based Zincke reaction with supported amines s. M. Eda, M.J. Kurth et al., J. Org. Chem. *65*, 5131-5 (2000).

Ethyldiisopropylamine   *i-Pr$_2$NEt*
Polymer-based synthesis of guanidines from prim. amines s. *60*, 147   NH$_2$ → N=C(N⊂)$_2$

Acylase
Kinetic resolution of amines ←
by asym. N-acylation with acylamines in aq. organic medium NH → NAc

148.

Phenylacetamide and 1.33 eqs. 2-amino-4-phenylbutane in 3:1 water/acetonitrile adjusted to pH 11 with 3 M H_2SO_4, penicillin acylase (*Alcaligenes faecalis*) added, and stirred for 2.5 h with pH control by addition of 2 M NaOH (pH 11) → (R)-N-phenylacetyl-2-amino-4-phenylbutane. Y 43.5% (e.e. 97%). This is a useful alternative to existing lipase-catalyzed routes for reaction with arylalkylamines. *E. coli* penicillin acylase displayed low enantioselectivity and reactivity. F.e.s. L.M. van Langen, R.A. Sheldon et al., Tetrahedron:Asym. *11*, 4593-600 (2000).

Polymer-based triarylphosphine ←
Aldimines from aldehydes and azides N_3 → N=CHR
with Ph_3P cf. *45*, 220; with a polystyrene-based triarylphosphine s. A.B. Charette et al., Org. Lett. *2*, 3777-9 (2000).

Magnesium perchlorate/chiral bis(Δ^2-oxazolines) ←
Asym. synthesis of 5-isoxazolidones from N-(α,β-ethyleneacyl)-2-pyrrolidones O

149.

Chiral 3-aryl-5-isoxazolidones. Startg. pyrrolidone added to a soln. of 0.3 eq. $Mg(ClO_4)_2$ and 0.3 eq. 1,1-bis[(3aS,8aR)-indano[1,2-*d*]oxazolin-2-yl]cyclopropane in dichloromethane under N_2, stirred for 30 min at room temp. before cooling to -60°, 1.1 eqs. N-benzylhydroxylamine added, the mixture stirred at the same temp. for 34 h, then quenched with water → product. Y 87% (e.e. 96%). Variations of the N-substituent of hydroxylamine, Lewis acid, and electronic properties of the β-substituent of the startg. N-(α,β-ethyleneacyl)-2-pyrrolidones were also studied. The products are precursors of chiral β-amino-β-arylcarboxylic acids. F.e.s. M.P. Sibi, M. Liu, Org. Lett. *2*, 3393-6 (2000).

Ferric chloride $FeCl_3$
Hydrazones from azides s. *60*, 155 CHN_3 → C=NN<

Rhodium(II) acetate $Rh_2(OAc)_4$
N-Alkylation with diazo compds. C=N_2 → CHN<
of α-amino-esters s. *52*, 164; N-protected amino(silyl)acetic from diazo(silyl)acetic acid esters and N-protected amines, also peptide synthesis, s. C. Bolm et al., Angew. Chem. Int. Ed. Engl. *39*, 2288-90 (2000).

Palladous acetate/triphenyl phosphite $Pd(OAc)_2/(PhO)_3P$
Regio- and stereo-specific N-allylation with N-allylbenzotriazoles NH → N-C-C=C
of amines cf. *57*, 159; of sulfonic acid amides with $(PhO)_3P$ as ligand s. A.R. Katritzky et al., J. Org. Chem. *65*, 8063-5 (2000).

Via intermediates v.i.
Nucleoside base exchange s. *60*, 121 ←

Halogen ↑   NC ↓↑ Hal

Without additional reagents   w.a.r.
N-Arylation with (η^6-fluoroarene)chromium(0) carbonyl complexes   NH → NAr
s. *8*, 563s57; polymer-based N-arylation of cyclic sec. amines (cf. *11*, 510s56) with polymer-based dicarbonyl(isonitrile)chromium-complexed ar. fluorides, and substitution with other nucleophiles (PhCH$_2$S-, MeO-) s. S. Maiorana, E. de Magistris et al., Tetrahedron Lett. *41*, 7271-5 (2000).

Microwaves s. under NaHCO$_3$ and Al$_2$O$_3$ ←

Sodium hydride   NaH
Regiospecific N-alkylation of the imidazole ring   NH → NR
s. *9*, 512s39; 9-subst. guanines from 2-amino-6-(4*H*-1,2,4-triazol-4-yl)purines via regiospecific N^9-alkylation s. K. Alarcon, M. Demeunynck et al., Tetrahedron Lett. *41*, 7211-5 (2000).

N-Arylation with diaryliodonium salts of 5-fluorouracil s. *53*, 162s60   NH → NAr

Sodium hydroxide   NaOH
Sym. tert. amines from halides   3 RHal → R$_3$N

150.   3 BnCl $\xrightarrow{H_2NCONH_2}$ Bn$_3$N

A mixture of urea, 6.4 eqs. 50% aq. NaOH, and 6.2 eqs. benzyl chloride pressurized to 60 psi at 20° with N$_2$ in a glass-lined Parr pressure vessel, then heated at 120° for 40 h → tribenzylamine. Y 82%. The optimum reaction temp. and time of this potentially industrial-scale process depended on the reactivity of the halide. Contaminating quaternary ammonium salts were produced if the ratio of halide to N exceeded 4:1. F.e.s. N. Sachinvala et al., J. Org. Chem. *65*, 9234-7 (2000).

Potassium hydroxide/tetraoctylammonium bromide   KOH/(C$_8$H$_{17}$)$_4$NBr
(E)-Alkoximes from oxo compds. via oximes   CO → C=NOR
One-pot procedure under phase transfer catalysis

151.

$$Ph\text{-CHO} \xrightarrow{H_2NOH \cdot HCl} \left[Ph\text{-CH=N-OH} \right] \xrightarrow{Br\text{-CH}_2\text{-C} \equiv \text{CH}} Ph\text{-CH=N-O-CH}_2\text{-C} \equiv \text{CH}$$

1.25 eqs. Powdered hydroxylamine hydrochloride and 25% aq. KOH soln. (1 part) added to a soln. of benzaldehyde and 2.5 mol% tetraoctylammonium bromide in toluene (0.6 part), the mixture refluxed for 2 h, 4 eqs. propargyl bromide added, and stirred at room temp. for 3 h → benzaldehyde O-propargyloxime. Y 79% (100% (E)). The products were usually obtained as single (E)-isomers (91-100% selectivity), exceptions being thienyl derivs. (58-87% E). Pyridyl-derivs. afforded low yields in the presence of MeI as a result of undesirable N-quaternization. F.e.s. E. Abele, E. Lukevics et al., Org. Prep. Proced. Int. *32*, 153-9 (2000).

Cesium hydroxide   CsOH
Partial and preferential mono-N-alkylation   NH → NR

152.

H$_2$N-CH$_2$CH$_2$-S-CH$_2$CH$_2$-NH$_2$ + Br-CH$_2$CH$_2$-NH$_2$ → H$_2$N-CH$_2$CH$_2$-S-CH$_2$CH$_2$-NH-CH$_2$CH$_2$-NH$_2$

H$_2$N-(CH$_2$)$_3$-NH-(CH$_2$)$_3$-NH$_2$ + BnCl $\xrightarrow{0°}$ H$_2$N-(CH$_2$)$_3$-NH-(CH$_2$)$_3$-NHBn

of di-*prim*-amines. 2.5 eqs. CsOH·H$_2$O and 1,5-diamino-3-thiapentane added to activated, powdered, dry 4 Å molecular sieves in anhydrous DMF, 1.5 eqs. 2-bromoethylamine hydro-

bromide added with vigorous stirring, stirred at 23° for 12 h, and quenched with 1 N NaOH →
product. Y 73%. Diamines possessing prim. and sec. amino groups reacted preferentially at the
primary nitrogen (steric control), and unsym. di-*prim*-amines having both primary and secondary
carbon atoms linked to nitrogen reacted *regiospecifically* with the amino group at the primary
site. No overalkylation and no racemization of chiral diamines was observed. Reaction with
Cs_2CO_3 was problematic. F.e. and partial N-tosylation s. R.N. Salvatore, K.W. Jung et al., Tetrahedron
Lett. *41*, 9705-8 (2000).

Potassium tert-butoxide KOBu-t
N-Arylation with ar. fluorides NH → NAr
with K_2CO_3 cf. *8*, 563s*53*; 1-arylation of pyrazoles (with KOBu-*t*) s. X. Wang et al., Org. Lett. *2*,
3107-9 (2000).

n-*Butyllithium/trimethylsilyl chloride* *BuLi/Me₃SiCl*
Regiospecific N-acylation of diamines via *in situ*-N-silylation NH → NAc
of *prim*-amino groups with retention of sec. cf. *58*, 154; mono-N-acylation of piperazines at the
more hindered site s. T. Wang, Z. Zhang et al., J. Org. Chem. *65*, 4740-2 (2000).

Sodium hydrogen carbonate/microwaves ←
N-Arylation of α-aminocarboxylic acids under microwave irradiation in water NH → NAr

153.

A soln. of startg. amino acid and 2 eqs. $NaHCO_3$ *in water* added to a quartz reaction vessel,
followed by 1 eq. 2,4-dinitrofluorobenzene, the vessel placed into the cavity of a focused mono-
mode microwave reactor (Synthewave 402), irradiated *for 35 sec,* and the product precipitated
by addition of 2 N HCl → (S)-N-(2,4-dinitrophenyl)alanine. Y 93%. There was no reaction under
conventional heating. 2,4-Dinitro-chlorobenzene and -bromobenzene were less reactive. F.e.s.
Y.-J. Cherng, Tetrahedron *56*, 8287-9 (2000).

Potassium carbonate K_2CO_3
Mono-N-alkylation of tetraamines NH → NR
via their tridentate tricarbonylchromium complexes (with Na_2CO_3) cf. *46*, 351; with simple and
long-chain alkyl halides via tris-N-formylation (with K_2CO_3) s. V. Boldrini, M. Sisti et al.,
Tetrahedron Lett. *41*, 6527-30 (2000).

*Cesium carbonate/tetra-*n-*butylammonium iodide* Cs_2CO_3/Bu_4NI
Urethans from amines and halides NH → NCOOR
with CO_2/Cs_2CO_3 s. *50*, 214; prepn. of polymer-based urethans with added Bu_4NI s. R.N. Salvatore,
K.W. Jung, Org. Lett. *2*, 2797-800 (2000).

Sodium azide s. under $FeCl_3$ NaN_3

Pyridine C_5H_5N
Carboxylic acid amides from carboxylic acid chlorides COCl → CON<
s. *9*, 524s*59*; improved work-up with a polymer-based tris(2-aminoethyl)amine as acid chloride
scavenger, also use of the resin as an isocyanate scavenger, s. A. Chesney et al., J. Comb. Chem.
2, 434-7 (2000).

Alumina/microwaves ←
N-Arylation under microwave irradiation NH → NAr
on a solid support without solvent s. *59*, 171; with ar. chlorides s. M. Kidwai et al., Synth.
Commun. *30*, 4479-88 (2000).

Trimethylsilyl chloride s. under BuLi *Me₃SiCl*

Polymer-based cyclic iminophosphoric acid amides
Regiospecific N⁹-alkylation of 2-aminopurines   NH → NR

154.

A soln. of 2-amino-6-benzyloxypurine and methyl iodide in acetonitrile treated with polymer-based BEMP (2-*tert*-butylimino-2-diethylamino-1,3-dimethylperhydro-1,3,2-diazaphosphorine) at room temp., and the mixture purified by filtration through Al_2O_3/H^+ → product. Y 75% (purity >99%). There was no contamination with the N⁷-alkyl deriv., this novel application of solid-supported reagents avoiding laborious regioisomer separation and work-up steps, and facilitating purine library synthesis in a parallel or high-throughput format ('chemoselective high-throughput purification'). F.e. and subsequent **N-acylation** (with 2 eqs. acyl chloride and the same reagent in THF with scavenging of excess of the electrophile using poly-trisamine or amine-derivatized silica) s. W. McComas, K. Kim et al., Tetrahedron Lett. *41*, 3573-6 (2000).

Carboxylic acid amides from acid chlorides s. *60*, 159   COCl → CONHR

p-Toluenesulfonyl chloride/triethylamine/sodium hydroxide/
 benzyltriethylammonium chloride
N,N-Disubst. cyanamides from amidoximes and halides s. *60*, 131   RHal → R(R')NCN

Calcium hypochlorite/molecular sieves/N-sodio compd.
N-Tosylaziridines from ethylene derivs.
and Bromamine-T with CuCl cf. *55*, 159s57; under mild conditions with $Ca(OCl)_2$ and 5 Å molecular sieves, also benzylic tosylamination with $Rh_2(OAc)_4$, s. B.M. Chanda et al., J. Org. Chem. *66*, 30-4 (2001).

Iron/ammonium chloride Fe/NH_4Cl
Urethans from azides N_3 → NHCOOR
with Me_3P cf. *48*, 348s58; with Fe/NH_4Cl, also combinatorial synthesis of urethans **and sulfonic acid amides** (with $PhSO_2Cl$), s. S. Chandrasekhar, Ch. Narsihmulu, Tetrahedron Lett. *41*, 7969-72 (2000).

Ferric chloride/sodium azide $FeCl_3/NaN_3$
N,N-Dimethylhydrazones from halides via azides   CHHal → C=NNMe₂

155.

One-pot procedure. A mixture of *p*-bromobenzyl bromide and 1.1 eqs. NaN_3 in acetonitrile stirred under argon for 4.5 h (monitored by TLC), filtered (through a disposable pipet with a cotton plug), 0.09 eq. $FeCl_3·6H_2O$ added to the filtrate and washings, the mixture purged with argon, 4.4 eqs. N,N-dimethylhydrazine added via syringe, the mixture heated to reflux for 14 h, cooled to ambient temp., satd. aq. Na-bicarbonate added, and the mixture stirred at ambient temp. for a further ca. 10 min → product. Y 89%. The method is simple, efficient, convenient, high-yielding and in one-pot. It can also be effected from prim. tosylates, while sec. tosylates were prone to eliminate. F.e.s. I.C. Barrett, M.A. Kerr, Synlett *2000*, 1673-5; **from the intermediate azides** s. idem., J. Org. Chem. *65*, 6268-9 (2000).

Cobalt(II) acetate/tetra-n-butylammonium iodide/potassium carbonate
N-Arylation with diaryliodonium salts   NH → NAr
with CuI/Na_2CO_3 cf. *53*, 162s59; of imidazoles with 5 mol% $Co(OAc)_2/Bu_4NI/K_2CO_3$ s. L. Wang,

Z.C. Chen, J. Chem. Res., Synop *2000*, 367-9; 1-arylation of 5-fluorouracil with NaH s. J.Z. You, Y.Z. Chen et al., Chin. Chem. Lett. *11*, 865-6 (2000).

Nickel/carbon/1,1'-bis(diphenylphosphino)ferrocene/lithium tert-*butoxide/* ←
n-*butyllithium*
Ar. amines from chlorides NH → NAr
Heterogeneous nickel-catalyzed N-arylation

156.

A slurry of 5 mol% nickel/charcoal (5%), 2.5 mol% 1,1'-bis(diphenylphosphino)ferrocene, and 1.2 eqs. LiOBu-*t* in dry toluene stirred at room temp. under argon for 90 min, 0.1 eq. *n*-BuLi (2.42 *M* in hexane) added dropwise, stirring continued for 30 min, solutions of 4-chlorobenzonitrile and 1.5 eqs. startg. amine in toluene added, and the mixture refluxed vigorously for 3.5 h (oil bath temp. 130°) → product. Y 92%. This method is viewed as a simple economical alternative to the standard homogeneous Pd-catalyzed route, and is generally applicable to condensing electron-rich or -deficient ar. chlorides with prim. and sec. amines. Reaction failed with hexamethyl-disilazane, imidazole and aza-crown ethers. F.e.s. B.H. Lipshutz, H. Ueda, Angew. Chem. Int. Ed. Engl. *39*, 4492-4 (2000).

Tris(dibenzylideneacetone)dipalladium/2-(di-tert-*butylphosphino)biphenyl/* ←
sodium tert-*butoxide*
Triarylamines from prim. ar. amines ArNH$_2$ → ArN(Ar')(Ar'')
and two different ar. halide molecules

157.

One-pot procedure. 0.5 Mol% Pd$_2$(dba)$_3$, 2 mol% 2-(di-*tert*-butylphosphino)biphenyl, and 2.1 eqs. Na-*tert*-butoxide charged to a Schlenk tube under argon, 1,3-phenylenediamine added along with 1 eq. 4-bromo-*tert*-butylbenzene, 1.05 eqs. 4-chlorotoluene and toluene, and the mixture stirred at 80° until GC analysis indicated complete consumption of the intermediate diarylamine → product. Y 95%. The reaction can also be carried out as a one-pot two-step process in which the final ar. halide is added after completion of the first N-arylation. In this way the second ar. halide can be a more reactive or available halide. F.e. and preparation of a library of 27 triarylamines s. M.C. Harris, S.L. Buchwald, J. Org. Chem. *65*, 5327-33 (2000).

Tris(dibenzylideneacetone)dipalladium/2,2'-bis(diphenylphosphino)- ←
1,1'-binaphthyl/sodium tert-*butoxide or cesium carbonate*
Ar. amines from halides NH → NAr
s. *52*, 171s59; scope and limitations s. J.P. Wolfe, S.L. Buchwald, J. Org. Chem. *65*, 1144-57 (2000); aminothiophenes with milder Cs$_2$CO$_3$ as base s. T.J. Luker et al., Tetrahedron Lett. *41*, 7731-5 (2000); mono-N-arylation of chiral diamines s. C.G. Frost, P. Mendonça, Tetrahedron: Asym. *10*, 1831-4 (1999); partial conversion of *o*-dibromobenzene (cf. *55*, 161s58), and subsequent reaction with a second chiral amine component to give chiral unsym. *o*-diamino-benzenes s. F.M. Rivas, S.T. Diver et al., ibid. *11*, 1703-7 (2000).

*Palladous chloride/tri-*o-*tolylphosphine/sodium* tert-*butoxide* ←
Ar. amines from halides
with Pd(dba)$_2$/(*o*-tol)$_3$P/NaOBu-*t* cf. *51*, 171s59; tert. ar. amines from sec. amines with PdCl$_2$ cf. V. Bavetsias, E.A. Henderson, J. Chem. Res., Synop *2000*, 418-9.

Palladous chloride/1,1'-bis(diphenylphosphino)ferrocene/cesium carbonate
**Palladium-catalyzed 3-component synthesis of ar. amidines
from ar. halides and amines**   ArC(=NR)N<

158.

1.5 eqs. *t*-Butyl isocyanide and 5 mol% PdCl$_2$ added to a mixture of 1.3 eqs. dry Cs$_2$CO$_3$, 10 mol% 1,1'-bis(diphenylphosphino)ferrocene, *p*-bromoanisole, and 5 eqs. pyrrolidine in dry, degassed toluene in a Schlenk tube under argon, the tube sealed, and heated in an oil bath at 109° for 20 h → product. Y 83%. This tin-free synthesis is generally applicable to electron-rich or -poor ar. or heteroar. halides, but is limited to *t*-butyl isocyanide and bidentate ligands. F.e. and **N-de-*tert*-butylation** (with concd. HCl), also with Pd(OAc)$_2$, s. C.G. Saluste, R.J. Whitby et al., Angew. Chem. Int. Ed. Engl. *39*, 4156-8 (2000).

Via intermediates   v.i.
Polymer-based synthesis of tert. amines via Hofmann degradation   >NH → >NR
using ester resins s. *51*, 172; using amide resins (with increased compatibility with Grignards or reducing agents or transesterification conditions) cf. M.J. Plater, D.C. Rees et al., J. Comb. Chem. *2*, 508-12 (2000).

**Polymer-based synthesis of carboxylic acid amides
from carboxylic acid halides and amines via urethans**   NH → NAc

159.

Startg. Merrifield resin-bound *p*-nitrophenyl carbonate in DMF added to 3 eqs. startg. amine hydrochloride and 6 eqs. *i*-Pr$_2$NEt, the mixture shaken at room temp. for 6 h, the resulting resin-bound carbamate washed with DMF (3x) and dichloromethane (3x), swelled with dichloromethane, treated with 0.5 eqs. dry ZnBr$_2$ and 3 eqs. acetyl bromide under argon, agitated at room temp. for 24 h, a soln. of 1.5 eqs. Et$_3$N in dichloromethane added dropwise, the resin filtered off, and the filtrate worked up → product. Y 81% (overall). The method is efficient, mild, simple and high-yielding. Benzyl esters were stable under these conditions, and no epimerization was observed. F.e., **also carboxylic acid esters from alcohols** (with DMAP) or phenols (with Cs$_2$CO$_3$) via polymer-based carbonates (without the need for Et$_3$N), s. W.-R. Li, Y.-S. Lin, Y.-C. Yo, Tetrahedron Lett. *41*, 6619-22 (2000); solution-phase procedure for heterocyclic amines with polymer-based 2-*tert*-butylimino-2-diethylamino-1,3-dimethylperhydro-1,3,2-diazaphosphorine as base, **also polymer-based N-acylation** of weakly nucleophilic heterocyclic amines **with supported *o*-acoxynitro compds.** (with scavenging of unreacted amine by Amberlite IRA-120) cf. K. Kim, K. Le, Synlett *1999*, 1957-9.

N-Acylureas from carboxylic acid chlorides   NH → NCOR
s. *14*, 516; mono-N-acylation of cyclic ureas via partial deacylation of cyclic N,N'-diacylureas with KN(SiMe$_3$)$_2$ s. S.P. Bew, S.G. Davies et al., Tetrahedron Lett. *40*, 7143-6 (1999).

Sulfur ↑ NC ↓↑ S

Without additional reagents   w.a.r.
Isothiocyanates from 2H-1,2,3-triazolium-1-imides under mild conditions ←

160.

15 eqs. Carbon disulfide added to a soln. of the startg. triazolium-1-imide in dry acetone, and the mixture stirred at room temp. for 30 min → phenyl isothiocyanate (Y 89%) and 2-phenyl-4,5,6,7-tetrahydro-2H-1,2,3-benzotriazole (Y 90%). Electron-withdrawing groups on the pendant N-aryl group lowered the yield. The mechanism involves a dipolar cycloaddition of the triazolium-1-imide to carbon disulfide followed by fragmentation. F.e.s. R.N. Butler, L.M. Wallace, J. Chem. Soc. Perkin Trans. 1 2000, 4335-8.

Triethylamine Et_3N
N-Thioacylation   NH → NC(S)R
with N-thioacyl-1,3,2-dioxaphosphorinane 2-sulfides s. 60, 199

N-Subst. carboxylic acid thioamides from dithiocarboxylic acid esters
with Et_3N s. 8, 529; N-(2-hydroxyalkyl)-α-phosphonylcarboxylic acid thioamides s. N. Leflemme, S. Masson et al., Synthesis 2000, 1143-7.

Sodium azide/mercuric chloride/triethylamine $NaN_3/HgCl_2/Et_3N$
5-Aminotetrazoles from thioureas   ○
via carbodiimides (not isolated) with HgO/HN_3 cf. 13, 366; mono-, di- and tri-subst. derivs. with $HgCl_2/NaN_3/Et_3N$, limitations, s. R.A. Batey, D.A. Powell, Org. Lett. 2, 3237-40 (2000).

Silver nitrate $AgNO_3$
Polymer-based synthesis of guanidines from N-thioacyltriazenes and amines ←
via amidinotriazenes

161.

A soln. of allylamine in acetonitrile added to a mixture of startg. polymer-based N-thioacyltriazene (loading 0.45 mmol/g) and 4.2 eqs. $AgNO_3$ in a closed vial, the resulting mixture agitated for 12 h at 45°, and the resin cleaved with 10% trifluoroacetic acid in dichloromethane at room temp. for 5 min → product. Y 88% (>90% purity as trifluoroacetate salt). This is part of a multistep polymer-based route **from two different amine molecules and isothiocyanates**. F.e.s. S. Dahmen, S. Bräse, Org. Lett. 2, 3563-5 (2000).

Benzyltriethylammonium permanganate $BnEt_3N^+MnO_4^-$
Guanidines from thioureas and amines $NH \rightarrow N(C=N)N<$

162.
PhNH-C(S)-NHPh + $H_2NC_6H_{11}$-c $\xrightarrow{BnEt_3N^+ MnO_4^-}$ PhNH-C(=NC_6H_{11}-c)-NPh

N,N''-Diaryl-derivs. A stirred mixture of 2 eqs. cyclohexylamine and N,N'-diphenylthiourea in THF at 5-10° treated portionwise over 15 min with 1 eq. $BnEt_3N^+MnO_4^-$, and stirring continued for 15 min → product. Y 92%. This method is simple, mild, provides high yields and avoids toxic by-products. F.e.s. N. Srinivasan, K. Ramadas, Tetrahedron Lett. *42*, 343-6 (2001).

Remaining Elements ↑ NC ↓↑ Rem

Cupric acetate $Cu(OAc)_2$
N-Arylation with aryllead tricarboxylates $NH \rightarrow NAr$
N-phenylation of anilines cf. *42*, 397; 1-arylation of imidazoles s. G.I. Elliot, J.P. Konopelski, Org. Lett. *2*, 3055-7 (2000).

Cupric acetate/triethylamine $Cu(OAc)_2/Et_3N$
Sec. ar. amines from prim. ar. amines and triarylbismuthines
with PhI(OAc)$_2$ cf. *40*, 286s55; sec. diarylamines with 1 eq. Cu(OAc)$_2$/Et$_3$N (with retention of N-subst. amide groups) s. R.J. Sorenson, J. Org. Chem. *65*, 7747-9 (2000); N-arylation of mono-, di- and tri-protected hydrazines s. O. Loog, U. Ragnarson et al., Synthesis *2000*, 1591-7.

Carbon ↑ NC ↓↑ C

Without additional reagents w.a.r.
Tert. amines from halides $NMe \rightarrow NR$
with cleavage of N-methyl groups cf. *28*, 413; tert. hetar. amines from hetar. chlorides s. A.I. Khalaf et al., Tetrahedron *56*, 8567-71 (2000).

Sodium azide/pyridinium tosylate $NaN_3/[C_5H_5NH][TsO]$
2-Azidoalcohols from 1,3-dioxolan-2-ones with inversion of configuration C
s. *52*, 175; from 1,3-dioxolane-2-thiones with added pyridinium tosylate s. L. He, R. Bittman et al., J. Org. Chem. *65*, 7627-33 (2000).

Potassium fluoride/bis(2-oxazolidon-3-yl)phosphoryl chloride/triethylamine ←
Peptide synthesis via N-decarbo-9-fluorenylmethoxylation $NFmoc \rightarrow NCOR$
cf. *50*, 241; dipeptides with KF/BOP-Cl/Et$_3$N, also N-acetyl-, N-benzoyl- and N-tosyl-α-amino esters, s. W.-R. Li, H.-H. Chou, Synthesis *2000*, 84-90.

Sodium iodide NaI
N-Subst. lactams from O-alkyllactims ←
s. *31*, 427; with allyl, benzyl, propargyl, or cinnamyl halides using NaI s. W.R. Bowman, C.F. Bridge, Synth. Commun. *29*, 4051-9 (1999).

Triethylamine Et_3N
α-Sulfamido- from α-amino-carboxylic acid esters $NH \rightarrow NHSO_2NHR$

163.
$MeO_2CC(Ph_2)NH_2$ + $HOCHPh_2$ $\xrightarrow{ClSO_2NCO}$ $MeO_2CC(Ph_2)NHSO_2NHCHPh_2$

A soln. of 2 eqs. benzhydrol in hexane added dropwise to chlorosulfonyl isocyanate in the same solvent, the mixture warmed slightly above room temp. until the white suspension disappeared and bubbling stopped, cooled to -78°, a soln. of 1 eq. startg. ester and 2 eqs. Et$_3$N added, stirred for 0.5 h at the same temp., then for 2 h at room temp., and quenched with water → N-di-phenylmethyl-N'-[1,1-diphenyl-1-(methoxycarbonyl)methyl]sulfamide. Y 95%. The improved procedure avoids the formation of side products. F.e. and cyclization to 1,2,5-thiadiazolid-3-one 1,1-dioxides s. Z. Xiao, J.W. Timberlake, J. Heterocycl. Chem. *37*, 773-7 (2000).

Zinc bromide/triethylamine $ZnBr_2/Et_3N$
Polymer-based synthesis of carboxylic acid amides   COHal → CON<
from acid halides and urethans s. *60,* 159

1,3-Dichloro-5,5-dimethylhydantoin s. under $K_2OsO_2(OH)_4$ ←

Carbamic acid silyl esters   >NCOOSi≤
Enamines from ketones and carbamic acid silyl esters   COCH → C=CN<

164.

A mixture of 3-pentanone and *2.5 eqs.* trimethylsilyl dimethylcarbamate allowed to react *at room temp.* for 4 h in the absence of solvent and with the exclusion of moisture → product. Y 85%. Although there are certain limitations (e.g. acetone and acetaldehyde undergo self-condensation), the method is advantageous in that it is mild and clean, and obviates handling volatile amines. One equivalent of the silyl carbamate serves as dehydrating agent. F.e. and partial conversion of acetylacetone s. F. Kardon, M. Mörtle, D. Knausz, Tetrahedron Lett. *41,* 8937-9 (2000).

Stannous bromide/triethylamine $SnBr_2/Et_3N$
Acylamines from carbobenzoxyamines under mild conditions   NCOOBn → NAc

165.

Acetylamines. 0.54 eq. $SnBr_2$ and 5 eqs. acetyl bromide added sequentially to a stirred soln. of N-Cbz-L-Phe-OMe in methylene chloride under argon at room temp., a soln. of 1.5 eqs. Et_3N in the same solvent added dropwise when the carbamate had been consumed, and stirred at room temp. with TLC monitoring (9 h) → Ac-L-Phe-OMe. Y 81%. Reaction with benzoyl chloride was slower (24-36 h). Esters, sulfides and the peptide bond were unaffected, and there was no noticeable racemization. cf. *13,* 518s48. F.e. and **from** [Merrifield] **polymer-based carbobenzoxyamines** with $ZnBr_2/Et_3N$, also comparison of reagents and solvents s. W.-R. Li et al., Tetrahedron *56,* 8867-75 (2000).

Pyridinium tosylate s. under NaN_3 $[C_5H_5NH][TsO]$

Palladous acetate/lithium bromide $Pd(OAc)_2/LiBr$
(E)-2-Ethylenetosylamines from 2-ethylenealcohols   ←
with double bond shift s. *60,* 130

Potassium osmate/1,4-bis(9-O-dihydroquinine)phthalazine/   ←
1,3-dichloro-5,5-dimethylhydantoin/sodium hydroxide
2-Oxazolidones from ethylene derivs.   O
via regiospecific asym. Sharpless oxyamination

166.

Chiral 4-aryl-2-oxazolidones. A soln. of 1 eq. startg. ethylene deriv. and 0.025 eq. $(DHQ)_2PHAL$ in *n*-propanol added to a mixture of 3.1 eqs. ethyl carbamate, 1.5 eqs. 1,3-dichloro-5,5-dimethylhydantoin (as substitute for *tert*-butyl hypochlorite) and 3 eqs. NaOH in 1:1 water/*n*-propanol at 20°, treated with a soln. of 0.02 eq. $K_2OsO_2(OH)_4$ and 0.08 eq. NaOH in water, stirred for 3 h, 3.1

eqs. NaOH added, stirred for a further 1 h, then quenched with 1.8 eqs. Na-sulfite → product. Y 81% (as a 5:1 mixture of regioisomers; e.e. 98% for the major isomer after column chromatographic separation). The use of (DHQD)$_2$PHAL resulted in the formation of the (R,R)-isomer. F.e. on both large and small scale, also with dichlorocyanuric acid Na-salt, s. N.S. Barta et al., Org. Lett. 2, 2821-4 (2000).

Elimination ⇑

Hydrogen ↑   NC ⇑ H

N-Bromoacetamide   AcNHBr
Cyclic amidinium salts from 1,1-diamines   ←
with NBS cf. 32, 413; Δ^2-imidazolinium salts from imidazolidines with AcNHBr s. A. Salerno, I.A. Perillo et al., Synth. Commun. 30, 3369-82 (2000).

Sodium dichloroisocyanurate/alumina   ←
Nitrile oxides from aldoximes   CH=NOH → C≡N→O
with PhICl$_2$ cf. 46, 397s47; rapid, heterogeneous procedure with Na-dichloroisocyanurate/alumina s. B. Syassi, M. Soufiaoui et al., Tetrahedron Lett. 40, 7205-9 (1999).

Iodine/potassium hydroxide   I$_2$/KOH
Pyridines from 1,4-dihydropyridines   ←
with CCl$_4$ cf. 25, 649s56; mild inexpensive method with I$_2$/KOH in methanol s. J.S. Yadav et al., Synthesis 2000, 1532-4.

Hydroxyapatite-bound ruthenium(III) complex/oxygen   ←
Nitriles from prim. amines   CH$_2$NH$_2$ → CN
Heterogeneous Ru-catalyzed aerobic oxidation

167.   n-C$_{11}$H$_{23}$CH$_2$NH$_2$ ⟶ n-C$_{11}$H$_{23}$CN | PhCN ⟶ PhCONH$_2$

p-Xylene and *n*-dodecylamine added successively to RuHAP (6.5 mol% Ru^{3+}; prepared by treating a calcium hydroxyapatite, Ca$_{10}$(PO$_4$)$_6$(OH)$_2$, with aq. RuCl$_3$ at 25° for 24 h, then filtering the slurry, and drying at 110° overnight), the mixture stirred at 125° under 1 atm. O$_2$ for 24 h, filtered, and the organic layer distilled → *n*-dodecanenitrile. Y 94%. The catalyst shows high catalytic activity and may be reused. The method is applicable to aliphatic, allylic and benzylic prim. amines, and amines were oxidized preferentially over prim. alcohols. Oxidation of dibenzylamine afforded N-benzylidenebenzylamine in 91% yield, but other sec. benzylamines gave mixtures of imines and benzaldehyde (formed by hydrolysis). Oxidation of tert. amines did not proceed. F.e., **also carboxylic acid amides from nitriles** (in water under N$_2$), and one-pot prepn. of nicotinamide from 3-aminomethylpyridine, s. K. Mori, K. Kaneda et al., Chem. Commun. 2001, 461-2.

Rhodium(II) acetate/phenyl iodosoacetate/magnesium oxide   ←
2-Oxazolidones from urethans by catalytic oxidative ring closure   ○

168.

with asym. induction. A. ca. 0.2 M soln. of startg. urethan in methylene chloride treated with 5 mol% Rh$_2$(OAc)$_4$, 1.4 eqs. PhI(OAc)$_2$, and 2.3 eqs. MgO (for removing liberated acetic acid) at 40° for 12 h → product. Y 82%. The process is based on inexpensive reagents (other bases being less efficient). However, reaction is limited to substrates having a benzylic or tertiary site. A concerted, metal-directed (non-nitrenoid) mechanism is invoked. F.e. and with Rh(II)-triphenylacetate s. C.G. Espino, J. Du Bois, Angew. Chem. Int. Ed. Engl. 40, 598-600 (2001).

Via intermediates v.i.
Δ³-Oxazolines from oxazolidines via 3-chlorooxazolidines ←
by N-chlorination with *t*-BuOCl and dehydrochlorination with KO₂/18-crown-6 cf. *48*, 392; with NaOCl and ethanolic KOH, respectively, s. S. Favreau, R. Fellous et al., Tetrahedron Lett. *41*, 9787-90 (2000).

Oxygen ↑   NC ↑ O

Indium(III) chloride $InCl_3$
Nitriles from aldoximes CH=NOH → CN

169.

0.13 eq. Anhydrous InCl₃ added to a soln. of benzaldoxime in dry acetonitrile (stirred previously for 2 min), and the mixture refluxed for 1.5 h → benzonitrile. Y 98%. The method is simple, mild and general, and avoids the use of toxic, expensive or strongly acidic or basic reagents. Functional groups such as phenolethers, phenols, and aromatic chlorides, amines and nitro compds. remained unaffected. F.e. incl. aliphatic, conjugated, aromatic and heteroaromatic (2-furyl-, 2-thienyl- and 2-pyridyl-)nitriles, **also carboxylic acid anilides** from ar. ketoximes **via Beckmann rearrangement** under the same conditions, s. D.C. Barman, J.S. Sandhu et al., Chem. Lett. *2000*, 1196-7.

Samarium s. under TiCl₄ Sm
Samarium diiodide SmI₂
Titanium tetrachloride/samarium TiCl₄/Sm
Cyclic azomethines from nitrooxo compds. ○
from nitroketones with TiCl₄/Zn cf. *13*, 539s*49*; 2H-1,4-benzoxazines from *o*-nitroaryloxyketones with TiCl₄/Sm s. Y.M. Ma, Z. Zhang, J. Chem. Res., Synop *2000*, 388-9; from nitroaldehydes with SmI₂ cf. A. Kamal, P.S.M.M. Reddy et al., Tetrahedron Lett. *41*, 8631-4 (2000).

Stannous phenylmercaptide/thiophenol/triethylamine ←
Cyclic hydroxamic acids from nitrocarboxylic acid esters
with Zn/NH₄Cl cf. *20*, 387s*47*; polymer-based synthesis of oligosaccharides with traceless removal of the support on formation of the 4-hydroxy-2H-1,4-benzoxazin-3-one ring with Sn(SPh)₂/PhSH/Et₃N s. S. Manabe, Y. Ito et al., Synlett *2000*, 1241-4.

2-Pyridinesulfonyl chloride RSO_2Cl
Nitriles from carboxylic acid amides CONH₂ → CN
α-Formamidinonitriles s. *5*, 301s*60*

Iron/acetic acid Fe/AcOH
Cyclic β-arylaminocarboxylic from α,β-ethylene(nitro)carboxylic acid esters ○
Reductive intramolecular Michael addition

170.

1,2,3,4-Tetrahydroquinolin-2-ylacetic acid esters. A mixture of ethyl (E)-5-(2-nitrophenyl)-2-pentenoate and 1.5 eqs. Fe powder (>100 mesh) in acetic acid heated with stirring at 115° under N₂ for 30 min → ethyl 1,2,3,4-tetrahydroquinolin-2-ylacetate. Y 98%. The procedure is simple, clean, inexpensive and efficient, yields being high regardless of the geometry or substitution at the Michael terminus (although Z-isomers reacted more quickly). F.e. incl. 3,4-dihydro-2H-1,4-benzoxazin-3-ylacetic and 1,2,3,4-tetrahydroquinoxalin-2-ylacetic acid esters, s. R.A. Bunce, M.L. Ackerman et al., J. Org. Chem. *65*, 2847-50 (2000).

*Tris(dibenzylideneacetone)dipalladium/(R)-2,2'-bis(diphenylphosphino)-
1,1'-binaphthyl/potassium carbonate*
**β,γ-Ethylenenitriles from cyclobutanone O-acyloximes
via palladium-catalyzed ring opening-β-elimination**

171.

A mixture of startg. (E)-oxime benzoate, 5 mol% Pd$_2$(dba)$_3$·CHCl$_3$, 7.5 mol% (R)-(+)-BINAP, and 1 eq. K-carbonate in THF refluxed for 9 h → product. Y 86% (with 8% γ,δ-ethylene isomer). Reaction is presumed to involve initial oxidative addition of Pd(0) to the N-O bond to give a cyclobutaniminopalladium(II) complex, followed by formation of an alkylpalladium(II) species which undergoes β-elimination. Substrates lacking a β-hydrogen afforded the corresponding **cyanocyclopropanes**. The (Z)-oxime benzoate gave a 37:48 mixture of the same products in refluxing THF but 83:8 in refluxing ether. F.e.s. T. Nishimura, S. Uemura, J. Am. Chem. Soc. *122*, 12049-50 (2000).

Nitrogen ↑   NC ⇈ N

Without additional reagents   w.a.r.
Richter-type cinnoline synthesis
s. *58*, 187; cinnolines from *o*-acetylenetriazenes under forcing conditions s. D.B. Kimball, M.M. Haley, Org. Lett. *2*, 3825-7 (2000).

Microwaves s. under KHSO$_5$ ←

Mercuric perchlorate/sodium tetrahydridoborate   HgClO$_4$/NaBH$_4$
**Cyclic sec. amines from ethyleneazides
via mercury-promoted intramolecular Schmidt reaction**

172.

A soln. of 1-azidododec-5-ene in THF added dropwise to a soln. of ca. 4.5 eqs. Hg(II)-perchlorate trihydrate in the same solvent at room temp., the mixture stirred for 45 min during which gas evolution was observed, a soln. of 13.4 eqs. NaBH$_4$ in 15% aq. NaOH added, and stirring continued for a further 2 h → 2-heptylpiperidine. Y 73%. This process is superior to acid-promoted Schmidt cyclization in that it is milder, tolerates acid-sensitive functionalities, and avoids some of the limitations of the carbocation method, such as rearrangement prior to the ring closure. F.e., incl. double ring closures s. W.H. Pearson et al., J. Org. Chem. *65*, 8326-32 (2000).

Potassium peroxymonosulfate/alumina/microwaves   ←
Nitriles from hydrazones   CH=NN< → CN
with H$_2$O$_2$ cf. *22*, 408; rapid procedure with Oxone under microwave irradiation on Al$_2$O$_3$ in the absence of solvent s. T. Ramalingam et al., Synth. Commun. *30*, 4507-12 (2000).

Ferrous sulfate/ammonia   FeSO$_4$/NH$_3$
Lactams from azidocarboxylic acid esters
Pyrrolo[2,1-*c*][1,4]benzodiazepines s. *52*, 13s60

Halogen ↑   NC ⇈ Hal

Sodium hydride   NaH
Intramolecular N-alkylation
N-carbalkoxy-5-isoxazolidones cf. *50*, 257; N-carbalkoxymorpholines s. T. Takemoto, T. Nishi et al., Tetrahedron Lett. *41*, 1785-8 (2000).

Formation of Hal-Hal Bond

Uptake ⇓

Addition to Halogen HalHal ⇓ Hal

Sodium persulfate/hydrogen chloride $Na_2S_2O_8/HCl$
Iododichlorides from iodides I → ICl_2
with CrO_3/HCl cf. *35*, 182s*57*; with $Na_2S_2O_8$/HCl, notably from deactivated ar. iodides, s. A. Baranowski et al., J. Chem. Res., Synop *2000*, 435-7.

Formation of Hal-C Bond

Uptake ⇓

Addition to Oxygen and Carbon HalC ⇓ OC

Lithium bromide/Amberlist 15 ←
1,2-Bromhydrins from epoxides ⌄o⌄ → C(Br)C(OH)
with Amberlyst 15 cf. *46*, 429s*47*, *48*; regio- and stereo-specific formation of 3-ethylene-2,1-bromhydrins from 3-ethyleneepoxides with Amberlite 15, **also 1,3-dienes** with LiI, s. R. Antonioletti, G. Righi et al., Tetrahedron Lett. *41*, 9315-8 (2000).

Silicon tetrachloride/chiral bicyclic o-methoxyphenylphosphonic acid diamides ←
1,2-Chlorhydrins from epoxides with desymmetrization ⌄o⌄ → C(Cl)C(OH)

A soln. of cyclooctene oxide in methylene chloride (1.2 mmol scale) treated at -78° with *freshly distilled* $SiCl_4$ in the presence of 10 mol% P_S-chiral bicyclic o-methoxyphenylphosphonic acid diamide for 4 h, and worked up with KF/KH_2PO_4 → product. Y 77% (e.e. >99%). Enantioselectivity is considerably higher than by prior art (cf. *44*, 403s*57*). The o-methoxy group appears critical for binding $SiCl_4$ prior to complexation with the epoxide. F.e.s. J.M. Brunel, G. Buono et al., Angew. Chem. Int. Ed. Engl. *39*, 2554-7 (2000).

Trihalogeno-vanadium or -chromium imide complexes ←
1,2-Halogenhydrins from epoxides ⌄o⌄ → C(Hal)C(OH)
with VCl_3 cf. *48*, 423; regio- and stereo-specific ring opening of benzyl 2,3- and 3,4-epoxyethers with [p-TolN=VCl_3] or [t-BuN=Cr(Hal)$_3$·DME] (Hal = Cl or Br) in stoichiometric amount or in catalytic amount (with Me_3SiCl or Me_3SiBr) s. W.-H. Leung, L.-L. Yeung et al., Synlett *2000*, 677-9.

Manganese(II) halides $MnHal_2$
Acoxyhalides from cyclic ethers C̃
4-acoxyiodides with GaI_3 cf. *43*, 402s*58*; 4-acoxy-chlorides, -bromides and -iodides with the appropriate Mn(II)-halides, also with metal halides, such as LiCl, $ZnCl_2$, $CuCl_2$, $CoCl_2$ or $FeBr_2$ s. J. Kang, S.-H. Kim, Bull. Korean Chem. Soc. *21*, 611-2 (2000).

Ferric trifluoroacetate/[quaternary] ammonium halides
1,2-Halogenhydrins from epoxides ← ; $\triangledown_O \to C(Hal)C(OH)$
with $FeCl_3/SiO_2$ cf. *50*, 55s52; with a little $Fe(OCOCF_3)_3$ and [quaternary] ammonium halides s. N. Iranpoor, H. Adibi, Bull. Chem. Soc. Jpn. *73*, 675-80 (2000).

Addition to Carbon-Carbon Bonds   HalC ⇓ CC

Without additional reagents w.a.r.
1,2-Dibromides from ethylene derivs. $C=C \to C(Br)C(Br)$
s. *3*, 3; with addition of allyloxy-substituted Perloza MT-100 resin as scavenger of the Br_2 excess s. A. Chesney, P.G. Steel et al., J. Comb. Chem. *2*, 434-7 (2000).

(E)-α,β-Ethylene-β-tosyloxyiodides $C\equiv C \to C(OTs)=C(I)$
from acetylene derivs. s. *55*, 183s60

Lithium acetate LiOAc
2-Ethylene-2,1-halogenhydrins from allenes $C=C=C \to C=C(Hal)C(OH)$
with $Ca(OCl)_2$ cf. *27*, 555; 2-ethylene-2,1-iodohydrins with I_2/LiOAc in aq. acetonitrile, regio- and stereo-specific preparation of (E)-α,β-ethylene-γ-iodosulfoxides, s. S. Ma et al., Org. Lett. *2*, 3893-5 (2000).

Sodium azide/sodium iodide/ammonium ceric nitrate ←
1,2-Azidoiodides from ethylene derivs. $C=C \to C(I)C(N_3)$
with NaN_3/CaF_2 cf. *43*, 456; regiospecific conversion under mild conditions with NaN_3/NaI/CAN, also preferential conversion of a dienone, s. V. Nair et al., Synlett *2000*, 1597-8; with $Me_3SiN_3/Et_4NI/PhI(OAc)_2$ s. A. Kirsching et al., J. Org. Chem. *64*, 6522-5 (1999).

Lithium chloride/acetic acid LiCl/AcOH
(Z)-2-(Alkylseleno))-1,3-enynes from 1-acetylene-1-selenides $CH=C(Cl)SeR$
via Sonogashira coupling with (Z)-1-chloreneselenides

A mixture of startg. selenide and 1.2 eqs. LiCl in acetic acid stirred at room temp. for 2 h, and quenched with water → (1Z)-1-chloro-1-(methylseleno)pent-1-ene (Y 84%; Z/E 100/0), added to 0.5 eq. $PdCl_2(PhCN)_2$ and 10 mol% CuI in piperidine at room temp. under N_2, stirred for 10 min, 2 eqs. startg. alkyne added, and worked up when TLC indicated completion of reaction → (4Z)-5-(methylseleno)dodec-4-en-6-yne (Y 86%). F.e.s. A. Sun, X. Huang, Synthesis *2000*, 1819-21.

Cuprous trifluoromethanesulfonate CuOTf
2-Chlorosulfonylamines from ethylene derivs. $C=C \to C(Cl)C(NHSO_2R)$
and N,N-dichloro-*p*-toluenesulfonamide/$ZnCl_2$ cf. *24*, 555s58; *anti*-2-chlorosulfonylamines with N,N-dichloro-2-nitrobenzenesulfonamide and a little Cu(OTf)·PhH s. G. Li et al., Org. Lett. *2*, 2249-52 (2000).

Ammonium ceric nitrate s. under NaN_3 $(NH_4)_2Ce(NO_3)_6$
Phenyl iodosoacetate s. under Me_3SiN_3 $PhI(OAc)_2$

N-Bromosuccinimide NBS
1,2-Bromhydrins from ethylene derivs. $C=C \to C(Br)C(OH)$
s. *8*, 272; regiospecific formation from γ,δ-ethylene-β-hydroxysulfoxides with asym. induction s. S. Raghavan et al., Chem. Commun. *1999*, 1845-6.

N-Iodosuccinimide NIS
2-Iodo-2-deoxyaldoses from glycals s. *60*, 64   C≡C → C(I)C(OH)

N-Chloromethyl-N'-fluoro-1,4-diazoniabicyclo[2.2.2]octane   ←
2-Functionalized fluorides from ethylene derivs.   ←
carbohydrate derivs. from glycals s. *49*, 407s*55*; 2-fluoro-2-deoxyglycosyl azides and conversion to α-(2-fluoro-2-deoxyglycosylamino)carboxylic acid derivs. s. M. Albert, K. Dax et al., Synlett *1999*, 1483-5.

Trimethylsilyl azide/phenyl iodosoacetate/tetraethylammonium iodide   ←
1,2-Azidoiodides from ethylene derivs. s. *43*, 456s*60*   C≡C → C(I)C(N₃)

Trimethylsilyl halides/methanol   Me₃SiHal/MeOH
Regio- and stereo-specific synthesis and reactions of 1-halogenenolethers   CH=C(Hal)OR
from alkoxyacetylenes

175.

Stille coupling with inversion of configuration. 0.99 eq. Me₃SiBr added to a soln. of startg. alkoxyacetylene and 0.99 eq. methanol in methylene chloride at -40° for 10 min → intermediate 1-bromoenolether (Y 99%), in acetonitrile refluxed with tributylvinyltin in the presence of a little PdCl₂(PPh₃)₂ until reaction complete → product (Y 91%). The procedure is notably of value for preparing β-*alkyl*-1-halogenenolethers (which are difficult to prepare by other means) and reliant on the controlled *in situ*-generation of HHal. These products can be stored in frozen benzene for a month and may be readily converted to **1-lithioenolethers as acyl carbanion equivalents** by treatment with *n*-BuLi or *t*-BuLi. These in turn can be quenched with electrophiles, or can be converted to lower-order, higher-order or mixed higher-order cuprates. F.e. and electrophiles, also further transition metal-catalyzed conversions, e.g. Sonogashira coupling (with retention of configuration) and carbonylation (with inversion) s. W. Yu, Z. Jin, J. Am. Chem. Soc. *122*, 9840-1 (2000).

Rearrangement   ∩

Nitrogen/Halogen Type   HalC ∩ NHal

Cuprous chloride   CuCl
Cyclic 1,2-chloramines from ethylene-N-chloramines   ○
with TiCl₄/TiCl₃ cf. *27*, 530; regio- and stereo-specific conversion with 10 mol% CuCl or 1 eq. CuCl₂, 3-chloro- and 2-α-chloro-piperidines, s. R. Göttlich, Synthesis *2000*, 1561-4.

Exchange   ↕

Hydrogen ↑   HalC ↕ H

Electrolysis   ⚡
Diaryliodonium salts from ar. iodides and arenes   Ar₂I⁺
with CrO₃/Ac₂O/H₂SO₄ cf. *51*, 204; electrochemical route s. M.J. Peacock, D. Pletcher, Tetrahedron Lett. *41*, 8995-8 (2000).

Potassium carbonate K_2CO_3
Bromoketones from cyclic tert. alcohols
Retro-Barbier fragmentation

176.

under mild conditions. 6 eqs. K_2CO_3 added in 1 eq. portions to a soln. of 1-methylcyclopentanol in chloroform, stirred at 0° for 10 min, 5 eqs. Br_2 added in 1 eq. portions, and stirring continued for 5 h → product. Y 97%. This is the first example of a *direct* retro-Barbier fragmentation. The method is clean, high-yielding, and generally applicable to aliphatic and aryl ketone derivs. F.e. and with Na_2CO_3, KOH or NaOH s. W.-C. Zhang, C.-J. Li, J. Org. Chem. 65, 5831-3 (2000).

Sodium hydrogen sulfite s. under $NaIO_4$ $NaHSO_3$

Cupric bromide $CuBr_2$
4-Halogeno-5-hydroxy-Δ³-2-pyrrolones from α-allenecarboxylic acid amides
Oxidative halogenolactamization

177.

A mixture of startg. alleneamide and 2 eqs. $CuBr_2$ in THF/H_2O (1:1) refluxed for 24 h, and quenched with satd. NH_4Cl → product. Y 94%. An unexpected facile oxidation follows the initial bromolactamization. F.e.s. S. Ma, H. Xie, Org. Lett. 2, 3801-3 (2000).

Boron fluoride s. under KI BF_3

1-Aryloxy-1,2-benziodoxol-3(1H)-ones ←
Ar. iodination H → I
with 1-tosyloxy-1,2-benziodoxol-3(1*H*)-one cf. 55, 183; comparison of 1-aryloxy-1,2-benz-iodoxol-3(1*H*)-ones (with special reference to the *p*-chloro-analog), also ar. chlorination and bromination with added LiCl or LiBr, and (E)-α,β-ethylene-β-tosyloxyiodides from acetylene derivs., s. T. Muraki, H. Togo et al., J. Org. Chem. 64, 2883-9 (1999).

N-Bromosuccinimide/hydrogen chloride NBS/HCl
Acid-catalyzed ar. bromination H → Br
in H_2SO_4/CF_3COOH cf. 20, 413s58; of activated arenes in acetone with 1 *M* HCl s. B. Andersh et al., Synth. Commun. 30, 2091-8 (2000).

N-Iodosuccinimide NIS
Halogeno-O-heterocyclics from ethylenealcohols O
iodo-O-heterocyclics s. 33, 477s51; 3-hydroxy-4-iodotetrahydrofurans from 3-ene-1,2-diols (or diol monoethers) with $I_2/NaHCO_3$, stereoselectivity, s. S.P. Beur, D.W. Knight et al., Tetrahedron Lett. 41, 4447-51 (2000); medium-sized 2-ethylene-α-iodolactolides (with NIS) s. H. Nogano et al., Synlett 2000, 1193-5; 2-subst. 3-iodotetrahydrofurans from 3-ethylenealcohols with $NaIO_4$/$NaHSO_3$ s. Y. Okimoto, Y. Ishii et al., Tetrahedron Lett. 41, 10223-7 (2000).

2,4,6,8-Tetraiodoglycoluril/sulfuric acid ←
Ar. iodination of deactivated arenes under mild conditions H → I

178.

4-Fluorobenzoic acid added in one portion to a well-stirred soln. of 0.5 eq. 2,4,6,8-tetraiodo-glycoluril in 90% sulfuric acid at 0°, and poured into ice-cold water after 1 h → product. Y 82%.

Deactivated arenes with electron-withdrawing substituents reacted readily and predictably with complete conversion. Other isomers were produced in some cases in minor quantities. F.e.s. V.K. Chaikovski, V.D. Filimonov et al., Tetrahedron Lett. *41*, 9101-4 (2000).

2,2-Dimethyl-α,α,α',α'-tetrakis(naphth-1-yl)-1,3-dioxolane-(4R,5R)-dimethanolato- ←
titanium dichloride/N'-chloromethyl-N-fluoro-1,4-diazoniabicyclo[2.2.2]octane
bis(fluoroborate)
Asym. α-halogenation of β-ketocarboxylic acid esters   H → F

179.

Asym. α-fluorination. 5 Mol% 2,2-dimethyl-α,α,α',α'-tetrakis(naphth-1-yl)-1,3-dioxolane-(4R,5R)-dimethanolatotitanium dichloride added to a soln. of the startg. β-ketoester in acetonitrile, the mixture stirred for 5 min, a soln. of 1.15 eqs. 0.145 *M* Selectfluor in acetonitrile added, stirring continued for 7 min, and worked up chromatographically → (+)-product. Y 82% (e.e. 70.8%). Asym. α-fluorination is achieved not by using a chiral fluorinating agent but by employing a chiral Lewis acid which induces formation of a chiral Ti(IV)-enolate for subsequent fluorination by the non-chiral reagent. Reaction is fast with good yields and enantioselectivity. F.e.s. L. Hintermann, A. Togni, Angew. Chem. Int. Ed. Engl. *39*, 4359-62 (2000); **asym. α-chlorination and α-bromination** with N-chloro- and N-bromo-succinimide in the presence of 5 mol% 2,2-dimethyl-α,α,α',α'-tetrakis(naphth-2-yl)-1,3-dioxolane-(4R,5R)-dimethanolatotitanium dichloride s. Helv. Chim. Acta *83*, 2425-35 (2000).

Chiral N-fluorosultams/lithium bis(trimethylsilyl)amide ←
Asym. α-fluorination
of ketones with chiral 3-cyclohexyl-2-fluoro-3-methyl-2,3-dihydrobenz[1,2-*d*]isothiazole 1,1-dioxide cf. *41*, 463s59; with (11S,12R,14R)-2-fluoro-14-methyl-11-(methylethyl)spiro[4*H*-benzo[*e*]-1,2-thiazine-3,2'-cyclohexane]-1,1-dione s. Z. Liu, Y. Takeuchi et al., J. Org. Chem. *65*, 7583-7 (2000).

N-Fluorobenzenesulfonimide ←
α-Fluorination
s. *39*, 458s58; of cyclic benzylphosphonic acid diamides and benzylphosphonamidates with asym. induction s. C.C. Kotoris, S.D. Taylor et al., J. Chem. Soc. Perkin Trans. 1 *2000*, 1271-81.

Iodine monochloride ICl
Halogenolactonization ○
update s. *33, 477s57*; 5-alkylidene-2(5*H*)-furanones from (2Z,4E)-2,4-dienecarboxylic acids via 5-α-halogeno-2(5*H*)-furanones (with ICl) s. S. Rousset et al., Synlett *2000*, 260-2.

Iodine monochloride/ferrocenium tetrakis[3,5-bis(trifluoromethyl)phenyl]borate/zinc oxide ←
Ar. iodination   H → I
with ICl cf. *1*, 419; at the 1.1-2 molar level under mild conditions with added ferrocenium tetrakis[3,5-bis(trifluoromethyl)phenyl]borate as Lewis acid (5 mol%) in the presence of DDQ or ZnO s. T. Mukaiyama et al., Tetrahedron Lett. *41*, 9383-6 (2000).

Iodic acid HIO_3
Sodium iodate or periodate/acetic acid/acetic anhydride/sulfuric acid ←
Ar. iodination
with $NaIO_3$ cf. *47*, 440; of both activated and deactivated arenes with I_2 and $NaIO_3$ or $NaIO_4$ in $AcOH/Ac_2O$ containing 98% H_2SO_4 s. P. Lulinski, L. Skulski, Bull. Chem. Soc. Jpn. *73*, 951-6 (2000); iodination of hydroxyacetophenones with I_2/HIO_3 (cf. *48*, 444) s. B.S. Dawane, Y.B. Vibhute, J. Indian. Chem. Soc. *77*, 299-300 (2000).

Sodium periodate/sodium hydrogen sulfite $NaIO_4/NaHSO_3$
3-Iodotetrahydrofurans from 3-ethylenealcohols s. *33*, 477s60 ○

Bis(collidine)bromine(I) hexafluorophosphate ←
Halogeno-N-heterocyclics by intramolecular halogenamination
s. *48*, 434s57; 3-bromo-1-tosylazetidines, stereoselectivity, s. S. Robin, G. Rouseau, Eur. J. Org. Chem. *2000*, 3007-11; N-condensed 3-iodo-1,2,3,4-tetrahydropyridines (with NIS or I_2/NaHCO$_3$) s. R. Lavilla et al., ibid. *1999*, 2997-3003; N-condensed bromo-subst. thiazolidines and tetrahydro-1,3-thiazines s. P. Wippich, M. Gütschow et al., Synthesis *2000*, 714-20; 3,4-dihydro-2,1-benzothiazine 2,2-dioxides from β-arylsulfonic acid amides with PhI(OAc)$_2$/I$_2$ under irradiation s. H. Togo, M. Yokoyama et al., J. Org. Chem. *65*, 926-9 (2000).

Oxygen ↑   HalC ↓↑ O

Electrolysis ↯
1,1-Dihalides from aldehydes s. *29*, 518s60   CHO → CHHal$_2$

Sodium iodide s. under CeCl$_3$ NaI
Potassium iodide s. under BF$_3$ KI

Boron fluoride/potassium iodide BF_3/KI
Cerium(III) chloride/sodium iodide $CeCl_3$/NaI
Iodides from alcohols under mild conditions   OH → I

180.   PhCH$_2$OH   ⟶   PhCH$_2$I

1.5 eqs. CeCl$_3$·7H$_2$O added to a stirred suspension of benzyl alcohol and 1.2 eqs. NaI in acetonitrile, stirred for 20 h at reflux, diluted with ether, and treated with 0.5 *N* HCl → benzyl iodide. Y 90%. This method, which is applicable to a wide range of alcohols, is simple, eco-friendly, and efficient. Furthermore, there was no elimination (of prim. or sec. iodides) or rearrangement (in contrast to the reaction of sec. alcohols with HI). However, tert. alcohols and β-hydroxycarbonyl compds. did undergo competing elimination. THP ethers, esters, and NO$_2$ groups were unaffected, but the procedure is not compatible with acetals, trialkylsilyl groups, and methoxymethyl and *p*-methoxybenzyl ethers. Significantly, substituted allyl alcohols (with an internal unsaturation) reacted *without* double bond shift or isomerization; however, allyl alcohols having a terminal double bond did undergo allyl shift. F.e.s. M. Di Deo, G. Bartoli et al., J. Org. Chem. *65*, 2830-3 (2000); allyl and benzyl iodides with BF$_3$-etherate/KI in dioxane s. B.P. Bandgar et al., Tetrahedron Lett. *42*, 951-3 (2001).

Dimethylformamide/carbon tetrahalides/copper/nickel ←
1,1-Dihalides from aldehydes   CHO → CHHal$_2$
1,1-dichlorides with DMF/SOCl$_2$ cf. *29*, 518s*34*; also 1,1-dibromides with DMF/CCl$_4$ or CBr$_4$ in the presence of Cu- and Ni-powder, and by electrolysis with a copper rod as anode and a concentric nickel grid cylinder as cathode, s. E. Léonel, J.-P. Paugam et al., Synth. Commun. *29*, 4015-24 (1999).

Carbon tetrahalides s. under DMF and HMPA   CHal$_4$

Bis(trichloromethyl) carbonate/pyridine $(Cl_3CO)_2CO$/C_5H_5N
Glycosyl chlorides from aldoses   OH → Cl

181.

0.4 eq. Bis(trichloromethyl) carbonate added to a soln. of startg. aldose (1 mmol) in dry THF, the mixture stirred at room temp. with exclusion of moisture, pyridine (0.1 mL) added in three portions, and the mixture stirred at room temp. for 2-4 h → product. Y 87% (α-anomer only).

The process is based on a commercially available reagent and easily available starting materials. It is compatible with acid-labile groups, such as isopropylidene acetals, as well as benzyl ethers, but is less efficient with partially protected acetates. F.e.s. R.M. Cicchillo, P. Norris, Carbohydr. Res. *328*, 431-4 (2000).

N-Chlorodiisopropylamine s. under Ph₃P	*i-Pr₂NCl*
N-Iodosuccinimide s. under HF-py	*NIS*
2-Chloro-1,3-dimethyl-Δ²-imidazolinium chloride/triethylamine	←
Replacement of hydroxyl groups by chlorine	OH → Cl

182.
$$CH_2=CH(CH_2)_7CH_2OH \longrightarrow CH_2=CH(CH_2)_7CH_2Cl$$
(via 2-chloro-1,3-dimethylimidazolinium chloride, 3 h) (Y 84%)

Prim. chlorides. A soln. of startg. prim. alcohol in methylene chloride treated with 1 eq. 2-chloro-1,3-dimethyl-Δ²-imidazolinium chloride [DMC] and 1 eq. Et₃N at room temp. for 21 h → product. Y 99%. There was no reaction with sec. alcohols so that **selective conversions** are feasible. Acid-sensitive Cbz groups, ethylenic unsaturations, and chlorine were unaffected. F.e., also **β-chloro-α,β-ethyleneketones from β-diketones**, and further reactions s. T. Isobe, T. Ishikawa, J. Org. Chem. *64*, 5832-5 (1999).

Triphenylphosphine/pyridine	*Ph₃P/C₅H₅N*
Triphenylphosphine/N-chlorodiisopropylamine	*Ph₃P/i-Pr₂NCl*
1,2-Bis(diphenylphosphino)ethane	*Ph₂PCH₂CH₂PPh₂*
Hexamethylphosphoramide/carbon tetrahalides	*HMPA/CHal₄*
Replacement of hydroxyl groups by halogen	OH → Hal

with Ph₃P/CCl₄ cf. *21*, 606; 5-chloro-5-deoxynucleosides with Ph₃P/i-Pr₂NCl s. Y.-S. Zhou, Y.-F. Zhao et al., Synlett 2000, 671-3; 3-bromooxepans with (n-C₈H₁₇)₃P/CBr₄ or dppe·2Br₂ s. K. Fujiwara, A. Murai et al., ibid. 1187-9; chiral 2,3-epoxyiodides with Ph₃P/py/I₂ s. Z. Liu, Y. Li et al., Tetrahedron:Asym. *9*, 3755-62 (1998); 6-chloro- and 6-bromo-inosine derivs. with HMPA/CCl₄ or CBr₄ (or NBS) s. E.A. Véliz, P.A. Beal, Tetrahedron Lett. *41*, 1695-7 (2000).

Sulfur tetrafluoride	*SF₄*
1,1,1-Trifluorides from carboxylic acids	COOH → CF₃

with SF₄/TiF₄ cf. *16*, 606; without TiF₄, hydrolysis of the products to ω-trifluoromethyl-α-functionalized carboxylic acids, s. H. Schedel, K. Burger et al., Synthesis 2000, 1681-8.

Hydrogen fluoride-pyridine/N-iodosuccinimide	←
Fluorides from xanthates	OC(S)SR → F

with p-MeC₆H₄IF₂ cf. *51*, 205; with HF-py/NIS s. K. Kanie, T. Hayama et al., Bull. Chem. Soc. Jpn. *73*, 471-84 (2000).

Nitrogen ↑   HalC ↓↑ N

Pyridine	*C₅H₅N*
N-(4-Chloro-1,2,3-dithiazol-5-ylidene)-1,1-chlorohydrazones from tetrazoles s. *60*, 205 C.	
Nitrosyl fluoroborate or hexafluorophosphate	*NOBF₄ or NOPF₆*
Balz-Schiemann reaction in ionic liquids	ArNH₂ → ArN₂⁺BF₄⁻ or PF₆⁻ → ArF

183.
$$\text{ArNH}_2 \xrightarrow{NOBF_4} [\text{ArN}_2^+ BF_4^-] \longrightarrow \text{ArF}$$
(mesityl example)

Ar. fluorides from prim. amines in one pot. Startg. aniline added slowly to 1 eq. NOBF₄ in 1-ethyl-3-methylimidazolium fluoroborate at 0° (ice bath), the mixture stirred at 0° for 0.5 h then at room temp. for 12 h (monitored by ^{1}H NMR), when diazotization complete the mixture heated until gas evolution observed, kept at this temp. (70°) for 1-3 h (monitored by ^{1}H NMR), 1 eq. i-Pr$_2$NEt added, and extracted with ether → product. Y 100% by ^{1}H NMR. This method overcomes many of the drawbacks of the classical method, giving high yields without the need for aq. work-up. The ionic medium may be recycled (with or without extraction of [i-Pr$_2$NHEt][BF$_4$]/[PF$_6$] with propyl acetate), which may mean the method is economical and environmentally friendly on the large scale. Use of [emim][CF$_3$CO$_2$], [emim][OTs] or [emim][OTf] gave the corresponding aryl trifluoroacetates or sulfonates. F.e. incl. reaction in the presence of ar. methoxy, halogen (Cl, Br), nitro or cyano groups with pre-formed diazonium salts s. K.K. Laali, V.J. Gettwert, J. Fluorine Chem. *107*, 31-4 (2001).

Ferrous chloride/trimethylsilyl chloride $\qquad$ *FeCl$_2$/Me$_3$SiCl*
4-α-Chloro-2-oxazolidones from 2-ethyleneazidoformic acid esters $\qquad$ ○
s. *59*, 195; (Z)-4-(α-chloroalkylidene)-2-oxazolidones from 2-acetyleneazidoformic acid esters s. T. Bach et al., Synlett *2000*, 1330-2.

Halogen ↑ $\qquad\qquad$ HalC ↓↑ Hal

Polymer-based cyclic iminophosphoric acid amides/benzoylquinine $\qquad$ ←
α-Halogenocarboxylic acid esters from carboxylic acid chlorides $\quad$ CHCOCl → C(Hal)COOR
by asym. α-halogenation-esterification

184.

A soln. of startg. acyl chloride in THF passed through an addition funnel containing polymer-based cyclic iminophosphoric acid amide at -78°, the resulting soln. of phenylketene added dropwise at the same temp. to 10 mol% benzoylquinine and 1 eq. hexachloro-2,4-cyclo-hexadienone, stirred at -78° for 4 h, quenched, and worked up chromatographically → (S)-product. Y 80% (e.e. 99%). Reaction is presumed to involve asym. halogenation of an intermediate zwitterionic enolate. The procedure is inexpensive. N-Halogenosuccinimides and alkyl hypochlorites were ineffective as halogen sources, and yields were lower with Proton Sponge as base. The enantiomeric product was obtained with 'pseudoenantiomeric' benzoylquinidine as chiral reagent. F.e. and chiral bromo derivs. with 2,4,4,6-tetrabromo-2,5-cyclohexadienone s. H. Wack, T. Lectka et al., J. Am. Chem. Soc. *123*, 1531-2 (2001).

Sulfur ↑ $\qquad\qquad$ HalC ↓↑ S

1,3-Dibromo-5,5-dimethylhydantoin s. under HF-py $\qquad$ ←

Chlorodiphenylsulfonium chloride [Ph$_2$SCl]$^+$Cl$^-$
Glycosyl chlorides from thioglycosides via glycosyl(chloro)sulfonium chlorides SR → Cl

185.

A soln. of 3 eqs. diphenyl sulfoxide in dry dichloromethane added dropwise to a stirred soln. of 2 eqs. oxalyl chloride in the same solvent at -78°, after 5 min (when the evolution of gas ceased), the resulting chlorodiphenylsulfonium chloride was placed under argon, treated dropwise with a soln. of startg. thioglycoside in the same solvent, then the mixture allowed to attain room temp. over a period of 1.5 h → β-chloride. Y 92%. This mild method avoids the handling of chlorine gas, while incorporation of the 4-chlorophenylthio group minimizes the generation of volatile sulfur-containing by-products. The procedure is amenable to scale-up affording a wide range of glycosyl chlorides in high yields. Groups unaffected by the reaction conditions include phthalimido, esters, acetals, azides and ethers. F.e. and stereoselectivity s. S. Sugiyama, J.M. Diakur, Org. Lett. *2*, 2713-5 (2000).

Iodine monohalides *IHal*
Glycosyl halides from thioglycosides SR → Hal
α-glycosyl bromides with IBr s. *17*, 630s*54*; also β-glycosyl chlorides with ICl, and epimerization of 'armed' thioglycosides with I$_2$, s. K.P.R. Kartha et al., Tetrahedron:Asym. *11*, 581-93 (2000).

Tetra-n-butylammonium dihydrogentrifluoride [Bu$_4$N][H$_2$F$_3$]
Hydrogen fluoride-pyridine/1,3-dibromo-5,5-dimethylhydantoin ←
Trifluoromethyl ethers from xanthates OCSSR → OCF$_3$
with HF-py/NBS s. *51*, 205s*53*; with HF-py/1,3-dibromo-5,5-dimethylhydantoin, also alkylthio-(difluoro)methyl ethers with [Bu$_4$N][H$_2$F$_3$], s. K. Kanie, T. Hayama et al., Bull. Chem. Soc. Jpn. *73*, 471-84 (2000).

Remaining Elements ↑ HalC ↓↑ Rem

N-Fluoro-O-(4-chlorobenzoyl)dihydroquininium fluoroborate ←
Asym. C-α fluorination of ketones via enoxysilanes C=C(OSi≤) → CFCO

186.

Startg. silyl enolether (readily obtained from 2-benzyl-1-indanone) added to a soln. of N-fluoro-O-(4-chlorobenzoyl)dihydroquininium fluoroborate (prepared *in situ* by stirring equimolar amounts of dihydroquinine 4-chlorobenzoate and Selectfluor in dry acetonitrile containing 3 Å molecular sieves at room temp. for 1 h) at -20° under N$_2$, and stirred overnight → product. Y 99% (e.e. 89%). Acyclic products were obtained with dihydroquinine acetate instead of the benzoate at -80°. F.e., also asym. α-fluorination of β-keto and α-cyano esters, s. N. Shibata, Y. Takeuchi et al., J. Am. Chem. Soc. *122*, 10728-9 (2000); asym. α-fluorination of cyclic ketones with N-fluorocinchonidinium fluoroborate/NaH cf. D. Cahard et al., Org. Lett. *2*, 3699-701 (2000).

Carbon ↑   HalC ↓↑ C

Phenyl iodosoacetate $PhI(OAc)_2$
Cyclic 2-acylamino-1,2-iodohydrins from α-acylaminocarboxylic acids   ←
Stereospecific conversion via cyclic N-acylimmonium salts

187.

A soln. of startg. chiral cyclic α-acylamino acid in acetonitrile treated with 2 eqs. phenyl iodoso-acetate and 2 eqs. I_2 at room temp. under N_2 with exposure to sunlight for 5 h, and poured into aq. $Na_2S_2O_3$ → (±)-product. Y 64%. Reaction is presumed to involve initial formation of a cyclic acylimmonium ion, followed by isomerization, iodination and attack of nucleophile (water). Only *trans*-2,3-disubst. derivs. were obtained since the bulky iodine atom hinders approach of water from the same face. F.e. incl. piperidine analogs and 2-methoxy derivs. (with methanol as nucleophile) s. A. Boto, E. Suárez et al., Tetrahedron Lett. *41*, 2495-8 (2000).

Hydrogen fluoride-pyridine $HF\text{-}C_5H_5N$
2-Halogenamines from 2-oxazolidones   C̰
2-bromamines with HBr/AcOH cf. *23*, 570; chiral 2-fluoramines with HF-py s. D. O'Hagan et al., Tetrahedron:Asym. *11*, 2033-6 (2000).

Formation of S-S Bond

Exchange   ↓↑

Hydrogen ↑   SS ↓↑ H

Sulfur dichloride SCl_2
Sym. disulfides from mercaptans $2\ RSH \rightarrow (RS)_2$
with S_2Cl_2 cf. *11*, 660; bis(2- and 3-pyrrolyl) disulfides with SCl_2 s. Q. Chen, D. Dolphin, Synthesis *2001*, 40-2.

Benzyltriphenylphosphonium dichromate   ←
Sym. disulfides from mercaptans
with nicotinium dichromate cf. *41*, 241; with benzyltriphenylphosphonium dichromate s. A.R. Hajipour et al., Org. Prep. Proced. Int. *31*, 335-40 (1999).

Sulfur ↑   SS ↓↑ S

Nitrogen monoxide/air NO/O_2
Disulfide interchange   RSSR'
with $PhI(OAc)_2$ cf. *47*, 470; with NO/air s. N. Tsutsumi, T. Itoh et al., Chem. Pharm. Bull. *48*, 1524-8 (2000).

Carbon ↑   SS ↓↑ C

Samarium diiodide/hexamethylphosphoramide   Sm/HMPA
Sym. disulfides from thiolic acid esters $2\ COSR \rightarrow (RS)_2$
with clay-supported NH_4NO_3 under microwave irradiation cf. *51*, 216s57; under mild, neutral conditions with SmI_2/HMPA s. B.W. Yoo et al., Synth. Commun. *30*, 4317-22 (2000).

Elimination ⇑

Carbon ↑   SS ↑ C

N-*Chlorosuccinimide/dimethyl sulfide* NCS/Me_2S
Polymer-based synthesis of cyclic disulfides from mercaptothioethers

188.

under mild conditions. 4 eqs. 0.08 M NCS and 5 eqs. dimethyl sulfide in methylene chloride allowed to react at 0° for 10 min, startg. Wang-supported mercaptothioether added, and allowed to react for a further 2 h at 0° then at room temp. for 2 h → product. Y 79% (from the Wang resin). The procedure is applicable to 6- to 14-membered cyclic disulfides and is compatible with the peptide bond. F.e.s. T. Zoller et al., Tetrahedron Lett. *41*, 9989-92 (2000).

Formation of S-Rem Bond

Uptake ⇓

Addition to Hydrogen and Remaining Elements   SRem ⇓ HRem

*Carbon oxide sulfide/di-*tert-*butyl hyponitrite* $COS/(t\text{-}BuO)_2N_2$
Silanethiols from organosilicon hydrides   ≽SiH → ≽SiSH

189.

1.9 eqs. Carbon oxide sulfide introduced as a slow steam of bubbles to a stirred soln. of α-naphthyldiphenylsilane and 5 mol% di-*tert*-butyl hyponitrite in dry dioxane, then the mixture heated in an oil bath at 60° for 2.5 h under argon → α-naphthyldiphenylsilanethiol. Y 83%. The method is mild, convenient and clean. Yields were improved in some instances by carrying out the reaction under a COS pressure ca. 2.5 bar above atmospheric, or by use of AIBN or lauroyl peroxide in place of the hyponitrite. F.e., also germanethiols from organogermanium hydrides, and application of organosilicon hydrides as relatively cheap, environmentally-friendly alternatives to Bu_3SnH for Barton-McCombie-type deoxygenations via xanthates, s. Y.D. Cai, B.P. Roberts, Tetrahedron Lett. *42*, 763-6 (2001).

Exchange ⇕

Hydrogen ↑   SRem ⇕ H

Chlorotris(triphenylphosphine)rhodium(I) $RhCl(PPh_3)_3$
Silyl sulfides from mercaptans and organosilicon hydrides   SH → SSi≼
s. *29*, 530; stable thiopolysiloxanes s. B.P.S. Chauhan, P. Boudjouk, Tetrahedron Lett. *41*, 1127-30 (2000).

Oxygen ↑ SRem ↓↑ O

Diphenyl phosphorochloridate/pyridine $(PhO)_2P(O)Cl/C_5H_5N$
Diphenyl phosphorochloridate/2,4,6-trichlorophenol/hexamethyldisilthiane ←
Nucleoside H-thiophosphonates P-OH → P-SH
with *t*-BuCOCl/quinoline/Me$_3$SiSSiMe$_3$ cf. *52*, 208; with (PhO)$_2$P(O)Cl/2,4,6-trichlorophenol, also nucleoside H-dithiophosphonates with added H$_2$S/py, s. J. Cieslak, J. Stawinski et al., J. Org. Chem. *65*, 7049-54 (2000).

Halogen ↑ SRem ↓↑ Hal

Pyridine C_5H_5N
Monothiolphosphonic acid esters PO(OR)(SR')
from phosphonous acid esters and sulfenyl chlorides cf. *31*, 520; from mercaptans, alcohols and phosphonic dichlorides, nucleoside 3'-monothiolphosphonates, and conversion to dinucleoside phosphonates, s. J. Pyzowski, W.J. Stec et al., Tetrahedron Lett. *41*, 1223-6 (2000); nucleoside thiolthionophosphonates s. A. Chworos, W.J. Stec et al., ibid. 1219-22.

Remaining Elements ↑ SRem ↓↑ Rem

Samarium diiodide/sodium salt SmI$_2$/Na$^+$
Selenosulfides from diselenides RSeSeR → RSeSR'

190.   n-C$_8$H$_{17}$SSO$_3$Na + PhSeSePh ⟶ n-C$_8$H$_{17}$SSePh

and thiosulfuric acid salts. Startg. diselenide in THF added in one portion to 2 eqs. SmI$_2$ (freshly prepared from 2 eqs. samarium powder and 2 eqs. iodine in THF under N$_2$ at 40°) at room temp., 2 eqs. startg. thiosulfate added after 4 h to the resulting yellow soln., and the mixture stirred for a further 4 h → product. Y 79%. The method is simple and mild with good yields. F.e., also from arylsulfenyl chlorides under the same conditions, s. H. Guo, Y. Zhang, J. Chem. Res., Synop. *2000*, 374-5.

Formation of S-C Bond

Uptake ⇓

Addition to Carbon-Carbon Bonds SC ⇓ CC

n-Butyllithium/(1R,2R)-1-dimethylamino-2-(2-methoxyphenoxy)-1,2-diphenylethane ←
Asym. Michael addition of arylmercaptans C=C → CHC(SAr)
to α,β-ethylenecarboxylic acid esters
via *anti*-protonation of a catalytically generated β-(arylthio)carboxylic acid ester enolate

191.

1 Mol% n-Butyllithium in hexane added to a soln. of 1.2 eqs. 2-(trimethylsilyl)benzenethiol in 1:1 toluene/hexane, followed by a soln. of 1.2 mol% (1R,2R)-1-dimethylamino-2-(2-methoxyphenoxy)-1,2-diphenylethane in the same solvent at 0°, stirred for 30 min at room temp., a soln. of startg. enoate in the same solvent added at -78°, stirred for 30 min, and quenched with water → product. Y 99% (e.e. 92%) with 99% recovery of the ligand. Reaction is characterized by a rate-determining enantiofacial differentiating addition of the mercaptan to the enoate mediated by the chiral ligand, followed by rapid protonation with the same mercaptan (serving as *achiral* proton source) of the resulting enolate before it undergoes a conformational change. This is the first such enolate route based on a catalytic amount of a metal cation. α-Alkyl-subst. enoates required a higher temp. and a longer reaction time. F.e and desulfurization to give **chiral carboxylic acid esters** s. K. Nishimura, K. Tomioka et al., Angew. Chem. Int. Ed. Engl. *40*, 440-2 (2001).

Benzyltriethylammonium tetrathiomolybdate   [BnNEt$_3$]MoS$_4$
Michael addition to α,β-ethyleneketones s. *60*, 198   C=C → CHC(SR)

Rearrangement

Hydrogen/Carbon Type   SC ∩ HC

Potassium tert-*butoxide*   KOBu-t
1,4-Dithiafulvenes from 1,4-dithiins
Base-induced ring contraction

192.

2,5-Diphenyl-1,4-dithiin treated with 5 eqs. KOBu-*t* in THF at reflux for 0.5 h → product. Y 75%. For the ring contraction to be synthetically useful, an alkoxide with pK_a >19 is required, and the ion pairs must be sufficiently dissociated in the reaction medium (bases with smaller cations such as NaOBu-*t* or LiOBu-*t* being ineffective in THF in the absence of crown ether). The reaction is proposed to follow an E1cB mechanism, and the equilibrium between the intermediate anions is driven to the right by the protonation of the anion of the product, which is highly favoured due to the weaker acidity of dithiafulvenes. Moreover, the dithioles are more thermodynamically stable than dithiins. F.e. incl. benzo-fused analogs s. R. Andreu, J. Garín et al., Tetrahedron Lett. *42*, 875-7 (2001); isomerization of 1,4,5,8-tetrathianaphthalene to tetrathiafulvalene s. R.L. Meline, R.L. Elsenbaumer, J. Chem. Soc. Perkin Trans. 1 *1998*, 2467-9; with *i*-Pr$_2$NLi cf. idem., Synthesis *1997*, 617-9; S. Nakatsuji et al., Liebigs Ann. *1997*, 729-32; idem., Chem. Commun. *1994*, 841-2; ring contraction with tetra-*n*-butylammonium hydroxide s. M.L. Anderson et al., Acta Chem. Scand *49*, 503-14 (1995).

Oxygen/Nitrogen Type   SC ∩ ON

Irradiation
Generation of, and ring closures via, iminyl radicals
from unsatd. acyloximes cf. *49*, 370; by irradiation of unsatd. ketoxime xanthates, 5-α-functionalized Δ^1-pyrrolines, s. F. Gagosz, S.Z. Zard, Synlett *1999*, 1978-80.

Carbon/Carbon Type   SC ∩ CC

Sodium hydride/dimethyl sulfoxide   NaOH/DMSO
6-(Vinylthio)-2,3-dihydro-4-thiopyrones from 2-(α,β-ethyleneacylene)-1,3-dithiolanes
Base-induced fragmentation-type ring closure

193.

1 eq. NaH (50%) added to DMSO, the mixture stirred at 70° for 1 h, a soln. of 1-(1,3-dithiolan-2-ylidene)-4-phenyl-3-buten-2-one in DMSO added to the formed dimsylsodium, the mixture

stirred at the same temp. for 1 h, then poured into cold water, and neutralized with dil. HCl → 2-phenyl-6-vinylthio-2,3-dihydro-4H-thiopyran-4-one. Y 75%. This simple method is applicable to a variety of alkenoyl ketenemercaptals. F.e. of 2- or 2,5-di-subst. thiopyrones incl. 2-styryl-derivs. s. R. Samuel, C.V. Asokan et al., Synlett *2000*, 1804-6.

Exchange ↕

Oxygen ↑   SC ↕ O

Without additional reagents   w.a.r.
Alkylation with isoureas   SH → SR
review s. *29*, 547s*35*; 2-(alkylthio)-benzoxazoles and -benzothiazoles s. A. Harizi et al., Tetrahedron Lett. *41*, 5833-5 (2000).

Microwaves s. under Lawesson's reagent   ←
Ammonium thiocyanate s. under Fe(OCOCF₃)₃ NH_4SCN

Potassium salt K^+
Thiolic from sulfonic acid esters OSO_2R → SAc
s. *11*, 666s*33*; chiral α-(acylthio)nitriles with inversion of configuration, also xanthate analogs, s. F. Effenberger, S. Gaupp, Tetrahedron:Asym. *10*, 1765-75 (1999).

Zinc iodide ZnI_2
Thiolic acid esters from thiolic acids and alcohols   COSH → COSR
s. *41*, 505; preparation of thioester resin linkers for solid-phase synthesis of peptide C-terminal thioacids s. A.S. Goldstein, M.H. Gelb, Tetrahedron Lett. *41*, 2797-800 (2000).

Dimethylaluminum chloride Me_2AlCl
Thiolic from carboxylic acid esters   COOR → COSR'
with Me_3Al cf. *29*, 553; solid-phase synthesis of peptide C-terminal thioesters with Me_2AlCl s. D. Swinnen, D. Hilvert, Org. Lett. *2*, 2439-42 (2000).

Indium(III) chloride or bromide $InCl_3$ or $InBr_3$
Mercaptals from aldehydes   CHO → $CH(SR)_2$
Selective Lewis acid-catalyzed conversion

194.   Ph–CH(=O)H + HS–CH₂CH₂–SH → Ph–CH(S–CH₂CH₂–S) (1,3-dithiolane)

1 eq. 1,2-Ethanedithiol and *10 mol%* $InBr_3$ added successively to a stirred soln. of startg. aldehyde *in methylene chloride* at 0° under N_2, stirred for 5 min, warmed to room temp. for 1 h, and worked up → product. Y 80%. Reaction is applicable to the condensation of acyclic mercaptans in *stoichiometric* amount with aliphatic or aromatic aldehydes, and generally leaves ketones unaffected. 1,3-Dithiolanes were obtained, however, from both aldehydes and ketones. It is not essential to eliminate water from the system; in fact reaction of ar. aldehydes and enals is efficient *in water* with excellent recovery of the catalyst for reuse up to ten times with no loss of activity. F.e.s. M.A. Ceschi, C. Peppe et al., Tetrahedron Lett. *41*, 9695-9 (2000); 1,3-dithiolanes with $InCl_3$, selectivity, s. S. Muthusamy et al., ibid. *42*, 359-62 (2001).

Lipase   ←
Thiolic acid esters from carboxylic acids and mercaptans   COOH → COSR

195. $Me(CH_2)_{10}CO_2H$ + $HSC_{12}H_{25}$ → $Me(CH_2)_{10}C(O)SC_{12}H_{25}$

in the absence of solvent. A mixture of lauric acid (80 mg), 1-dodecanethiol (242 mg), and 100 mg Novozyme (RTM: immobilized *Candida antarctica* lipase) stirred at 40° and 500 mbar for 96 h → dodecyl thiollaurate. Y 93% (98% purity). The reduced pressure serves to remove water so that molecular sieves (or other dehydrating agents) are not required. F.e.s. N. Weber, K.D. Mukherjee et al., German patent DE-19905962.

Formic acid   *HCOOH*
N-Protected α-aminosulfones from aldehydes and sulfinic acids   CHO → CH(NHR)SO$_2$R'
N-carbalkoxy derivs. cf. *20*, 450; N-sulfonyl derivs. with formic acid s. F. Chemla, J.-F. Normant et al., Synthesis *2000*, 75-7.

Amberlite IRA 400 hydrogen sulfide/trifluoroacetic anhydride   ←
Mercaptans from alcohols under mild conditions   OH → SH
One-pot procedure via trifluoroacetoxy compds.

Preferential conversion. 1.2 eqs. Trifluoroacetic anhydride added to a soln. of (-)-menthol (e.e. 99%) in methylene chloride, stirred for 15 min at room temp., the liberated trifluoroacetic acid and solvent removed, the residue diluted with acetonitrile, treated with 1.2 eqs. Amberlite IRA 400 (SH$^-$ form), stirred for 5 min at 25°, the resin filtered off, and the solvent removed under reduced pressure → (+)-menthyl mercaptan. Y 88% (e.e. 98.8%). The procedure is rapid, simple, and high-yielding, and is not accompanied by dialkyl sulfide formation. It is generally applicable to various prim., sec., tert., acyclic and cyclic alcohols, diols, and benzyl alcohols, and leaves carbonyl, methoxy, methylenedioxy, C=C, ester, THP ethers, isopropylidenedioxy, Boc, Fmoc and phenolic groups unaffected. Furthermore, reaction with sec. alcohols proceeds **with inversion of configuration.** F.e. and chemoselectivity s. B.P. Bandgar et al., Chem. Lett. *2000*, 1304-5.

Amberlite IRA 400 dithiocarbamate   ←
Dithiocarbamic acid esters from alcohols   OH → SC(S)N<
and Zn-dimethyldithiocarbamate/Ph$_3$P/EtO$_2$CN=NCO$_2$Et cf. *38*, 506s*42*; from prim. or sec. alcohols via trifluoroacetates with Amberlite IRA 400 (dithiocarbamate form), also preferential conversion of a sec./tert. 1,3-diol, s. B.P. Bandgar et al., J. Chem. Res., Synop *2000*, 450-1.

Silica chloride   ←
Mercaptals from acetals   C(OR)$_2$ → C(SR)$_2$
with MgBr$_2$ cf. *44*, 460; under heterogeneous conditions with 'silica chloride' (prepared by refluxing silica gel with SOCl$_2$ for 48 h) s. H. Firouzabadi et al., Synlett *2000*, 263-5.

2-Mercapto-5,5-dimethyl-1,3,2-dioxaphosphorinane 2-sulfide
Replacement of carbonyl oxygen by sulfur   CO → CS
2-Thioacylthio-1,3,2-dioxaphosphorinane 2-sulfides s. *60*, 199

2,4-Bis(p-methoxyphenyl)-1,3,2,4-dithiadiphosphetane-2,4-disulfide/hydrochlorides/   ←
microwaves
Carboxylic acid thioamides from amides
s. *34*, 525; rapid procedure from N-acyl-1,2-diamine hydrochlorides under microwave irradiation without solvent s. R. Olsson et al., Tetrahedron Lett. *41*, 7947-50 (2000); thiocoumarins from coumarins s. J.N. Gadre et al., Indian J. Chem. *35B*, 60-2 (1996).

Phosphorus pentasulfide   *P$_2$S$_5$*
Dithiocarboxylic acid esters from carboxylic acids and mercaptans   COOH → CSSR

A mixture of benzoic acid, 1 eq. 2-thienyl mercaptan, and 20 mol% P$_4$S$_{10}$ in toluene refluxed (110°) for 10 h → product. Y 91%. The reagents are readily available and inexpensive, yields are high, and the method avoids the use of basic conditions or high temp. required by previous procedures. The reaction failed with *p*-hydroxybenzoic acid, *p*-acetoxybenzoic acid and pyridinecarboxylic acids. F.e., **also from alcohols,** s. A. Sudalai, B.C. Benicewicz et al., Org. Lett. *2*, 3213-6 (2000).

Phosphorus pentasulfide/hexamethyldisiloxane $P_2S_5/(Me_3Si)_2O$
1,2-Dithiole-3-thiones from β-ketocarboxylic acid esters
with Lawesson's reagent cf. *36*, 539; with $P_2S_5/S/(Me_3Si)_2O$ s. T.J. Curphey, Tetrahedron Lett. *41*, 9963-6 (2000).

Camphorsulfonic acid RSO_3H
3,1-Benzoxathian-4-ones from o-mercaptocarboxylic acids and oxo compds.
with Et_3N cf. *35*, 68; also from acetals with camphorsulfonic acid s. R. Siedlecka, J. Skarzewski, Pol. J. Chem. *74*, 1369-74 (2000).

Nafion-H ←
S-Acetylation s. *35*, 105s60 $SH \rightarrow SAc$

Hydrogen chloride/sodium cyanide HCl/NaCN
1,3-Di(thiocyanates) from 1,2-dithiolane 1-oxides

198.

2 eqs. aq. NaCN soln. added dropwise over 1 h to a soln. of 4-dimethylamino-1,2-dithiolane 1-oxide in toluene at 5° while maintaining the pH at <7 using 5 N HCl, allowed to stand for 1 h at the same temp., and the pH adjusted to 10 with 4 N NaOH → product. Y 84%. F.e.s. Japanese patent JP-2000086619 (Takeda Chem. Ind. Ltd.).

Ferric trifluoroacetate/ammonium thiocyanate $Fe(OCOCF_3)_3/NH_4SCN$
Thiiranes from epoxides
with Fe(III)-porphyrin complex cf. *55*, 205s57; with $Fe(OCOCF_3)_3$ s. N. Iranpoor, H. Adibi, Bull. Chem. Soc. Jpn. *73*, 675-80 (2000).

Fe(III)-exchanged montmorillonite ←
Sulfones from arenes $H \rightarrow SO_2R$
and sulfonyl chlorides with Fe(III)-exchanged montmorillonite cf. *57*, 207; **also from sulfonic acids or anhydrides**, and with zeolite β, regioselectivity, s. B.M. Choudary et al., J. Chem. Soc. Perkin Trans. 1 *2000*, 2689-93.

Nitrogen ↑ SC ↓↑ N

Sodium salt Na^+
Ar. thioethers from diazonium salts and mercaptans $N_2^+ \rightarrow SR$
diaryl sulfides s. *41*, 511; also alkyl aryl sulfides from diazonium o-benzenedisulfonimide salts s. M. Barbero, I. Degani et al., J. Org. Chem. *65*, 5600-8 (2000).

Halogen ↑ SC ↓↑ Hal

Without additional reagents w.a.r.
Reactions of diaryliodonium salts with S-nucleophiles ←
s. *48*, 510; aryl dithiocarbamates from polymer-based diaryliodonium salts s. D.J. Chen, Z.C. Chen, J. Chem. Res., Synop *2000*, 352-3.

Sodium methoxide/crown ether NaOMe/crown
Thioglycosides from 1-thiosugars $SH \rightarrow SR$
and ar. fluorides with K_2CO_3 cf. *50*, 322; S-glycosylcysteine derivs. with NaOMe/15-crown-5 s. L. Jobron, G. Hummel, Org. Lett. *2*, 2265-7 (2000).

Triethylamine Et_3N
S-Acylation of cyclic dithiophosphoric acid O,O-diesters s. *60*, 199 $SH \rightarrow SAc$

Anion exchanger-supported tetrahydridoborate BH_4^-
Thioethers from mercaptans $Hal \rightarrow SR$
s. *53*, 228; S-alkylation of 2-mercaptobenzimidazoles s. G. Lim, D. Kim, N. Yoon, World Intellectual Property Organization patent WO-200027841 (Dong-A Pharm. Co. Ltd.)

Benzyltriethylammonium tetrathiomolybdate [BnEt$_3$N]$_2$MoS$_4$
3-Ketothioethers from α,β-ethyleneketones and halides

199.

Intramolecular conversion. 2.2 eqs. Benzyltriethylammonium tetrathiomolybdate added to a soln. of startg. allyl bromide in acetonitrile, and the mixture stirred at room temp. for 6 h → product. Y 98%. This **one-pot process** is clean (avoiding handling of mercaptans), and involves conversion of the halide to disulfide, followed by reduction to the thiol and Michael addition. Both inter- and intra-molecular applications were demonstrated. F.e. and intermolecular reaction with α,β-ethylene-nitriles and -esters s. K.R. Prabhu, S. Chandrasekaran et al., Angew. Chem. Int. Ed. Engl. *39*, 4316-9 (2000).

Sulfur ↑   **SC ↓↑ S**

Triethylamine   Et$_3$N
Preparation of, and thioacylation with,   ←
2-thioacylthio-1,3,2-dioxaphosphorinane 2-sulfides

200.

Selective S-thioacylation. Startg. acid chloride added to 1 eq. startg. dithiophosphoric acid in benzene, treated with 1 eq. pyridine or Et$_3$N with iced water-cooling, and worked up after 15 min → intermediate acyl dithiophosphate (Y 96%), and 2 eqs. of the above dithiophosphoric acid in benzene refluxed for 4 h → intermediate thioacyl dithiophosphate (Y 91%), as a soln. treated dropwise with 2-hydroxyethyl mercaptan (1 part) in benzene and Et$_3$N (1.1 parts) at room temp., filtered, and the filtrate worked up → product (Y 97%). This is a simple and efficient method from cheap starting materials (avoiding the use of dithiocarboxylic acid salts). The thioacylating agents are stable (storable for several months without noticeable change), and even those derived from aliphatic acids may be handled without special precautions. The 2nd and 3rd step may be combined in one pot. Product isolation is very simple. However, there was no reaction with methanol. F.e. incl. **N-thioacylation** of amines (with alcohols or phenols unaffected) or hydroxylamines s. L. Doszczak, J. Rachon, Chem. Commun. *2000*, 2094-5.

Zinc/aluminum chloride   Zn/AlCl$_3$
Thioethers from halides and disulfides   SSR → SR'
with Mg cf. *12*, 686; alkyl aryl sulfides from diaryl disulfides with Zn/AlCl$_3$ s. B. Movassagh et al., J. Chem. Res., Synop *2000*, 350-1.

Samarium s. under TiCl$_4$ Sm

Samarium diiodide   SmI$_2$
Titanium tetrachloride/samarium   TiCl$_4$/Sm
Reductive ring closure with *o*-nitrodisulfides
benzothiazolines s. *8*, 657s*59*; 2,3-dihydro-1,5-benzothiazepines from α,β-ethyleneketones s. W. Zhong, Y. Zhang et al., J. Chem. Res., Synop *2000*, 386-7; with TiCl$_4$/Sm, also benzothiazoles, s. Synth. Commun. *30*, 4451-60 (2000).

Remaining Elements ↑   SC ⇅ Rem

Cupric acetate/ pyridine $Cu(OAc)_2/C_5H_5N$
Ar. thioethers from arylboronic acids and mercaptans $B(OH)_2 \rightarrow SR$

201.

A soln. of startg. mercaptan, 2 eqs. startg. arylboronic acid, 3 eqs. pyridine, 1.5 eqs. Cu(OAc)$_2$, and 75 wt% 4 Å molecular sieves *in DMF* refluxed under argon for 3 h → product. Y 88%. The procedure is generally applicable to prim. and sec. aliphatic and aryl mercaptans, and is tolerant of any substitution in the aryl ring (e.g. NO$_2$, CN, OMe). However, reaction times were longer with *o*-subst. arylboronic acids for steric reasons. F.e. incl. reaction of sec. thiols with retention of sulfur chirality, and S-aryl cysteine derivs. s. P.S. Herradura, R.K. Guy et al., Org. Lett. 2, 2019-22 (2000).

Carbon ↑   SC ⇅ C

Microwaves s. under Montmorillonite   ←
Cuprous iodide s. under $PdCl_2(PPh_3)_2$ *CuI*

Magnesium   *Mg*
Asym. synthesis of sulfoxides from Grignard compds. $RMgHal \rightarrow RS(O)R'$
with a phosphonylmethyl carbanion as leaving group cf. *58*, 213; chiral methyl sulfoxides from aryl methyl sulfoxides with an aryl carbanion as leaving group s. M. Annunziata, F. Naso et al., J. Org. Chem. *65*, 2843-6 (2000).

Montmorillonite KSF/microwaves   ←
Protection of carbonyl groups as 1,3-oxathiolanes   O
under microwave irradiation in the absence of solvent

202.

by **transthioacetalation.** Phenylacetaldehyde, 1.5 eqs. 2,2-dimethyl-1,3-oxathiolane, and montmorillonite KSF (0.1 g per mmol substrate) introduced with stirring in the reactor of a Synthewave 402 microwave apparatus, subjected to irradiation (the temp. being raised to 90° within 1 min), kept at the same temp. for a further 14 min by continuous adjustment of the power between 0 and 15 W, cooled to room temp., filtered, and the product purified by low-pressure distillation → 2-benzyl-1,3-oxathiolane. Y 86% (100% conversion). The procedure is low-energy, and allows product isolation by simple filtration and distillation, thereby avoiding the need to use a solvent. F.e. and **protection as 1,3-dithiolanes** using 2,2-dimethyl-1,3-dithiolane and Amberlyst 15 s. B. Pério, J. Hamelin, Green Chem. *2000*, 252-5.

Polymer-based phosphonium iodide   ←
Trans-S-alkylation of thiocyanates with halides $RSCN \rightarrow R'SCN$

203.   n-BuI + MeSCN ⟶ n-BuSCN

Startg. alkyl iodide (0.2 mmol) and 5 eqs. methyl thiocyanate in toluene heated for 12 h at 90° in the presence of 50 mmol of polystyrene-supported tributylphosphonium iodide (prepared by reaction of Bu$_3$P with a chloromethylated cross-linked polystyrene) → product. Y 99%. This method is useful where anhydrous conditions are required. Alkyl chlorides gave much lower yield. F.e.s. N. Ohtani et al., J. Chem. Soc. Perkin Trans. 2 *2000*, 1851-6.

Dichlorobis(triphenylphosphine)palladium(II)/cuprous iodide/triethylamine
Diaryl sulfides from arylthio-2-acetylenes and ar. iodides   I → SAr

204.

A mixture of *m*-chloroiodobenzene in THF containing 3 mol% PdCl$_2$(PPh$_3$)$_2$, 40 mol% CuI, and 4 eqs. Et$_3$N stirred at room temp. under argon for 0.5 h, 3-[2-(N-*p*-toluenesulfonyl)aminophenylthio]prop-1-yne added, stirring continued at room temp. for 8 h, then refluxed for 20 h → product. Y 62%. Both CuI and PdCl$_2$(PPh$_3$)$_2$ are essential as no product is formed if either is omitted. Optimum yields were achieved with 40 mol% CuI. The method is general for aromatic and heteroaromatic iodides, except for *p*-anisyl and 2-thienyl iodides where disubstituted alkynes were also formed. F.e. incl. thienyl and pyrimidinyl derivs. s. B. Nandi, N.G. Kundu et al., Tetrahedron Lett. *41*, 7259-62 (2000).

Elimination   ⇑

Oxygen ↑   SC ⇑ O

Acetic acid/microwaves   ←
2-Amino-1,3,4-thiadiazoles from N-acylthiosemicarbazides   O
with AcCl cf. *12*, 690; 2-acylamino-1,3,4-thiadiazoles from 1,4-diacylthiosemicarbazides with AcOH under microwave irradiation in the absence of solvent s. Z. Li, X. Wang, Y. Da, Synth. Commun. *30*, 3971-83 (2000).

Sulfur ↑   SC ⇑ S

Triphenylphosphine   Ph$_3$P
2-Cyano-1,3,4-thiadiazoles from tetrazoles   ∞
via N-(4-chloro-1,2,3-dithiazol-5-ylidene)-1,1-chlorohydrazones

205.   Ar = p-MeOC$_6$H$_4$

Startg. tetrazole allowed to react with 1.1 eqs. Appel's salt and pyridine in dichloromethane at room temp. → intermediate (Y 94%), in dichloromethane treated with 2 eqs. PPh$_3$ at room temp., and allowed to react for 15 min → product (Y 99%). F.e.s. V.-D. Le, C.W. Rees et al., Tetrahedron Lett. *41*, 9407-11 (2000).

Carbon ↑   SC ⇑ C

Without additional reagents   w.a.r.
Generation of reactive sulfines by elimination   ←
α,β-ethylenesulfines cf. *49*, 507; cyano- and formyl-sulfine by retro-Diels-Alder reaction under flash vacuum pyrolysis s. N. Pelloux-Léon, Y. Vallée et al., Eur. J. Org. Chem. *1999*, 3041-5.

Titanium tetrachloride   TiCl$_4$
Δ^2-Thiazolines from 2-acylaminothioethers   O
with PCl$_5$ cf. *34*, 896s57; with TiCl$_4$ s. P. Raman, J.W. Kelly et al., Org. Lett. *2*, 3289-92 (2000).

Trifluoromethanesulfonic acid
Stereospecific ring expansion of 2,4-dithia-6,8-diazabicyclo[3.2.2]nonanes via sulfur group migration

CF_3SO_3H

206.

with 2,3-dihydrobenzofuran ring closure. Startg. bridged heterocycle treated with anhydrous trifluoromethanesulfonic acid in acetonitrile for 8 h at room temp. → product. Y 61%. This is part of a multistep route to aspirochlorine with high stereospecificity and in high yield. F.e.s. Z. Wu, S.J. Danishefsky et al., Angew. Chem. Int. Ed. Engl. *39*, 3866-8 (2000).

Formation of Rem-C Bond

Uptake ⇓

Addition to Nitrogen and Carbon   RemC ⇓ NC

Without additional reagents   w.a.r.
α-Aminophosphorus(V) compds. from azomethines $C{=}N \to C(NH)P(O){<}$
with Me_3SiCl/Et_3N cf. *41*, 556s*46*; α-aminophosphonic acid monoesters via [instantaneous] diselective addition of spirocyclic tetraoxyphosphoranes s. K. Vercruysse, G. Eteastereomad-Moghadam et al., Eur. J. Org. Chem. *2000*, 281-9.

1,5,7-Triazabicyclo[4.4.0]dec-5-ene   ←
α-Aminophosphonic acid esters from azomethines s. *60*, 226 $C{=}N \to C(NH)PO(OR)_2$

Dichloro(1,5-cyclooctadiene)platinum(II) $Pt(cod)Cl_2$
α-(Borylamino)boronic acid esters from aldimines $C{=}N \to C(B(OR)_2)N(B(OR)_2)$

207.

3 Mol% $Pt(cod)Cl_2$, N-benzylidene-2,6-dimethylaniline and 1.08 eqs. bis(catecholato)diborane in benzene stirred for 3 h under N_2 in a screw-capped vial, the latter placed in a -5° freezer, solvent removed by sublimation under vacuum, and the dark brown solid worked up → product. Y 78%. Reaction is selective for ar. N-arylaldimines (preferably bulky substrates with an electron-donating group on the N-aryl residue). Poor yields were obtained with aliphatic aldimines. The presence of tert. phosphines or coordinating solvents (e.g. THF) reduced the catalytic activity. F.e. and **1,2-diborylation** of terminal alkenes, styrenes and alkynes (cf. *49*, 518s*52,54,56*) with the same catalyst s. G. Mann, R.T. Baker et al., Org. Lett. *2*, 2105-8 (2000).

Addition to Halogen and Carbon   RemC ⇓ HalC

Magnesium/ethyl bromide/ferrous chloride/magnesium chloride ←
Transition metal-catalyzed generation of organomagnesium chlorides   RMgCl

208.   MeO-N-Cl →(Mg/MgCl$_2$, FeCl$_2$) [MeO-N-Fe(MgCl)$_{0-1}$] → MeO-N-MgCl

1.33 eqs. Magnesium powder in THF treated with 10.9 mol% EtBr, the mixture stirred for 1 h, 5.1 mol% FeCl$_2$ added followed by a soln. of 5 mol% MgCl$_2$ in the same solvent, stirring continued for 3 min, startg. chloro compd. added dropwise over 30 min (exothermic from 22-45°), and stirred for a further 23.5 h → product. Y 96%. The conversion of chlorides is normally very difficult to achieve because of their lack of reactivity. F.e. incl. arylmagnesium chlorides s. B. Bogdanovic, M. Schwickardi, Angew. Chem. Int. Ed. Engl. **39**, 4610-2 (2000); functionalized arylmagnesium bromides with Rieke magnesium cf. J. Lee, R.D. Rieke et al., J. Org. Chem. **65**, 5428-30 (2000).

Addition to Carbon-Carbon Bonds   RemC ⇓ CC

Without additional reagents   w.a.r.
2-Alkoxymonoorganomercury chlorides from ethylene derivs.   C=C → C(OR)C(HgCl)
s. **16**, 681; regio- and stereo-specific oxymercuration of trisubst. alkoxy- and siloxy-2-ethylenes s. Tetrahedron Lett. **41**, 3713-6 (2000).

Irradiation   ⋀⋀⋀
Generation of silyl radicals from silylboranes   ←

209.   C$_6$H$_{13}$CH=CH$_2$ + [(i-Pr$_2$N)$_2$Ḃ + ṠiMe$_2$Ph] ──→ C$_6$H$_{13}$CH$_2$CH$_2$SiMe$_2$Ph

(i-Pr$_2$N)$_2$BSiMe$_2$Ph

Regiospecific hydrosilylation of ethylene derivs. A soln. of 1-octene and 1.2 eqs. of dimethylphenylsilylbis(diisopropylamino)borane in hexane irradiated under N$_2$ for 1.5 h at room temp. using a high-pressure Hg-lamp → dimethyl(*n*-octyl)phenylsilane. Y 56%. Regiospecific addition of the silane took place at the terminal carbon or preferentially at the β-carbon in the case of methyl crotonate. F.e. incl. photochemical silylative ring closure of 1,6-dienes, and prepn. of the silylboranes s. A. Matsumoto, Y. Ito, J. Org. Chem. **65**, 5707-11 (2000).

Silver trifluoromethanesulfonate/bromine   AgOTf/Br$_2$
Regiospecific asym. oxyselenation of ethylene derivs.   C=C → C(OR)C(SeR′)
with *o*-α-(alkylthio)diselenides cf. **49**, 515s59; with a 2,2′-bis(isonitroso)diselenide s. T.G. Back, Z. Moussa, Org. Lett. **2**, 3007-9 (2000); chiral 2-alkoxy-3-(arylseleno)tetrahydropyrans with (S,S)-hydrobenzoin/PhSeCl/Et$_3$N s. K.S. Kim et al., J. Chem. Soc. Perkin Trans. 1 **2000**, 1341-3.

Diethylzinc/N,N,N′,N′-tetramethylethylenediamine   Et$_2$Zn/TMEDA
β-Nitrophosphonic acid esters from 1,1-nitroethylene derivs.   C=C → CHCPO(OR)$_2$
Asym. Michael addition

210.   [chiral phosphite structure] + Ar⌒NO$_2$ → [product structure]   Ar = *p*-MeC$_6$H$_4$

Startg. chiral phosphite (prepared from (R,R)-4,5-bis(diphenylhydroxymethyl)-2,2-dimethyl-1,3-dioxolane and PCl$_3$) added under argon at 0° to a soln. of 1 eq. TMEDA in THF, 1 eq. diethylzinc added dropwise followed by a soln. of the startg. nitroethylene in the same solvent, the resulting mixture stirred at -78° for 12 h, then quenched at 0° with satd. aq. NH_4Cl → product. Y 86% (d.e. 96%). The intermediate, reactive phosphite-diethylzinc adduct, which is normally reactive but insoluble, is solubilized by TMEDA, thus making addition possible at low temperatures with increased diastereoselectivity. F.e., also hydrolysis to yield **chiral β-nitrophosphonic acids**, s. D. Enders et al., Angew. Chem. Int. Ed. Engl. 39, 4605-7 (2000).

Mercuric acetate chloride/ytterbium triflate $\qquad$ $Hg(Cl)OAc/Yb(OTf)_3$
4-Chloromercuri-1,3-dioxanes from 3-ethylenealcohols and oxo compds.
from aldehydes with $Hg(OAc)_2$/NaCl cf. 59, 220; also from ketones with 5 mol% $Yb(OTf)_3$ as catalyst s. S.D. Dreher, J.L. Leighton et al., Org. Lett. 2, 3197-9 (2000).

Mercuric triflate/sodium chloride $\qquad$ $Hg(OTf)_2/NaCl$
Intramolecular oxymercuration of ethylenealcohols
with $Hg(OAc)_2$ cf. 11, 224; 5-, 6- and 7-membered O-heterocyclics with $Hg(OTf)_2$/NaCl s. H. Imagawa et al., Chem. Pharm. Bull. 46, 1341-2 (1998).

Sodium tetrahydridoborate $\qquad$ $NaBH_4$
Enetellurides from acetylene derivs. $\qquad$ C≡C → CH=C(TeR)
(Z)-α,β-ethylene-β-(organotelluro)phosphonic acid esters cf. 50, 343s58; also (Z)-β-aryltelluro-α,β-ethylenesulfones from α,β-acetylenesulfones s. X. Huang et al., J. Org. Chem. 66, 74-80 (2001).

Bromine s. under AgOTf $\qquad$ Br_2

Lithium tri-n-butylmanganate $\qquad$ $Li[MnBu_3]$
(Z)-Enegermanes $\qquad$ C≡C → CH=C(Ge≤)
from acetylene derivs. and organogermanium hydrides
radical addition with Et_3B cf. 43, 517; with Li-tri-*n*-butylmanganate as catalyst, also intramolecular carbogermylation of dienes, and hydrogermylation of olefins, s. H. Kinoshita, K. Oshima et al., Bull. Chem. Soc. Jpn. 73, 2159-60 (2000); enegermanes from terminal alkynes **via (Z)-2-germylenestannanes** with 1,2-germyl group migration cf. T. Nakano et al., Chem. Lett. 2000, 1408-9; regio- and stereo-specific hydrostannylation of 2-acetylenealcohols with dilithium trialkyl-(stannyl)manganates s. Synlett 1999, 1417-9.

Rhodium(II) acetate $\qquad$ $Rh_2(OAc)_4$
Hydrosilylation of ethylene derivs. $\qquad$ C≡C → CHC(Si≤)
with $RhCl(PPh_3)_3$ cf. 29, 603s57; N-protected 1,1-aminosilanes from enamines with $Rh_2(OAc)_4$ s. G.W. Hewitt, S.McN. Sieburth et al., Tetrahedron Lett. 41, 10175-9 (2000).

Bis(η³-allylpalladium chloride)/(R)-2-bis[3,5-bis(trifluoromethyl)phenyl]- $\qquad$ ←
phosphino-1,1'-binaphthyl
Asym. hydrosilylation of ethylene derivs.
with (R)-3-diphenylphosphino-3'-methoxy-4,4'-biphenanthryl as ligand cf. 47, 542s52; regio-specific asym. hydrosilylation of styrenes with (R)-2-bis[3,5-bis(trifluoromethyl)phenyl]-phosphino-1,1'-binaphthyl s. T. Hayashi et al., Chem. Lett. 2000, 1272-3.

(Ethylene)bis(triphenylphosphine)platinum $\qquad$ $Pt(CH_2=CH_2)(PPh_3)_2$
Catalyst-dependent regio- and stereo-specific silaboration of alkylidenecyclopropanes

211.

While Pd complexes favour silaboration of alkylidenecyclopropanes by *distal* cleavage of the cyclopropane ring, Pt complexes induce *proximal* cleavage. **E: α-Alkylidene-γ-silylboronic acid esters**. A mixture of 2-(dimethylphenylsilyl)-4,4,5,5-tetramethyl-1,3,2-dioxaborolane, 1.2-2 eqs. startg. cyclopropane, and 2 mol% (ethylene)bis(triphenylphosphine)platinum in toluene heated at 110° for 24 h → product. Y 75%. With Pd(dba)$_2$/triethyl phosphite as catalyst, the corresponding **β-alkylidene-γ-silylboronic acid esters** were obtained. Interestingly, with benzylidenecyclopropane, proximal cleavage took place with either catalyst, the Pt complex favouring the (Z)-product while the Pd complex gave the (E)-isomer. Given this E/Z selectivity, it is possible that the Pt-catalyst induces initial oxidative addition of the proximal C-C bond to the transition metal to give an alkylidenemetallacyclobutane prior to reductive elimination, whereas the Pd complex favours a direct oxidative addition mechanism. F.e.s. M. Suginome, Y. Ito et al., J. Am. Chem. Soc. *122*, 11015-6 (2000).

Dichloro(1,5-cyclooctadiene)platinum(II)   *Pt(cod)Cl$_2$*
Addition of tetraalkoxydiboranes to multiple bonds   ←
with Pt(PPh$_3$)$_4$ s. *49*, 518s56; 1,2-diborylation of terminal alkenes, styrenes and alkynes with Pt(cod)Cl$_2$ s. G. Mann et al., Org. Lett. *2*, 2105-8 (2000).

Via intermediates   *v.i.*
Enegermanes from terminal acetylene derivs.   C≡CH → C(Ge≤)=CH$_2$
via (Z)-2-germylenestannanes s. *43*, 517s60

Rearrangement   ↻

Oxygen/Remaining Elements Type   RemC ↻ ORem

Lithium diisopropylamide   *i-Pr$_2$NLi*
1,3-O→C- and 1,3-S→C-Phosphinyl group migration with retention of P-chirality   ←

Chiral 1,1'-bi-2-naphthol-3,3'-di(phosphine oxides). A soln. of startg. (S$_P$,R,S$_P$)-BINOL bis-(phosphinate) in THF added to 10 eqs. LDA [from 1:1 *n*-BuLi (2.5 *M* in hexane)/*i*-Pr$_2$NH] under N$_2$ at -78°, and the mixture stirred at -78° to room temp. for 12 h → product. Y 88% (e.e. >98.5%). F.e. incl. *o*-hydroxy- and *o*-mercapto-phosphine oxides (or sulfides), also 1,2-N→C-[thio]-phosphinyl group migration, s. T.-L. Au-Yeung, R.K. Haynes et al., Tetrahedron Lett. *42*, 457-60 (2001).

Exchange   ↕

Hydrogen ↑   RemC ↕ H

Palladium-carbon or -alumina   *Pd-C or Pd/Al$_2$O$_3$*
Ar. deuteriation   H → D
with Pd-C cf. *39*, 555s41; of imidazoles and imidazolium salts [with D$_2$O], also with Pd/Al$_2$O$_3$, s. C. Hardacre et al., Chem. Commun. *2001*, 367-8.

Cyclooctadiene(β-diketonato)iridium(I) complexes   ←
Ar. deuteriation
with di(acetone)dihydridobis(triphenylphosphine)iridium fluoroborate cf. *39*, 555s47; *o*-deuteriation with cyclooctadiene(β-diketonato)iridium(I) complexes s. L.P. Kingston, W.J.S. Lockley et al., Tetrahedron Lett. *41*, 2705-8 (2000).

Oxygen ↑   RemC ↓↑ O

Zirconocene dichloride/n-butyllithium $Cp_2ZrCl_2/BuLi$
(E)-Vinylzirconium compds. from enolethers   ←

213.

Ethylene derivs. A soln. of 3.3 eqs. 1.6 M n-BuLi in hexane added slowly to a soln. of 1.5 eqs. Cp_2ZrCl_2 in dry THF at -78°, stirred for 1 h, 1 eq. startg. enolether (E:Z 75/25) added, allowed to warm to room temp., stirred for 2.5-5 h, cooled to -20°, and hydrolyzed with acid [1 N HCl] → 1-undecene. Y 90%. (E)-Vinylzirconium compds. were obtained irrespective of the geometry of the enolether or the bulk of the alkoxy group. The method was also applied successfully to enoxysilanes, but vinyl bromides and vinyl carbonates failed to react. F.e. and reactions of the product s. A. Liard, I. Marek, J. Org. Chem. *65*, 7218-20 (2000).

Halogen ↑   RemC ↓↑ Hal

Irradiation s. under KOBu-t   ///

Potassium tert-*butoxide/irradiation*   KOBu-t////
Arylphosphonic acid esters from ar. halides   Hal → $PO(OR)_2$
and dialkyl phosphites with KOBu-t/FeSO$_4$ cf. *40*, 540; direct phosphonylation of mono- and dihalogenanilines under irradiation s. N. Defacqz et al., Synthesis *1999*, 1368-72.

n-Butyllithium/(-)-sparteine   ←
Asym. synthesis of ferrocenes   ←
via *o*-metalation cf. *49*, 536s*54*; asym. *o,o'*-disubstitution of ferrocene-1,1'-dicarboxamide with BuLi/(-)-sparteine s. R.S. Laufer, V. Snieckus et al., Org. Lett. *2*, 629-31 (2000).

Trimethylsilyl chloride/ethyldiisopropylamine $Me_3SiCl/i-Pr_2NEt$
Unsym. phosphinic acids from halides   >P(O)H → >P(O)R
s. *40*, 424; from two different halide molecules and $H_2P(O)ONH_4$ with *i*-Pr$_2$NEt as base s. E. Soulier, J.-C. Clément et al., Eur. J. Org. Chem. *2000*, 3497-503.

Palladous acetate/triphenylphosphine/triethylamine $Pd(OAc)_2/Ph_3P/Et_3N$
Palladium-catalyzed P-arylation   >PH → >PAr
aryldialkylphosphines s. *38*, 584s*59*; N-protected *p*-phosphino-D- and -L-phenylalanines s. H.-B. Kraatz, A. Pletsch, Tetrahedron:Asym. *11*, 1617-21 (2000).

Palladous acetate/2-(dicyclohexylphosphino)biphenyl/triethylamine   ←
Arylboronic acid esters from ar. halides   ArHal → $ArB(OR)_2$
by B-arylation using $PdCl_2(dppf)/Et_3N$ cf. *54*, 244; from hindered ar. halides using $Pd(OAc)_2$/2-(dicyclohexylphosphino)biphenyl s. O. Baudoin et al., J. Org. Chem. *65*, 9268-71 (2000).

Palladous acetate/1,1'-bis(diphenylphosphino)ferrocene/propylene oxide   ←
1-Acetylenephosphonic acid esters $CH=CBr_2 → C≡CPO(OR)_2$
from 1,2-ethylene-1,1-bromides

214.

and phosphorous acid diesters. A soln. of startg. 1,1-dibromoalkene in DMF, 3 eqs. propylene oxide, and 2 eqs. dimethyl phosphite added to a mixture of 0.2 eq. Pd(OAc)$_2$ and 0.4 eq. dppf in the same solvent under dry N$_2$ at room temp., the mixture heated at 80° for 14 h, then cooled to room temp. → product. Y 76%. The reaction tolerates a range of functional groups in both the dibromoalkene and H-phosphonate, e.g. carboxylic acid esters, phenolethers, ar. nitro compds. and furans. With 1,1-dibromo-2-(furan-2-yl)ethylene, tris(2-furyl)phosphine was the preferred ligand. F.e.s. M. Lera, C.J. Hayes, Org. Lett. *2*, 3873-5 (2000).

Tetrakis(triphenylphosphine)palladium(0)/triethylamine Pd(PPh$_3$)$_4$/Et$_3$N
Arylphosphonous acids from ar. halides Hal → PH(O)OH

215.

2 Mol% Pd(PPh$_3$)$_4$ added to a soln. of iodobenzene, 3 eqs. Et$_3$N and 1.1 eqs. anilinium hypophosphite in anhydrous DMF, and heated under N$_2$ at 85° for 2-6 h → phenylphosphinic acid. Y 89% (96% by ^{31}P NMR). The anilinium reagent is a safer alternative to anhydrous H$_3$PO$_2$, and is cheap, non-hygroscopic, and stable in air and moisture. Various palladium catalysts are useful, as well as solvents (especially acetonitrile or DMF), and pyridine may be used in place of Et$_3$N. The reaction has broad scope with applicability to ar. iodides, bromides and triflates, as well as benzyl chlorides, and shows high selectivity for monosubstitution. Reducible functions are compatible, although reaction times may need to be controlled in the presence of excess hypophosphite. F.e.s. J.-L. Montchamp, Y.R. Dumond, J. Am. Chem. Soc. *123*, 510-1 (2001).

Remaining Elements ↑   RemC ↓↑ Rem

Azodiisobutyronitrile RN=NR
Enestannanes from eneselenides C=CScR → C=CSn≤

216.

2 eqs. Tri-*n*-butyltin hydride and 0.19 eq. AIBN added to a soln. of startg. (E)-vinyl selenide in toluene under argon, and the resulting mixture stirred at 85° for 3 h → (E)-product. Y 83%. This new deselenative stannylation reaction is highly stereospecific and proceeds without simple reduction or reductive ring closure. This is part of a 5-step **asym. synthesis of α-amino-α-vinyl-carboxylic acids**. F.e. and enantiomers s. D.B. Berkowitz et al., J. Am. Chem. Soc. *122*, 11031-2 (2000).

Tri-n-butyltin hydride/azodiisobutyronitrile Bu$_3$SnH/RN=NR
Dithiophosphonic acid esters from ethylene derivs. C=C → CHCPO(SR)$_2$
Regiospecific radical addition

217.

A degassed soln. of 4 eqs. tributyltin hydride and 0.5 eq. azodiisobutyronitrile in benzene added over 10 h to a refluxing degassed soln. of 3 eqs. startg. dithiaphosphinane and α-methylstyrene in the same solvent, and the mixture refluxed for an additional 4 h → product. Y 75%. No addition product could be isolated with *n*-butyl vinyl ether. F.e. incl. addition to 3-methylenefuranoses s. C. Lopin, S.R. Piettre et al., Tetrahedron Lett. *41*, 10195-200 (2000).

Tris(dibenzylideneacetone)dipalladium/diphenyl-2-pyridylphosphine/potassium carbonate ←
Bis(allylpalladium chloride)/2-(diphenylphosphino)phenol/sodium hydroxide ←
Silanes from halides Hal → Si≼
with Pd(PPh$_3$)$_4$ cf. *31*, 592s*37*; **arylsilanes** with [PdCl(C$_3$H$_5$)]$_2$/*o*-diphenylphosphinophenol and 1.2 eqs. NaOH s. E. Shirakawa, T. Hiyama et al., Chem. Commun. *2000*, 1895-6; with Pd$_2$(dba)$_3$/diphenyl-2-pyridylphosphine/K$_2$CO$_3$ [or 2-(di-*tert*-butylphosphino)biphenyl/KF for electron-poor substrates] cf. L.J. Goossen, A.-R.S. Ferwanah, Synlett *2000*, 1801-3.

Dichloro[1,1'-bis(diphenylphosphino)ferrocene]palladium(II) PdCl$_2$(dppf)
Arylboronic acids from diazonium fluoroborates N$_2^+$ → B(OH)$_2$

218.

Anhydrous, deoxygenated methanol added via syringe to a mixture of startg. diazonium fluoroborate, 1 eq. bis(pinacolato)diboron, and 3 mol% PdCl$_2$(dppf) under N$_2$, allowed to react at room temp. for 1 h, a further 0.5 eq. of the diazonium salt and 0.75 mol% catalyst added in 1 h intervals, and heated at 40° in a sand bath until reaction complete (3 h) → product. Y 96%. This is the first example of C-B bond formation by displacement of a nitrogen function. Conditions are mild and the procedure is based on a benign solvent with *no added base*. Furthermore, ar. chlorine and bromine were unaffected. F.e.s. D.M. Willis, R.M. Strongin, Tetrahedron Lett. *41*, 8683-6 (2000).

Carbon ↑ RemC ↓↑ C

Cesium fluoride CsF
Silylacetylenes from terminal acetylene derivs. H → Si≼

219.

A mixture of startg. alkyne, 2 mol% CsF, and 1.2 eqs. trimethylsilyltrifluoromethane in THF stirred at room temp. for 30 min, the catalyst filtered off, and the filtrate evaporated → N-(2-propenyl)-N-[3-(trimethylsilyl)-2-propynyl]tosylamide. Y 100%. Reaction is thought to involve intermediate formation of a pentacoordinate silicate in a catalytic cycle. Acetals, esters, iodides, silyl, amino, and amido groups, and ketones (not aldehydes) were tolerated. F.e. incl. triethylsilyl derivs. (with 10 mol% CsF), also **acetylenestannanes** (with Bu$_3$SnCF$_3$), and with a little KF in DMF, s. M. Ishizaki, O. Hoshino, Tetrahedron *56*, 8813-9 (2000).

Formation of C-C Bond

Uptake ⇓

Addition to Oxygen and Carbon CC ⇓ OC

Lithium/naphthalene Li/C$_{10}$H$_8$
Generation and trapping of ω-lithioalkoxides from O-heterocyclics C
s. *45*, 368s*52*; from benzannelated cyclic ethers, also cleavage of N-heterocyclic analogs, s. U. Azzena et al., Tetrahedron *56*, 8375-82 (2000); branched-chain functionalized carbohydrates from carbohydrate spiro-epoxides and -oxetanes s. T. Soler, M. Yus et al., Tetrahedron:Asym. *11*, 493-517 (2000).

*Lithium/naphthalene or 4,4'-di-*tert-*butylbiphenyl* ←
Synthesis of alcohols from oxo compds. CO → C(OH)R
by reductive lithiation s. *47*, 576s*55*; 3-methylene-1,5-diols from 3-chloro-2-(chloromethyl)prop-1-ene and conversion to 1,6-dioxaspiro[3.4]octanes (with Ag_2O/I_2) s. F. Alonso, M. Yus et al., Synthesis *2000*, 949-52; β-hydroxycarboxylic acids from chloroacetic acid/LDA, also γ-hydroxy- from β-chloro-carboxylic acids, and their conversion to γ-lactones with TsOH, s. I.M. Pastor, M. Yus, Tetrahedron Lett. *41*, 5335-9 (2000).

Sodium hydride NaH
Ketocarboxylic acids from dicarboxylic acid anhydrides C
with organozinc compds. cf. *1*, 697; regiospecific ring opening of α-acoxysuccinic anhydrides with sodio compds., also conversion to 5-(carboxymethyl)tetronic acids s. C.A. Mitsos et al., J. Org. Chem. *65*, 5852-3 (2000).

Organolithium compds./chiral 2-aminoalcohols ←
Asym. synthesis of sec. alcohols from aldehydes CHO → CH(OH)R
with chiral pyrrolidine-derived 2-aminoalcohols cf. *35*, 439; with chiral 1-amino-1,2-diphenyl- ethanols s. M. Schön, R. Naef, Tetrahedron:Asym. *10*, 169-76 (1999).

Methyllithium MeLi
α'-Iodolactols from lactones CO → C(OH)CH$_2$I

220.

1-Iodo-1-deoxy-2-uloses. 1.2 eqs. ca. 1.6 M MeLi in ether added slowly under argon (dropwise in ca. 5 min) to a stirred soln. of startg. lactone and 1.5 eqs. methylene iodide in dry toluene at -70°, the mixture kept between -70° to -60° for 15 min, satd. aq. NH$_4$Cl added at the same temp., and warmed to room temp. → product. Y 75%. This method is direct and complementary to an earlier two-step process (C. Wilcox, G.W. Long, H. Sugh, Tetrahedron Lett. *25*, 395-8 (1984), and D. Noort et al., Synlett, *1990*, 205-6). It was applied successfully to several representative γ- and δ-lactones with protecting groups such as ether, acetal, silyl ether. F.e.s. B. Bessieres, C. Morin, Synlett *2000*, 1691-3.

n-*Butyllithium/potassium* tert-*butoxide/(+)-diisopinocampheyl(methoxy)borane/* ←
boron fluoride
anti-**3-Ethylene-2-silylalcohols from aldehydes and 2-ethylenesilanes** CH(OH)C(Si≤)C=C
via asym. allylboration

221.

1.1 eqs. startg. allylsilane added via syringe in one portion to a soln. of 1 eq. KOBu-*t* in dry THF at -78°, treated dropwise with 1 eq. *n*-BuLi in hexanes over 5 min, the mixture stirred at the same temp. for 10 min, warmed to -45° in a dry ice-acetone bath, stirred for 2 h, cooled to -78°, 1 eq. freshly prepared (+)-diisopinocampheyl(methoxy)borane added slowly via cannula, stirred for 30 min, 1.33 eqs. BF$_3$-etherate added dropwise followed slowly by 1.2 eqs. of the startg. aldehyde, the mixture stirred for 4 h at the same temp., and worked up with 1 M KH$_2$PO$_4$-KOH buffer (pH

6) and 30% H_2O_2 → product. Y 86% (e.e. 95%). This method is highly diastereo- and enantio-selective with various alkyl and aryl aldehydes, while ester, ether and silyl groups remained unaffected. F.e.s. W.R. Roush et al., Tetrahedron Lett. *41*, 9413-7 (2000).

n-Butyllithium/(-)-sparteine/boron fluoride
Asym. synthesis of glycol monoethers from 1,3-dioxolanes

A soln. of 4 eqs. *n*-BuLi in hexane (1.6 *M*) added to a mixture of 4 eqs. 2-bromo(ethyl)benzene and 2 eqs. (-)-sparteine in ether at -30°, stirred for 1 h, cooled to -78°, startg. 1,3-dioxolane added, followed slowly by 3 eqs. BF_3-etherate, stirring continued at the same temp. for 1 h, then hydrolyzed with satd. aq. NH_4Cl → product. Y 88% (e.e. 81%). Enantioselectivity was highest for aryllithiums and 2-aryl-1,3-dioxolanes possessing *o*-substituents, but decreased with bulky *tert*-butyl or phenyl groups. Selectivity was low with alkyllithiums. This is the first example of an asym. nucleophilic substitution at a single prochiral centre. F.e. and asym. nucleophilic substitution of dimethyl acetals (e.e. 5-40%) s. P. Muller, P. Nury, Org. Lett. *2*, 2845-7 (2000).

sec-Butyllithium/N,N,N′,N′-tetramethylethylenediamine   s-BuLi/TMEDA
***p*-Metalation of 3,5-dichloroarylcarboxylic acid amides**

A soln. of 1.2 eqs. TMEDA in anhyd. ether added to a soln. of 3,5-dichloro-N,N-diethylbenzamide in the same solvent under N_2, the mixture cooled to -78°, 1.2 eqs. *sec*-BuLi (in cyclohexane) added dropwise, stirred at the same temp. for 30 min, a soln. of 1.5 eqs. 4-methoxybenzaldehyde in ether added dropwise, the mixture allowed to warm slowly to room temp., and stirred for a further 2 h before quenching with water → 3,5-dichloro-N,N-diethyl-4-[hydroxy(4-methoxy-phenyl)methyl]benzamide. Y 56%. No *ortho*-metalation was observed suggesting that regio-control in other substrates may possibly be based on steric and electronic properties of the substituents. F.e.s. M. Demas, L.M. Bradley et al., J. Org. Chem. *65*, 7201-2 (2000).

Lithium diisopropylamide   i-Pr_2NLi
β-Hydroxycarboxylic acid esters from oxo compds.   CO → C(OH)C-COOR
and lithium ester enolates s. *25*, 487s*34*; condensation with chiral 2-acyl-1,3-dithiane 1-oxides with asym. induction s. J.L. Garcia Ruano, P.C. Bulman Page et al., J. Org. Chem. *65*, 6027-34 (2000).

Lithium diisopropylamide/hexamethylphosphoramide   i-Pr_2NLi/HMPA
β-Hydroxynitriles from nitriles   CO → C(OH)C-CN
anti-products from ar. aldehydes cf. *29*, 959; *syn*-products with added HMPA (6 eqs.) under thermodynamic control s. P.R. Carlier et al., Org. Lett. *2*, 2443-5 (2000).

tert-*Butyllithium* s. under MeMgBr   t-BuLi

Cesium fluoride   CsF
2-Siloxy-1,1,1-trifluorides from oxo compds.   CO → C(OSi≤)CF$_3$
Synthesis with addition of one C-atom

224.

ca. 10 Mol% dry CsF added to a soln. of benzophenone and 2 eqs. 4-[2,2,2-trifluoro-1-(trimethylsiloxy)ethyl]morpholine in DME, and heated at 80° for 5 h → product. Y 75%. The reagent is stable and more easy to prepare (from fluoroform and N-formylmorpholine) than CF$_3$SiMe$_3$ (cf. *44*, 577), and the procedure is environmentally friendly. High yields were obtained from non-enolizable ketones and aldehydes, even in the heterocyclic series. However, reaction with enolizable ketones resulted in α-(trifluoromethyl)methylenation. F.e.s. T. Billard, B.R. Langlois et al., Org. Lett. *2*, 2101-3 (2000); with N-(2,2,2-trifluoro-1-siloxyethyl)piperazine cf. Tetrahedron Lett. *41*, 8777-80 (2000); with tris(trimethylsilyl)amine and fluoroform in the presence of Me$_4$NF, also trifluoromethyl thioethers and selenides, s. J. Org. Chem. *65*, 8848-56 (2000).

Sodium iodide s. under Zn   NaI

1,8-Diazabicyclo[5.4.0]undec-7-ene   DBU
2,2,2-Trichloroalcohols from oxo compds.   CO → C(OH)CCl$_3$
Synthesis with addition of one C-atom under mild conditions

225.

in the absence of solvent. 1 eq. DBU added dropwise to a mixture of 2-nitrobenzaldehyde and 2 eqs. chloroform under N$_2$, and the mixture stirred for 10 min → product. Y 98%. Reaction is very straightforward and generally applicable to ar. and aliphatic aldehydes or ketones. There were no Cannizzaro by-products (even from electron-deficient aryloxo compds.). However, hindered aldehydes and ketones gave lower yields. F.e. and with DBN s. V.K. Aggarwal, A. Mereu, J. Org. Chem. *65*, 7211-2 (2000).

1,5,7-Triazabicyclo[4.4.0]dec-5-enes   ←
Catalytic Henry reaction under mild conditions   CO → C(OH)CH$_2$NO$_2$

226.

in the absence of solvent. Cyclohexanone and nitromethane (0.5 ml per mmol ketone) stirred without solvent in the presence of 10 mol% 1,5,7-triazabicyclo[4.4.0]dec-5-ene for 1 h at 0° → product. Y 82%. Reaction is generally applicable to aliphatic and aromatic aldehydes, as well as alicyclic ketones and α,β-unsatd. oxo compds (not aliphatic ketones or acetophenone). Formation of Nef by-products and dehydration of the product is effectively avoided. Reaction is fast, and in many instances superior, to that with tetramethylguanidine. F.e. and with the 7-methyl-deriv. or polymer-based 1,5,7-triazabicyclo[4.4.0]dec-5-ene, also reactions of dialkyl phosphites, incl. α-aminophosphonic acid esters from azomethines, s. D. Simoni et al., Tetrahedron Lett. *41*, 1607-10 (2000).

3-Hydroxyquinuclidine
Baylis-Hillman reaction ←
with aq. Me$_3$N cf. *39*, 593s*59*; soluble polymer-based conversion in ethanol with a PEG-based acrylate using 3-hydroxyquinuclidine as reagent, and combinatorial synthesis of 1,4-oxazepin-7-ones, s. R. Räcker, O. Reiser et al., J. Org. Chem. *65*, 6932-9 (2000).

CHO → CH(OH)C(=CH$_2$)CO

Cupric trifluoromethanesulfonate/2,2′-isopropylidenebis(4(S)-tert-butyl-Δ^2-oxazoline) ←
α-Aryl-α-hydroxy- from α-oxo-carboxylic acid esters and arenes
via regiospecific asym. Friedel-Crafts hydroxyalkylation

H → C(OH)COOR

227.

A mixture of 10 mol% Cu(OTf)$_2$ and 11 mol% 2,2′-isopropylidenebis(4(S)-*tert*-butyl-Δ^2-oxazoline) in dry THF stirred in a Schlenk tube for 0.5-1 h, 1.5 eqs. ethyl glyoxylate and 1 eq. N,N-dimethylaniline added, and the mixture stirred at 0° for 36 h → product. Y 82% (e.e. 94%). Reaction occurs on the aromatic ring of anilines, cyclic amines such as indoles, naphthylamines, furans and alkoxybenzenes. F.e., catalysts and solvents, also with alkyl trifluoropyruvates, s. N. Gathergood, K.A. Jørgensen et al., J. Am. Chem. Soc. *122*, 12517-22 (2000).

Magnesium/cuprous bromide-dimethyl sulfide
Synthesis of β-allenecarboxylic acids from β-alk-1-ynyl-β-lactones
with asym. induction

Mg/CuBr-Me$_2$S

228.

A soln. of startg. chiral β-lactone and isopropylmagnesium bromide in THF treated with 10 mol% CuBr·Me$_2$S at -78° until reaction complete → product. Y 92% (e.e. 92%). This is part of an operationally simple 2-step procedure **from α,β-acetylenealdehydes.** Reaction is generally applicable to the synthesis of chiral 1,3-di- and 1,1,3-tri-subst. allenes, and takes place by S$_N$2′ ring opening with the anticipated complete transfer of lactone chirality. The addition is insensitive to the steric environment of the nucleophilic carbon atom, being applicable to straight-chain, branched and aryl Grignards. F.e. and addition of unbranched nucleophiles [with added CuCN·2LiBr], and of zinc ester enolates s. Z. Wan, S.G. Nelson, J. Am. Chem. Soc. *122*, 10470-1 (2000).

Magnesium/cerium(III) chloride
Synthesis of 1,1-hydroxysilanes from acylsilanes
s. *44*, 545; 2-ethylene-1,1-hydroxysilanes from acylsilanes with addition of 2 C-atoms from vinylmagnesium bromide with 3 eqs. CeCl$_3$·7H$_2$O, also asym. variants, s. B.F. Bonini, A. Ricci et al., Synlett *2000*, 1688-90.

Mg/CeCl$_3$
COSi≤ → C(OH)R(Si≤)

Zinc (s.a. under VOCl$_3$)
Barbier-type synthesis of 3-ethylenealcohols with activated zinc
from ZnCl$_2$ and Na in liq. NH$_3$ cf. *34*, 614s*51*; with activated zinc prepared electrochemically (from ZnCl$_2$/NH$_4$Cl) *under ultrasonication,* also Reformatskii synthesis and reductive dehalogenation, and sonoelectrochemical preparation of Zn,Cu alloys, s. A. Durant, V. Libert et al., Eur. J. Org. Chem. *1999*, 2845-51.

Zn
CO → C(OH)C-C=C

Zinc/ammonium iodide/sodium iodide
Barbier-type synthesis of 3-ethylenealcohols
with Zn/NH$_4$Cl cf. *40*, 567s*51*; *18*, 736s*58*; γ-hydroxy-α-methylenesulfoxides from aldehydes and (S$_S$)-3-chloro-2-(*p*-tolylsulfinyl)-1-propene with asym. induction in the presence of NH$_4$I and NaI s. F. Márquez, A. Delgado et al., Org. Lett. *2*, 547-9 (2000).

Zn/NH$_4$I/NaI

Zinc/dibromo[1,2-bis(diphenylphosphino)ethane]nickel(II) Zn/Ni(dppe)Br$_2$
Synthesis of sec. benzylalcohols from aldehydes CHO → CH(OH)Ar
with Zn/CrCl$_3$/Me$_3$SiCl cf. *52*, 255; with Zn and 0.1 eq. Ni(dppe)Br$_2$ s. K.K. Majumdar, C.-H. Cheng, Org. Lett. *2*, 2295-8 (2000).

Methylmagnesium bromide/tert-butyllithium MeMgBr/t-BuLi
Synthesis of sec. benzylalcohols from aldehydes
with *i*-PrMgBr/*n*-BuLi cf. *59*, 250; *o*-sulfamidobenzyl alcohols with MeMgBr/*t*-BuLi, also conversion to 3,4-dihydro-1*H*-2,1,3-benzothiadiazine 2,2-dioxides, s. J. Agejas, C. Lamas et al., Tetrahedron Lett. *41*, 9819-23 (2000).

Dialkylzinc/(1S,2R)-N,N-dibutylnorephedrine ←
Asym. synthesis of sec. alcohols from aldehydes in the absence of solvent CHO → CH(OH)R

229.

3 eqs. Diethylzinc added via cannula to 5 mol% (1S,2R)-N,N-dibutylnorephedrine at 0° under argon, the mixture stirred for 10 min, *p*-tolualdehyde added slowly, stirring continued for 2 h, excess of diethylzinc removed at reduced pressure, then the residue quenched with satd. aq. NH$_4$Cl → product. Y 99% (e.e. 90%). Reaction is faster than by the traditional solution-phase procedure (*42*, 616s*60*) without decrease in enantioselectivity. F.e. and chiral aminoalcohols s. I. Sato, K. Soai et al., Chem. Commun. *2000*, 2471-2.

Dialkylzinc/(R)-α-deuteriobenzyl alcohol R$_2$Zn/PhCH(D)OH
Asym. induction with chiral α-deuteriobenzyl alcohols ←

230.

A soln. of 0.075 mmol diisopropylzinc (1 *M*) in toluene added at 0° to a soln. of 0.025 mmol (R)-α-deuteriobenzyl alcohol (>95% e.e.) in the same solvent, a soln. of 0.025 mmol of the startg. aldehyde in the same solvent added dropwise via syringe at the rate of 1 drop/30 sec, the mixture stirred at 0° for 12 hrs., a second portion of diisopropylzinc (0.2 mmol), aldehyde (0.1 mmol) and solvent added, stirring continued for 3.5 h, a third portion of diisopropylzinc (0.8 mmol) and aldehyde (0.4 mmol) added, the mixture stirred for 2.5 h, finally a fourth portion of diisopropylzinc (2 mmol) and aldehyde (1 mmol) added, stirring continued for 2 h, then quenched with 1 *M* HCl → product. Y 98% (e.e. 95%). This is the first highly enantioselective synthesis mediated by an isotopic chiral inducer. In this instance, reaction benefits from **asym. automultiplication** (cf. *51*, 271): the chiral deuterioalcohol initially reacts with diisopropylzinc to form a chiral isopropylzinc alkoxide which induces a small level of face selectivity on formation of the product (also as an isopropylzinc alkoxide); this in turn serves as an enhanced chiral inducer so that the e.e. of the product increases to a high level as reaction proceeds. F.e.s. I. Sato, K. Soai et al., J. Am. Chem. Soc. *122*, 11739-40 (2000).

Dialkylzinc/chiral aminoalcohols or thiazolidines or 2-aminodisulfides ←
Asym. synthesis of sec. alcohols from aldehydes ←
update s. *42*, 616s*59*; with D-altritol-derived 2-aminoalcohols s. B.T. Cho et al., Tetrahedron:Asym. *11*, 2149-57 (2000); with chiral N-(9-phenylfluoren-9-yl)-2-aminoalcohols s. M.R. Paleo, F.J. Sardina et al., J. Org. Chem. *65*, 2108-13 (2000); with camphor-derived 4-aminoalcohols s. M. Knollmuller et al., Tetrahedron:Asym. *10*, 3969-75 (1999); with chiral 3,5-dihydro-4*H*-dinaphth-[2,1-*c*:1',2'-*e*]azepine-derived 2-aminoalcohols s. N. Arroyo, M. Widhalm et al., ibid. *11*, 4207-19 (2000); with chiral (C$_2$-symmetric) aziridine- and piperidine-derived 2-aminoalcohols s. M. Shi et al., ibid. 4923-33; with chiral 2,4-bis(1-hydroxycyclopentyl)azetidines s. J. Wilken, J.

Martens et al., ibid. 2143-8; with further chiral azetidines s. M. Shi, J.-K. Jiang, ibid. *10*, 1673-9 (1999); with chiral pyrrolidine-derived 2-aminoalcohols s. X. Yang, R. Wang et al., ibid. 133-8; with chiral TADDOL-derived 4-aminoalcohols s. C.-T. Qian et al., ibid. *11*, 1733-40 (2000); with camphor- or fenchone-derived 4-aminoalcohols s. N. Hanyu, T. Fujita et al., ibid. 4127-36; with further fenchone derivs. s. B. Goldfuss et al., J. Org. Chem. *65*, 77-82 (2000); with chiral pyridyl- and bipyridyl-alcohols s. P. Collomb, A. von Zelewsky, Tetrahedron:Asym. *9*, 3911-7 (1998); with (S)-3-methyl-2-(2-pyridylmethylamino)-1,1-diphenylbutan-1-ol s. H. Yun, Y. Wu et al., Tetrahedron Lett. *41*, 10263-6 (2000); s.a. Tetrahedron:Asym. *11*, 3543-52 (2000); with fructose-derived pyridylalcohols s. Y.-G. Zhou et al., Chin. J. Chem. *18*, 121-3 (2000); with chiral 3-indolyl-based 2-aminoalcohols s. W.-M. Dai et al., Tetrahedron:Asym. *11*, 2315-37 (2000); with chiral ferrocenylaminoalcohols s. S. Bastin, L. Pelinski et al., Tetrahedron Lett. *41*, 7303-7 (2000); s.a. Tetrahedron:Asym. *10*, 1647-51 (1999); O. Delacroix, J. Brocard et al., ibid. 4417-25; with chiral thiazolidine-4-carboxylic acid esters s. Q. Meng, Y. Guan et al., ibid. *11*, 4255-61 (2000); with L-cysteine-derived 2-aminodisulfides s. ibid. *10*, 1733-8 (1999); with $Cr(CO)_3$-complexed indan-derived aminoalcohols s. ibid. *9*, 2595-610 (1998); kinetic study of product inhibition s. T. Rosner, D.G. Blackmond et al., Org. Lett. *2*, 2511-3 (2000).

Dialkylzinc/chiral 5-carbamyl-3-pyridylcarbinols ←
Catalytic asym. automultiplication ←
s. *51*, 271s*58*; high asym. induction (e.e. 86-8%) on dialkylzinc addition to 5-carbamyl-3-pyridinecarboxaldehydes with a little chiral 5-carbamyl-3-pyridylcarbinols of low e.e. (4%) s. S. Tanji, K. Soai et al., Tetrahedron:Asym. *11*, 4249-53 (2000); asym. amplification on addition of diethylzinc to benzaldehyde with a little chiral *o*-hydroxybenzylamine as ligand s. ibid. 3361-73.

Dialkylzinc/titanium tetraisopropoxide/chiral soluble or polymeric or polymer-based ←
  1,1'-bi-2-naphthols or o,o'-dihydroxybiphenyls or 1,2-di(sulfonylamines)
Asym. synthesis of sec. alcohols from aldehydes CHO → CH(OH)R
s. *44*, 565s*59*; with camphor-derived di(sulfonylamines) s. C.-D. Hwang, B.-J. Uang, Tetrahedron: Asym. *9*, 3979-84 (1998); s.a. O. Prieto, M. Yus et al., ibid. *11*, 1629-44 (2000); preparation and catalytic screening of peptidyl di(sulfonylamines) s. A.J. Brouwer, R.M.J. Liskamp et al., J. Org. Chem. *65*, 1750-7 (2000); conformational analysis of chiral N,N'-bis(arenesulfonyl)-1,2-diaminocyclohexanes s. J. Balsells, P.J. Walsh et al., Angew. Chem. Int. Ed. Engl. *39*, 3428 (2000); with bridged polymeric (R)-BINOLs as ligand cf. C. Dong, Z. Yu et al., Tetrahedron:Asym. *11*, 2449-54 (2000); s.a. B.H. Lipshutz, Y.J. Shin, Tetrahedron Lett. *41*, 9515-21 (2000); with a polymer-based (S)-BINOL-3,3'-dicarboxamide s. X.-W. Yang, R. Wang et al., J. Org. Chem. *65*, 295-6 (2000); with a bis(steroidal) *o,o'*-dihydroxybiphenyl cf. K. Kostova et al., Tetrahedron:Asym. *11*, 3253-6 (2000); with (R)-5,6,7,8-tetrahydro-1,1'-bi-2-naphthol as ligand s. X. Shen, K. Ding et al., ibid. 4321-7.

Dialkylzinc/titanium tetraisopropoxide/fluorous chiral 1,1'-bi-2-naphthol ←
Asym. synthesis of sec. alcohols from aldehydes
in a fluorous 2-phase medium s. *58*, 235; with (S)-4,4',6,6'-tetrakis(perfluorooctyl)-1,1'-bi-2-naphthol as ligand s. Y. Tian, K.S. Chan, Tetrahedron Lett. *41*, 8813-6 (2000).

Diethylzinc s.a. under Ar_2Zn and $RhCl(PPh_3)_3$ Et_2Zn

Diethylzinc/(R,R)-2,6-bis[2-(diphenylhydroxymethyl)-1-pyrrolidinylmethyl]-p-cresol/ ←
  triphenylphosphine sulfide/molecular sieves
Catalytic asym. aldol condensation of ketones with aldehydes CHO → CH(OH)C-CO

231.

A soln. of 10 mol% Et_2Zn (1 *M* in hexane) added under argon to a soln. of 5 mol% (R,R)-2,6-bis[2-(diphenylhydroxymethyl)-1-pyrrolidinylmethyl]-*p*-cresol in THF at room temp., the mixture stirred for 30 min, added to a suspension of the startg. aldehyde, 15 mol% triphenylphosphine sulfide, 4 Å molecular sieves and 10 eqs. acetophenone in THF, stirring continued for 2 days at 5°, then quenched with 1 N HCl → product. Y 79% (e.e. 99%). In this *direct* asym. aldol condensation, the liberation of 4 equivalents of ethane would suggest that the catalytic cycle involves a *bimetallic* zinc alkoxide wherein the ketone is bound to one of the zinc atoms as an enolate prior to *re*-face selective addition to the aldehyde, itself coordinated via the carbonyl group to the second, proximal zinc atom. F.e.s. B.M. Trost, H. Ito, J. Am. Chem. Soc. *122*, 12003-4 (2000).

Diisopropylzinc   i-Pr_2Zn
Dialkylzinc-mediated Reformatskii-type synthesis   CO → C(OH)C-COOR
via zinc enolate equivalents

232.

β-Hydroxycarboxylic acid esters. 1.5 eqs. Ethyl iodoacetate added under argon at room temp. to a soln. of 2.1 eqs. 1 M diisopropylzinc in hexane, stirred for 4 h, 1 eq. benzaldehyde added at 0°, stirred for 20 h, and quenched with satd. aq. NH_4Cl → product. Y 98%. Reaction is generally applicable to aliphatic and ar. aldehydes as well as aryl ketones and enones. However, there was no reaction with diethylzinc and ethyl iodoacetate. F.e.s. I. Sato, K. Soai et al., Bull. Chem. Soc. Jpn. *73*, 2825-6 (2000).

Diarylzinc compds./1-diphenylhydroxymethyl-2-(4(S)-isopropyl-Δ^2-oxazolin-2-   ←
yl)ferrocene/diethylzinc
Asym. synthesis of sec. benzylalcohols from aldehydes   CHO → CH(OH)Ar
with (-)-3-*exo*-(dimethylamino)isoborneol as ligand cf. *55*, 239; improved procedure for chiral diarylcarbinols with 1-diphenylhydroxymethyl-2-(4(S)-isopropyl-Δ^2-oxazolin-2-yl)ferrocene/ diethylzinc as ligand s. C. Bolm et al., Angew. Chem. Int. Ed. Engl. *39*, 3465-7 (2000).

Bis(iodozincio)methane $CH_2(ZnI)_2$
cis-**Cyclopropane-1,2-diol O-derivs. from α-diketones**

233.

A soln. of 1.2 eqs. bis(iodozincio)methane (prepared in THF from zinc and diiodomethane in the presence of $PbCl_2$) added dropwise at 25° to a soln. of the startg. diketone in the same solvent, the mixture stirred for 30 min, 2.4 eqs. trimethylsilyl chloride added dropwise, stirring continued for 30 min, then quenched with satd. aq. NH_4Cl → product. Y 97%. The *cis*-isomer is produced stereoselectively. Reaction is thought to involve initial coordination of the zinc reagent to the *s-cis*-form of the diketone. followed by generation of an intermediate α-(iodozinciomethyl)ketone prior to alkoxide-mediated ring closure and quenching with the chlorosilane. By using acetic anhydride instead of trimethylsilyl chloride, the diacetate was formed. F.e.s. K. Ukai, S. Matsubara et al., J. Am. Chem. Soc. *122*, 12047-8 (2000).

Zinc triflate/(-)-N-methylephedrine/triethylamine   ←
Asym. synthesis of 2-acetylenealcohols   CHO → CH(OH)C≡C
from terminal acetylene derivs. and aldehydes under mild conditions
s. *58*, 236; chiral 2-acetylene-1,4-diols, also degradation to sec. propargyl alcohols, s. D. Boyall, E.M. Carreira et al., Org. Lett. *2*, 4233-6 (2000).

Magnesium bromide/magnesium
Synthesis of α-bromoglycols from α,β-epoxyaldehydes
Regio- and stereo-specific conversion

$MgBr_2/Mg$

234.

5 eqs. $MgBr_2·Et_2O$ added to a soln. of (2S*,3R*)-2,3-epoxyhexanal in methylene chloride at -50°, stirred for 12 h, 1.5 eqs. 2 M MeMgBr in THF added, and quenched with satd. NH_4Cl soln. after 12 h → (2S*,3R*,4S*)-4-bromo-2,3-heptanediol. Y 67%. *anti,syn*-Isomers were obtained exclusively under chelation control. The method is also effective in the presence of bulky groups at C-3 or in the Grignard. F.e. and conversion to *syn,syn*-α-aminoglycols s. G. Righi et al., Eur. J. Org. Chem. **2000**, 3127-31.

Indium
Barbier-type synthesis of 3-ethylenealcohols
in aq. medium s. **40**, 567s58; 10-allyl-10-hydroxyanthracen-9-ones s. S. Kumar et al., J. Chem. Res., Synop **2000**, 314-5; 3-allyl-3-hydroxycephams **with asym. induction** s. J.E. Lee, Y.S. Cho et al., Synth. Commun. **30**, 4299-308 (2000); chiral N-protected δ-amino-γ-hydroxy-α-methylenecarboxylic acid esters s. M.D. Chappell, R.L. Halcomb, Org. Lett. **2**, 2003-5 (2000); chain extension of mannose to **sialic acids** (in aq. HCl) s. M. Warwell, W.-D. Fessner, Synlett **2000**, 865-7.

In
$CO → C(OH)C-C≡C$

3,4-Unsatd. 2,2-difluoroalcohols from aldehydes
3-ethylene-2,2-difluoroalcohols s. **40**, 567s53; details s. Tetrahedron **56**, 8275-80 (2000); C-silyl protected 3-acetylene-2,2-difluoroalcohols or 1,1-difluoroallenes s. Z.G. Wang, G.B. Hammond, J. Org. Chem. **65**, 6547-52 (2000).

←

Diisobutylaluminum hydride s. under Cp_2TiCl_2

$i\text{-}Bu_2AlH$

Trialkylalanes
3-Acetylenealcohols from epoxides and ynalanes
s. **32**, 231; 3-acetylene-3'-hydroxythioethers from 2,3-epoxythioethers, regio- and stereo-specific ring opening via thiiranium salts with **double inversion of configuration**, s. M. Sasaki, M. Miyashita et al., Tetrahedron Lett. **40**, 9267-70 (1999).

R_3Al
$→ C(OH)C\text{-}C≡C$

Trialkylalanes/titanium tetraisopropoxide/chiral α,α,α',α'-tetraphenyl-1,3-dioxolane-4,5-dimethanols
Asym. synthesis of sec. alcohols from aldehydes
with (R)-1,1'-bi-2-naphthol as ligand cf. **53**, 268; with TADDOLs (at an optimum ratio of TADDOL/Ti(IV)/AlR_3 at 1:25:25) s. J.-F. Lu, H.-M. Gau et al., Tetrahedron:Asym. **11**, 2531-5 (2000).

←
$CHO → CH(OH)R$

Diisopinocampheyl(methoxy)borane s. under BuLi
R_2BOMe

Dicyclohexylborinyl triflate/triethylamine
Asym. aldol condensation with 1,4-dioxan-2-ones via their boron enolates

R_2BOTf/Et_3N
←

235.

2.5 eqs. Dry Et_3N added dropwise over 5 min to a stirred soln. of (S,S)-5,6-diphenyl-1,4-dioxan-2-one in dry dichloromethane at -78°, the mixture stirred for another 5 min, 3 eqs. dicyclohexylborinyl triflate in hexane added dropwise over 10 min by syringe, the mixture stirred for 2-3 h at the same temp., quenched with pH 7 buffer, followed by methanol and 30%

aq. H_2O_2, and the mixture stirred while warming to room temp. → *anti*-product. Y 86% (d.e. 92%). The *anti*-selectivity is a consequence of the enolate necessarily being locked in the (E)-form. However, yields and stereoselectivities were lower with aromatic and unsatd. aldehydes. F.e. and conversion to **chiral *anti*-α,β-dihydroxycarboxylic acids** s. M.B. Andrus et al., Org. Lett. *2*, 3035-7 (2000).

Dicyclohexylborinyl chloride/triethylamine R_2BCl/Et_3N
Aldol condensation via enol borinates   CHO → CH(OH)C-CO
via (E)-enol borinates cf. *32*, 614s49; *syn*-β-hydroxyketones with asym. induction via (Z)-enol borinates (with dicyclohexylborinyl chloride), also conversion to chiral *syn*-α-methyl-β-hydroxyesters, s. M. Carda, J.A. Marco et al., Tetrahedron:Asym. *11*, 3211-20 (2000).

Chiral 2-bromo-1,3-disulfonyl-1,3,2-diazaborolidines/ethyldiisopropylamine   ←
Asym. aldol condensation
s. *45*, 402; with (phenylthio)acetic acid esters s. E.J. Corey, S. Choi, Tetrahedron Lett. *41*, 2769-72 (2000).

Boron fluoride s. under BuLi BF_3

Chiral chloroaluminum bis(phosphinylalkoxides) $ClAl(OR)_2$
Cyanohydrins from oxo compds. under bifunctional asym. catalysis   CO → C(OH)CN
from aldehydes with a chiral chloroaluminum bis(phosphinyl)BINOLate cf. *57*, 238; also from acylophenones with carbohydrate-based chloroaluminum bis(phosphinylalkoxides) s. M. Kanai, M. Shibasaki et al., Tetrahedron Lett. *41*, 2405-9 (2000).

Samarium/ammonium chloride or hydrogen chloride Sm/NH_4Cl or HCl
Pinacolization of aryl ketones   2 CO → C(OH)C(OH)
with Sm/allyl bromide cf. *53*, 22s59; with Sm/NH_4Cl under ultrasonication s. M.K. Basu, F.F. Becker et al., J. Chem. Res., Synop *2000*, 406-7; with Sm in 5:1 aq. HCl/THF, also from ar. aldehydes, s. S. Talukdar, J.M. Fang, J. Org. Chem. *66*, 330-3 (2001).

Chiral yttrium salen complex   ←
Chiral 1,3-diol monoesters by catalytic asym. aldol-Tishchenko reaction   ←

236.

A soln. of benzaldehyde and isobutyraldehyde in methylene chloride treated with 2 mol% pentameric yttrium alkoxide, $Y_5O(OPr-i)_{13}$, and 13 mol% chiral salen complex in the presence of 4 Å molecular sieves → product. Y 70% (>15:1 mixture of regioisomers; S/R 87:13). The two regioisomers are formed in similar enantiopurity suggesting a non-selective intramolecular acyl migration after formation of the initial aldol-type adduct. The transition state is thought to involve a hexacoordinate *monomeric* yttrium salen complex wherein the isopropyl group of the intermediate hemiacetal occupies a pseudoequatorial position in readiness for the rate-determining hydride transfer. Details s. C.M. Mascarenhas, J.P. Morken et al., Angew. Chem. Int. Ed. Engl. *40*, 601-3 (2001).

Scandium(III) triflate $Sc(OTf)_3$
α,β-Ethylene-δ-hydroxyaldehydes from 3-ethyleneepoxides s. *60*, 431

Lanthanide(III) triflates $Ln(OTf)_3$
***o*-α-Hydroxyalkylation of phenols**   CO → C(OH)Ar
of phenols with EtMgBr cf. *43*, 563; with Yb(III)-, Nd(III)-, Dy(III)-, La(III)- or Sc(III)-triflate, also regiospecific α-hydroxyalkylation of phenolethers and heteroarenes, s. W. Zhang, P.G. Wang, J. Org. Chem. *65*, 4732-5 (2000).

Ytterbium(III) triflate/chiral bis(Δ^2-oxazolines)
Asym. ene reaction   ←
   $CO \rightarrow C(OH)C-C\equiv C$
with Yb(OTf)$_3$/(S)-6,6'-dibromo-1,1'-bi-2-naphthol cf. *54, 261*; with a chiral bis(2-oxazoline) (Ph-pybox) as ligand s. C. Qian, L. Wang, Tetrahedron:Asym. *11*, 2347-57 (2000).

Cerium(III) chloride s. under Mg $CeCl_3$

Samarium diiodide SmI_2
1,2-*trans*-C-Glycosides from silyl glycosides via glycosyl iodides   I →R

237.

1,2-*trans*-C-(1-Hydroxyalkyl) glycosides from oxo compds. A soln. of startg. per-silylated sugar in methylene chloride treated with 1.1 eqs. Me$_3$SiI at 25° for 30 min, solvent removed, the residue treated successively with 2 eqs. cyclohexanone and a soln. of 2.2 eqs. SmI$_2$ in THF at room temp. (25°) for ca. 1.5 h (until the blue colouration disappeared), 3:1 methanol/1 *M* HCl added at 25° (to effect desilylation), and acetylated with Ac$_2$O in the presence of *2,6-lutidine* (pyridine itself undergoing ring opening) *at 25°* → product. Y 85% (after silica gel column chromatography; 100% α). There was no competing β-elimination, as takes place with the corresponding gluco- and galacto-pyranosyl sulfones (cf. *50, 531s55*). Reaction also takes place with benzyl-protected silyl glycosides. F.e. incl. C-disaccharides, also with added NiCl$_2$ as catalyst, s. N. Miquel, J.-M. Beau et al., Chem. Commun. *2000*, 2347-8.

Decarboxylase   ←
Asym. benzoin condensation $2 ArCHO \rightarrow ArCH(OH)COAr$
with chiral thiazolium salts cf. *26, 675s40*; with benzoylformate decarboxylase s. A.S. Demir, M. Muller et al., Tetrahedron:Asym. *10*, 4769-74 (1999).

Chiral titanium(IV) 1,1'-bi-2-naphthoxides/o,o'-dihydroxybiphenyls   ←
Asym. ene reaction $CO \rightarrow C(OH)C-C\equiv C$
with glyoxylic acid esters using chiral Ti(IV)-BINOLate complexes cf. *44, 568s55*; enhanced enantioselectivity with an added racemic *o,o'*-dihydroxybiphenyl s. M. Chavarot, Y. Vallee et al., Tetrahedron:Asym. *9*, 3889-94 (1998); with bis(acetonitrile)[(S)-2,2'-bis(di-*p*-tolylphosphino)-1,1'-binaphthyl]palladium(II) hexafluoroantimonate s. J. Hao, K. Mikami et al., Org. Lett. *2*, 4059-62 (2000).

Di-tert-butoxytitanium(IV) 1,1'-bi-2-naphthoxide/(R)-mandelic acid   ←
Asym. titanium(IV)-catalyzed aldol condensation $CHO \rightarrow CH(OH)C-CO$
mediated by chiral α-hydroxycarboxylic acids

238.

1.5 eqs. Benzaldehyde added to a suspension of 1 eq. di-*tert*-butoxytitanium(IV) 1,1'-bi-2-naphthoxide in toluene under argon, the mixture stirred for 30-60 min at room temp., 1 eq. (R)-mandelic acid added, stirred for 15 min, 3-pentanone added, stirred again for 5-8 h at the same temp., and quenched with satd. aq. NaHCO$_3$ → product. Y 72% (*syn*-selectivity 73%; e.e. 94% for the major isomer). Asym. induction benefits from ligand exchange between the BINOLate complex and the chiral α-hydroxy acid to generate *in situ* a chiral titanium(IV) alkoxide to induce the required face selectivity. There was no added advantage in using a chiral BINOL ligand. F.e. and ligand optimization s. R. Mahrwald, Org. Lett. 2, 4011-2 (2000).

Chiral cyclopentadienyl(1,3-dioxolane-4,5-dimethanolato-O,O')titanium chloride/ ←
magnesium
Asym. synthesis of 3-ethylenealcohols from aldehydes CHO → CH(OH)C-C≡C
s. 45, 382s47; chiral 1,5-enyn-4-ols from α,β-acetylenealdehydes and a preformed chiral allyl-(cyclopentadienyl)titanium TADDolate complex s. S. BouzBouz, J. Cossy et al., Tetrahedron Lett. 41, 8877-80 (2000); **chiral 5-ene-1,3-diols** s. Org. Lett. 2, 501-4 (2000); chiral 4-ene-1,2-diol 1-mono-ethers and -esters from O-protected α-hydroxyaldehydes s. Tetrahedron:Asym. 10, 3859-62 (1999).

Titanocene dichloride/diisobutylaluminum hydride/trimethylsilyl triflate ←
Isocyclic alkoxy-3-ethylenes from cyclic (1,3-diene)acetals ∞
via intramolecular allyltitanation s. 57, 321; cyclo-propane, -butane and -pentane analogs s. N. Thery et al., Eur. J. Org. Chem. 2000, 1483-8.

Zirconocene dichloride/lithium diisopropylamide $Cp_2ZrCl_2/i\text{-}Pr_2NLi$
Asym. aldol condensation with carboxylic acid amides CHO → CH(OH)C-CON<
with N-acyl-β-ketoamides using $TiCl_4/i\text{-}Pr_2NEt$ cf. 45, 383; with simple amides based on (S,S)-pseudoephedrine via Zr(IV)-enolates (using $Cp_2ZrCl_2/i\text{-}Pr_2NLi$) s. J.L. Vicario et al., J. Org. Chem. 65, 3754-60 (2000); with (S)-2-(pyrrolidin-2-yl)propan-2-ol as chiral auxiliary cf. E. Hedenström et al., J. Chem. Soc. Perkin Trans. 1 2000, 1513-8.

Tin/trimethylsilyl chloride Sn/Me_3SiCl
Sym. glycol (pinacol) ethers from aldehydes 2 CHO → CH(OR)CH(OR)

239. 2 Ph-CHO + 2 $HOCH_2CH_2Cl$ ⟶ $ClCH_2CH_2O$-CH(Ph)-CH(Ph)-OCH_2CH_2Cl

A soln. of benzaldehyde and 1.5 eqs. 2-chloroethanol in dry THF stirred with 0.75 eq. tin powder and 2.5 eqs. Me_3SiCl at room temp. for 10 h with exclusion of moisture → 1,8-dichloro-4,5-diphenyl-3,6-dioxaoctane. Y 89% (dl:meso 1:1). This provides a straightforward, high-yielding route to **sym. 4,5-diaryl-1,8-dihalogeno-3,6-dioxaoctanes**, which are useful precursors of O-macroheterocyclics such as crown ethers or [2]catenananes. F.e.s. H. Tong, Z. Ding et al., Synth. Commun. 30, 4097-105 (2000); f. sym. glycol ethers (without formation of by-products such as pinacols or sym. alkenes) s. Acta Chimica Sinica 57, 1152 (1999).

Vanadyl chloride/zinc $VOCl_3/Zn$
Sym. glycol (pinacol) esters from aldehydes 2 CHO → CH(OAc)CH(OAc)

240. 2 Ph-CHO + 2 Ac_2O ⟶ AcO-CH(Ph)-CH(Ph)-OAc

Benzaldehyde treated with 3 mol% $VOCl_3$, 2 eqs. Ac_2O and 2 eqs. Zn in DME at 80° for 24 h under argon → product. Y 85% (dl/meso 80:20).This catalytic method gives good diastereoselectivity and the products are easily handled. However, reaction did not proceed under these conditions when benzoyl chloride was used as acylating agent. F.e. and with $TiCl_4$ (10 mol%)/Al (2 eqs.)/AcCl (2 eqs.) at room temp. (with complementary diastereoselectivity in some cases) s. T. Hirao et al., Synlett 2000, 1658-60.

Chiral quaternary ammonium fluorides ←
Asym. Henry reaction CHO → CH(OH)CH_2NO_2
with chiral rare earth BINOLate complexes cf. 48, 600; with a chiral azabicyclic quaternary ammonium fluoride (cf. 34, 610s51) s. E.J. Corey, F.-Y. Zhang, Angew. Chem. Int. Ed. Engl. 38, 1931-4 (1999).

Cobalt carbonyl/triphenylphosphine $Co_2(CO)_8/Ph_3P$
β-Hydroxyaldehydes from epoxides via hydroformylation

241. ⟨epoxide⟩ + CO/H₂ $\xrightarrow{Co_2(CO)_8}$ HO–CH₂–CH₂–CHO

A 1:1 mixture of CO and H_2 introduced at 100 bar into a steel autoclave containing a soln. of 0.4 mol% $Co_2(CO)_8$ and 1.6 mol% Ph_3P in toluene, heated at 100° for 1 h, a mixture of ethylene oxide and diglyme at 0° added dropwise, heated at 100° for 3 h, and worked up at 5° → 3-hydroxypropanal. Y 65%. F.e.s. R. Weber, European patent EP-1000921 (Degussa-Huels AG); details and further P- or P,O-ligands s. Chem. Commun. *2000*, 1419-20.

Dibromo[1,2-bis(diphenylphosphino)ethane]nickel(II) s. under Zn $Ni(dppe)Br_2$
Chlorotris(triphenylphosphine)rhodium(I)/diethylzinc $RhCl(PPh_3)_3/Et_2Zn$
Rhodium-catalyzed Reformatskii-type synthesis $CO \rightarrow C(OH)C\text{-}COOR$
under mild conditions

242. Br–CH₂–CO₂Et $\xrightarrow{[Rh]}$ [Br[Rh]–CH₂–CO₂Et] $\xrightarrow{Et_2Zn}$ CH₂=C(OZnEt)(OEt) $\xrightarrow{PhCHO}$ Ph–CH(OH)–CH₂–CO₂Et

Ethyl bromoacetate, benzaldehyde, and 2.2 eqs. ca. 1 M Et_2Zn in hexane added to a stirred soln. of 5 mol% $RhCl(PPh_3)_3$ in THF at 0°, stirred for *5 min,* and quenched with satd. aq. $NaHCO_3 \rightarrow$ product. Y 82%. The procedure is rapid and generally applicable to ar. and aliphatic aldehydes and ketones. Significantly, there was no addition of the dialkylzinc to the aldehyde group. Reaction is initiated by oxidative addition of Rh(I) to the carbon-halogen bond, followed by transmetalation with Et_2Zn to give a reactive ethylzinc enolate with liberation of Rh(I) to complete the catalytic cycle. F.e. and intramolecular conversion s. K. Kanai, T. Honda et al., Org. Lett. *2*, 2549-51 (2000).

Bis(acetonitrile)[(S)-2,2′-bis(di-p-tolylphosphino)-1,1′-binaphthyl]- ←
palladium(II) hexafluoroantimonate
Asym. ene reaction $CO \rightarrow C(OH)C\text{-}C\!=\!C$
with glyoxylic acid esters s. *44,* 568s60

Addition to Nitrogen and Carbon CC ⇓ NC

Without additional reagents w.a.r.
β-Ketocarboxylic acid amides $>\!N\text{-}C\!=\!CH \rightarrow\, >\!N\text{-}C\!=\!C\text{-}CONHR$
from enamines and isocyanates via α,β-ethylene-β-aminocarboxylic acid amides
s. *20,* 502; polymer-based synthesis with a supported enamine s. F. Aznar, C. Valdes, M.-P. Cabal, Tetrahedron Lett. *41,* 5683-7 (2000).

Lithium/naphthalene $Li/C_{10}H_8$
Ring opening of N-heterocyclics via reductive lithiation s. *45,* 368s60

Sodium/liq. ammonia or naphthalene ←
2-Subst. 1,2-dihydropyridine from pyridine ring $C\!=\!N \rightarrow C(R)NH$
with Li cf. *3,* 568; *4,* 768; from electron-deficient pyridines by Birch-type reductive alkylation with Na/liq. NH_3 or Na/naphthalene s. T.J. Donohoe et al., Org. Lett. *2,* 3861-3 (2000).

n-Butyllithium BuLi
Phthalimidine ring by cyclocarbonylation
of 2-(*o*-bromoaryl)-Δ^2-oxazolines under dual metal catalysis cf. *52,* 422; from 2-aryl-Δ^2-oxazolines via *o*-lithiation with *n*-BuLi s. K. Iwamoto, S. Murai et al., J. Org. Chem. *65,* 7944-8 (2000).

N-Protected 3-silyl-2-pyrrolidones from aziridines
Lithium silylethynolates as nucleophilic silylketene equivalents

243.

1.1 eqs. n-BuLi (1.5 M in hexane) added via syringe under N_2 at -78° to a soln. of 0.9 eq. trimethylsilyldiazomethane in dry THF, the mixture stirred 1 h, then again under a CO atmosphere for 2 h, a soln. of the startg. aziridine in THF added at -78°, the mixture warmed to 20°, stirred for 12 h, then quenched with satd. aq. NH_4Cl at -78° → product. Y 77% (77:23 mixture of diastereoisomers). The reaction is analogous to that with epoxides except that trimethylaluminum is not required (cf. *52*, 380). F.e., also **3-alkylidene-2-pyrrolidones** by addition of the aldehyde before the aqueous quench, s. K. Iwamoto, S. Murai et al., J. Org. Chem. *66*, 169-74 (2001).

n-Butyllithium/diethylzinc   *BuLi/Et$_2$Zn*
β-Aminocarboxylic acid esters from aldimines   C=N → C(NH)CH$_2$COOR
Asym. synthesis with addition of two C-atoms

244.

A soln. of (R R)-2,6-bis(2-isopropylphenyl)-3,5-dimethylphenyl acetate in THF treated with 1 eq. BuLi at -78° for 30 min, 1 eq. Et_2Zn and 1 eq. startg. aldimine added, and worked up after 18-48 h → (S)-product. Y 82% (d.r. 95.5:4.5). The *o*-methoxy group (or *o*-fluorine) is critical for coordination of the lithium cation, reaction failing with the corresponding N-phenylaldimine. The Lewis acid was essential to activate the imine function. F.e. and conversion to the corresponding β-*prim*-aminocarboxylic acid methyl esters with 99% recovery of the chiral auxiliary s. S. Saito et al., Org. Lett. *2*, 1891-4 (2000).

Aryllithium compds./(R,R)-1,2-dimethoxy-1,2-diphenylethane   ←
Asym. synthesis of sec. amines from azomethines   C=N → C(R)NH
with chiral 2-aminoethers as ligand cf. *39*, 612s53; addition of aryllithiums with (R,R)-1,2-dimethoxy-1,2-diphenylethane as ligand, and conversion to chiral α-acylamino acids, s. M. Hasegawa, K. Tomioka et al., Tetrahedron *56*, 10153-8 (2000).

Lithium diisopropylamide   *i-Pr$_2$NLi*
Asym. synthesis of tosylamines from N-tosylaziridines   ⊂
and Grignards with chiral Cu(II)-salen complexes cf. *58*, 255; chiral γ-tosylaminohydrazones from SAMP-hydrazones (with LDA) s. D. Enders et al., Synlett *2000*, 641-3.

Lithium bis(trimethylsilyl)amide   *LiN(SiMe$_3$)$_2$*
Asym. synthesis of N-[(β-phosphinylamino)acyl]sultams   C(NHP(O)<)C-CON<
from N-acylsultams and N-phosphinylaldimines

245.

A soln. of 1.07 eqs. N-propionyl-(2S)-bornane-10,2-sultam in anhydrous THF at -78° under inert atmosphere treated dropwise with 1.13 eqs. 1 M LHMDS, the resulting yellow soln. stirred for ca. 30 min, a soln. of P,P-diphenyl-N-(phenylmethylene)phosphinic amide in THF added, and stirring continued for ca. 3-4 h (monitored by TLC) before quenching with satd. NH$_4$Cl soln. → (2'R,3'R)-(+)-product. Y 73% (syn:anti >95:<5; e.r. >95:<5). This method gives virtually complete diastereo- and enantio-control in acceptable to good yields. Since the chiral auxiliary may be cleaved easily without racemization (with LiOH in THF/water), and the diphenylphosphinyl is also cleaved easily, it provides an operationally simple route to α-methylated β-amino acids from readily-available startg. m. F.e.s. A.B. McLaren, J.B. Sweeney, Synlett 2000, 1625-7.

Cuprous hexafluorophosphate/(R)-2,2'-bis(di-p-tolylphosphino)- ←
1,1'-binaphthyl
N-Protected arylglycinates from electron-rich arenes   C≡N → C(Ar)NH
Regiospecific asym. synthesis with addition of two C-atoms

246.

A soln. of ethyl (carbethoxyimino)acetate and N,N-dimethylaniline in THF treated with 5 mol% CuPF$_6$/(R)-Tol-BINAP at -78° until reaction complete → (R)-product. Y 75% (e.e. 96%). Only p-subst. products were isolated. Reaction was also facile with N-carbo-*tert*-butoxy-, N-carbomethoxy-, N-carbobenzoxy- and N-tosyl-derivs., the (S)-enantiomers being obtained from the bulkier N-Boc and N-Ts derivs. F.e. and electron-rich arenes, also N-decarbalkoxylation, s. S. Saaby, K.A. Jorgensen et al., Angew. Chem. Int. Ed. Engl. 39, 4114-6 (2000).

Cupric trifluoromethanesulfonate s. under R$_2$Zn   Cu(OTf)$_2$
Magnesium/zirconocene dichloride   Mg/Cp$_2$ZrCl$_2$
Synthesis of sec. amines from azomethines   C=N → C(R)NH
s. 4, 643; chiral (C$_2$-symmetric) 1,2-diamines s. S. Roland, P. Mangeney, Eur. J. Org. Chem. 2000, 611-6; f. method (by addition of RLi) s. G. Martelli, D. Savoia, Tetrahedron 56, 8367-74 (2000); addition of EtMgBr to normally unreactive azomethines with added Cp$_2$ZrCl$_2$ as catalyst, also 1,2-addition of Cp$_2$Zr(CH$_2$=CH$_2$) in the presence of Grignard reagents, s. T. Takahashi et al., Chem. Commun. 2001, 31-2.

Zinc/ammonium chloride   Zn/NH$_4$Cl
Sym. 1,2-diamines from azomethines   2 C=N → C(NH)C(NH)
with Zn in 10% NaOH cf. 47, 614s55; from ar. azomethines with Zn-powder and NH$_4$Cl (or L-tyrosine) **in water** without organic solvent s. T. Tsukinoki et al., Green Chem. 2, 117-22 (2000).

Dialkylzinc/1(S)-[9-(piperidin-1-yl)fluoren-9-yl]ethanol/triisopropylsilyl chloride   ←
Asym. synthesis of phosphinic acid amides from N-phosphinylimines   CH(R')NHP(O)R$_2$
s. 48, 625s58; enhanced enantioselectivity under dual catalysis with 1(S)-[9-(piperidin-1-yl)-fluoren-9-yl]ethanol and *i*-Pr$_3$SiCl s. C. Jimeno, M.A. Pericàs et al., Org. Lett. 2, 3157-9 (2000); with dendritic 2-aminoalcohols as ligand cf. I. Sato, K. Soai et al., Tetrahedron:Asym. 11, 2271-5 (2000).

Dialkylzinc/cupric trifluoromethanesulfonate/(S)-2-diphenylphosphinomethyl-   ←
4,4-dibenzyl-N-pivaloylpyrrolidine
Asym. synthesis of sulfonylamines from N-sulfonylimines   C=NSO$_2$R → C(R')NHSO$_2$R

247.

A soln. of 1.3 mol% (S)-2-diphenylphosphinomethyl-4,4-dibenzyl-N-pivaloylpyrrolidine in toluene added to a suspension of 1 mol% Cu(OTf)$_2$ in the same solvent, the mixture stirred for 1 h at room temp., 2 eqs. diethylzinc (soln. in hexane) added, stirring continued for 20 min before cooling to 0°, a soln. of the startg. aldimine in toluene added, the mixture stirred at 0° for 4 h, then quenched with aq. 10% HCl → product. Y 97% (e.e. 94%). The catalyst is superior to simpler phosphines in terms of catalytic activity, reaction selectivity (addition vs. reduction) and enantioselectivity. Sulfonyl substituents having electron-withdrawing groups or steric bulk had a deleterious effect on reactivity possibly due to the reduced coordinating ability of the sulfonyl group with the presumed intermediate zinc cuprate. F.e., also conversion to **chiral prim. amines** by removal of the sulfonyl group s. H. Fujihara, K. Tomioka et al., J. Am. Chem. Soc. *122*, 12055-6 (2000).

Dialkylzinc/zirconium tetraisopropoxide/N-(o-hydroxybenzylidene)dipeptide amides ←
Zr(IV)-catalyzed asym. synthesis of sec. amines from aldimines   CH=N → CH(R)NH

248.

Chiral sec. benzylamines. A mixture of 1 mol% N-(*o*-hydroxybenzylidene)-Val-Phe-NHBu-*n*, 20 mol% Zr(OPr-*i*)$_4$·HOPr-*i*, and the startg. aldimine in toluene stirred under N$_2$ at 22° over 24 h, and quenched with satd. aq. NH$_4$Cl → (S)-product. Y 81% (conversion >98%; e.e. 95%). The ligand and transition metal salt were optimized by screening of parallel libraries. F.e. and conversion to chiral prim. benzylamines by oxidative cleavage of the N-protective group s. J.R. Porter, M.L. Snapper, A.H. Hoveyda et al., J. Am. Chem. Soc. *123*, 984-5 (2001).

Diethylzinc s. under BuLi   Et$_2$Zn

Zinc chloride/trimethylsilyl cyanide   ZnCl$_2$/Me$_3$SiCN
α-Aminonitriles from azomethines   C=N → C(NH)CN
with ZnI$_2$ cf. *31*, 616; hexahydro-1,4-diazepin-2-one analogs with asym. induction using ZnCl$_2$ s. K. Namba, Y. Ohfune et al., J. Am. Chem. Soc. *122*, 10708-9 (2000).

Indium   In
Indium-mediated nucleophilic 1,2-allylation of hydrazones and nitrones   ←

249.

in aq. organic medium. An equimolar amount of startg. hydrazone in DMF added to a soln. of allylindium reagent (prepared *in situ* from equimolar amounts of indium powder and allyl bromide in 3:1 DMF/H$_2$O), stirred for 11 h at room temp., and quenched with 10% aq. NH$_4$Cl → product. Y 80%. Nitrones reacted similarly to give the corresponding 3-ethylenehydroxylamines. Reaction is generally applicable to tosyl- and aryl-hydrazones derived from ar. aldehydes and ketones, and to nitrones derived from ar. aldehydes. Phenolethers, ar. halides, phenols, ar. nitro groups and acyclic acetals remained unaffected. F.e.s. H.M.S. Kumar et al., Tetrahedron Lett. *41*, 9311-4 (2000).

Synthesis of 1-carbalkoxy-1,2-dihydropyridines   ←
via Grignard addition to 1-carbalkoxypyridinium salts cf. *26*, 684; 2-allyl-1-carbaryloxy-1,2-dihydropyridines from 1-carbaryloxypyridinium salts and β,γ-ethylenebromides with In s. T.-P. Loh et al., Tetrahedron Lett. *41*, 7779-83 (2000).

Diethylaluminum cyanide/isopropanol
α-Sulfinylaminonitriles from N-sulfinylimines
Asym. Strecker synthesis

Et_2AlCN/i-$PrOH$
C=NS(O)R → C(CN)NHS(O)R

250.

β-Alkoxy-derivs. A soln. of 1.5 eqs. diethylaluminum cyanide in THF cooled to 0° under argon, a small volume of *isopropanol* added via syringe, the mixture stirred at the same temp. for 10-15 min, added via cannula to a soln. of the startg. sulfinylimine in THF at -78°, the resulting mixture allowed to warm to room temp., stirred for 8 h, cooled to -78°, and quenched with satd. aq. NH_4Cl → (R_S,2S,3S)-(N-*p*-toluenesulfinyl)-2-amino-(O-benzyloxy)-3-isovaleronitrile. Y 94% (d.e. 74%). The overall method allows access to the four stereoisomers of phenylserine. F.e. and conversion to **chiral α-amino-β-hydroxycarboxylic acids** s. F.A. Davis et al., J. Org. Chem. 65, 7663-6 (2000); ketimine adducts s. ibid. 8704-8; aldimine adducts s.a. ibid. 61, 440-1 (1996).

Boron fluoride-acetic acid
Arylcarboxylic acid thioamides from isothiocyanates
$BF_3·AcOH$
N=C=S → NHC(S)Ar
with $AlCl_3$ cf. 20, 510; with $BF_3·2AcOH$, regioselectivity, s. J. Sosnicki et al., J. Heterocycl. Chem. 36, 1033-41 (1999).

Samarium diiodide
Syntheses via regiospecific reductive ring opening of N-acyllactams
SmI_2

251.

ω-Acylamino-α′-hydroxyketones from oxo compds. A mixture of startg. N-acyl-2-pyrrolidone and 1.25 eqs. cyclobutanone in THF added quickly to 2 eqs. 0.1 M SmI_2 in THF in a Schlenk tube under argon at 20°, the mixture stirred for 10 min, then quenched with 0.1 M HCl → product. Y 85%. N-Methyl-succinimide or -glutarimide failed to react. F.e., also ω-acylaminoketones with 4 eqs. SmI_2 and 3 eqs. EtOH (at 20° for 3.17 h), **sym. ω,ω′-di(acylamino)-α-diketones** with 1 eq. SmI_2 in the absence of an oxo compd., and **ω-acylaminoaldehydes** with 2 eqs. EtOH (also in the absence of an oxo compd.), s. S. Farcas, J.-L. Namy, Tetrahedron Lett. 41, 7299-302 (2000).

Samarium diiodide/hexamethylphosphoramide
2-Hydrazinoalcohols from hydrazones and oxo compds.
$SmI_2/HMPA$
C(OH)C(NHN<)
with SmI_2/t-BuOH s. 46, 612s51; cyclic analogs by reductive ring closure with $SmI_2/HMPA$ s. D. Riber, T. Skrydstrup et al., J. Org. Chem. 65, 5382-90 (2000).

Trifluoroacetic acid/pyridine
α-Hydroxycarboxylic acid amides from aldehydes and isonitriles
CF_3COOH/C_5H_5N
CHO → CH(OH)CONH
Passerini reaction with H_2SO_4 or HCl cf. 20, 511; chiral N-protected β-amino-α-hydroxy amides with CF_3COOH/py in CH_2Cl_2 s. J.E. Semple et al., Org. Lett. 2, 2769-72 (2000).

Trimethyl cyanide s.a. under $ZnCl_2$
Me_3SiCN

Trimethylsilyl cyanide/(R,R)-tartaric acid/chiral diamine
Asym. Strecker synthesis
←
CH=NR → CH(NHR)CN
with chiral N-(2-thioureidocyclohexyl)-*o*-hydroxyaldimines cf. 55, 253; from (R)-N-(1-phenyl-ethyl)aldimines by addition of Me_3SiCN in the presence of (R,R)-tartaric acid and a chiral diamine for enhancement of diastereoselectivity s. E. Leclerc, P. Mangeney et al., Tetrahedron: Asym. 11, 3471-4 (2000).

Zirconium tetraisopropoxide s. under R_2Zn
$Zr(OPr-i)_4$

Zirconocene dichloride s. under Mg Cp_2ZrCl_2

Titanium tetrachloride/ethyldiisopropylamine $TiCl_4/i\text{-}Pr_2NEt$
Stereospecific synthesis of α-alkoxy-β-aminocarboxylic acid esters C(NH)C(OR)COOR'
from aldimines

252.

via chlorotitanium(IV) enolates. A soln. of 1.1 eqs. TiCl$_4$ in dichloromethane added dropwise to a stirred soln. of methyl methoxyacetate in the same solvent at -78° under N$_2$, after 5 min 1.1 eqs. i-Pr$_2$NEt added dropwise [affording a violet soln. indicative of enolate formation], after a further 1 h the enolate soln. transferred via cannula to a stirred soln. of 0.5 eq. startg. aldimine in dichloromethane at -78° for 1 h, and quenched by addition of aq. 1 M HCl → product. Y 95% (*anti:syn* 95:5). Non-enolizable aliphatic imines were completely unreactive under these conditions. The predominant *anti*-selectivity is thought to arise by addition to a *monodentate* Ti(IV)-enolate through a chair-like transition state. F.e.s. J.C. Adrian Jr. et al., J. Org. Chem. *65*, 6264-7 (2000).

Zwitterionic cyclooctadienerhodium(I) tetraphenylborate/triphenyl phosphite/ ←
    carbon monoxide
(2Z,6E)-4H-1,4-Thiazepin-5-ones from 2-alk-1-ynylthiazoles ○
Regio- and stereo-specific cyclohydrocarbonylative ring expansion

253.

A mixture of 2 mol% zwitterionic cyclooctadienerhodium(I) tetraphenylborate, 8 mol% triphenyl phosphite, and the startg. thiazole in dichloromethane flushed with CO in a glass-lined autoclave, pressurized to 10.5 atm. with CO, H$_2$ introduced to a total pressure of 21 atm., then heated in an oil bath at 110° for 18 h → product. Y 90%. This process, involving a zwitterionic rhodium catalyst in a complex catalytic cycle, is simple and general, tolerating such functional groups as ether, ester and chloro. The ligand is essential to the reaction. F.e.s. B.G. Van den Hoven, H. Alper, J. Am. Chem. Soc. *123*, 1017-22 (2001).

Addition to Remaining Elements and Carbon   CC ⇓ RemC

Zinc bromide $ZnBr_2$
1,2-Oxasilacyclopentanes from siliranes and carbonyl compds. ○
with CuBr$_2$ cf. *56*, 265; reversal of regioselectivity (ring opening at the *less* subst. site) for ring expansion with methyl formate, α,β-unsatd. carbonyl and oxo compds., s. A.K. Franz, K.A. Woerpel, Angew. Chem. Int. Ed. Engl. *39*, 4295-9 (2000).

Addition to Carbon-Carbon Bonds   CC ⇓ CC

Without additional reagents *w.a.r.*
Asym. Michael addition with chiral azomethines C=C → CHC(R)
s. *43*, 607s*48*; generation of chiral quaternary hydrocarbon groups by asym. Michael addition to 3-acetoxyacrylonitrile **as acetaldehyde equivalent** s. L. Keller, J. d'Angelo et al., Tetrahedron Lett. *42*, 381-3 (2001); asym. addition to nitroalkenes s. C. Thominiaux, J. d'Angelo et al.,

Tetrahedron:Asym. *10*, 2015-21 (1999); with chiral α-(arylthio)azomethines/enamines s. ibid. *11*, 995-1002 (2000).

Diels-Alder reaction with acetylene derivs.
s. *22*, 761s*47*; homobarrellenes from 2*H*-cyclohepta[*b*]furan-2-ones s. V. Nair et al., J. Chem. Soc. Perkin Trans. 1 *2000*, 3795-8; N-protected 1,2,3,4-tetrahydroisoquinolines s. S. Kotha, N. Sreenivasachary, Chem. Commun. *2000*, 503-4.

Diels-Alder reaction with α,β-ethylenesulfones
s. *29*, 792s*35*; with 2-(trifluoroacetyl)vinyl sulfones, also conversion to 1-trifluoroacetyl-1,3-cyclohexadienes (with DBU) and trifluoroacetophenones, s. A.L. Krasovsky, V.G. Nenajdenko et al., Tetrahedron *57*, 201-9 (2001).

Diels-Alder reaction with α-methylenealdehydes generated *in situ* from 1,3-dioxins

254.

2 eqs. Startg. diene added to a soln. of the startg. dioxin in toluene, and the mixture refluxed under argon for 5 h → 2-methoxy-1-triisopropylsilylethynyl-4-trimethylsilyloxy-3-cyclohexene-1-carboxaldehyde. Y 100% (d.r. 4:1). Sensitive acroleins, incl. α-acyl derivs., can thus be generated *in situ* under mild, neutral conditions. Silyl groups, acetylene derivs., ethers, ketones and carboxylic acid amides remained unaffected. F.e. incl. intramolecular cycloaddition s. S.P. Fearnley, R.L. Funk et al., Tetrahedron *56*, 10275-81 (2000); generation of 2-acoxyacroleins s. J. Org. Chem. *61*, 2598-9 (1996).

Asym. Diels-Alder reaction with 3-siloxy-1,3-dienamines
s. *53*, 288; asym. cycloaddition of enals to 1-(2-oxazolidon-3-yl)-3-siloxy-1,3-dienes, and conversion to chiral 4-hydroxymethyl-2-cyclohexenones s. J.M. Janey, V.H. Rawal et al., J. Org. Chem. *65*, 9059-68 (2000); stereospecific cycloaddition to enones s. R. Paczkowski, M.E. Maier et al., Org. Lett. *2*, 3967-9 (2000).

Asym. Diels-Alder reaction with α,β-ethylenesulfoxides
review s. *49*, 849s*50*; with chiral 1-(1-sulfinylvinyl)cyclohexenes s. M.C. Aversa et al., Tetrahedron: Asym. *10*, 3907-17 (1999); with chiral sulfinyl-*p*-quinones under thermal conditions and with ZnBr$_2$ for reversal of regioselectivity s. M.C. Carreño et al., ibid. *11*, 4279-96 (2000); s.a. J. Org. Chem. *65*, 453-8 (2000); with naphthoquinone analogs s. ibid. 4355-63; with 4-sulf(i,o)nyl-5-(*l*-menthyloxy)furan-2(5*H*)-ones s. J.L. Garcia Ruano et al., Tetrahedron:Asym. *11*, 4737-52 (2000).

Diels-Alder reaction with anthracenes
s. *27*, 694; **asym. induction** with chiral 9-α-alkoxyanthracenes, also with a soluble PEG-based substrate, and under Lewis acid catalysis, s. A. Sanyal, J.K. Snyder, Org. Lett. *2*, 2527-30 (2000).

High-pressure Diels-Alder reaction
with maleic anhydride derivs. s. *27*, 696s*42*; decahydro-*as*-indacene skeleton from maleic anhydride **with asym. induction** s. M. Banwell et al., J. Chem. Soc. Perkin Trans. 1 *2000*, 3555-8; *endo*-selectivity with a chiral α,β-ethylenesulfoxide ((Z)-3-*p*-tolylsulfinylacrylonitrile) (cf. *49*, 849s*50*), also with ZnBr$_2$ as catalyst (cf. *36*, 667s*58*), s. J.C. Garciá-Ruano et al., J. Org. Chem. *65*, 7938-43 (2000).

High-pressure 1,3-dipolar cycloaddition
s. *43*, 249; (1→2)-linked pseudo-aza-C-disaccharides by enhanced *exo-anti*-addition of enantiopure nitrones to glycals s. F. Cardona, A. Goti et al., Synlett *1998*, 1444-6.

Bicyclo[3.n.1]alk-2-en-(n+6)-ones
from chromium α,β-ethylene(alkoxy)carbene complexes and isocyclic enamines s. *60*, 419.

Cycloaddition with 4-hydroxythiazolium betaines ←
review s. *43*, 943s*49*; 2,3-dihydrothiophene-2-carboxylic acid amides from enazo derivs. s. M. Avalos et al., J. Org. Chem. *65*, 5089-97 (2000); 4-nitro-derivs. from 1,1-nitroethylene derivs. s. Chem. Commun. *1995*, 2213-4

Irradiation (s.a. under Chiral bicyclic lactams and t-BuSSBu-t)   ⫷
Stereospecific radical 3-component synthesis ←
with acetylene derivs., diselenides and ethylene derivs. cf. *57*, 257; ζ-arylseleno-α-(arylseleno-methylene)-γ,δ-ethylenecarboxylic acid esters from vinylcyclopropanes s. A. Ogawa et al., J. Org. Chem. *65*, 7682-5 (2000).

Stereospecific silylative ring closure of 1,6-dienes   ○
with Me$_3$SiH under Pd(II)-catalysis cf. *55*, 299; silylmethyl-cyclopentanes, -tetrahydrofurans and -pyrrolidines with dimethylphenylsilylbis(diisopropylamino)borane under irradiation s. A. Matsumoto, Y. Ito, J. Org. Chem. *65*, 5707-11 (2000).

**Photo-induced electron-transfer ring closure
of ω-phthalimidoalcohols with ethylene derivs.
Medium- and large-ring N-condensed O,N-heterocyclics**

255.

A deaerated soln. of startg. phthalimidoalcohol (0.025 M) and excess of α-methylstyrene in benzene irradiated at room temp. (λ >300 nm) until TLC indicated completion of reaction → product. Y 78% (d.r. 2.5:1). The procedure is most effective for medium (eight- to nine-membered) rings. F.e. incl. intramolecular insertion s. J. Xue, J.-H. Xu et al., Tetrahedron Lett *41*, 8553-7 (2000).

Pyrrolidines from aziridines   ○
by anionic cycloaddition with BuLi cf. *33*, 640; 2-(2-cyanovinyl)pyrrolidines from electron-deficient alkenes under irradiation s. K. Ishii et al., J. Chem. Soc. Perkin Trans. 1 *2000*, 3022-4.

Sodium hydride   NaH
3-Methylenetetrahydrofurans from ethynylcarbinols   ○
and electron-deficient ethylene derivs. with *n*-BuLi/Pd(OAc)$_2$/Ph$_3$P cf. *53*, 291; β-alkylidene-γ-lactolides from enolethers with NaH, 2-alkoxy-3-alk-1-ynyl-4-methylene-2-perfluoroalkyl-3-sulfonyltetrahydrofurans, s. M. Yoshimatsu, J. Murase, J. Chem. Soc. Perkin Trans. 1 *2000*, 4427-31.

Cesium hydroxide   CsOH
(E)-β,γ-Ethylenenitriles from acetylene derivs. s. *51*, 337s*60*   C≡C → CH=C(R)

Potassium tert-*butoxide*   KOBu-t
Catalytic regiospecific 1,2-addition of CH-acidic compds.   C=C → CHC(R)

256.

A mixture of cyclohexanecarbonitrile and 1.33 eqs. styrene added to a soln. of *0.33 eq.* *t*-BuOK in N-methyl-2-pyrrolidone under argon, and stirred at room temp. for 15 h → product. Y 91%. The procedure was extended to imines and ketones (with DMSO as the preferred solvent for the latter). F.e. incl. an intramolecular addition s. A.L. Rodriguez, P. Knochel et al., Org. Lett. *2*, 3285-7 (2000).

Organolithium compds./aluminum tris(2,6-diphenylphenoxide)   RLi/Al(OAr)₃
1,4(6)-Addition to arylcarboxylic acid chlorides ←

257.

A soln. of 1.1 eqs. aluminum tris(2,6-diphenylphenoxide) in toluene treated with benzoyl chloride at -78°, 3 eqs. methyllithium added, and worked up with concd. HCl at -78° to room temp. → product. Y 99% (2.6:1 mixture of 1,6- and 1,4-adducts). Interestingly, lower organometallics, e.g. MeLi and enolates (e.g. methyl acetate lithium enolate), react predominantly with ar. aldehydes by 1,2-addition, although reaction of bulky nucleophiles with both acyl chlorides and ar. aldehydes (cf. *51*, 300) takes place exclusively or predominantly by 1,6-addition. F.e.s. S. Saito, H. Yamamoto et al., J. Am. Chem. Soc. *122*, 10216-7 (2000).

Organolithium compds./ferric acetoacetonate   RLi/Fe(acac)₃
Regio- and stereo-specific synthesis of functionalized ethylene derivs. ←
from acetylene derivs. via iron-catalyzed carbolithiation

258.

3 eqs. 1 M n-BuLi in hexane added to a soln. of startg. alkyne and 10 mol% Fe(acac)₃ in toluene under argon at -40°, warmed immediately to -20°, stirred for 2 h, and quenched with 1 N HCl → product. Y 97%. The choice of solvent and Fe salt is critical. The procedure is inexpensive and suitable for carbolithiation of alkynes bearing alkoxy or amino functions, the Lewis basic heteroatom activating the alkyne bond towards attack by the organolithium agent. There was no reaction with unfunctionalized alkynes. F.e., also tetrasubst. alkynes by carbolithiation-electrophile capture s. M. Hojo, A. Hosomi et al., Angew. Chem. Int. Ed. Engl. *40*, 621-3 (2001).

Methyllithium   MeLi
Stereospecific Michael addition-aldol condensation ←
α-(arylthiomethyl)-β-hydroxy esters with *n*-BuLi cf. *55*, 261; amide derivs., also α-(arylselenomethyl) analogs, s. A. Kamimura et al., J. Chem. Soc. Perkin Trans. 1 *2000*, 4499-504.

n-Butyllithium s.a. under Zn   BuLi

n-Butyllithium/(-)-sparteine ←
Asym. 1,4-addition of N-(2-ethylene)urethans to 1,1-nitroethylene derivs. C=C → CHC(R)

259.

1.02 eqs. (-)-sparteine added under N₂ to a soln. of the startg. protected allylamine in toluene, the mixture cooled to -78°, 1.02 eqs. *n*-BuLi (1.5 M in hexane) added, stirring continued for 1 h, a soln. of 1.3 eqs. startg. nitroethylene deriv. in toluene added dropwise via syringe pump over 1 h at the same temp., the mixture stirred for a further 10 min, then quenched with methanol → product. Y 83% (diastereoselectivity >99%). High yields and diastereo- and enantio-selectivities were achieved with a variety of such allylamines and nitroalkenes. F.e., also conversion to **chiral N-protected piperidin-3-ylcarbinols**, and to **chiral 2-piperidones** via δ-nitrocarboxylic acid esters, s. T.A. Johnson, P. Beak et al., J. Am. Chem. Soc. *123*, 1004-5 (2001).

Lithium diisopropylamide i-Pr_2NLi
Asym. Michael addition with carboxylic acid amides C≡C → CHC(R)
chiral glutaric acid amide esters from chiral amides cf. *42*, 638; chiral glutaric acid diamides from chiral N-(α,β-ethyleneacyl)sultams, *anti*-addition, s. B. Liang, M.M. Joulié et al., Org. Lett. *2*, 4157-60 (2000).

Lithium bis(trimethylsilyl)amide $LiN(SiMe_3)_2$
Michael addition with sulfoxides
s. *18*, 965s*26*; *22*, 835s*23*; *38*, 238; stereospecific addition of heteroaryl alkyl sulfoxides to α,β-ethylenecarbonyl compds. with $LiN(SiMe_3)_2$ s. M. Casey et al., Synlett *2000*, 1721-4.

Lithium chloride s. under BH_3-THF	LiCl
Lithium bromide s. under $Pd(OAc)_2$	LiBr
Lithium iodide s. under RCu	LiI
Sodium iodide s. under $Pd(OAc)_2$	NaI

Polymer-based quaternary ammonium hydroxide ←
Michael addition in aq. medium
with NaOH/hexadecyltrimethylammonium chloride cf. *52*, 247; with an amphiphilic polymer-based quaternary ammonium hydroxide in water (cf. *54*, 288) for addition of cyclic β-keto esters s. K. Shibatomi, Y. Uozumi et al., Synlett *2000*, 1643-5.

Piperidine R_2NH
Michael addition-intramolecular aldol condensation ○
with Cs_2CO_3 s. *44*, 602; bicyclo[3.3.1]nonenols with piperidine or Bu_4NF s. K. Aoyagi, Y. Yamamoto et al., J. Org. Chem. *64*, 4148-51 (1999).

2,2′-Isopropylidenebis((4S)-4-tert-butyl-Δ²-oxazoline)copper(II) bistriflate ←
Catalytic asym. Friedel-Crafts alkylation with electron-deficient ethylene derivs. H → R

260.

Chiral γ-aryl-α-ketocarboxylic acid esters. 1 eq. Startg. enone added to a soln. of 5 mol% $Cu(OTf)_2$(2,2′-isopropylidenebis((4S)-4-*tert*-butyl-Δ²-oxazoline) in ether under N_2, stirred at room temp. for 15 min, cooled to -78°, 1 eq. 5-methoxyindole added, and the mixture stirred at the same temp. for 1 h → (R)-product. Y 95% (e.e. 99.5%). This is the first highly enantioselective, catalytic Friedel-Crafts alkylation of arenes or heteroarenes with such β,γ-ethylene-α-ketoesters. The γ-substituent may be alkyl, aryl or benzyloxymethyl. F.e. and comparison of Lewis acid complexes s. K.B. Jensen, K.A. Jørgensen et al., Angew. Chem. Int. Ed. Engl. *40*, 160-3 (2001); asym. alkylation of 5-membered heteroarenes with arylidenemalonic acid esters s. Chem. Commun. *2001*, 347-8.

6-Acyl-2,2,4,4-tetraalkoxytetrahydropyrans from α-ketocarbonyl compds. ○
Catalytic asym. [2+2+2]-cycloaddition

261.

The startg. α-ketocarboxylic acid ester and 3 eqs. ketene diethyl acetal added to a soln. of 20 mol% $Cu(OTf)_2$(2,2′-isopropylidenebis((4S)-4-*tert*-butyl-Δ²-oxazoline) in ether under N_2 at -78°, and stirred overnight at the same temp. → product. Y 74% (e.e. 83%). The reaction (for which a sequential aldol addition is involved) is generally applicable to α-ketoesters and sym. or unsym. α-diketones; it is also regiospecific and proceeds with good to high isolated yields and high enantioselectivities. F.e. and hydrolysis to **chiral 6-acyl-5,6-dihydro-2-pyrones** with formic acid in pentane/methylene chloride s. H. Audrain, K.A. Jørgensen, J. Am. Chem. Soc. *122*, 11543-4 (2000).

Lithium diorganocuprates/zinc chloride $\qquad$ $LiCuR_2/ZnCl_2$
1,4-Addition-aldol condensation $\qquad$ ←
with $LiCuR_2/Bu_3P$ cf. *38*, 673; *33*, 807s40 (review); with $LiCuR_2/ZnCl_2$ s. J. Méndez-Andino, L.A. Paquette, Org. Lett. *2*, 4095-7 (2000).

Lithium dialkylcuprates/trimethylsilyl chloride $\qquad$ $LiCuR_2/Me_3SiCl$
1,4-Addition $\qquad$ C=C → CHC(R)
to enones s. *41*, 638s*45*; with lithium bis(methylenecyclopropyl)cuprate s. G. Peron, J.D. Kilburn, Tetrahedron Lett. *42*, 347-9 (2001).

Lithium silyl(cyano)cuprates $\qquad$ $LiCu(Si \leqslant)CN$
Syntheses via silylcupration of carbon-carbon multiple bonds $\qquad$ ←
s. *36*, 825s*58*; synthesis of β-arylsilanes from styrenes by silylcupration-electrophile trapping s. V. Liepins, J.-E. Bäckvall, Chem. Commun. *2001*, 265-6.

Organocopper(I) compds./lithium iodide/dimethylaluminum chloride $\qquad$ ←
Asym. 1,4-addition to 3-(α,β-ethyleneacyl)-2-oxazolidones $\qquad$ C=C → CHC(R)
with Grignard compds./CuBr cf. *37*, 657s*50*; also with RCu·LiI/Me$_2$AlCl for asym. 1,4-addition to *tert*-leucine-based derivs. s. C. Schneider, O. Reese, Synthesis *2000*, 1689-94.

Silver acetate s. under RhCl(PPh$_3$)$_3$ $\qquad$ *AgOAc*
Cuprous cyanide s. under BH$_3$-THF $\qquad$ *CuCN*
Cuprous chloride s. under Cp$_2$ZrCl$_2$ $\qquad$ *CuCl*

Cuprous iodide/tri-n-butylphosphine/tert-butyllithium $\qquad$ ←
1,4-Addition with organocopper(I) compds.
s. *43*, 616; β-(2-ethoxyvinyl)lactones from *cis*-1-bromo-2-ethoxyethylene, also conversion to β-propargyl derivs., s. S. Bennabi, M.A. Ciufolini et al., Tetrahedron Lett. *41*, 8873-6 (2000).

Cupric chloride s. under Fe(CO)$_5$ and PdCl$_2$ $\qquad$ $CuCl_2$

Gold trichloride $\qquad$ $AuCl_3$
2-γ-Ketofurans from α,β-acetyleneketones $\qquad$ O
and α-alleneketones s. *60*, 74

Magnesium $\qquad$ *Mg*
Asym. synthesis of 2,6-disubst. 3,6-dihydro-2H-pyrans $\qquad$ ←
from 2,6-dialkoxydihydro-η^3-pyranylmolybdenum complexes
by sequential nucleophilic substitution

262.

Chiral 2,3,6-trisubst. 5,6-dihydro-2H-pyrans. Startg. chiral 2,6-dialkoxydihydro-η^3-pyranyl-molybdenum complex (readily prepared by treatment of the corresponding (-)-η^3-pyranyl-molybdenum complex [e.e. 98%] with 1.1 eqs. Br$_2$ at -78°, followed by quenching with 40%

methanolic NaOMe) in methylene chloride treated with 1 eq. trityl hexafluorophosphate at -78°, warmed to 0° over 5 min, the intermediate cationic diene precipitated with *tert*-butyl methyl ether, redissolved in THF, treated with ethylmagnesium bromide at -78° for 15 min, the mixture poured through a pad of silica gel (treated with 5% Et$_3$N/hexane) with ether, concentrated, the second intermediate cationic diene precipitated with HBF$_4$ in THF/*tert*-butyl methyl ether at 0° during 5 min, redissolved in THF, and treated with Bu$_4$NCN at -78° → intermediate 2,6-disubst. complex (Y 85%; e.e. 98%), treated with 2 eqs. acetic acid in methylene chloride under irradiation for 24 h → product (Y 78% as a mixture with 8% of the double bond isomer). A variety of nucleophiles participated in the reaction (Grignards, enolates, cyanide ion), and there was no racemization. F.e.s. J. Yin, L.S. Liebeskind et al., J. Am. Chem. Soc. *122*, 10458-9 (2000).

Zinc s.a. under Cp$_2$TiCl$_2$ and Cp$_2$VCl$_2$	Zn
Zinc/n-butyllithium/chiral Δ^2-oxazolines	←
Asym. allylzincation of cyclopropenone ketals	

s. *43*, 770s53; regioselectivity, **also under high pressure** (more rapidly), s. M. Nakamura, E. Nakamura et al., Org. Lett. *2*, 2193-6 (2000).

Zinc/trifluoroacetic acid	Zn/CF$_3$COOH
β-Perfluoroalkyl-α,β-ethyleneiodides	C≡C → CI=C(Rf)
from acetylene derivs. and perfluoroalkyl iodides	

with Na$_2$S$_2$O$_4$/NaHCO$_3$ cf. *47*, 668; (E)-isomers from terminal alkynes with a little Zn-powder (10 mol%) and 20 mol% CF$_3$COOH s. M.P. Jennings, P.V. Ramachandran et al., J. Org. Chem. *65*, 8763-6 (2000).

Organomagnesium halides s. under Titanium complexes	RMgHal
Dialkylzinc/cupric trifluoromethanesulfonate/chiral bis(phosphites),	←
phosphoromonoamidites, 3-hydroxythioethers or N-(o-phosphinoarylidene)dipeptides	
Dialkylzinc/nickel(II) acetoacetonate/chiral 2-hydroxythioethers	←
Asym. 1,4-addition of dialkylzinc to α,β-ethyleneketones	C=C → CHC(R)

s. *52*, 297s55; with carbohydrate-based cyclic 1,1'-bi-2-naphthol bis(phosphites) as ligand cf. O. Pàmies, M. Dieguez et al., Tetrahedron:Asym. *10*, 2007-14 (1999); *11*, 4377-83 (2000); with a chiral cyclic 1,1'-bi-2-naphthol phosphoromonoamidite (BIBAPHOSHQUIN) (cf. *56*, 275) s. C.G. Arena, F. Baraone et al., ibid. *11*, 2387-92 (2000); with carbohydrate-based 3-hydroxy-thioethers s. O. Pàmies et al., ibid. 871-7; asym. 1,4-addition to cyclic enones with modular N-(*o*-phosphinoarylidene)dipeptides (cf. *59*, 288) s. S.J. Degrado, A.H. Hoveyda, J. Am. Chem. Soc. *123*, 755-6 (2001); asym. 1,4-addition to chalcone with Ni(acac)$_2$ as catalyst and a chiral 2-hydroxythioether as ligand s. Y. Yin, T.-K. Yang, Tetrahedron:Asym. *11*, 3329-33 (2000).

Diethylzinc s. under Ni(acac)$_2$	Et$_2$Zn
Diisopropylzinc s. under BH$_3$-THF	i-Pr$_2$Zn
Lithium trialkylzincates/ethyl dichlorovanadate	LiZnR$_3$/VO(OEt)Cl$_2$
Nucleophilic 1,4-addition-alkylation of cyclic α,β-ethyleneketones	←
with lithium trialkylzincates	

263.

Novel nucleophilic *vic*-dialkylation at *both* the α- and β-positions of cyclic enones is reported. E: 1.2 eqs. Startg. mixed lithium trialkylzincate in hexane added to a soln. of startg. ethyleneketone in THF, allowed to react at -78° under argon for 1 h, 2 eqs. ethyl dichlorovanadate added, and the mixture stirred for 20 h at 0° to room temp. → product. Y 51% (*trans/cis* 71/29). The procedure involves initial 1,4-addition with generation of a dialkylzinc enolate, followed by oxidative nucleophilic α-alkylation by the oxovanadium reagent so that two different alkyl groups are introduced. Such 1,4-addition-alkylation (incorporating the same alkyl group at the adjacent sites) was also achieved with dialkylzincs (or trialkyl-alanes or -boranes) in the presence of the same reagent acting as both Lewis acid and oxidant. F.e.s. T. Hirao, T. Takada, H. Sakurai, Org. Lett. 2, 3659-61 (2000).

Magnesium-aluminum tert-butoxide hydrotalcite
Heterogeneous Michael addition ←
  C=C → CHC(R)
with activated montmorillonite K10 cf. *38*, 668s54; with Mg-Al hydrotalcite having *t*-BuO⁻ between layers for enhanced basicity, also Knoevenagel condensation (cf. *57*, 310s59), s. B.M. Choudary et al., Tetrahedron *56*, 9357-64 (2000).

Zinc triflate s. under Bu₃SnH $Zn(OTf)_2$
Zinc chloride s. under LiCuR₂ $ZnCl_2$

Borane-tetrahydrofuran/diisopropylzinc/cuprous cyanide/lithium chloride ←
Regio- and stereo-specific syntheses via hydroboration-1,2-boryl migration ←

264.

of tetrasubst. ethylene derivs. 3 eqs. 1 M BH₃-THF in THF added to a soln. of 2-ethyl-1,1-diphenyl-1-butene in THF at 25°, stirred for 10 min at the same temp. then for 4 h at 50°, cooled to 0°, solvent and excess of borane removed *in vacuo* (0.1 mmHg for 60 min), 2.6 eqs. 2.5 M *i*-Pr₂Zn in THF added at 25°, after 2 h (the colouration having turned to dark grey) the mixture was stirred for a further 45 min, excess of *i*-Pr₂Zn removed *in vacuo* at 0°, the residue diluted with THF, cooled to -78°, a soln. of 0.2 eq. 1 M CuCN·2LiCl in THF added slowly, stirred for 15 min, 3 eqs. allyl bromide added, warmed to 25°, stirred for 1 h, and quenched with 3 M aq. HCl → (4R*,5S*)-5-benzhydryl-4-methyl-1-heptene. Y 62%. The procedure generates up to three stereogenic centres in one step with 100% diastereoisomeric purity. If there is a choice, migration of boron takes place towards the larger alkyl group. F.e. and conversion of the rearranged borane to the corresponding **alcohols and sec. amines** s. L.O. Bromm, P. Knochel et al., J. Am. Chem. Soc. *122*, 10218-9 (2000).

9-Borabicyclo[3.3.1]nonane s. under PdCl₂(dppf) R_2BH

Trimethylaluminum/(R)-3,3'-diphenyl-1,1'-bi-2-naphthol ←
Asym. 1,3-dipolar cycloaddition of nitrones to ethylene derivs. ◯
with Mg(ClO₄)₂/chiral bis(Δ²-oxazolines) cf. *54*, 296s59; chiral 5-alkoxyisoxazolidines from enolethers with Me₃Al/(R)-3,3'-diphenyl-1,1'-bi-2-naphthol s. K.B. Jensen, K.A. Jørgensen et al., J. Org. Chem. *65*, 9080-4 (2000).

Catecholborane s. under Pd(PPh₃)₄ ←
Aluminum tris(2,6-diphenylphenoxide) s. under RLi $Al(OAr)_3$
Hydrotalcites s. Magnesium-aluminum tert-*butoxide hydrotalcite* ←

Boron fluoride BF_3
2-Alkylthio-3,6-dihydro-2H-thiopyrans from 1,3-dienes
s. *28*, 637; 2-alkylthio-3,6-dihydro-2*H*-thiopyran-2-ylphosphonic acid esters from methyl diisopropylphosphonodithioformate **under Lewis acid catalysis** with BF₃-etherate or ZnCl₂ s. B. Heuzé, S. Masson et al., Tetrahedron Lett. *41*, 7327-31 (2000).

N-Sulfonyl-pyrrolidines from -aziridines and ethylene derivs. ◯

265.

A soln. of 1 eq. BF₃-etherate in dichloromethane added dropwise under argon to a soln. of the startg. 2-phenylaziridine and 1.4 eqs. methylenecyclopentane in the same solvent at -78°, the mixture stirred for 20 min, and quenched with water → product. Y 80%. The presumed intermediate zwitterionic 1,3-dipole is stabilized by the aromatic ring and the sulfonyl group, and is sufficiently electron-deficient to react even with *non-activated* alkenes. This formal **1,3-dipolar cycloaddition** did not, however, work well with cyclopentene and cyclohexene. F.e.s. I. Ungureanu, A. Mann et al., Angew. Chem. Int. Ed. Engl. *39*, 4615-7 (2000).

Dimethylaluminum chloride s. under RCu   *Me$_2$AlCl*

Diethylaluminum chloride   *Et$_2$AlCl*
Asym. Diels-Alder reaction with α,β-ethylenecarboxylic acid esters   ○
with carbohydrate-based esters s. *47*, 651; with Et$_2$AlCl as Lewis acid s. M.L.G. Ferreira, V.F. Ferreira et al., Tetrahedron:Asym. *9*, 2671-80 (1998).

Indium(III) chloride   *InCl$_3$*
***endo*-Selective ionic Diels-Alder reaction with α,β-ethyleneacetals**

266.

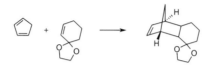

20 Mol% anhyd. InCl$_3$ added to a stirred soln. of startg. ethyleneketal and 3 eqs. cyclopentadiene in nitromethane, and allowed to react at 15-20° for 1 h → product. Y 75% (*endo:exo* 94:6). The method is applicable to acyclic or cyclic ketals or acetals. Reaction failed with Yb(OTf)$_3$ and Sc(OTf)$_3$. F.e.s. B.G. Reddy, Y.D. Vankar et al., Tetrahedron Lett. *41*, 10333-6 (2000).

Bis(η^5-pentamethylcyclopentadienyl)(tetrahydrofuran)samarium   *Cp*$^*_2$Sm·THF*
Regio- and stereo-specific ring closure of dienes
cycloalkylcarbinols via Zr-catalyzed carboalumination cf. *50*, 405; from 1,5- and 1,6-dienes via hydroboration with 1,3-dimethyl-1,3,2-diazaborolidine in the presence of Cp*$_2$Sm·THF (cf. *50*, 402s59) s. G.A. Molander, D. Pfeiffer, Org. Lett. *3*, 361-3 (2001).

((R,S)-2,2-Bis(4′-tert-butyl-2′-(trimethylsilyl)cyclopentadien-1′-yl)-   ←
   1,3-dioxa-2-siladinaphtho[2,1-d:1,2-f]cycloheptane)yttrium complex
Regiospecific intramolecular carbosilylation of dienes
with Me(Cp*)$_2$Y(THF) cf. *50*, 402; 5- and 6-membered silylmethylated heterocyclics and carbocyclics **with asym. induction** (e.e. 5 to 50%) using ((R,S)-2,2-bis(4′-*tert*-butyl-2′-(trimethylsilyl)cyclopentadien-1′-yl)-1,3-dioxa-2-siladinaphtho[2,1-*d*:1,2-*f*]cycloheptane)yttrium complex s. A.R. Muci, J.E. Bercaw, Tetrahedron Lett. *41*, 7609-12 (2000).

Scandium(III) triflate   *Sc(OTf)$_3$*
[4+3]-Cycloaddition via oxyallyl cations
review s. *39*, 883s*41*,*42*; 2-siloxy-4-cycloheptenone ring from cyclic 1,3-dienes and 2-(triisopropylsilyloxy)acrolein with 10 mol% Sc(OTf)$_3$ s. M. Harmata, U. Sharma, Org. Lett. *2*, 2703-5 (2000).

Ytterbium(III) triflate   *Yb(OTf)$_3$*
Lewis acid-catalyzed high-pressure Diels-Alder reaction
with ZnBr$_2$ cf. *44*, 610; with Yb(OTf)$_3$, notably for reaction of 1,3-cyclohexadiene with electron-deficient ketodienophiles s. A.C. Kinsman, M.A. Kerr, Org. Lett. *2*, 3517-20 (2000).

1,4-Cyclohexadiene   ←
Bergman cycloaromatization
s. *48*, 665s59; naphthalenes from *o*-diynes in sealed melting point tubes s. N. Choy, K.C. Russell et al., Tetrahedron Lett. *41*, 6955-8 (2000).

Azodiisobutyronitrile (s.a. under (Me$_3$Si)$_3$SiH)   *AIBN*
Radical addition of iodides to terminal ethylene derivs.   C=C → C(I)C(R)
of CF$_3$CH$_2$I cf. *47*, 658, and of ClCH$_2$I (with AIBN) cf. *36*, 668; γ-iodophosphonic from α-iodophosphonic acid esters, also β,γ-ethylene-γ-iodophosphonic acid esters from acetylene derivs. s. P. Balczewski et al., Tetrahedron Lett. *41*, 3687-90 (2000).

Radical ring closures of enynes with bisfunctionalization
s. *49*, 669s*53*; 2-homoallyl-1-(stannylmethylene)cyclopentanes and O- or N-heterocyclic analogs from 1,6-enynes and allylstannanes s. K. Miura, A. Hosomi et al., J. Org. Chem. *65*, 8119-22 (2000).

Chiral bicyclic lactams/irradiation
Photochemical asym. [2+2]-cycloaddition
review s. *19*, 764s*40*; of 4-methoxycarbostyrils with ethylene derivs. in toluene **via hydrogen-bonding** complexation with a bicyclic lactam as host molecule s. T. Bach, H. Bergmann, J. Am. Chem. Soc. *122*, 11525-6 (2000).

N-Hydroxyphthalimide s. under Co(OAc)₂

L-Proline/trans-2,5-dimethylpiperazine
Asym. Michael addition of aliphatic nitro compds. to cyclic α,β-ethyleneketones

$C=C \rightarrow CHC(R)$

267.

A mixture of 2-cyclohexen-1-one, 2.1 eqs. 2-nitropropane, 1.02 eqs. *trans*-2,5-dimethylpiperazine, and 3.8 mol% L-proline in chloroform (previously passed through a bed of Brockmann 1 grade basic alumina) stirred for 62 h at room temp. → product. Y 88% (e.e. 93%). This is the first example of a catalytic asym. conjugate addition of a nitroalkane to a cyclic enone with high enantioselectivity. Piperazines appear to be the optimum additive, the pronounced non-linear effects of reagent *vs.* product enantioselectivities indicating the formation *in situ* of a complex multicomponent chiral catalytic system. F.e. and with certain acyclic enones s. S. Hanessian, V. Pham, Org. Lett. *2*, 2975-8 (2000).

m-Chloroperoxybenzoic acid s. under Ti(OPr-i)₄   *ArCOO₂H*

Di-tert-butyl disulfide/irradiation   *t-BuSSBu-t/⫯⫯*
cis-3-Thiabicyclo[3.3.0]octanes from 1,6-dienes
Double radical ring closure

268.

A soln. of startg. diene and di-*tert*-butyl disulfide in hexane irradiated in a water-cooled quartz photochemical cell for 5 h at 10° → product. Y 73%. Reactions were much slower and less efficient through Pyrex (40% yield for this example after 48 h), but were accelerated in the presence of Et₃B. Reaction is presumed to involve thiyl radical addition to one of the alkene groups, followed by cyclization through a chair-like transition state, and terminating by homolytic substitution at sulfur. Heteroatoms in the linking chain were tolerated. F.e.s. D.C. Harrowven et al., Tetrahedron Lett. *41*, 9345-9 (2000).

Tris(trimethylsilyl)silane/azodiisobutyronitrile   *(Me₃Si)₃SiH/AIBN*
γ-Ketocarbonyl compds. from α,β-ethylenecarbonyl compds. $C=C \rightarrow CHC-COR$
via radical carbonylation in supercritical carbon dioxide

269.

1-Iodooctane, 1.2 eqs. acrylonitrile, and 1.5 eqs. (Me₃Si)₃SiH added via syringe to 0.3 eq. AIBN in a stainless steel autoclave under argon, CO introduced to a pressure of 50 atm., followed by

CO_2 to a total pressure of 220 atm., the mixture heated to 80° with more CO_2 being added to achieve a final pressure of 310 atm., the mixture held at the same temp. for 5 h, cooled to ca. -78°, the gases vented off, the residue warmed to 25°, diluted with ether, and treated with 1,8-diazabicyclo[5.4.0]undecene and a soln. of I_2 in ether → product. Y 90%. This method avoids handling such less friendly solvents as benzene. F.e.s. Y. Kishimoto, T. Ikariya, J. Org. Chem. 65, 7656-9 (2000).

Bis(trimethylsilyl) peroxide s. under Zr(OBu-t)$_4$	(Me$_3$SiO)$_2$
Trimethylsilyl cyanide s. under Zr(OBu-t)$_4$	Me$_3$SiCN
Trimethyl azide s. under Pd(dba)$_2$	Me$_3$SiN$_3$
Titanium tetraisopropoxide/isopropylmagnesium chloride	Ti(OPr-i)$_4$/i-PrMgCl
2-Ethylenealcohols from acetylene derivs. and oxo compds.	CO → CH≡C-C(OH)

s. 52, 310s59; α,β-ethylene-β'-hydroxy-β-silylcarboxylic from α,β-acetylene-β-silylcarboxylic acid esters and aldehydes **with asym. induction** s. D. Suzuki, F. Sato et al., Angew. Chem. Int. Ed. Engl. 39, 3290-2 (2000).

3-Hydroxysilanes from enesilanes and aldehydes CHO → CH(OH)C-CH(Si≤)

270.

Nucleophilic 2-silylethylation. A soln. of 3 eqs. isopropylmagnesium chloride in ether (1.5 M) added to a mixture of 1.5 eqs. Ti(IV)-isopropoxide and 1.5 eqs. trimethyl(vinyl)silane in the same solvent at -78°, warmed to -50° over 30 min, stirred at -50 to -55° for 2 h, cyclohexanecarboxaldehyde added at the same temp., stirred for 2 h, then quenched with 1 N HCl → product. Y 87%. The intermediate silylethylene-titanium complex also reacts smoothly with azomethines, a second enesilane molecule, and with acetylene derivs. F.e. and electrophiles (other than H⁺), also *trans*-**2-silylcyclopropanols** from carboxylic acid esters, s. R. Mizojiri, F. Sato et al., J. Org. Chem. 65, 6217-22 (2000).

Ring closures of diynes via titanacyclopentadienes ○
s. 58, 273; 3,4-di(alkylidene)-2-pyrrolidones from α,β-acetylenecarboxylic acid acid alk-2-ynylamides, s. H. Urabe, F. Sato, Org. Lett. 2, 3481-4 (2000).

Titanium tetraisopropoxide/titanium tetraiodide/m-chloroperoxybenzoic acid	←
α-Alkoxy-β-hydroxyketones from aldehydes	CHO → CH(OH)CH(OR)COMe
Synthesis with addition of two C-atoms	

271.

2 eqs. Methoxyallene and 2 eqs. *m*-chloroperoxybenzoic acid in dichloromethane stirred for 30 min at 0° under argon, the resulting mixture added to the startg. aldehyde and 1.5 eqs. each of TiI$_4$ and Ti(OPr-*i*)$_4$ in the same solvent at -78°, stirred at -78 to -20°, and quenched with satd. aq. NaHCO$_3$, 5% aq. NaHSO$_3$, and Et$_3$N → product. Y 71% (*anti:syn* 96:4). *syn*-α,β-**Dialkoxyketones** were obtained **from acetals**. Phenolethers, ar. halides, ar. nitro, and ethylenes remain unaffected. F.e.s. R. Hayakawa, M. Shimizu, Org. Lett. 2, 4079-81 (2000).

*Zirconium tetra-*tert-*butoxide/1,1,4,4-tetraphenylbutane-1,4-diol/* ←
triphenylphosphine oxide/bis(trimethylsilyl) peroxide/trimethylsilyl cyanide
β-Hydroxynitriles from ethylene derivs. via stereospecific epoxidation   C(OH)C(CN)

272.

One-pot conversion. 10 Mol% Zr(OBu-*t*)$_4$ added at 0° to a mixture of 10 mol% 1,1,4,4-tetraphenylbutane-1,4-diol and 10 mol% triphenylphosphine oxide in dichloromethane, stirred for 10 min, 2 eqs. bis(trimethylsilyl) peroxide, 2 eqs. trimethylsilyl cyanide, and the startg. ethylene deriv. added, and the mixture stirred at 50° for 4 h → product. Y 100%. Trimethylsilyl cyanide was necessary for both the epoxidation and the subsequent formation of β-cyanohydrin, while the zirconium catalyst assumes the triple role of oxidant, Lewis acid and nucleophile. With unsym. ethylene derivs. the cyano group is introduced *regioselectively* at the least hindered carbon atom in a *trans*-fashion. F.e.s. S. Yamasaki, M. Shibasaki et al., J. Am. Chem. Soc. *123*, 1256-7 (2001).

Bis(2-methoxyethyl)zirconocene dihydride   ←
Ring closures of enynes   ○

with Cp$_2$ZrCl$_2$/BuLi cf. *50*, 428; of *terminal* enynes with bis(2-methoxyethyl)zirconocene dihydride, also reductive coupling of ethylene derivs. and ring closures of dienes s. P. Wipf, X. Wang, Tetrahedron Lett. *41*, 8237-41 (2000).

Trimethylsilyl chloride   Me$_3$SiCl
1,2,3,4-Tetrahydroquinolines via hetero-Diels-Alder reaction

from arylaminomethyl ethers with TiCl$_4$ cf. *38*, 760; from 1,3,5-triarylhexahydro-1,3,5-triazines by cycloaddition with 1,2-bis(trimethylsiloxy)cyclobutene using Me$_3$SiCl s. H.-J. Ha et al., J. Org. Chem. *65*, 8384-6 (2000).

Titanocene dichloride/zinc or manganese   Cp$_2$TiCl$_2$/Zn or Mn
Titanium(III)-mediated radical ring closures of acetyleneepoxides   OC̈

s. *51*, 318; bicyclo[n.3.0] systems with Zn or Mn as reductant, diastereoselectivity, **also cyclopentylcarbinol ring from ethyleneepoxides**, and tetrahydrofuran analogs, s. A. Gansäuer, M. Pierobon, Synlett *2000*, 1357-9.

*Titanocene dichloride/*n-*butylmagnesium chloride*   Cp$_2$TiCl$_2$/BuMgCl
Regiospecific 3-component radical synthesis of silanes   ←
from ethylene derivs., halides and chlorosilanes

273.

A soln. of ca. 2.2 eqs. *n*-BuMgCl (0.9 *M* in THF) added to a mixture of styrene, 1.1 eqs. Et$_3$SiCl, and 5 mol% Cp$_2$TiCl$_2$, then 1.1 eqs. *t*-BuBr added via syringe at 0° over 10 min under N$_2$, the mixture stirred 40 min, 0.1 *N* HCl added at the same temp., allowed to warm to 20°, then quenched with satd. aq. NH$_4$Cl → 3,3-dimethyl-1-phenyl-1-(triethylsilyl)butane. Y 95%. The method is efficient from terminal aryl- and silyl-ethenes, but sluggish with simple terminal and internal alkenes. It is thought that alkyl radicals are initially generated by electron transfer from a titanocene ate complex; radical addition to the alkene then takes place, followed by generation of a benzylmagnesium halide which is quenched with the chlorosilane. F.e., **also 2-ethylenesilanes from 1,3-dienes**, s. S. Nii, N. Kambe, J. Terao, J. Org. Chem. *65*, 5291-7 (2000).

Zirconocene dichloride/n-butyllithium/cuprous chloride $Cp_2ZrBu_2/CuCl$
9,10-Dihydroanthracene ring from *o*-di(propargyl)arenes and acetylene derivs.
[2+2+2]-Cycloaddition via zirconacyclopentadienes

274.

The startg. diyne added at -78° to a soln. of Cp_2ZrBu_2 (prepared from 1.2 eqs. Cp_2ZrCl_2 and 2.4 eqs. *n*-BuLi (2.48 *M*)) in THF, the mixture warmed to room temp., stirred for 1 h, 3 eqs. dimethyl acetylenedicarboxylate and 2 eqs. CuCl added at the same temp., stirred for a further 1 h, then quenched with 3 *N* HCl → product. Y 78%. DDQ oxidation affords the corresponding **linearly condensed polyarenes** (up to 7 rings). F.e.s. T. Takahashi et al., J. Am. Chem. Soc. *122*, 12876-7 (2000).

Titanium tetrachloride (s.a. under [CpRu(MeCN)$_3$]PF$_6$) $TiCl_4$
3-Silylcyclopentyl ketones from α,β-ethyleneketones and 2-ethylenesilanes
with allyldimethyltritylsilane cf. *50*, 407s57; with allylic benzhydryldimethylsilanes for subsequent facile Tamao oxidation s. Z.-H. Peng, K.A. Woerpel, Org. Lett. *2*, 1379-81 (2000).

Titanium tetrachloride/triethylamine $TiCl_4/Et_3N$
Titanium tetrachloride or bromide/dimethyl sulfide $TiCl_4$ or $TiBr_4/Me_2S$
Enhancement of the Baylis-Hillman reaction   ←
with α,β-ethyleneketones via β′-chloro-β-hydroxyketones

275.

While the standard dabco-catalyzed Baylis-Hillman reaction is often slow and limited in scope, the conversion can be accomplished smoothly and more generally *in two steps* with isolation of an intermediate chlorine adduct. E: *1.4 eqs.* $TiCl_4$ added to a soln. of *0.2 eq.* Et_3N in methylene chloride at -78°, stirred for 5 min, a soln. of *p*-nitrobenzaldehyde in the same solvent and 3 eqs. methyl vinyl ketone added at the same temp., kept at -78° for 12 h, and quenched with satd. aq. $NaHCO_3$ → *anti*-product. Y 81%. The latter was readily converted to the desired Baylis-Hillman adduct by treatment with 2 eqs. Et_3N or DBU in methylene chloride (or on purification by TLC). The proportion of the two reagents is critical, and pre-complexation is thought beneficial in providing a reactive nucleophilic chloride source. Reaction is also facile with acrylonitrile, but methyl acrylate was effectively unreactive. F.e.s. M. Shi et al., Org. Lett. *2*, 2397-400 (2000); **α,β-ethylene-β-halogeno-β′-hydroxy- from α,β-acetylene-carbonyl compds.** with $TiCl_4$ or $TiBr_4$ and Me_2S s. T. Kataoka et al., Angew. Chem. Int. Ed. Engl. *39*, 2358-60 (2000).

Titanium tetraiodide s. under Ti(OPr-i)$_4$ TiI_4

*Tri-*n*-butyltin hydride/zinc triflate/chiral bis(Δ^2-oxazolines)/*   ←
4,4-diphenyl-2-oxazolidone/triethylborane
Catalytic asym. radical 1,4-addition   C=C → CHC(R)
s. *55*, 277; enhancement of enantioselectivity with added 4,4-diphenyl-2-oxazolidone s. M. Murakata, O. Hoshino et al., Org. Lett. *3*, 299-302 (2001).

Tri-n-butyltin hydride/azodiisobutyronitrile Bu₃SnH/AIBN
Radical ring closure of 2,n-dienecarbonyl compds. ○

276.

3-β-Ketotetrahydrofurans. 1.1 eqs. Tri-*n*-butyltin hydride and 0.3 eq. AIBN added slowly over 1 h to a refluxing benzene soln. of the startg. ketone → product. Y 82% (*cis/trans* 1:1.8). F.e. and cyclic alcohols from ethyleneoxo compds. (cf. *45*, 413) s. D. Bebbington, A.F. Parsons et al., Tetrahedron Lett. *41*, 8941-5 (2000); **2-piperidone ring** from 3-aza-1,5-dien-4-ones s. S. Lesniak, J. Fisinska, Synthesis *2001*, 135-9.

Stannous chloride s. under PdCl₂(PPh₃)₂ SnCl₂

Tris(triphenylsilyl) vanadate VO(OSiPh₃)
(Z)-α,β-Ethylene-β'-hydroxyketones ←
from 2-acetylenealcohols and aldehydes

277.

Benzaldehyde and 1.2 eqs. startg. propargyl alcohol added via syringe to a suspension of 5 mol% tris(triphenylsilyl) vanadate in 1,2-dichloroethane in a reaction vial, and the mixture heated at 80° for 20 h → product. Y 95% ((Z)-selectivity 98%). This is a useful atom-economical alternative to the Baylis-Hillman reaction (which normally requires a β-unsubst. acceptor). The thermodynamically less stable (Z)-isomers are formed predominantly, but (Z)-selectivity was lower with aliphatic aldehydes. Reaction is thought to proceed via **aldol-type addition to a vanadyl allenolate**. F.e.s. B.M. Trost, S. Oi, J. Am. Chem. Soc. *123*, 1230-1 (2001).

Vanadocene dichloride/zinc/trimethylsilyl chloride Cp₂VCl₂/Zn/Me₃SiCl
1-Amino-5-cyanocyclopentenes from two α,β-ethylenenitrile molecules ○
with Sm/Me₃SiCl/*t*-BuOH cf. *55*, 284s*58*; dimerization of arylidenemalononitriles with 5 mol% Cp₂VCl₂ and 2 eqs. Zn/Me₃SiCl s. L. Zhou, T. Hirao, Tetrahedron Lett. *41*, 8517-21 (2000).

Ethyl dichlorovanadate s. under LiZnR₃ VO(OEt)Cl₂

Lithium trifluoromethanesulfonate LiOTf
Diels-Alder reaction
with LiClO₄ cf. *46*, 659s*47,50,59*; with safer LiOTf s. J. Auge et al., Synlett *2000*, 877-9.

Polymer-based arenedicarbonylchromium(0) complexes ←
[6+2]-Cycloaddition
with asym. induction cf. *46*, 631s*51*; using a little, recyclable, poly[(4-diphenylphosphinomethylstyrene)chromium(0) (η⁶-benzene)dicarbonyl] complex s. J.H. Rigby et al., Org. Lett. *2*, 3917-9 (2000).

Lithium perchlorate LiClO₄
1,2,3,4-Tetrahydroquinolines from N-arylaldimines and ethylene derivs.
via hetero-Diels-Alder reaction
with BF₃-etherate cf. *21*, 739; from cyclic ethylene derivs. under *neutral* conditions with 5 *M* LiClO₄ in ether s. J.S. Yadav et al., Synlett *2001*, 240-2.

5(S)-Benzyl-2,2,3-trimethyl-4-imidazolidone hydroperchlorate ←
Chiral amine-catalyzed asym. 1,3-dipolar cycloaddition ←
of α,β-ethylenealdehydes with nitrones

278.

Chiral isoxazolidine-4-carboxaldehydes. 4 eqs. Crotonaldehyde added to a soln. of startg. nitrone, 20 mol% 5(S)-benzyl-2,2,3-trimethyl-4-imidazolidone hydroperchlorate, and 6 eqs. water in nitromethane at -20°, and kept at the same temp. under aerobic conditions until TLC indicated completion of reaction (35-160 h) → product. Y 70% (*endo:exo* 99:1; e.e. 99%). This is the first example of an organo-catalyzed 1,3-dipolar cycloaddition, which is operationally simple, economical, and a useful alternative to Lewis acid-catalyzed procedures. Significantly, reaction is generally applicable to enals which are poor substrates in metal-catalyzed conversions. Cycloaddition is facilitated by *in situ*-conversion of the aldehyde to a more reactive immonium salt. The asym. induction was also high from the TfOH salt. F.e.s. W.S. Jen, D.W.C. MacMillan et al., J. Am. Chem. Soc. *122*, 9874-5 (2000).

Manganese s. under Cp_2TiCl_2 Mn

Manganese(III) acetate $Mn(OAc)_3$
3-Hydroxy-1,2-dioxanes from ketones and ethylene derivs.
electrocatalytic procedure cf. *42*, 636; 8-acyl-1-hydroxy-2,3-dioxa-8-azabicyclo[4.4.0]decane-6-carboxylic acid esters with $Mn(OAc)_3$ s. R. Kumabe, H. Nishino et al., Tetrahedron Lett. *42*, 69-72 (2001).

Manganese(II) acetate/cobalt(II) acetate/oxygen $Mn(OAc)_2/Co(OAc)_2/O_2$
Cocatalytic redoxidative radical C-α-alkylation of ketones with ethylene derivs. H → R

279.

1-Octene added to a soln. of 10 eqs. cyclohexanone, 0.5 mol% $Mn(OAc)_2$ and 0.1 mol% $Co(OAc)_2$ in acetic acid under a balloon of O_2, and heated at 80° for 10 h → 2-octylcyclohexanone. Conversion 87% (selectivity 62%). Reaction is generally applicable to cyclic and aliphatic ketones, α-monosubst. derivs. reacting preferentially at the secondary site. A dual catalytic cycle is invoked, wherein O_2 oxidizes Co(II) to Co(III), which in turn oxidizes Mn(II) to Mn(III). F.e.s. T. Iwahama, Y. Ishii et al., Chem. Commun. *2000*, 2317-8.

Iron carbonyl/sodium tetrahydridoborate/acetic acid/cupric chloride ←
Biscarbonylation of acetylene derivs. ←
cyclobutenediones cf. *54*, 312; **succinimides** in the presence of prim. amines s. C. Rameshkumar, M. Periasamy, Synlett *2000*, 1619-21.

Ferric acetoacetonate s. under RLi $Fe(acac)_3$

(1,3,5-Cycloheptatriene)(1,5-cyclooctadiene)iron ←
Benzene ring from acetylene derivs. by cyclotrimerization s. *42*, 676s60 ○

Cobalt/charcoal Co/C
Heterogeneous Pauson-Khand reaction
of enynes with cobalt-on-mesoporous silica cf. *58*, 281; with inexpensive, readily prepared, and

reusable cobalt-on-charcoal for both inter- and intra-molecular conversions s. S.U. Son, Y.K. Chung et al., Angew. Chem. Int. Ed. Engl. *39*, 4158-60 (2000).

Cobalt carbonyl $Co_2(CO)_8$
Benzene ring from diynes and acetylene derivs.
with Ni(acac)$_2$/Ph$_3$P/*i*-Bu$_2$AlH cf. *57*, 300s*58* and with RhCl(PPh$_3$)$_3$ cf. *33*, 658s*57*; cofacial indan-subst. porphyrins with Co$_2$(CO)$_8$ (stoichiometric) by metal-templated cycloaddition s. J.T. Fletcher, M.J. Therien, J. Am. Chem. Soc. *122*, 12393-4 (2000).

Intramolecular Pauson-Khand reaction-Diels Alder reaction

280.

2 eqs. 2,3-Dimethyl-1,3-butadiene added to a soln. of startg. diyne and 5 mol% Co$_2$(CO)$_8$ in dichloromethane under CO in a stainless steel reactor, pressurized to 30 atm. with CO, and heated at 110° for 18 h → product. Y 90%. The diyne may contain an oxygen tether. A methoxy-substituted diene gave the Pauson-Khand adduct only, while silyl- and *t*-butyl-substituted diynes failed to react. F.e.s. S.U. Son, Y.K. Chung et al., J. Org. Chem. *65*, 6142-4 (2000).

Cobalt carbonyl/cyclohexylamine or tert. amine N-oxides $Co_2(CO)_8/c$-$C_6H_{11}NH_2$ or R_3NO
Pauson-Khand reaction
s. *40*, 475s*59*; tetracyclic N-condensed indoles from indolylenynes with Me$_3$NO as promotor s. L. Perez-Serrano, J. Perez-Castells et al., Synlett *2000*, 1303-5; 2,3-disiloxybicyclo[4.3.0]non-1(9)-en-8-ones with asym. induction using cyclohexylamine as promotor s. C. Mukai, M. Hanaoka et al., J. Org. Chem. *65*, 6654-9 (2000); regioselective reaction of subst. norbornenes and diazabicyclo[2.2.1]heptenes with terminal alkynes s. V. Derdau, S. Laschat et al., Eur. J. Org. Chem. *2000*, 681-9; 3-aza-4-oxabicyclo[4.3.0]non-6-en-8-ones by intramolecular reaction s. S.G. Koenig, D.J. Austin et al., Tetrahedron Lett. *41*, 9393-6 (2000); intermolecular reaction with Co$_2$(CO)$_6$-complexed terminal ynamines at low temp. s. J. Balsells, M.A. Pericàs et al., J. Org. Chem. *65*, 7291-302 (2000); ultrasonic enhancement and reactions of desymmetrized cobalt alkyne complexes s. Vol. *59* (p.158 under *40*, 675).

Cobalt carbonyl/polymer-based tert. amine N-oxides   ←
Pauson-Khand reaction using a polymer-based N-oxide as promotor
s. *58*, 286; improved procedure with a *high-loading* supported N-methylmorpholine N-oxide s. D.S. Brown, W.J. Kerr et al., Synlett *2000*, 1573-6.

Cobalt carbonyl/chiral bicyclic 2-phosphino-1,3-oxathianes/N-methylmorpholine N-oxide   ←
Asym. Pauson-Khand reaction
Improved procedure with chiral P,S-ligands

281.

Chiral P,S-ligands (e.g. PuPHOS) are more effective in asym. Pauson-Khand reaction than chiral [bis]phosphines (cf. *49*, 674s*51*), performing the unprecedented multiple role of (a) enhancing diastereoselectivity in their coordination to the prochiral dicobalt complex by thermodynamic equilibriation, (b) increasing reactivity and (c) controlling stereoselection by directing the reaction to the cobalt atom where sulfur is coordinated. E: A soln. of PuPHOS-complexed ethynyltrimethylsilane (prepared as a crystalline solid in 68% yield by heating a mixture of (ethynyltrimethylsilane)-$Co_2(CO)_6$ with PuPHOS·BH_3 in toluene with 2 eqs. DABCO at 80° for 17 h, followed by crystallization of the resulting 3:1 mixture of diastereoisomers) in methylene chloride containing N-methylmorpholine N-oxide (NMO) allowed to react with 10 eqs. norbornadiene at room temp. for 3 days in a Schlenk flask under N_2 → product. Y 93% (e.e. 97%). The close proximity of the S and O atoms in the ligand is crucially important in order to effect the necessary bridging between the two cobalt atoms, the stereospecificity being controlled by selective cleavage of the sulfur residue to liberate a coordinatively unsaturated cobalt atom to bind the incoming olefin. NMO does not affect the rate significantly but is believed to improve the stereoselectivity by preventing CO-promoted complex epimerization. F.e.s. X. Verdaguer, M.A. Pericàs, A. Riera et al., J. Am. Chem. Soc. *122*, 10242-3 (2000).

Dodecacarbonyltetracobalt/cyclohexylamine $Co_4(CO)_{12}/c$-$C_6H_{11}NH_2$
Catalytic Pauson-Khand reaction
s. *50*, 415s55; with added cyclohexylamine as promotor (30 mol%) s. M.E. Krafft, L.V.R. Boñaga, Angew. Chem. Int. Ed. Engl. *39*, 3676-80 (2000).

[η⁵-(4-Hydroxybutyryl)cyclopentadiene](1,5-cyclooctadiene)cobalt(I)   ←
Pyridines from nitriles and two acetylene deriv. molecules
s. *31*, 658s34 (review); under mild conditions with a little [η⁵-(4-hydroxybutyryl) cyclopentadiene](1,5-cyclooctadiene)cobalt(I) *in aq. ethanol* under mild conditions s. A.W. Fatland, B.E. Eaton, Org. Lett. *2*, 3131-3 (2000).

Cobaltous acetate s.a. under Mn(OAc)₂ $Co(OAc)_2$

Cobaltous acetate/N-hydroxyphthalimide
Cyclic β-hydroxyacetals   C=C → C(OH)C-C(OR)$_2$
from electron-deficient ethylene derivs. and cyclic acetals
Regiospecific oxidative radical addition

282.

2-β-Hydroxy-1,3-dioxolanes. Methyl acrylate added under a balloon of O_2 to a soln. of 5 eqs. 2-methyl-1,3-dioxolane, 5 mol% N-hydroxyphthalimide, and 0.05 mol% $Co(OAc)_2$, and vigorously stirred at room temp. for 3 h → product. Y 81%. Reaction takes place in a catalytic cycle initiated by Co(II)-catalyzed abstraction of hydrogen atom from the acetal by phthalimide N-oxyl. F.e. and nitrile analogs s. K. Hirano, Y. Ishii et al., Chem. Commun. *2000*, 2457-8.

Cobaltous chloride/hexamethyldisilthiane $CoCl_2/(Me_3Si)_2S$
Hetero-Diels-Alder reaction with thioacylsilanes
s. *43*, 597; generation of thioacylsilanes from 1-benzotriazol-1-yl-1,1-silylthioethers with $CoCl_2/(Me_3Si)_2S$, and further catalysts, s. A. Degl'Innocenti, A.R. Katritzky et al., J. Org. Chem. *65*, 9206-9 (2000).

Bis(1,5-cyclooctadiene)nickel(0) $Ni(cod)_2$
β,γ-Ethylene-δ-stannylketones from 1,3-dienes and acylstannanes COC-C=C-C(Sn≤)
Stereospecific 1,4-acylstannylation

283.

A mixture of 3 eqs. startg. diene and 1 eq. startg. acylstannane *in toluene* treated with 5 mol% $Ni(cod)_2$ at 50° for 10 min → (Z)-1-phenyl-5-trimethylstannyl-3-penten-1-one. Y 72%. 2,3-Disubst. 1,3-dienes also gave a single isomer, but 2-subst. 1,3-dienes gave a mixture of regioisomers. Decarbonylation of the acylstannane was the predominant reaction in polar media (e.g. DMF, THF). This is a rare example of a transition metal-catalyzed carbometalation of a 1,3-diene at the 1,4-position. It is thought to involve initial oxidative insertion of Ni(0) into the C-Sn bond, followed by insertion of the diene to give a π-allyl complex, which then undergoes reductive elimination. F.e. and subsequent regiospecific synthesis of **β,γ-ethylene-ε-hydroxyketones from aldehydes** with Bu_2SnCl_2 s. E. Shirakawa, T. Hiyama et al., J. Am. Chem. Soc. *122*, 9030-1 (2000); **β,γ-ethylene-β-stannylketones from allenes** (cf. *58*, 287) s. Chem. Commun. *2001*, 263-4.

Bis(1,5-cyclooctadiene)nickel(0)/tri-n-butylphosphine/triethylborane ←
Regio- and stereo-specific synthesis of 2-ethylenealcohols C≡C → CH=C-CH(OH)R
from acetylene derivs. and aldehydes

284.

0.2 eq. Tri-*n*-butylphosphine added via syringe under argon to a mixture of 0.1 eq. $Ni(cod)_2$ in degassed toluene in a Schlenk flask, the mixture stirred for 5 min at 23°, a soln. of 2 eqs. Et_3B (1 *M* in hexane) added via syringe, stirring continued for 10-15 min, a soln. of the startg. alkyne and 1 eq. octanal in toluene added dropwise over 1 min, the mixture stirred at room temp. for 18 h, and quenched with satd. aq. NH_4Cl and 1 *N* HCl → product. Y 89% (regioselectivity >98%). The method is highly regioselective using commercially available reagents with no reduction being observed in the acetylene or aldehyde. Reaction is regio- and stereo-specific, and generally applicable to coupling internal or terminal alkynes with aromatic or aliphatic aldehydes. F.e.s. W.-S. Huang, T.F. Jamison et al., Org. Lett. *2*, 4221-3 (2000).

Bis(1,5-cyclooctadiene)nickel(0)/triphenylphosphine $Ni(cod)_2/Ph_3P$
1,2-Dialkylidenecyclobutanes from allenes ☐
Regiospecific nickel(0)-catalyzed cyclodimerization under mild conditions

285.

A soln. of (perfluorohexyl)allene in dry toluene added to a mixture of 0.1 eq. $Ni(cod)_2$ and 0.4 eq. Ph_3P in the same solvent under argon, and stirred for 30 min at room temp. → 1,2-bis-(2,2,3,3,4,4,5,5,6,6,7,7,7-tridecafluoroheptylene)cyclobutane. Y 81%. The codimerization proceeded efficiently in the presence of a wide range of electron-withdrawing groups, e.g. carbalkoxy, as well as the perfluorohexyl group. However, no reaction occurred with 1-(perfluorohexyl)-1-methylallene even at 100°, and the reaction with 1-(perfluorohexyl)-3-methylallene gave a complex mixture at room temp. The high regioselectivity is explained in terms of the preferred formation of a nickelacyclopentane, controlled by the electronic and steric effects of the allene substituent. Reaction failed with bidentate ligands, indicating that ligand dissociation precedes reductive elimination of the metal in the catalytic cycle. F.e.s. S. Saito, Y. Yamamoto et al., J. Am. Chem. Soc. *122*, 10776-80 (2000).

Nickel(II) acetoacetonate/diethylzinc/triphenylphosphine/phenol ←
Benzene ring from acetylene derivs. by cyclotrimerization ○
1,2,4-trisubst. benzenes with a Ru-carbene complex cf. *42*, 676s57; with Ni(acac)$_2$/Et$_2$Zn/Ph$_3$P/PhOH s. N. Mori, S. Ikeda et al., Chem. Commun. *2001*, 181-2; with (1,3,5-cycloheptatriene)(1,5-cyclooctadiene)iron(0) s. C. Breschi et al., J. Organomet. Chem. *607*, 57-63 (2000); **under heterogeneous conditions** with a stable, recyclable bis(pyridyl)palladium(II) dichloride complex supported on polysiloxanes *with controllable solubility* and degree of cross-linking, and with improved regioselectivity, s. Y.-S. Fu, S.J. Yu, Angew. Chem. Int. Ed. Engl. *40*, 437-40 (2001).

Dodecacarbonyltriruthenium $Ru_3(CO)_{12}$
N-Heteroarene-directed acylation with ethylene derivs. under carbonylation H → COR
s. *48*, 681s57; *o*-acylation of 2-arylpyridines s. Y. Ie, S. Murai et al., J. Org. Chem. *65*, 1475-88 (2000).

3,4-Dihydro-2-pyridones from cyclopropyl ketimines ○
via ruthenium-catalyzed carbonylative ring expansion

286.

Cyclopropyl phenyl N-*tert*-butylketimine and 2 mol% Ru$_3$(CO)$_{12}$ in toluene charged to a stainless steel autoclave, the mixture flushed with CO and pressurized to 2 atm., then heated in an oil bath at 160° for 60 h → 1-(*tert*-butyl)-3,4-dihydro-6-phenyl-2-pyridone. Y 76%. The substituent on the imine nitrogen is limited to *tert*-butyl and cyclohexyl. Coordination of nitrogen to ruthenium facilitates conversion to a ruthenacyclic which undergoes carbonyl insertion prior to reductive elimination of the metal. The reaction failed at higher temp. and/or higher pressure. F.e.s. A. Kamitani, S. Murai et al., J. Org. Chem. *65*, 9230-3 (2000).

Tris(acetonitrile)(cyclopentadienyl)ruthenium(II) hexafluorophosphate/ ←
  titanium tetrachloride
Cyclic 2-ethyleneamines from ω-alleneamines and electron-deficient ethylene derivs. ○
Cyclic δ-amino-γ-methyleneketones

287.

A soln. of the startg. alleneamine in DMF added to 10 mol% CpRu(CH$_3$CN)$_3$PF$_6$ in a pressure tube, 1.5 eqs. methyl vinyl ketone and 15 mol% TiCl$_4$ (1 *M* in dichloromethane) added, the tube capped, heated at 60° for 2 h, cooled to room temp., 1 eq. pyrrolidine added, and the mixture stirred at room temp. for 30 min → product. Y 90%. Interestingly, there was no poisoning of the catalyst by the amine, nor undesirable Michael addition of the latter. Co-catalytic TiCl$_4$ (as Lewis acid) is essential for high yields. Reaction is thought to involve intermediate formation of a ruthenacycle. N-Benzyl, N-*p*-methoxybenzyl and carbonyl groups were tolerated. F.e. and piperidines s. B.M. Trost et al., J. Am. Chem. Soc. *122*, 12007-8 (2000).

Chloro(1,5-cyclooctadiene)(η⁵-pentamethylcyclopentadienyl)ruthenium(II)   Cp*RuCl(cod)
**Ruthenium-catalyzed [2+2]-cycloaddition of 5-subst. norbornenes
with acetylene derivs.
Effect of remote substituents on regioselectivity**

288.

A mixture of 5 eqs. startg. norbornene and 1 eq. acetylene in THF added via cannula to an oven-dried screw-capped vial containing 0.07 eq. Cp*RuCl(cod) under N_2, and the mixture stirred in the dark *at 25°* for 67 h → product. Y 96% (7.5:1 isomer ratio). Regioselectivity was higher with 5-*exo*-substituents than 5-*endo*-groups, this being evidence for the preferred formation of an intermediate ruthenacycle. F.e.s. R.W. Jordan, W. Tam, Org. Lett. *2*, 3031-4 (2000).

Chloro(η⁵-pentamethylcyclopentadienyl)bis(triphenylphosphine)ruthenium(II)   ←
Addition of polyhalides to ethylene derivs.   ←
with $RuCl_2(PPh_3)_3$ cf. *18*, 776s38,46,57; addition of CCl_4 and $CHCl_3$ *at 40°* with air-stable, accessible chloro(η⁵-pentamethylcyclopentadienyl)bis(triphenylphosphine)ruthenium(II) s. F. Simal, A. Demonceau et al., Tetrahedron Lett. *41*, 6071-4 (2000).

Ruthenium carbene complex   ←
Intramolecular ene-yne metathesis-Diels-Alder reaction   ○

289.

2-Sulfonyl-2,3,3a,4,5,6-hexahydro-1*H*-pyrrolo[3,4-*c*]pyridazines. A soln. of startg. enyne and 1.1 eqs. startg. dienophile in dichloromethane added to a soln. of 0.05 eq. Grubb's catalyst ($PhCH=RuCl_2(PCy_3)_2$) in the same solvent, and the mixture stirred at room temp. for 50 h under N_2 → product. Y 78%. Grubb's catalyst was stable even in the presence of strong Lewis acids (such as BCl_3 or $AlCl_3$) which are sometimes required for Diels-Alder reactions. The yield was higher than that obtained by a stepwise procedure. F.e. and hexahydroindenes s. D. Bentz, S. Laschat, Synthesis *2000*, 1766-73.

Soluble or polymer-based ruthenium dichloro(imidazolidin-2-ylidene) complexes   ←
1,3-Dienes by ene-yne metathesis $CH=CHC=CH_2$
with $PhCH=RuCl_2(PCy_3)_2$ s. *54*, 320; metathesis with α-propargylglycinates *en route* to subst. phenylalanines s. S. Kotha et al., Synlett *2000*, 853-5; 2-α-acoxy-1,3-butadienes from propargyl esters at elevated ethylene pressure s. J.A. Smulik, S.T. Diver, J. Org. Chem. *65*, 1788-92 (2000); 2-α-hydroxy-1,3-butadienes with a soluble ruthenium dichloro(imidazolidin-2-ylidene) complex under 60 psi ethylene, functional group compatibility, s. Org. Lett. *2*, 2271-4 (2000); with a polymer-based complex cf. S.C. Schurer, S. Blechert et al., Angew. Chem. Int. Ed. Engl. *39*, 3898-901 (2000).

Ruthenium (arylseleno)carbene complex   ←
Ring opening-cross metathesis of bicyclic ethylene deriv. with eneselenides   C

290.

2 Mol% $PhSeCH=RuCl_2(PCy_3)_2$ added to a soln. of phenyl vinyl selenide and 1 eq. of the startg. 7-oxanorbornene in dichloromethane, and the mixture stirred at room temp. for 20 h → product.

Y 99% ((E)-selectivity 84%). High regioselectivity is explained by the enhanced thermodynamic stability of the catalyst which is regenerated exclusively in the catalytic cycle on addition of the reactive eneselenide (i.e. there is no concurrent self-metathesis of the cyclic alkene). Grubb's catalyst, PhCH=RuCl$_2$(PCy$_3$)$_2$, is similar in activity but must be combined with the chalcogenide prior to addition of the 7-oxanorbornene deriv. F.e.s. H. Katayama, F. Ozawa et al., Angew. Chem. Int. Ed. Engl. 39, 4513-5 (2000).

Chiral rhodium(I) bis(1,3,2-diazaphospholidine) complexes ←
Regiospecific asym. hydroformylation of enolesters   C=C(OAc) → CHC(OAc)CHO

291.

0.5 Mol% [Rh(acac)(CO)$_2$] and 0.75 mol% chiral *bidentate* bis(1,3,2-diazaphospholidine) complex [ESPHOS] in toluene flushed with CO/H$_2$ in a mini-autoclave, stirred at 60° under <8 atm. CO/H$_2$, a soln. of vinyl acetate in the same solvent injected, the CO/H$_2$ pressure raised to 8 atm., and stirring continued under these conditions at 500 rpm for 5 h → (S)-product. Y 90.3% (conversion 98.9%; branched:linear aldehyde ratio 94.5:5.5; e.e. 89%). Regioselectivity in favour of the branched aldehyde is higher than with (R,S)-BINAPHOS (cf. 49, 683), and a lower pressure is required. Furthermore, yields and enantioselectivity were lower with monodentate 1,3,2-diazaphospholidines as ligand. However, the e.e. was 0% with styrene, indicating that a *coordinating group* (OAc) in the substrate is essential for high asym. induction. Interestingly, the catalyst also shows activity for aldehyde hydrogenation. F.e. and comparison of chiral ligands s. S. Breeden, M. Wills et al., Angew. Chem. Int. Ed. Engl. 39, 4106-8 (2000).

Dicarbonyl(chloro)rhodium(I) dimer   [Rh(CO)$_2$Cl]$_2$
1,4-Cycloheptadienes from unactivated vinylcyclopropanes and acetylene derivs.
via regiospecific rhodium-catalyzed [5+2]-cycloaddition

292.

5 Mol% [Rh(CO)$_2$Cl]$_2$ added to a soln. of startg. vinylcyclopropane in 1,2-dichloroethane containing 5% 2,2,2-trifluoroethanol in a Schlenk flask, the mixture sparged with argon for 1 min, 2.8 eqs. methyl propiolate added via syringe, the flask sealed, heated in an oil bath at 80° for 1 h, then cooled to room temp. → product. Y 95% (100% conversion). The C$_1$-position of the vinylcyclopropane must bear a substituent bulky enough to force the molecule into the *s-cis* conformation which leads to formation of the desired cyclic *cis*-alkene. The presence of 2,2,2-trifluoroethanol was found to improve the yield and reaction rate. F.e.s. P.A. Wender et al., J. Am. Chem. Soc. 123, 179-80 (2001).

Chloro(1,5-cyclooctadiene)rhodium(I) dimer   [Rh(cod)Cl]$_2$
Cyclic α-aminoketones from amines   H → Ac
via rhodium-catalyzed carbonylation at an sp^3-hybridized carbon

293.

A mixture of the startg. amine, isopropanol, and 4 mol% [RhCl(cod)]$_2$ in a stainless steel autoclave pressurized with ethylene gas to 5 atm. and with CO to a total pressure of 10 atm., heated in an oil bath at 160° for 40 h, cooled to room temp., and vented → product. Y 84%. The reaction is regiospecific and limited to cyclic amines, such as pyrrolidine and 1,2,3,4-tetrahydroisoquinoline, with a 2-pyridyl residue on nitrogen. Reaction possibly proceeds via **pyridyl nitrogen-directed oxidative addition** of rhodium to the *sp³-hybridized* carbon-hydrogen bond (there being no reaction with N-alkyl- or N-phenyl-analogs). Electron-withdrawing groups, as well as groups at C_6 of the pyridine ring, resulted in low yields. The reaction cannot be applied to benzo-fused pyrrolidines. F.e. and regioselectivity s. N. Chatani, S. Murai et al., J. Am. Chem. Soc. *122*, 12882-3 (2000).

Rhodium phosphine complexes/imidazolium hexafluorophosphates ←
Hydroformylation in ionic liquid solvents C═C → CHC-CHO
in phosphonium tosylates cf. *4,* 667s*56*; in a 2-phase alkene/[BMIM]PF_6 medium with guanidinium-modified bis(diarylphosphino)xanthenes as ligand for good linear selectivity and catalyst recovery s. P. Wasserscheid et al., Chem. Commun. *2001,* 451-2.

Dendritic rhodium phosphine complexes ←
Heterogeneous hydroformylation
on silica cf. *4,* 667s*59*; bead-supported catalysts s. P. Arya, H. Alper et al., J. Org. Chem. *65,* 1881-5 (2000); with dendritic polyhedral oligomeric phosphine-functionalized silsesquioxanes for enhanced linear selectivity s. L. Ropartz, D.J. Cole-Hamilton et al., Chem. Commun. *2001,* 361-2.

Rhodium bis(phosphite) complexes ←
Hydroformylation
update s. *4,* 667s*59;* with chelated calix[4]arene-based bis(phosphites) for controlled regioselectivity *at low pressure* s. R. Paciello, M. Röper et al., Angew. Chem. Int. Ed. Engl. *38,* 1920-3 (1999).

Rhodium cyclic phosphonite complexes ←
Isomerizing hydroformylation
linear aldehydes with 4,5-bis(phosphino)xanthenes as ligand cf. *56,* 296; with oxy-functionalized tricyclic monodentate phosphonite complexes under mild conditions s. D. Selent, A. Börner et al., Angew. Chem. Int. Ed. Engl. *39,* 1639-41 (2000).

Chlorotris(triphenylphosphine)rhodium(I) $RhCl(PPh_3)_3$
o-**Alkylation of ar. ketimines with ethylene derivs.** s. *60,* 360 H → R

Chlorotris(triphenylphosphine)rhodium(I)/silver acetate/triethylamine ←
Stereospecific [2+2+2]-cycloaddition-1,3-dipolar cycloaddition O

294.

A soln. of startg. diyne in dry THF added dropwise over 2 h to a stirred mixture of 2 eqs. N-prop-2-ynylmaleimide, and 5 mol% $RhCl(PPh_3)_3$ in *dry THF* under N_2 at room temp., stirred for 15 h,

2.1 eqs. startg. imine, 2.1 eqs. AgOAc, and 2.4 eqs. Et₃N added, and stirring continued for 72 h → product. Y 44%. Three rings, five bonds and 4 stereocentres are formed stereospecifically in one operation. F.e. (all 3-(5-indanylmethyl)-3,7-diazabicyclo[3.3.0]octane-2,4-diones) and with isolation of the intermediate s. R. Grigg et al., Tetrahedron 56, 8967-76 (2000).

Palladous acetate $Pd(OAc)_2$
cis-β-Hetaryl-α,β-ethylenecarboxylic C≡C → CH=C(R)
from α,β-acetylenecarboxylic acid esters and hetarenes
Palladium-catalyzed regio- and stereo-specific 1,4-addition under mild conditions

295.

Ethyl phenylpropiolate added to a soln. of 5 mol% Pd(OAc)₂ and 3 eqs. N-methylpyrrole in acetic acid, and the mixture stirred at room temp. for 24 h → product. Y 83%. *cis*-Adducts were obtained exclusively as a result of *trans*-carbopalladation of the alkyne, reaction with pyrroles and furans taking place at the 2- or 5-position. However, indoles react at C₃ unless this position is blocked. Bis-addition can occur when the substituent on the acetylene is small or linear, such as methyl. Catalyst/solvent combinations were studied with Pd(OAc)₂ in acetic acid being the optimized choice. F.e.s. W. Lu, Y. Fujiwara et al., Org. Lett. 2, 2927-30 (2000).

Palladous acetate/lithium bromide $Pd(OAc)_2 /LiBr$
Ring closures of α,β-acetylenecarboxylic acid derivs. via halogenopalladation ○
α-(1-halogenalkylidene)-γ-lactones from allyl esters s. 48, 692s53; 3-(1-bromoalkylidene)-2-pyrrolidones from N-allylamides s. X. Xie, X. Lu, Synlett 2000, 707-9.

Palladous acetate/sodium iodide $Pd(OAc)_2/NaI$
Intramolecular aminopalladation-1,4-addition

296.

N-Protected γ-(2-oxazolidon-4-ylidene)oxo compds. Startg. carbamyloxy-2-acetylene reacted with 15 eqs. acrolein in THF in the presence of 0.05 eq. Pd(OAc)₂ and 2 eqs. NaI at room temp. → product. Y 85%. F.e. and 2-imidazolidone and 2-pyrrolidone analogs s. A. Lei, X. Lu, Org. Lett. 2, 2699-702 (2000).

Palladous acetate/triethylamine $Pd(OAc)_2/Et_3N$
β-Aryl- from α,β-ethylene-ketones and ar. halides C=C → CHC(Ar)
with Pd(OAc)₂(PPh₃)₂/HCOOH/Et₃N cf. 39, 640; from β-subst. enones **in supercritical carbon dioxide** with Pd(OAc)₂/Et₃N s. S. Cacchi et al., Synlett 2000, 650-2.

Palladous acetate/triphenylphosphine/triethylamine $Pd(OAc)_2/Ph_3P/Et_3N$
Telomerization of 1,3-dienes ←
using Pd(OAc)₂ cf. 27, 724s32; with phenols using Pd(OAc)₂/Ph₃P/Et₃N s. A. Krotz, M. Beller et al., Chem. Commun. 2001, 195-6.

Palladous acetate/tris(2,6-dimethoxyphenyl)phosphine/palladous bis(trifluoroacetate) ←
4-Alkylidene-3,4-dihydro-2H-pyrans ○
from 3-acetylenealcohols and terminal acetylene derivs.
Regiospecific carbopalladation-cycloisomerization under sequential catalysis

297.

A soln. of 2 mol% Pd(OAc)$_2$ and 2 mol% tris(2,6-dimethoxyphenyl)phosphine in benzene stirred at room temp. for 20 min, solns. of 1 eq. of startg. acetylenealcohol and 1.1 eqs. *t*-butylacetylene in benzene added, stirring continued until complete consumption of the alcohol, 3 mol% Pd(OCOCF$_3$)$_2$ added, and the mixture stirred until completion → product. Y 61% (with <5% by-product). Satisfactory yields were also obtained by using a higher loading of catalyst and ligand. However, attempts to force the reactions by heating resulted in an increase in lactone by-product. The terminal alkyne may possess a *prim*- or *sec*-hydroxyl group, but *tert*-hydroxyl or bulky groups had a detrimental effect on the rate. Esters, alcohols, silyl ethers, PMB ethers, nitriles and alkyl chlorides were tolerated. F.e. and reaction at a higher loading of ligand with Pd(OAc)$_2$ as the sole catalyst s. B.M. Trost, A.J. Frontier, J. Am. Chem. Soc. *122*, 11727-8 (2000).

Palladous acetate/chiral (o-phosphinoaryl)-Δ^2-oxazolines ←
1,3-Enynes from internal and terminal acetylene derivs. CH≡C-C≡C
with Ru-vinylidene complex cf. *56*, 295; regioselective cross-coupling of mono- and di-subst. alkynes with Pd(OAc)$_2$/2-(o-phosphinoaryl)-Δ^2-oxazolines, **also with kinetic resolution** of propargyl alcohols using a chiral ligand s. U. Lücking, A. Pfaltz, Synlett *2000*, 1261-4.

Bis(dibenzylideneacetone)palladium(0)/potassium acetate/trimethylsilyl azide ←
2-Aryl-2-ethyleneazides from allenes and ar. iodides C=C=C → C=C(Ar)C(N$_3$)
Regioselective conversion

298.

A mixture of 5 mol% Pd(dba)$_2$, 1.5 eqs. Me$_3$SiN$_3$, 4-iodoacetophenone, 2 eqs. 1,1-dimethylallene, and 1.6 eqs. KOAc in DMF stirred at 70-75° for 24 h under N$_2$ → product. Y 96% (by NMR; 84:16 mixture of regioisomers). The regioselectivity depends largely on the nature and size of the substituent on the allene bond (*tert*-butylallene exhibiting the highest regioselectivity). Ar. iodides possessing electron-donating groups gave the highest yield. F.e. and reduction [with Ph$_3$P/H$_2$O] to the corresponding **2-aryl-2-ethyleneamines** as a *single* isomer (via rapid 1,3-azido group shift), s. H.-M. Chang, C.-H. Cheng, J. Chem. Soc. Perkin Trans.1 *2000*, 3799-807.

Tris(dibenzylideneacetone)dipalladium $Pd_2(dba)_3$
1-Stannyl-1,4-dienes from acetylene derivs. and 2-ethylenestannanes ←
with ZrCl$_4$ cf. *52*, 305; with Pd$_2$(dba)$_3$, regio- and stereo-selectivity, s. E. Shirakawa, T. Hiyama et al., Org. Lett. *2*, 2209-11 (2000).

Tris(dibenzylideneacetone)dipalladium/chiral ferrocenylmonophosphines
3-Alkylidene-4-vinylcyclohexenes from 1,2,4-trienes and unactivated 1,3-dienes
Regiospecific asym. [4+2]-cycloaddition

An asym. version of *53*, 326 is reported. E: 10 eqs. 4.4 M 1,3-butadiene in methylene chloride and startg. vinylallene added successively to 2.5 mol% $Pd_2(dba)_3 \cdot CHCl_3$ and 6 mol% chiral ferrocenylmonophosphine ligand in the same solvent at room temp. under N_2, and worked up after 10 min → product. Y 85% (e.e. 83%). Reaction is thought to involve an intermediate square planar palladium(II) complex in which three of the four coordination sites are occupied by the cycloaddition partners, and wherein *mono*dentate chiral ligands are more effective than bidentate ligands (e.g. DuPHOS, BINAP). Unlike Lewis acid-catalyzed methods, it is also applicable to unactivated substrates (without coordinating heteroatoms). F.e. and optimization of the ligand s. M. Murakami, Y. Ito et al., Chem. Commun. *2000*, 2293-4.

Tris(dibenzylideneacetone)dipalladium/tris(2,6-dimethoxyphenyl)phosphine $Pd_2(dba)_3/Ar_3P$
Arylacetylenes from 1,3-enynes and 1,3-diynes
with $Pd(PPh_3)_4$ cf. *54*, 325s*59*; N-Boc-protected *p*-aminoarylacetylenes from 2-amino-1,3-enynes and 1,3-diynes with $Pd_2(dba)_3$/tris(2,6-dimethoxyphenyl)phosphine s. S. Saito, Y. Yamamoto et al., J. Org. Chem. *65*, 4338-41 (2000).

Tetrakis(triphenylphosphine)palladium(0) $Pd(PPh_3)_4$
Styrenes from two 1,3-enyne molecules
s. *51*, 335s*58*; cross-cycloaddition under reactivity control s. S. Saito, Y. Yamamoto et al., Org. Lett. *2*, 3853-5 (2000).

Regio- and stereo-specific intramolecular carbopalladation of bis(allenes)

with **1,4-silylstannylation**. 1.2 eqs. Trimethyl(tributylstannyl)silane added to a mixture of startg. bis(allene) and 5 mol% $Pd(PPh_3)_4$ in THF, and refluxed for 3 h → product. Y 78%. By contrast, **1,4-distannylation** takes place with distannanes to give *cis*-divinyl derivs. (as opposed to *trans*-products by ring closure of bis(dienes)). F.e. and Pd catalyst, **also exocyclic 1,3-dienes** with Bu_3SnH, s. S.-K. Kang et al., J. Am. Chem. Soc. *122*, 11529-30 (2000).

Tetrakis(triphenylphosphine)palladium(0)/catecholborane/sodium hydroxide
Regio- and stereo-specific hydroboration-Suzuki coupling s. *49*, 836s*60*

Tetrakis(triphenylphosphine)palladium(0)/tri-n-butyltin hydride/microwaves
Regio- and stereo-specific hydrostannylation-Stille coupling
with $PdCl_2(PPh_3)_2/Bu_3SnH$ cf. *56*, 307; under microwave enhancement with $Pd(PPh_3)_4$ as catalyst s. R.E. Maleczka Jr. et al., Org. Lett. *2*, 3655-8 (2000).

Palladous chloride/cupric chloride
β-Chloro-α,β-ethylene-α-(organoseleno)carboxylic acid esters
from 1-acetylene-1-selenides and alcohols
Regio- and stereo-specific carbonylation under mild conditions

$PdCl_2/CuCl_2$
C(Cl)=C(SeR)COOR'

301.

Ph—≡—SeMe + MeOH $\xrightarrow[CuCl_2]{CO}$ Ph\\C=C/SeMe, Cl/COOMe

Methanol and startg. acetyleneselenide added under 1 atm. CO at room temp. to 5 mol% $PdCl_2$ and 3 eqs. $CuCl_2$ in benzene, and worked up after 1.5 h → methyl (E)-3-chloro-2-methylseleno-3-phenylacrylate. Y 95% (E/Z >98/2). Reaction is presumed to involve regio- and stereo-specific 1,2-chloropalladation, followed by carbonyl insertion and elimination of Pd(0) prior to its reoxidation to Pd(II) with $CuCl_2$ to complete the cycle. Aliphatic 1-acetylene-1-selenides, however, gave the (Z)-isomer, although the reason for the change in stereoselectivity is not apparent. F.e. s. X. Huang, A. Sun, J. Org. Chem. *65*, 6561-5 (2000).

Dichlorobis(triphenylphosphine)palladium(II)/stannous chloride/hydrogen chloride ←
Regiospecific synthesis of 3-ethylenealcohols
from allenes and aldehydes

CHO → CH(OH)C-CH=C

302.

Ph-C(=O)-H + CH₂=C=CMe₂ → Ph-CH(OH)-C(Me)₂-CH=CH₂

in aq. organic medium. A mixture of benzaldehyde, 2-3 eqs. 1,1-dimethylallene, 2.5-3 eqs. $SnCl_2$, 2 mol% $PdCl_2(PPh_3)_2$, and HCl in aq. DMF allowed to react at 25° for 24 h → product. Y 95%. High chemo-, regio- and stereo-selectivity can be achieved by appropriate choice of the reaction conditions. Reaction is thought to involve initial hydrostannylation of the allene, followed by nucleophilic allylation of the aldehyde. F.e.s. H.M. Chang, C.H. Cheng, Org. Lett. *2*, 3439-42 (2000).

Dichloro[1,1'-bis(diphenylphosphino)ferrocene]palladium(II)/ ←
9-borabicyclo[3.3.1]nonane/potassium phosphate
Hydroboration-Suzuki coupling ←
N-protected 2-arylamines cf. *49*, 836s58; chiral N-protected α-amino-δ-arylcarboxylic acid esters, also coupling with ene-bromides and -iodides, s. P.N. Collier, R.J.K. Taylor et al., Tetrahedron Lett. *41*, 7115-9 (2000); (E)-β-sulfonylaminostyrenes and 3-indolyl derivs. with $Pd(PPh_3)_4$/catecholborane s. B. Witulski, U. Bergsträsser et al., Tetrahedron *56*, 8473-80 (2000).

Chloro(methyl)[2-(2-pyridyl)-4(R)-isopropyl-Δ²-oxazoline]palladium(II)/
sodium tetrakis[3,5-bis(trifluoromethyl)phenyl]borate
Regiospecific asym. silylative ring closure of 1,6-dienes ○
chiral (silylmethyl)cyclopentanes with Et_3SiH cf. *58*, 300; with $Me_2Si(H)OSiPh_2Bu-t$ for a subsequent facile Tamao oxidation to the corresponding chiral cyclopentylcarbinols s. T. Pei, R.A. Widenhoefer, Tetrahedron Lett. *41*, 7597-600 (2000).

Polysiloxane-supported palladium(II) dichloride complex ←
Benzene ring from acetylene derivs.
Heterogeneous cyclotrimerization s. *42*, 676s60

Chloro(cyclooctadiene)iridium(I) dimer/ ←
(S)-2,2'-bis(di-o-tolylphosphino)-1,1'-binaphthyl
Iridium-catalyzed asym. Pauson-Khand reaction

303.

[structure: enyne → Ir complex intermediate → bicyclic enone product]
[OC>Ir<P/Cl P] P)P = (S)-tolBINAP

10.5 Mol% (S)-tolBINAP and 4.93 mol% $[Ir(cod)Cl]_2$ in toluene stirred at 40° under 1 atm. CO, a soln. of startg. enyne in the same solvent added to the light-yellow soln., refluxed for 24 h, the mixture passed through a small pad of silica gel (eluting with 3:1 hexane/ethyl acetate), and purification effected by TLC on silica gel → product. Y 75% (e.e. 91%). Other P-ligands were less effective. The enyne may be all-carbon or possess a N- or O-atom in the tether, and be substituted by an alkyl or aryl group on the alkyne residue. Other P-chiral ligands were less effective. A bicyclic iridacyclopentene is thought to be involved in the catalytic cycle. F.e. and a highly enantioselective (but low-yielding) intermolecular conversion s. T. Shibata, K. Takagi, J. Am. Chem. Soc. *122*, 9852-3 (2000).

Hexacarbonyldiplatinum(I) bis(hydrogen sulfate)/sulfuric acid   [Pt(CO)$_3$]$_2$[HSO$_4$]$_2$/H$_2$SO$_4$
Tert. carboxylic acids from ethylene derivs. ←
via carbonylative Wagner-Meerwein rearrangement
with Ag$_2$O/H$_2$SO$_4$ cf. *31*, 637; with the homoleptic, dinuclear complex, [Pt(CO)$_3$]$_2$[HSO$_4$]$_2$, s. Q. Xu et al., J. Org. Chem. *65*, 8105-7 (2000).

Platinum chloride   PtCl$_2$
Regiospecific intramolecular carboplatination-solvation of enynes   O

Exocyclic alkoxy-3-ethylenes. A mixture of startg. enyne and 5 mol% PtCl$_2$ in methanol refluxed for 14 h → product. Y 88%. Reaction is presumed to involve *anti*-attack of the alkene group onto an intermediate (η^2-alkyne)platinum complex, followed by quenching of the incipient cation by the solvent (alcohol, AcOH or water). Substrates with a stereo-defined internal alkene group reacted with high stereoselectivity. The reaction does not proceed in the presence of proton donors like HCl or HI. F.e.s. M. Méndez, A.M. Echavarren et al., J. Am. Chem. Soc. *122*, 11549-50 (2000).

Via intermediates   v.i.
Regiospecific synthesis of (E)-3-ethylene- from 3-acetylene-alcohols   C≡C → CH=C(R)
One-pot conversion via intramolecular hydrosilylation s. *60*, 406

Rearrangement   ∩

Hydrogen/Carbon Type   CC ∩ HC

Irradiation s. under 1,4-Dicyanobenzene   ///

Cesium hydroxide   CsOH
Anionic cycloisomerization of acetylene derivs.   O
with *n*-BuLi cf. *51*, 337s*57*; isocyclic β-alkylidenenitriles, also (E)-selective intramolecular addition, s. C. Koradin, P. Knochel et al., Synlett *2000*, 1452-4.

Potassium tert-butoxide   KOBu-t
Alleneamines from 2-acetyleneamines   C≡C-CH → CH=C=C
N-allenylcarbazoles with KOH/DMSO cf. *21*, 749; cyclic N-acylalleneamines, e.g. N-allenyl-2-oxazolidones, -2-imidazolones, and -lactams with KOBu-*t*, s. L.-L. Wei, R.P. Hsung et al., Tetrahedron *57*, 459-66 (2001).

2,3-Dihydrothiophenes from 1-acetylene-1-thioethers

305.

2-Aryl-2,3-dihydrothiophenes. 2-Bromobenzyl 1-propynyl sulfide in dry acetonitrile added to a soln. of 2 eqs. KOBu-*t* in the same solvent at 0°, stirred for 24 h, and quenched with water → 2-(2-bromophenyl)-2,3-dihydrothiophene. Y 74%. Reaction is subject to an interesting *o*-**halogen effect** as the cyclization is sluggish or fails in the absence of the halogen. It may involve initial formation of a benzylic carbanion, prior to 5-*endo-dig* ring closure, or proceed via [2.3]-thia-Wittig rearrangement. F.e. and double cycloisomerization s. L.K. McConachie, A.L. Schwan, Tetrahedron Lett. *41*, 5637-41 (2000).

Lithium diisopropylamide i-*Pr₂NLi*
Intramolecular Michael addition with asym. induction
with DBU cf. *33*, 662s55; of α,β-ethylenesulfoxides to α,β-ethylenecarboxylic acid esters with LDA s. N. Maezaki, T. Tanaka et al., Org. Lett. *3*, 29-31 (2001).

1,4-Dicyanobenzene/biphenyl/collidine/irradiation ←
Photochemical migration of carbon-carbon double bonds ←
s. *20*, 533s53; 3-aryl- from 1-aryl-cyclohexenes in the presence of 1,4-dicyanobenzene/biphenyl/collidine s. D. Mangion, D.R. Arnold et al., Org. Lett. *3*, 45-8 (2001).

Organosilicon hydrides s. under (η³-Allyl)Pd(Cl)PCy₃ R_3SiH

Nafion/silica composite ←
Migration of carbon-carbon double bonds s. *33*, 8898s59 ←

Perchloric acid/acetic anhydride $HClO_4/Ac_2O$
2-Cyclopentenones from cross-conjugated dienones ◯
with protic acid s. *24*, 726s29; with dilute HClO₄ and Ac₂O in ethyl acetate s. A.F. Mateos et al., Tetrahedron *57*, 1049-57 (2001).

[1,2-Bis((2R,5R)-2,5-dimethylphospholan-1-yl)benzene]nickel(II) diiodide/ ←
lithium hydridotriethylborate
Asym. isomerization of O-allyl to O-vinyl derivs. RO-C≡C-CH
4,5-dihydro-1,3-dioxepins with a chiral dihalogenonickel(II) phosphine complex s. *34*, 668s55; improved enantioselectivity with NiI₂((R,R)-MeDuPHOS)/LiBHEt₃, also determination of abs. configuration, s. H. Frauenrath et al., Angew. Chem. Int. Ed. Engl. *40*, 177-9 (2001).

Carbonyl(dihydrido)tris(triphenylphosphine)ruthenium(II) $RuH_2(CO)(PPh_3)_3$
Regio- and stereo-specific ruthenium-catalyzed cycloisomerization of enynes ◯
with [CpRu(MeCN)₃]PF₆ cf. *58*, 306; 1,2-di(alkylidene)carbapenams with RuH₂(CO)(PPh₃)₃ s. M. Mori et al., Org. Lett. *2*, 3245-7 (2000).

[1,2-Bis(diphenylphosphino)ethane]rhodium(I) triflate [Rh(dppe)]OTf
4-Cyclooctenones from δ-cyclopropyl-γ,δ-ethylenealdehydes ◯
via rhodium-catalyzed intramolecular hydroacylation

306.

A soln. of startg. vinylcyclopropane in 1,2-dichloroethane added via cannula to a soln. of 20 mol% Rh(dppe)OTf in the same solvent under argon, and the mixture stirred at 65° for 10 h →

product. Y 63%. Reaction is thought to involve initial oxidative addition of Rh(I) to the aldehydic C-H bond, followed by rapid intramolecular hydrometalation, ring expansion to a 9-membered rhodacyclic, and reductive elimination. The formation of 2-vinylcyclopentanes was minimal under these conditions. *tert*-Butyldiphenylsilyl protective groups remained unaffected while *tert*-butyldimethylsilyl groups were cleaved. The preferred catalyst in some instances was Rh(dppe)ClO$_4$ under an ethylene atmosphere, while neutral Rh-complexes were ineffective. F.e. and from *cis*-substrates s. A.D. Aloise, M.D. Shair et al., J. Am. Chem. Soc. *122*, 12610-1 (2000).

Chloro[(2R,2'R)-bis(diphenylphosphino)-(1R,1'R)-dicyclopentane]-
  rhodium(I) dimer/silver hexafluoroantimonate
Asym. cycloisomerization of 1,6-enynes

The first highly enantioselective Rh-catalyzed enyne cycloisomerization is reported as an asym. variant of *59*, 321. **E: Chiral 3-methylene-4-vinyltetrahydrofurans.** A soln. of startg. enyne and 5 mol% [{Rh((R,R,R,R)-BICP)Cl}$_2$] in 1,2-dichloroethane stirred for 1 min in a Schlenk tube (in a glove box), 5 mol% AgSbF$_6$ added, and stirred at room temp. for 1 h with TLC monitoring → product. Y 67% (e.e. 98%). The catalytic activity and enantioselectivity are markedly dependent on the nature of the chiral ligand as well as on subtle variations in substrate substitution: (R,R)-Me-DuPhos and (R,R,R,R)-BICP were the most active ligands for O-tethered enynes, while (R,R,R,R)-BICPO was the ligand of choice for N-tethered enynes. With arylalkyne derivs., reaction was favoured by electron-withdrawing groups in the benzene ring. F.e. and comparison of ligands s. P. Cao, X. Zhang, Angew. Chem. Int. Ed. Engl. *39*, 4104-6 (2000).

Palladium/carbon   Pd-C
Migration of carbon-carbon double bonds
with Pd-C s. *12*, 931; *22*, 737s*25* (review); 2-ethylene-1,1-di(silanes) from ene-1,1-di(silanes) s. D.M. Hodgson et al., Chem. Commun. *2001*, 153-4.

Palladous acetate/trifluoroacetic acid   Pd(OAc)$_2$/CF$_3$COOH
Cycloisomerization of acetylene derivs.
carbostyrils from α,β-acetylenecarboxylic acid anilides with PPA cf. *19*, 788; with Pd(OAc)$_2$/CF$_3$COOH, also coumarins from α,β-acetylenecarboxylic acid aryl esters, s. C. Jia, Y. Fujiwara et al., J. Org. Chem. *65*, 7516-22 (2000).

Palladous trifluoroacetate/(R)-2,3:2',3'-bis(methylenedioxy)-
  6,6'-bis(diphenylphosphino)biphenyl
Asym. palladium-catalyzed cycloisomerization of enynes
chiral N-protected 3-methylene-4-vinylpyrrolidines with (S,S)-(R,R)-*p*-CF$_3$C$_6$H$_4$-TRAP as ligand cf. *49*, 699s*51*; chiral 3-methylene-4-vinyltetrahydrofurans having a quaternary carbon centre with Pd(OCOCF$_3$)$_2$/(R)-2,3:2',3'-bis(methylenedioxy)-6,6'-bis(diphenylphosphino)biphenyl s. M. Hatano, K. Mikami et al., Angew. Chem. Int. Ed. Engl. *40*, 249-53 (2001).

(Phenanthroline)methylpalladium(II) chloride/
  sodium tetrakis[3,5-bis(trifluoromethyl)phenyl]borate
Palladium-catalyzed cycloisomerization of dienes
of 1,6-dienes with PdCl$_2$(MeCN)$_2$/AgBF$_4$ cf. *38*, 706s*56*; cyclopentenes from 1,5- or 1,6-dienes with (phenanthroline)methylpalladium(II) chloride s. P. Kisanga, R.A. Widenhoefer et al., J. Org. Chem. *66*, 635-7 (2001).

η^3-Allyl(chloro)(tricyclohexylphosphine)palladium(II)/sodium tetrakis-
[3,5-bis(trifluoromethyl)phenyl]borate/organosilicon hydrides
Cyclopentenes from 1,6-dienes

Dimethyl diallylmalonate and 1.5 eqs. triethylsilane added sequentially via syringe to a soln. of 4.7 mol% (η^3-allyl)Pd(Cl)PCy$_3$ and 4.7 mol% Na-tetrakis[3,5-bis(trifluoromethyl)phenyl]borate in methylene chloride at 0°, and stirred at room temp. for 20 min → 4,4-dicarbomethoxy-1,2-dimethylcyclopentene. Y 89%. The mechanism of the reaction in unclear, but it is thought that the silane (required in at least stoichiometric amount) serves, firstly, to activate the π-allyl palladium precatalyst for the initial cycloisomerization to an alkylidenecyclopentane, and, secondly, to stabilize an intermediate palladium hydride complex which catalyzes the subsequent double bond shift. Reaction is most facile with terminal 1,6-dienes (a large excess of the silane being required for 1,6-dienes possessing internal olefin groups). The substrate may possess a homoallylic ester, acyl, acetoxy, pivaloyloxy or sulfonyl group, and allylic methyl and phenyl groups are tolerated. However there was no reaction with 1,7-dienes. F.e.s. P. Kisanga, R.A. Widenhoefer, J. Am. Chem. Soc. *122*, 10017-26 (2000).

Oxygen/Carbon Type

n-*Butyllithium*
Asym. [1,2]-Wittig rearrangement
of benzyl phenolethers cf. *53*, 335; of aryl lactolides s. P. Gärtner et al., Tetrahedron:Asym. *11*, 1003-13 (2000).

sec-*Butyllithium/(-)-sparteine*
Asym. Wittig rearrangement
s. *53*, 336s56; of crotyl furfuryl ethers s. M. Tsubuki, T. Honda et al., Tetrahedron:Asym. *11*, 4725-36 (2000).

sec-*Butyllithium/(-)-sparteine/boron fluoride*
Isopropyllithium/(-)-α-isosparteine
Asym. transannular ring opening of cyclooctene oxides

Chiral bicyclo[5.1.0]oct-5-en-2-ols. 2 eqs. (-)-Sparteine in ether treated dropwise under argon with a soln. of 2 eqs. *sec*-BuLi in hexane at -90°, stirred for 30 min, a soln. of 1.5 eqs. BF$_3$-etherate in ether added slowly at the same temp. over 20 min, then *meso*-cycloocta-1,5-diene oxide in the same solvent added dropwise over 20 min, and after a further 10 min the reaction quenched with methanol and 5% aq. H$_2$SO$_4$ → (1S)-bicyclo[5.1.0]oct-5-en-2-ol. Y 72% (e.e. 71%). Without BF$_3$, (R)-2,5-cyclooctadienol was formed. Cyclooctene oxide itself afforded the chiral *cis*-fused bicyclo[3.3.0]octan-2-ol with PhLi as base, while cycloocta-1,3-diene oxide gave the racemic bicyclo[3.3.0]octenol with *s*-BuLi/BF$_3$-etherate. F.e.s. A. Alexakis et al., J. Chem. Soc. Perkin Trans. 1 *2000*, 3354-5; chiral 5,6-disiloxybicyclo[3.3.0]octan-2-ols with *i*-PrLi/(-)-α-isosparteine s. D.M. Hodgson, I.D. Cameron, Org. Lett. *3*, 441-6 (2001).

Cupric trifluoromethanesulfonate
Claisen rearrangement
of α-allyloxy-α,β-ethylenecarboxylic acid esters s. *60*, 311

Gold(III) chloride AuCl₃
Condensed phenols from furylacetylenes
via regiospecific gold-catalyzed intramolecular Diels-Alder reaction

310.

2 mol% AuCl₃ added to a mixture of startg. furan in acetonitrile at 20° → product. Y 97%. The reaction is highly regiospecific, obviates the need to exclude air or water, and proceeds without formation of paramagnetic species or gold precipitation. Reaction is facilitated by electrophilic activation of the triple bond by coordination to the metal. F.e. with 6 mol% AuCl₃ s. A.S.K. Hashmi et al., J. Am. Chem. Soc. *122*, 11553-4 (2000).

Zinc bromide ZnBr₂
Cyclic β-hydroxyketones from 2,3-epoxyalcohols with ring expansion
with BF₃ cf. *22*, 742; with ZnCl₂ or ZnBr₂ for diastereoselective generation of quaternary hydrocarbon groups s. Y.Q. Tu, C.A. Fan et al., J. Chem. Soc. Perkin Trans. 1 *2000*, 3791-4.

Boron fluoride (s.a. under s-BuLi) BF₃
Ketones from epoxides with rearrangement
s. *45*, 436s*59*; from uncondensed 2-aryloxiranes **with anomalous aryl migration** s. Y. Kita et al., Tetrahedron *57*, 815-25 (2001).

Scandium(III) triflate/8-ethyl-1,8-diazabicyclo[5.4.0]undec-7-enium triflate
2,3-Dihydrobenzofurans from allyl phenolethers
with 40 mol% Mo(CO)₆ cf. *52*, 334; with 5 mol% Sc(OTf)₃ *in ionic liquids* (e.g. 8-ethyl-1,8-diazabicyclo[5.4.0]undec-7-enium triflate) s. F. Zulfiqar, T. Kitazume, Green Chem. *2*, 296-7 (2000).

Ytterbium(III) triflate Yb(OTf)₃
Ytterbium(III)-catalyzed Claisen rearrangement

311.

of α-allyloxy-α,β-ethylenecarboxylic acid esters. 0.075 eq. Yb(OTf)₃ and activated 3 Å molecular sieves added to a soln. of startg. allyl vinyl ether (Z:E of enolether 97:3) in dry dichloromethane, and the mixture stirred under argon for 24 h at 25° → product. Y 99% (*syn/anti* 96:4). The stereospecificity and reactivity strongly depend on the substrate structure, diastereoselectivity being poor with substrates possessing an (E)-configured allylic residue. The enhanced *syn*-selectivity is interpreted by a stepwise rearrangement involving a metal-coordinated ion pair. F.e. and with Cu(OTf)₂ s. M. Hiersemann, L. Abraham, Org. Lett. *3*, 49-52 (2001).

Trimethylsilyl triflate Me₃SiOTf
2-δ-Hydroxytetrahydropyrans from 2-homoallyloxytetrahydropyrans

312.

1 eq. Trimethylsilyl triflate added dropwise via syringe to a stirred soln. of 2-(3'-methylbut-3'-enyloxy)tetrahydropyran in methylene chloride at -10° under argon, stirred at the same temp. for 30 min, diluted with ether, and washed with pH 7.5 phosphate buffer → intermediate (Y 96%; 52:31:13 mixture of double bond isomers), in methanol treated at room temp. with Pd-on-

carbon (30% by wt), degassed, and hydrogenated at room temp. for 24 h with stirring → 4-(4′-methyltetrahydropyran-2′-yl)butan-1-ol (Y 100%; d.r. 30:1). F.e. incl. bicyclic analogs, and ring closures with trifluoromethanesulfonic acid s. D.J. Dixon, S.V. Ley et al., J. Chem. Soc. Perkin Trans. 1 *2000*, 1829-36.

Chloro(cyclooctadiene)iridium(I) dimer/tricyclohexylphosphine/cesium carbonate ←
γ,δ-Ethylenealdehydes from 4-oxa-1,7-dienes via Claisen rearrangement ←

313.

The startg. allyl ether added to a mixture of 1 mol% [Ir(cod)Cl]$_2$, 2 mol% tricyclohexylphosphine, and 1 mol% Cs$_2$CO$_3$ in toluene under argon, stirred at 100° for 1 h, then quenched with wet ether → product. Y 81% ((E)-selectivity 67%). Reaction presumably involves initial double bond migration prior to rearrangement, and clearly obviates the need to prepare less manageable allyl vinyl ethers. F.e.s. T. Higashino, Y. Ishii et al., Org. Lett. *2*, 4193-5 (2000).

Nitrogen/Carbon Type CC ∩ NC

Without additional reagents w.a.r.
2-α-Carboxypyridines via regiospecific intramolecular Diels-Alder reaction with α-(O-acyloximino)nitriles ○

314.

A 0.005 M soln. of startg. acyloxime in dry toluene stirred at reflux for 38 h → intermediate adduct (Y 74%), in dry DMF treated with 3.4 eqs. Cs$_2$CO$_3$, and stirred at room temp. for 17 h → product (Y 71%). Reaction tolerates nitriles and carboxylic acid esters. F.e. and via high pressure cycloaddition s. D.C. Bland, S.M. Weinreb et al., Org. Lett. *2*, 4007-9 (2000).

Halogen/Carbon Type CC ∩ HalC

Cuprous halides/N,N,N′,N′-tetramethylethylenediamine or 2,2′-bipyridyl ←
Ring closures with halogen atom transfer ○
with copper(I) complexes s. *40*, 493s*58*; 1-amino-2-pyrrolidones from N-allyl-N-acylhydrazines with chlorine atom transfer with CuCl/TMEDA s. F. Ghelfi, A.F. Parsons, J. Org. Chem. *65*, 6249-53 (2000); 3,3-difluoro-4-halogenomethyl-2-pyrrolidones via iodine or bromine atom transfer with CuI or CuBr/bipy s. H. Nagashima et al., ibid. *66*, 315-9 (2001).

Dilauroyl peroxide (RCOO)$_2$
Ring closures with iodine atom transfer
with Et$_3$B cf. *42*, 695s*53*; with dilauroyl peroxide, and further radical reactions s. C. Ollivier, P. Renaud et al., Synthesis *2000*, 1598-602.

Remaining Elements/Carbon Type CC ∩ RemC

Triethylborane/oxygen Et$_3$B/O$_2$
Radical aryl migration ←
1,5-Si→C-aryl migration with Bu$_3$SnH/AIBN cf. *56*, 490; 3-arylstannanes from ar. 3-iodostannanes via 1,4-Sn→C-aryl migration with Et$_3$B/O$_2$ s. K. Wakabayashi, K. Oshima et al., Org. Lett. *2*, 1899-901 (2000).

Carbon/Carbon Type CC ∩ CC

Without additional reagents w.a.r.
Regio- and stereo-specific intramolecular Diels-Alder reaction with tethered 1,3-dienes ○
s. 46, 696s48,53; 34, 693s47,48; with 1,3-dienesulfonic acid amides for the preparation of hexahydrobenzo[d]isothiazole 1,1-dioxides and hexahydro-2H-benzo[e][1,2]thiazine 1,1-dioxides s. I.R. Greig, M.J. Tozer et al., Org. Lett. 3, 369-71 (2001).

Intramolecular Diels-Alder reaction with allenes
s. 38, 723s42; 41, 690; with α-allenesulfones, regio- and stereo-selectivity, s. J.R. Bull, R. Hunter et al., J. Chem. Soc. Perkin Trans. 1 2000, 3129-39.

Irradiation ⚡
Photochemical skeletal rearrangement ←
with ring contraction and cyclopropane ring closure s. 25, 549; di-π-methane rearrangement of homobarrellene derivs. s. V. Nair et al., J. Chem. Soc. Perkin Trans. 1 2000, 3795-8.

Chloro[1,4-bis(diphenylphosphino)butane]rhodium(I) dimer/ [Rh(dppb)Cl]$_2$/AgSbF$_6$
silver hexafluoroantimonate
2,3-Fused 1,4-cycloheptadienes by intramolecular [5+2]-cycloaddition ○
with [CpRu(MeCN)$_3$]PF$_6$ cf. 58, 311; regioselectivity s. B.M. Trost, H.C. Shen, Org. Lett. 2, 2523-5 (2000); with [Rh(dppb)Cl]$_2$/AgSbF$_6$, also intramolecular [4+2]-cycloaddition (cf. 45, 439s55), s. B. Wang, P. Cao, X. Zhang, Tetrahedron Lett. 41, 8041-4 (2000).

Exchange ↕

Hydrogen ↑ CC ↕ H

Microwaves s. under CuCl$_2$ ←

Potassium hydroxide/oxygen KOH/O$_2$
Bis(acetonitrile)(α,α'-bis[(N)-methyl-2-pyridyl)ethylamino]-2-fluoro- ←
m-xylene)dicopper(I)/oxygen
Cupric chloride hydroxide/N,N,N',N'-tetramethylethylenediamine/oxygen
Aerobic dimerization of phenols 2 ArH → Ar-Ar
with CuCl/TMEDA cf. 31, 719s53; hindered sym. o,o'-dihydroxydiaryls with bis(acetonitrile)(α,α'-bis[(N)-methyl-2-pyridyl)ethylamino]-2-fluoro-m-xylene)dicopper(I), also **diphenoquinones**, s. R. Gupta, R. Mukherjee, Tetrahedron Lett. 41, 7763-7 (2000); sym. dihydroxydiaryls with KOH s. Japanese patent JP-11292813 (Chang Chung Artificial Resin Co. Ltd.); with Cu(II)-naphthenates s. JP-11292814 (Dainippon Ink & Chem. Co.); oxidative dimerization of simple phenols and natural phenanthrols with CuCl(OH)/TMEDA (cf. 27, 761s58) s. P.L. Majunder et al., J. Indian Chem. Soc. 77, 389-93 (2000).

Cupric chloride/potassium fluoride-alumina/microwaves ←
Sym. 1,3-diynes from terminal acetylene derivs. 2 C≡CH → C≡C-C≡C
with CuCl$_2$/NaOAc cf. 16, 780s58; with CuCl$_2$ and KF-on-Al$_2$O$_3$ under microwave irradiation in the absence of solvent s. G.W. Kabalka et al., Synlett 2001, 108-10.

Aluminum chloride AlCl$_3$
Ring closures with oxalyl chloride ○
s. 24, 755; carbazole-3,4-quinones from 2-vinylindoles with AlCl$_3$ s. A. Aygün, U. Pindur, Synlett 2000, 1757-60.

Ammonium ceric nitrate (NH$_4$)$_2$Ce(NO$_3$)$_6$
2,3-Dihydrofurans from ethylene derivs.
s. 50, 449s54; 5,6-dihydrofuro[2,3-d]pyrimidine-2,4(1H,3H)-diones s. K. Kobayashi et al., Synth. Commun. 30, 4277-91 (2000).

Carbon tetrabromide s.a. under PdI$_2$ CBr$_4$

Carbon tetrabromide-aluminum bromide
Superelectrophilic carbalkoxylation of hydrocarbons via regiospecific carbonylation

CBr_4-$AlBr_3$
H → COOR

315.

Cyclic tert. carboxylic acid esters. Methylcyclohexane added under CO to 1 eq. carbon tetrabromide and 2 eqs. anhydrous $AlBr_3$ in methylene bromide *at -40°*, the mixture stirred under a positive CO pressure at the same temp. for 0.5 h, an excess of isopropanol added carefully, allowed to warm to room temp., and stirred for a further 30 min → product. Y 67%. Reaction was also applicable to ethylcyclohexane, and to cycloheptane and cyclooctane (with ring contraction). Side reactions involving the intermediate carbocationic species were largely suppressed under these conditions. F.e.s. I. Akhrem et al., Tetrahedron Lett. *41*, 9903-7 (2000).

Titanium tetrachloride/nitromethane
Lead tetraacetate/boron fluoride
Oxidative dimerization of arenes

$TiCl_4$/$MeNO_2$
$Pb(OAc)_4$/BF_3
2 ArH → Ar-Ar

of phenolethers with ar. nitro compds. cf. *7*, 761; with $TiCl_4$ in $MeNO_2$ for dimerization of naphthalenes with electron-donating substituents s. J. Doussot, C. Ferroud et al., Tetrahedron Lett. *41*, 2545-7 (2000); regiospecific oxidative dimerization of naphthylisoquinoline alkaloids with $Pb(OAc)_4$/BF_3-etherate s. G. Bringmann et al., Synthesis *2000*, 1843-7.

Manganese(III) acetate
2,3-Dihydrofurans from ethylene derivs.

$Mn(OAc)_3$

s. *47*, 705s56; **with asym. induction** for preparing chiral 2,3-dihydrofuran-4-carboxylic acid esters s. F. Garzino, P. Brun et al., Tetrahedron Lett. *41*, 9803-7 (2000); 2,3-dihydrofuro[3,2-c]-carbostyrils s. G. Bar, A.F. Parsons et al., ibid. 7751-5.

5-Acoxy-2(5H)-furanones from acetylene derivs.

316.

5-Acoxy-5-aryl-2(5H)-furanones. 4 eqs. Anhydrous $Mn(OAc)_3$ added to a 0.2 M soln. of startg. alkyne in 1:1 acetic acid/acetic anhydride, refluxed for ca. 20 min, cooled, and quenched with water and ether → 5-acetoxy-5-phenyl-2(5H)-furanone. Y 78%. There was no reaction with alkylacetylenes, while silylacetylenes gave low yields. A 4-electron oxidation is involved, reaction being initiated by regiospecific addition of carboxymethyl radical to the alkyne. F.e.s. P.C. Montevecchi, M.L. Navacchia, Tetrahedron *56*, 9339-42 (2000).

Ferric chloride
Oxidative dimerization of arenes

$FeCl_3$
2 ArH → Ar-Ar

of phenolethers with $FeCl_3$/SiO_2 cf. *27*, 761s36; dibromotetraalkoxybiphenyls with $FeCl_3$ s. N. Boden, R.J. Bushby et al., Tetrahedron Lett. *41*, 10117-20 (2000).

Palladous chloride/triphenylphosphine
α-Aryl-α,β-ethylenecarboxylic acid esters from styrenes by regiospecific carbonylation

$PdCl_2$/Ph_3P
H → COOR

317.

A mixture of 500 g *p*-methylstyrene, 7.5 g $PdCl_2$ and 22 g triphenylphosphine in 1017 g methanol stirred at room temp. for 2 h, heated to 90° under 3.0 MPa carbon monoxide, an additional 1980 g startg. styrene added over 14 h at the same pressure, then stirred for 10 h → product. Y 55 wt% (at 92.7 weight% conversion). F.e., also α-aryl-α,β-ethylenecarboxylic acids by hydrocarboxylation, s. Japanese patent JP-2000086571 (Nippon Petrochemicals Co. Ltd.)

Palladous iodide-thiourea/carbon tetrabromide/cesium carbonate ←
Benzofuran-3-carboxylic acid esters from *o*-hydroxyarylacetylenes ○
Intramolecular oxypalladation-carbonylation

318.

Improved procedure. A soln. of startg. alkyne in methanol treated with a little PdI$_2$-thiourea, carbon tetrabromide (as reoxidant for Pd(0) to Pd(II)), and Cs$_2$CO$_3$ at 40° under a balloon of CO for <30 min → product. Y 84%. Reaction is generally applicable to both electron-rich *and* -deficient substrates, and an improvement on prior art (yields being low from electron-deficient substrates with CuCl$_2$ as oxidant). Significantly, silyl groups (e.g. TBS ethers) were unaffected so that **polymer-based syntheses** can be undertaken with a silyl linker. F.e.s. Y. Nan, H. Miao, Z. Yang, Org. Lett. *2*, 297-9 (2000).

Oxygen ↑   CC ↓↑ O

Without additional reagents   *w.a.r.*
Alkylidenemalononitriles from oxo compds.   CO → C=C(CN)$_2$
uncatalyzed procedure in ethanol cf. *15*, 571; condensation with aliphatic and [het]aryl aldehydes **in distilled water** (at 65°) s. F. Bigi et al., Green Chem. *2*, 101-3 (2000); uncatalyzed Knoevenagel condensation with barbituric acids under IR irradiation without solvent s. G. Alcerreca, F. Delgado et al., Synth. Commun. *30*, 1295-301 (2000).

Polymer-based Ugi 4-component condensation ←
s. *51*, 380; *56*, 330; *57*, 304; *17*, 809s56; with a supported galactosylamine as amine component s. K. Oertel, H. Kunz et al., Angew. Chem. Int. Ed. Engl. *39*, 1431-3 (2000).

Nenitzescu indole ring synthesis ○
s. *8*, 782; polymer-based synthesis with a supported β-amino-α,β-ethylenecarboxamide s. D.M. Ketcha, D.E. Portlock et al., Tetrahedron Lett. *41*, 6253-7 (2000).

Polymer-based Hantzsch 1,4-dihydropyridine synthesis
s. *52*, 352; with a supported β-ketoester, also preparation of a 272-member library of 1,4-dihydropyridines, s. J.G. Breitenbucher, G. Figliozzi, Tetrahedron Lett. *41*, 4311-5 (2000).

Polymer-based synthesis of 3,4-dihydro-2-pyridones
from α,β-ethylenecarboxylic acid anhydrides and enamines - 3,4-dihydro-2-pyridone-5-carboxamides s. *60*, 124

Microwaves (s.a. under AcOH and Pd(OAc)$_2$) ←
C-Dimethylaminomethylenation   CH$_2$ → C=CHNMe$_2$
s. *26*, 247; under microwave irradiation s. A.-K. Pleier, W.R. Theil et al., Synthesis *2001*, 55-62.

Δ^2-Oxazolines from aldehydes
Regiospecific 1,3-dipolar cycloaddition under microwave irradiation without solvent

319.

4-Carbalkoxy-4-cyano-Δ^2-oxazolines. A mixture of ethyl 2-cyano-2-[(1-ethoxy ethylidene)-amino]acetate **as nitrile ylid equivalent** and 1 eq. freshly distilled benzaldehyde contained in a cylindrical quartz tube irradiated in a Synthewave L 402 Prolabo focused microwave reactor (2450 MHz) at 70° (computer monitored) for 1 h → product. Y 93% (single regioisomer; *trans:cis* 87:13). The procedure is mild, short, simple, low-waste, and eco-friendly. F.e. and comparison with the thermal conversion, also a 4-cyano-Δ^2-imidazoline from an aldimine, s. J. Fraga-Dubreuil, J.P. Bazureau et al., Green Chem. *2000*, 226-9.

Sodium hydride   NaH
Glutarimides from carboxylic acid amides and α,β-ethylenecarboxylic acid esters via Michael addition-intramolecular N-acylation

320.

2-Sulfonylglutarimides via N,C-dianions. Startg. α-sulfonylcarboxylic acid amide treated with 2 eqs. NaH in THF, startg. α,β-ethylenecarboxylic acid ester added, and the mixture stirred at 25° until reaction complete → product. Y 74%. F.e., also synthesis of the aromatase inhibitor AG-1 s. M.-Y. Chang, N.-C. Chang et al., Tetrahedron Lett. *41*, 10273-6 (2000).

Sodium hydroxide   NaOH
Solid-state Claisen-Schmidt condensation   CHO → CH=C
with KOH cf. *1*, 549s59; with NaOH s. C.L. Raston, J.L. Scott, Green Chem. *2*, 49-52 (2000).

Potassium hydroxide   KOH
Friedländer quinoline ring synthesis
s. *2*, 651; alkyl-subst. 1,10-phenanthrolines, incl. chiral derivs., s. S. Gladiali, R.P. Thummel et al., J. Org. Chem. *66*, 400-5 (2001).

Potassium tert-butoxide   KOBu-t
Stobbe condensation   ←
in *tert*-butanol cf. *2*, 647; *without solvent* at room temp. s. K. Tanaka, F. Toda et al., Green Chem. *2*, 303-4 (2000).

Lithium diisopropylamide/acetic acid   i-Pr$_2$NLi/AcOH
(E)-α-Nitro-β,γ-ethylenesulfones from aldehydes   CHCHO → C=CHCH(NO$_2$)SO$_2$R
Synthesis with addition of one C-atom

321.

ca. 2 eqs. *n*-BuLi (2.5 *M* in hexanes) added to a soln. of 2 eqs. diisopropylamine in THF at dry ice temp., the mixture allowed to warm to -20° before recooling in dry ice, a soln. of 0.83 eq. phenylsulfonylnitromethane in THF added dropwise over 10 min, the mixture stirred for 1 h, allowed to warm to -40° before recooling in dry ice, a soln. of 1.2 eqs. propionaldehyde in THF added dropwise over 30 min, the mixture stirred for 30 min, glacial acetic acid added dropwise, then warmed to room temp. → (E)-product. Y 88%. The process requires formation of the dianion of phenylsulfonylnitromethane and reaction with unbranched aldehydes, affording an equilibrium mixture of α,β- and β,γ-ethylene-α-nitrosulfones favoring the latter at ca. pH 7. F.e., also Michael additions to the intermediate α-nitro-α,β-ethylenesulfones, s. P.A. Wade et al., J. Org. Chem. 65, 7723-30 (2000).

Lithium bis(trimethylsilyl)amide $LiN(SiMe_3)_2$
Polymer-based synthesis of 2-azetidinones from carboxylic acid esters and aldimines

322.

trans-**3-Acylamino-2-azetidinones.** 2.2 eqs. $LiN(SiMe_3)_2$ added to startg. triazene-linked ester resin in THF at -78°, stirred for 1.5 h, 3 eqs. startg. imine added, warmed to 0° within 16 h, stirred at room temp. for 7 h, the resulting resin washed and removed, suspended in 5% trifluoroacetic acid in methylene chloride, filtered after 1 min, the process repeated three times, the resulting diazonium salt dissolved in 5:2 THF/DMF, and heated to 60° for 15 min → product. Y 54% (d.e. ≥96%; 89% pure). This is the first example of a polymer-based synthesis of 2-azetidinones from ester enolates. The *trans*-configuration was favoured to the same degree by the solution-phase route. F.e.s. S. Schunk, D. Enders, Org. Lett. 2, 907-10 (2000).

Pyrrolidine
Xanthones from 2-vinyl-4-chromones and ketones via Diels-Alder reaction with enamines

323.

2-β-Styryl-4-chromone refluxed in acetone with a drop of pyrrolidine for 4 h with exposure to air → 1-methyl-3-phenyl-9H-xanthen-9-one. Y 76%. F.e. and with 1-methylene-1,2,3,4-tetrahydro-9H-xanthen-9-ones as by-products s. A.S. Kelkar, G.D. Brown et al., J. Chem. Soc. Perkin Trans. 1 2000, 3732-41.

Piperidine/acetic anhydride R_2NH/Ac_2O
2H-Pyran ring from α,β-ethylenealdehydes via α,β-ethylenealdiminium salts
s. 57, 312; 2H,5H-pyrano[3,2-c]benzopyran-5-ones s. G. Cravotto et al., Synthesis 2001, 49-51.

Triethylenediamine   *dabco*
3-Chromenes from *o*-hydroxyaldehydes and ethylene derivs.
3-Nitro-3-chromenes

324.

A mixture of 4-10 eqs. salicylaldehyde, 1 eq. 1-nitro-2,2-dimethylethylene, and 0.5-1 eq. dabco heated at 40° in the absence of solvent for 37 h → product. Y 99%. Reaction with sterically hindered 2,2-disubst. 1-nitroalkenes at 40° under the same conditions gave the intermediate *trans*-3-nitro-4-hydroxychromans. However, by increasing the temperature to 90°, the expected 2,2-disubst. 3-nitrochromenes were obtained. The method is simple and efficient. F.e. s. M.-C. Yan, C.-F. Yao et al., Tetrahedron Lett. *42*, 2717-21 (2001).

Morpholine/silica gel R_2NH/SiO_2
1,1-Nitroethylene derivs. from aldehydes $CHO \rightarrow CH{=}C(NO_2)$
under heterogeneous conditions
with gel-entrapped KOH cf. *4*, 702s*59*; under heterogeneous conditions with morpholine/silica gel in acetonitrile s. B.P. Bandgar et al., Montash. Chem. *131*, 949-52 (2000).

Allylcopper compds.   ←
Regiospecific synthesis of ethylene derivs. from sulfonyloxy-2-ethylenes $C(R)C{=}C$
s. *42*, 733s*58*; 3-arylthio-1,5-dienes from arylthio-2-ethylenes via thioether-stabilized allylcopper compds. (with *n*-BuLi/CuCN·2LiCl) s. X.-Z. Wang, Y.-L. Wu et al., J. Org. Chem. *65*, 8146-51 (2000).

Lithium dialkylcuprates $LiCuR_2$
Synthesis of 3,6-dihydro-2-pyrones
from α-allene-δ-hydroxycarboxylic acid esters via 1,4-addition

325.

4 eqs. Methyllithium (1.6 *M* in ether) added to 2 eqs. CuI in ether under N_2 at 0°, the suspension stirred for 2 min, cooled to -78°, a soln. of the startg. allenecarboxylic acid ester in ether added, stirred for 45 min, and quenched with aq. NH_4Cl/ammonia/methanol → 4,5-dimethyl-3-phenyl-3,6-dihydro-2-pyrone. Y 56%. $NaBH_4$ reduction of the allenecarboxylic acid esters led to β,γ-unsaturated esters having an E-configuration, but their lactonization failed. F.e. and prepn. of startg. allenes s. J.G. Knight et al., J. Chem. Soc. Perkin Trans. 1 *2000*, 3188-90.

Lithium dialkylcuprates/lithium cyanide/tri-n-butylphosphine   ←
Synthesis of 1,2,4-trienes from 5-acoxy-3-en-1-ynes $RC{=}C{=}C{-}C{=}C$
with remote asym. induction

326.

(60 : 40)

4 eqs. Tri-*n*-butylphosphine added to a suspension of 2 eqs. CuCN in ether at 0°, the mixture warmed to room temp., cooled to -30°, 12 eqs. *n*-butyllithium added, stirred for 15 min at the same temp., cooled to -80°, a soln. of startg. acetate (e.e. 99%) in ether added to the formed $LiCuBu_2$·LiCN, the mixture stirred at the same temp. for 1 h, and quenched with satd. NH_4Cl → product. (E:Z 60:40; e.e. 99% for each isomer). The phosphine ligand minimizes undesirable racemization of the substrate. F.e. and with $(EtO)_3P$ s. N. Krause, M. Purpura, Angew. Chem. Int. Ed. Engl. *39*, 4355-6 (2000).

Cupric trifluoromethanesulfonate $Cu(OTf)_2$
α,β-Ethyleneketones from methyl ketones and aldehydes $CHO \rightarrow CH{=}C{-}CO$
with a (2,2'-bipyridyl)zinc complex cf. *29*, 752s*36*; (E)-γ-aryl-β,γ-ethylene-α-ketocarboxylic acid esters with $Cu(OTf)_2$ (and added $(MeO)_3CH$ for electron-rich aldehydes) s. G. Dujardin, E. Brown et al., Synthesis *2001*, 147-9.

Copper(II)-catalyzed Friedel-Crafts reactions ←

327.

75 : 25

Startg. mesylate and 1 eq. naphthalene added with stirring to 10 mol% Cu(OTf)$_2$ suspended in dry 1,2-dichloroethane, and heated at 80° for 4 h → product. Y 97%. Cu(OTf)$_2$ is an efficient catalyst for Friedel-Crafts alkylation, acylation (of electron-rich arenes), benzoylation and sulfonylation, although Sn(OTf)$_2$ was preferred for the latter. However, there was no reaction with prim. mesylates or alkyl tosylates. F.e.s. R.P. Singh, V.K. Singh et al., Tetrahedron 57, 241-7 (2001).

Magnesium/calcium bis(tetrahydridoborate)   Mg/Ca(BH$_4$)$_2$
Synthesis of 4-ethylenealcohols   COOR → CH(OH)CH$_2$CH$_2$CH=CH$_2$
from carboxylic acid esters and two vinylmagnesium bromide molecules

328.

A soln. of *4 eqs.* vinylmagnesium bromide in THF added dropwise at room temp. under N$_2$ to a soln. of 0.25 eq. ca. 0.15 *M* Ca(BH$_4$)$_2$ in THF, the mixture stirred for 30 min, startg. ester added, stirring continued for a further 2 h, then hydrolyzed with satd. aq. NH$_4$Cl → product. Y 78%. F.e. and with Zn(BH$_4$)$_2$, **also synthesis of sec. alcohols** from Grignard compds. (cf. *7, 841*) s. S. Hallouis, R. Amouroux et al., Synth. Commun. *30*, 313-24 (2000).

Magnesium/titanium tetraisopropoxide   Mg/Ti(OPr-i)$_4$
Regio- and stereo-specific synthesis of ethylene derivs. from alkoxy-2-ethylenes ←
with Mg/NiCl$_2$(PPh$_3$)$_2$ cf. *51, 372*; with ethyl bromide (3-4 eqs.) and 1 eq. Ti(OPr-i)$_4$ via titanacyclopropanes, **also from 2-ethylenealcohols** or tetrahydropyran-2-yloxy-2-ethylenes, s. O.G. Kulinkovich et al., Synlett *2001*, 49-52.

Zinc/aluminum chloride   Zn/AlCl$_3$
Reductive dimerization of oxo compds.   2 CO → C=C
of ketones with Zn/1,2-bis(chlorodimethylsilyl)ethane cf. *36, 754s48*; also of aldehydes with Zn/AlCl$_3$, and cross-coupling, s. D.K. Dutta, D. Konwar, Tetrahedron Lett. *41*, 6227-9 (2000).

Diethylzinc s. under *Pd(OAc)$_2$*   Et$_2$Zn

Magnesium-aluminum tert-butoxide hydrotalcite   ←
Knoevenagel condensation s. *57, 310s60*   CO → C=C

Indium/indium(III) chloride or *Indium(I) chloride*   In/InCl$_3$ or InCl
β,γ-Ethyleneketones from α,β-ethyleneketones and aldehydes   CHO → CH=C-CHCO

329.

A soln. of *p*-fluorobenzaldehyde, 2 eqs. In, 0.5 eq. InCl$_3$, and 3 eqs. methyl vinyl ketone in 1:2 THF/water stirred at room temp. for 6 h, acidified with 1 *N* HCl, and stirred for 30 min → 5-(4-fluorophenyl)-4-penten-2-one. Y 79%. A radical mechanism is invoked wherein an initially-generated enone radical anion undergoes cyclopropanation with the aldehyde prior to dehydrative quenching with proton. The reaction is generally applicable to ar. and hetar. aldehydes, but cannot be extended, however, to acrolein, acrylonitrile, ethyl acrylate or acrylic acid. F.e. and with 2.5 eqs. InCl s. S. Kang, Y. Kim et al., Org. Lett. 2, 3615-7 (2000).

Calcium bis(tetrahydridoborate) s. under Mg $Ca(BH_4)_2$

Lithium tetrakis(pentafluorophenyl)borate/magnesium oxide $[LiB(C_6F_5)_4]/MgO$
Friedel-Crafts benzylation   H → Bn
with benzyl mesylates s. *11*, 894s59

Hydrotalcites s. Magnesium-aluminum tert-butoxide hydrotalcite   ←

Boron fluoride BF_3
Asym. synthesis of 3-silylpyrrolidines from (E)-2-ethylenesilanes via 1,2-silyl group migration

330.

2 eqs. BF$_3$-etherate added to a soln. of 1.1 eqs. benzaldehyde dimethyl acetal and 1.1 eqs. *tert*-butyl carbamate in dry dichloromethane *at -78°* under N$_2$, the dry ice/acetone bath removed, the mixture stirred for 10 min, cooled to -100°, the startg. silane in the same solvent added dropwise, stirred overnight at -85°, then quenched with a satd. soln. of NaHCO$_3$ → (2S,3R,4S,5S)-3-(dimethylphenylsilyl)-4-methyl-5-phenylpyrrolidine-2-acetic acid methyl ester. Y 78% (d.r. 30:1). Methyl carbamate reacted similarly with ar. aldehydes or acetals to give N-carbomethoxy derivs. Higher reaction temperatures promoted ring opening of the product to chiral N-protected 3-ethyleneamines. F.e.s. J.V. Schaus, J.S. Panek et al., Tetrahedron 56, 10263-74 (2000).

Boron fluoride/magnesium sulfate $BF_3/MgSO_4$
3-Acyltetrahydrothiophenes from 3-ethylene-2-hydroxymercaptans and oxo compds. Stereospecific conversion via pinacol-type rearrangement

331.

1 eq. BF$_3$-etherate added dropwise to a stirred mixture of startg. mercaptan, 1 eq. *p*-methylbenzaldehyde, 2 eqs. MgSO$_4$, and methylene chloride at -20°, kept at the same temp. until TLC indicated completion of reaction, and quenched with satd. aq. NH$_4$Cl → product. Y 71%. Reaction is generally applicable to aliphatic, α,β-unsatd. and ar. aldehydes, as well as ketones, but is limited in that the hydroxymercaptans must be substituted at the internal alkene carbon and the terminal alkene carbon must be unsubstituted. A single stereoisomer was obtained. It is thought that a complex protic acid, formed by reaction of BF$_3$-etherate with liberated water, is the key promotor. F.e. **and with asym. induction, also from** the intermediate (detectable) **5-vinyl-1,3-oxathiolanes**, s. A.M. Ponce, L.E. Overman, J. Am. Chem. Soc. 122, 8672-6 (2000).

Boron fluoride/trifluoroacetic acid BF_3/CF_3COOH
Asym. Diels-Alder reaction with *in situ*-generated azomethines
with *in situ*-generated methyleneammonium chlorides cf. *40*, 477; with chiral *in situ*-generated imines derived from glyoxylic acid amides with BF$_3$/CF$_3$COOH, also conversion to enantiopure bicyclic diamines, s. S.A. Modin, P.G. Andersson, J. Org. Chem. 65, 6736-8 (2000).

Aluminum chloride s. under Zn $AlCl_3$
Indium(I) iodide s. under Pd(PPh$_3$)$_4$ InI

Indium(III) chloride (s.a. under In) $InCl_3$
Lanthanum trichloride/hydrogen chloride $LaCl_3/HCl$
Catalytic Biginelli synthesis of 3,4-dihydro-2(1*H*)-pyrimidinones   ○
with Yb(OTf)$_3$ cf. *55*, 337s59; with LaCl$_3$ and a little HCl (notably from *aliphatic* aldehydes) s. J. Lu et al., Tetrahedron Lett. *41*, 9075-8 (2000); with InCl$_3$, **also 3,4-dihydropyrimidine-2(1*H*)-thiones** using thiourea, s. B.C. Ranu et al., J. Org. Chem. *65*, 6270-2 (2000).

Indium(III) chloride/p-toluenesulfonic acid/magnesium sulfate   ←
α-Amino-γ-lactones from ethylene derivs.

332.

Freshly distilled methyl glyoxylate, 1.1 eqs. 2,4-dichloroaniline, 1.2 eqs. 2-ethyl-1-butene, 1.5 eqs. TsOH, and 2 eqs. MgSO$_4$ added successively to a suspension of 1 eq. InCl$_3$ in methylene chloride at room temp., stirred at reflux overnight, and quenched with satd. aq. NaHCO$_3$ soln. and brine → product. Y 76%. The procedure is mild and applicable to both cyclic and acyclic alkenes, but failed with simple monosubst. alkenes and was suppressed with bulky alkyl glyoxylates. F.e. and electronic effects, also with 10-50 mol% Sc(OTf)$_3$, s. T. Huang, C.-J. Li, Tetrahedron Lett. *41*, 9747-51 (2000).

Lanthanum triisopropoxide/trimethylsilyl cyanide $La(OPr-i)_3/Me_3SiCN$
α-Siloxymalononitriles from acyloximes   COON=C → C(OSi≤)(CN)$_2$

333.

2.5 eqs. Me$_3$SiCN and startg. acyloxime added to a Schlenk tube containing a soln. of 10 mol% La(OPr-*i*)$_3$ in THF, stirred at room temp. for 15 h under argon, the catalyst precipitated with wet diisopropyl ether, and the filtrate worked up → 1-trimethylsiloxypentane-1,1-dinitrile. Y 83%. Reaction is thought to involve addition of *in situ*-generated La(CN)$_3$ to *in situ*-generated acyl cyanide in a catalytic cycle. Other lanthanide catalysts and Zr(OPr-*i*)$_4$ gave poor yields or were inactive. Furthermore, low yields were obtained from enolesters or carboxylic acid chlorides. F.e. **and from carboxylic acid anhydrides** (at 50°) s. A. Fujii, S. Sakaguchi, Y. Ishii, J. Org. Chem. *65*, 6209-12 (2000).

Ytterbium(III) tris(perfluoroalkanesulfonyl)methides   ←
Friedel-Crafts acylation in a fluorous medium   H → COR
with Sc(OTf)$_3$ cf. *58*, 378; with readily recyclable Yb(III)-tris(perfluoroalkanesulfonyl)methides (10 mol%) in perfluoromethyldecalin s. A.G.M. Barrett et al., Synlett *2000*, 847-9.

Yttrium(III) triflate $Y(OTf)_3$
C-Aminomethylation of the pyrrole ring   H → CH$_2$N<
of indoles with AcOH cf. *11*, 843; 2-aminomethylation of N-carbalkoxypyrroles with Y(OTf)$_3$ (13 mol%) s. C. Zhang, R. Li et al., Tetrahedron Lett. *42*, 461-3 (2001).

Samarium diiodide SmI_2
ω-Acylaminoketones from N-acyllactams and oxo compds. s. *60*, 251   ○

Samarium diiodide/(-)-camphor-2,10-sultam   ←
γ-Lactones from ketones and α,β-ethylenecarboxylic acid esters   ○
with asym. induction
s. *41*, 723s54; improved enantioselectivity with isosorbide as stoichiometric chiral auxiliary and 1 eq. (-)-camphor-2,10-sultam (chiral proton source) s. M.-H. Xu, G.-Q. Lin et al., Org. Lett. *2*, 2229-32 (2000).

Trifluoroacetic anhydride $(CF_3CO)_2O$
5-Subst. 2,3-dihydrobenzofurans from *p*-hydroxysulfoxides and ethylene derivs.
via ar. Pummerer-type rearrangement and *ipso*-substitution of ar. sulfinyl groups

334.

A soln. of startg. sulfoxide added to a soln. of ca. 1.3 eqs. *p*-methoxystyrene and 1.5 eqs. trifluoroacetic anhydride in anhydrous acetonitrile at -40°, stirred at the same temp. for 30 min, and quenched with satd. $NaHCO_3$ → intermediate 5-phenylthio deriv. (Y 88%), re-oxidized to the sulfoxide with *m*-chloroperoxybenzoic acid in the normal way → intermediate 5-phenylsulfinyl-deriv. (Y 94%), in THF treated with 5 eqs. *n*-BuLi at -78°, 5 eqs. DMF added immediately, stirred at -78° for 30-60 min, and quenched with satd. $NaHCO_3$ → product (Y 61%). This provides a convergent route to diversely substituted benzofuran neolignans. 2,3-Disubst. derivs. were obtained as *trans*-adducts exclusively, even from E/Z mixtures. The sulfinyl group has a dual role, facilitating both the rearrangement and the subsequent *ipso*-substitution with C-electrophiles. F.e. s. S. Akai, Y. Kita et al., Org. Lett. *2*, 2279-82 (2000).

Amberlite IRA 900 ←
2-Iminocoumarins via Knoevenagel condensation with nitriles
with LiBr cf. *52*, 358; with Amberlite IRA 900 (1 eq. OH⁻) s. C. Mhiri et al., Synth. Commun. *29*, 3385-99 (1999).

Acetic acid AcOH
Pyridines from aldehydes and two ketone molecules
One-pot procedure via aldol condensation-Michael addition in the absence of solvent

335.

Ar = p-BuOC$_6$H$_4$

Terpyridyls. 2 eqs. 4-Acetylpyridine, freshly distilled 4-butoxybenzaldehyde, and 2 eqs. NaOH pellets crushed with a pestle and mortar until a pale yellow powder was formed (ca. 15 min), the latter added to a stirred soln. of NH_4OAc (excess) in glacial acetic acid, heated at reflux for 2 h, and worked up → 4'-(4-butoxyphenyl)-4,2':6',4''-terpyridyl. Y 76%. In this simple, mild, economical and environmentally friendly method, there is a dramatic improvement in the yield for a range of compds. not accessible using conventional methods in organic solvents. Furthermore, no crystallization and/or chromatographic purification steps were required. F.e. incl. unsym. pyridines from two different ketone molecules, bipyridyls, and 2,4,6-triaryl-derivs. s. G.W.V. Cave, C.L. Raston, Chem. Commun. *2000*, 2199-200.

Acetic acid/microwaves ←
Biginelli synthesis of 3,4-dihydro-2(1H)-pyrimidinones under microwave irradiation in the absence of solvent
s. 55, 337s59; with added AcOH (0.3 eq.) s. J.S. Yadav et al., J. Chem. Res., Synop 2000, 354-5.

L-Proline or (R)-5,5-Dimethylthiazolidine-4-carboxylic acid ←
N-Protected β-aminoketones $COCH \rightarrow COC\text{-}CH(NHR)$
from ketones, aldehydes and N-protected amines
Catalytic asym. Mannich reaction

336.

A *direct*, highly enantioselective, catalytic asym. Mannich 3-component condensation has evolved without pre-formation of an enolate equivalent or imine. **E:** 35 Mol% L-proline, *p*-nitrobenzaldehyde, and 1.1 eqs. *p*-anisidine in 1:4 acetone/DMSO allowed to react for 12 h → (S)-product. Y 50% (e.e. 94%). Reaction is presumed to involve *in situ*-formation of a chiral enamine [from the ketone and proline], which then adds to an *in situ*-generated aldimine with excellent enantioselectivity. The procedure is applicable to both α-subst. and α-unsubst. aldehydes; it is also inexpensive, the auxiliary can be readily retrieved by filtration, and the N-protective group easily removed. F.e. incl. regiospecific asym. synthesis of N-protected β-amino-α-hydroxyketones s. B. List, J. Am. Chem. Soc. *122*, 9336-7 (2000); with (R)-5,5-dimethylthiazolidine-4-carboxylic acid cf. W. Notz, C.F. Barbas III et al., Tetrahedron Lett. *42*, 199-201 (2001).

Trifluoroacetic acid CF_3COOH
Polymer-based Pictet-Spengler ring closure ○
s. *8*, 823s55; 9*H*-pyrid[3,4-*b*]indoles with on-resin aromatization using cyanuric chloride s. J.X. Zhang et al., Chin. Chem. Lett. *11*, 955-6 (2000); with N-acylation and subsequent resin cleavage via piperazine-2,5-dione ring closure s. H. Wang, A. Ganesan, Org. Lett. *1*, 1647-9 (1999).

m-Chloroperoxybenzoic acid s. under Ti(OPr-i)$_4$ $ArCOO_2H$

Cyanuric chloride/potassium carbonate ←
α-Diazoketones from carboxylic acids $COOH \rightarrow COCN_2$
with dicyclohexylcarbodiimide cf. *26*, 806; diazomethyl ketones with cyanuric chloride/K_2CO_3 s. D.C. Forbes et al., Tetrahedron Lett. *41*, 9943-7 (2000).

Silica gel (s.a. under Morpholine) SiO_2
β-*tert*-Amino-α-methyleneketones $CH_3 \rightarrow C(=CH_2)CH_2N<$
from methyl ketones and sec. amines
Double Mannich reaction in the absence of solvent

337.

Startg. methyl ketone mixed with silica gel in a mortar, 3 eqs. formaldehyde (37% in water) and 2 eqs. startg. dialkylamine added, and stirred for 5 h in a capped flask at room temp. → product. Y 68%. The method is simple, reliable and environmentally safe. F.e.s. M.M. Mojtahedi et al., J. Chem. Res., Synop 2000, 380-1; with dimethyl(methylene)ammonium chloride cf. U. Girreser, M. Schütt et al., Synlett *1998*, 715-7.

Trimethylsilyl cyanide s. under La(OPr-i)$_3$ and Pd(PPh$_3$)$_4$ Me_3SiCN

Trimethylsilyl azide
Tetrazolo[1,5-*a*]piperazin-6-ones
from aldehydes, prim. amines and α-isocyanocarboxylic acid esters
Double ring closure by azide-modified Ugi 4-component condensation

Me_3SiN_3

338.

Equivalent amounts of 3-phenylpropionaldehyde, 3-phenylpropylamine, methyl isocyanoacetate, and trimethylsilyl azide (each 0.1 *M* in methanol) stirred together at reflux for 48 h, worked up, the crude residue redissolved in 1:1 THF/dichloromethane, treated with polymer-based isocyanate (to scavenge excess of amine), the suspension shaken for 15 h at room temp., filtered, and the filtrate worked up → product. Y 60%. Reaction is applicable to a range of aldehydes and prim. amines, and yields were highest with methyl isocyanoacetate. The procedure was also adapted to the solution-phase preparation of an **80-member library** by varying the three points of molecular diversity. F.e.s. T. Nixey, C. Hulme et al., Tetrahedron Lett. *41*, 8729-33 (2000).

Titanium tetraisopropoxide s.a. under Mg $Ti(OPr\text{-}i)_4$

Titanium tetraisopropoxide/titanium tetraiodide/m-chloroperoxybenzoic acid   ←
α,β-Dialkoxyketones from acetals $C(OR)_2 \rightarrow C(OR)CH(OR')CO$
Synthesis with addition of 2 C-atoms s. *60*, 271

Stannous triflate $Sn(OTf)_2$
Friedel-Crafts reactions s. *60*, 327   ←

Triphenylphosphine/diethyl azodicarboxylate/sodium hydride   ←
3-Aminofurans from α-cyanoketones and alcohols
Mitsunobu reaction-Thorpe cyclization

339.

3-Aminofuran-2-carboxylic acid esters. A soln. of 1.4 eqs. Ph$_3$P in anhydrous THF treated at 0° with 1.4 eqs. DEAD, 1.4 eqs. ethyl glycolate, and 4,4-dimethyl-3-oxopentanenitrile, allowed to warm to room temp., stirred for 15 h, 2.8 eqs. NaH added, and stirring continued for 5 h → ethyl 3-amino-5-*tert*-butylfuran-2-carboxylate. Y 83%. The method is limited to the synthesis of 5-alkyl, 5-aryl and 4,5-fused bicyclic furans. F.e. incl. 2-aryl derivs., and with isolation of the intermediate s. A.M. Redman, W.J. Scott et al., Org. Lett. *2*, 2061-3 (2000).

Dodecylbenzenesulfonic acid RSO_3H
Mannich reaction in an aq. colloidal dispersion $H \rightarrow CH(R)N{<}$

340.

Equivalent amounts of aniline and benzaldehyde, and 5 eqs. cyclohexanone, added successively at 23° to a soln. of *1 mol%* dodecylbenzenesulfonic acid *in water*, stirred for 1 h, and treated with satd. aq. NaHCO$_3$ and brine before work-up → product. Y 97% (74:26 mixture of diastereoisomers). The reagent functions as *both Brønsted acid* (to activate the aldimine by protonation prior to nucleophilic addition of the ketone) *and surfactant*, effectively 'solubilizing' the organic reactants in the resulting turbid aq. colloidal dispersion. Reaction does not require prior formation of a stabilized ketone enolate (e.g. an enoxysilane), and is generally applicable to aromatic, heteroaromatic and enolizable aliphatic aldehyde (when added slowly!). With 2-butanone, reaction takes place regioselectively at the less substituted site. There was no reaction with TsOH. F.e.s. K. Manabe, S. Kobayashi, Org. Lett. *1*, 1965-7 (1999).

Camphorsulfonic acid   *RSO$_3$H*
Pictet-Spengler ring closure with asym. induction   ○
with CF$_3$COOH cf. *8*, 823s*48*; chiral 1,2,3,4-tetrahydro-9*H*-pyrid[3,4-*b*]indoles from N-sulfinyl-tryptamines with camphorsulfonic acid s. C. Gremmen, G.-J. Koomen et al., Org. Lett. *2*, 1955-8 (2000).

p-*Toluenesulfonic acid s. under InCl$_3$*   *TsOH*

Iodine   *I$_2$*
2,4-Dialkoxychromans from *o*-hydroxyaldehydes under mild conditions

341.

A soln. of *o*-hydroxybenzaldehyde, 2 eqs. 2,2-dimethoxypropane, and 0.02 eq. I$_2$ in dry dichloromethane stirred under N$_2$ at room temp. for 0.5 h → product. Y 90%. The method is simple and general, the reagent and startg. materials are inexpensive and readily available, and the mild conditions tolerate phenolethers, and ar. bromides, chlorides, methylenedioxy compds., and nitro compds. The method, however, appears limited to the use of 2,2-dimethoxypropane (diethyl acetals of aldehydes being deprotected, instead), while *o*-hydroxyketones failed to react. HCl, HBr or HI were not suitable as catalyst, and only re-sublimed I$_2$ was effective under strictly moisture free conditions. F.e.s. J.S. Yadav et al., J. Chem. Soc. Perkin Trans. 1 *2000*, 3082-4.

Hydrogen chloride   *HCl*
Quinolines from α,β-ethylenealdehydes
and anilines in concd. HCl cf. *26*, 823; large-scale procedure from anilines or acetanilides in a 2-phase aq. organic medium s. M. Matsugi et al., Tetrahedron Lett. *41*, 8523-5 (2000).

Ferric chloride   *FeCl$_3$*
Pyridines from α,β-ethyleneoximes via Michael addition

342.

Pyridine-3-carboxylic acid esters. Ethyl acetoacetate and 5 mol% FeCl$_3$·6H$_2$O added to 1 eq. startg. oxime, heated with vigorous stirring at 150-60° for 2-4 h, unreacted β-keto ester distilled off under reduced pressure, the residue taken up in *tert*-butyl methyl ether, extracted with 1 *M* HCl, and the aq. extracts adjusted to pH 9 with aq. NH$_3$ → product. Y 81%. The procedure is simple and general, notably for the preparation of tetra- and penta-subst. pyridines. Reaction presumably involves initial Michael addition, but the course of the ensuing cyclodehydration is as yet unclear. Yields were lower with NiCl$_2$, CuCl$_2$ and CoCl$_2$. F.e.s. A.M. Chibiryaev, N. de Kimpe et al., Tetrahedron Lett. *41*, 8011-3 (2000).

Tricarbonyldichlororuthenium(II) dimer/triethylamine   [RuCl$_2$(CO)$_3$]$_2$/Et$_3$N
2-Cyclopentenones from ethylene derivs. and 2-ethylenecarbonic acid esters
Carbonylation via ruthenium η^3-allyl complexes

343.

A soln. of startg. allyl carbonate and 1.1 eqs. 2-norbornene in THF heated with 2.5 mol% [RuCl$_2$(CO)$_3$]$_2$ and 10 mol% Et$_3$N at 120° for 5 h under 3 atm. CO → *exo*-4-methyltricyclo-[5.2.1.0^{2,6}]dec-4-en-3-one. Y 80% (100% *exo*). Reaction is thought to involve initial generation of a ruthenium η^3-allyl complex, followed by *cis*-carboruthenation of the alkene, carbonyl insertion, intramolecular 1,2-acylruthenation, β-elimination and isomerization. Phenyl-subst. allylic carbonates, however, gave the more stable α-benzylidenecyclopentanones. The base was essential (Et$_3$N being optimal) and the CO pressure critical (3 atm. being optimal). F.e. and intramolecular conversion, also with (η^3-C$_3$H$_5$)RuBr(CO)$_3$/Et$_3$N, s. Y. Morisaki, T. Mitsudo et al., Org. Lett. 2, 949-52 (2000).

Carbonyl(chloro)[1,3-bis(diphenylphosphino)propane]rhodium(I) dimer   [Rh(CO)Cl(dppp)]$_2$
s. under Pd$_2$(dba)$_3$

Chlorotris(triphenylphosphine)rhodium(I)/trimethyl phosphite/sodium hydride   ←
Regio- and stereo-specific C-α-allylation with 2-ethylenecarbonic acid esters   H → C-C=C
cf. 55, 357; with chiral, non-racemic 2-ethylenecarbonates (without double bond shift and with retention of configuration at the chiral centre), notable for generating chiral quaternary hydrocarbon groups, s. P.A. Evans, L.J. Kennedy, Org. Lett. 2, 2213-5 (2000).

Palladous acetate/triphenylphosphine/diethylzinc   ←
Asym. synthesis of 3-acetylenealcohols   CH(OH)C-C≡C
from aldehydes and mesyloxy-2-acetylenes
with Pd(dppf)Cl$_2$/InI cf. 57, 317; from lactic aldehyde ethers s. J.A. Marshall et al., J. Org. Chem. 65, 8357-60 (2000); with Pd(OAc)$_2$/Ph$_3$P/Et$_2$Zn cf. Org. Lett. 2, 2897-900 (2000).

Palladous acetate/1,3-bis(diphenylphosphino)propane/triethylamine/microwaves   ←
Heck reactions with diol monovinyl ethers   ←
cyclic acetophenone ketals from aryl triflates cf. 54, 374; 2-vinyl-1,3-dioxolanes from enol triflates with microwave enhancement, also 2-alkoxy-1,3-dienes from enol triflates and enolethers, s. K.S.A. Vallin, A. Hallberg et al., J. Org. Chem. 65, 4537-42 (2000).

Palladous acetate/(R)- or (S)-7,7'-dimethoxy-2,2'-bis(diphenylphosphino)-   ←
1,1'-binaphthyl
Asym. Heck arylation with aryl triflates s. 46, 738s60   H → Ar

Palladous acetate/1,1'-bis(diphenylphosphino)ferrocene/potassium carbonate   ←
or triethylenediamine or microwaves
Regiospecific Heck arylation of 2-ethylene-*tert*-amines

344.

with aryl triflates. A mixture of startg. aryl triflate, 5 eqs. N,N-dimethylallylamine, 6 mol% Pd(OAc)$_2$, 26 mol% dppf, and 1.5 eqs. K$_2$CO$_3$ in acetonitrile heated at 80° for 20 h under N$_2$ in a heavy-walled Pyrex tube sealed with a Teflon stopcock → product. Y 71% (>99.5% conversion; β:γ >99.5:0.5). The very high regioselectivity is explained by *chelation-controlled* formation of a **cationic 5-membered σ-palladacyclic**, favoured by electron-poor triflates. In certain instances, there was a significant amount of deoxygenation of the triflate with Et$_3$N as base, and there was no reaction with vinyl triflates. F.e., steric effects, and functional group tolerance, also rate enhancement under microwave irradiation with electron-rich and -neutral triflates (although with reduced yield), s. K. Olofsson, A. Hallberg et al., J. Org. Chem. *65,* 7235-9 (2000); f. **2-aryl-2-ethylene-*tert*-amines** with DABCO as base cf. J. Wu et al., Tetrahedron Lett. *42,* 159-62 (2001); arylation of N-protected 2-ethyleneamines under microwave irradiation s. K. Olofsson, A. Hallberg et al., J. Org. Chem. *66,* 544-9 (2001).

Tris(dibenzylideneacetone)dipalladium/chiral phosphino-Δ^2-oxazolines/triethylamine ←
Asym. Heck vinylation with enol triflates H → C═C
s. *46,* 738s*59*; with hydroxyproline-derived (S,S,S)-2-[1-(*tert*-butoxycarbonyl)-4-(diphenylphosphino)pyrrolidin-2-yl]-4-isopropyl-Δ^2-oxazoline as ligand, also asym. Heck arylation, s. S.R. Gilbertson et al., Tetrahedron Lett. *42,* 365-8 (2001); with a chiral 2-(*o*-phosphinophenyl)-Δ^2-oxazoline derived from (1R,2S)-2-amino-3,3-dimethyl-1-indanol s. Y. Hashimoto et al., Tetrahedron:Asym. *11,* 2205-10 (2000).

Tris(dibenzylideneacetone)dipalladium/chiral 2-(o-phosphinophenyl)-Δ^2- ←
  oxazoline or (S)-2,2'-bis(diphenylphosphino)-3,3'-bibenzo[b]thiophene/base
Asym. Heck arylation with aryl triflates H → Ar
of 2,3-dihydrofurans with (R)-BINAP as ligand s. *46,* 738; with a chiral 2-(*o*-phosphinophenyl)-Δ^2-oxazoline (based on (1R,2S)-2-amino-3,3-dimethyl-1-indanol) s. Y. Hashimoto et al., Tetrahedron:Asym. *11,* 2205-10 (2000); with Pd(OAc)$_2$ and (R)- or (S)-7,7'-dimethoxy-2,2'-bis-(diphenylphosphino)-1,1'-binaphthyl as ligand s. D. Che, B.A. Keay et al., ibid. 1919-25; regiospecific asym. Heck arylation of N-carbalkoxy-Δ^2-pyrrolines with (S)-2,2'-bis-(diphenylphosphino)-3,3'-bibenzo[*b*]thiophene as ligand s. L.F. Tietze, K. Thede, Synlett *2000,* 1470-2.

Tris(dibenzylideneacetone)dipalladium/carbonyl(chloro)[1,3-bis(diphenylphosphino)- ←
  propane]rhodium(I) dimer/1,4-bis(diphenylphosphino)butane/
  N,O-bis(trimethylsilyl)acetamide
Bicyclo[3.3.0]oct-1-en-3-one-7,7-dicarboxylic ○
from γ,δ-acetylenemalonic acid esters and acoxy-2-ethylenes
Dual catalytic C-allylation-intramolecular Pauson-Khand reaction

A mixture of startg. alkyne and 2 eqs. allyl acetate added to a soln. of 1.5 mol% Pd$_2$(dba)$_3$·CHCl$_3$, 3 mol% dppb, 1.2 eqs. N,O-bis(trimethylsilyl)acetamide, and 7 mol% [RhCl(CO)(dppp)]$_2$ in toluene, the reaction vessel evacuated, charged with 1 atm. CO, and heated at 110° for 25 h → product. Y 92%. The procedure is simple, with minimal use of solvent and reagents, and minimal production of waste. The efficiency depends on the right combination of catalysts (preferably an electron-rich Pd(0) catalyst and a Lewis acidic Rh complex) and their relative proportion. N-Propargylsulfonamides reacted similarly to give the corresponding **7-aza-analogs**, but propargyl alcohols failed to react. F.e. and details [in Supporting Information] s. N. Jeong et al., J. Am. Chem. Soc. *122,* 10220-1 (2000).

Tetrakis(triphenylphosphine)palladium(0)   *Pd(PPh$_3$)$_4$*
4-Oxa-1,8-dienes from electron-deficient ethylene derivs.   C≡C → C(OR)C-C-C≡C
1,2-Alkoxyallylation-transetherification

346.

A soln. of the startg. dicyanoethylene, 1 eq. allyl ethyl carbonate, 10 eqs. allyl alcohol, and 5 mol% Pd(PPh$_3$)$_4$ in anhyd. THF stirred at room temp. under N$_2$ for 24 h → product. Y 96%. F.e.s. R.L. Xie, J.R. Hauske, Tetrahedron Lett. *41*, 10167-70 (2000).

Tetrakis(triphenylphosphine)palladium(0)/indium(I) iodide   *Pd(PPh$_3$)$_4$/InI*
Regiospecific synthesis of 3-ethylenealcohols from aldehydes   CHO → CH(OH)C-C≡C
by nucleophilic ('umpulonged') allylation with palladium π-allyl complexes
with Pd(PPh$_3$)$_4$/Zn cf. *43*, 699; via allylindium(III) compds. with InCl, InBr, InI or In for allylation with a wide range of allyl derivs. (allyl alcohols, acetates, carbonates, ethers, halides and thioethers) with double bond shift s. S. Araki et al., Org. Lett. *2*, 847-9 (2000).

Tetrakis(triphenylphosphine)palladium(0)/trimethylsilyl cyanide   *Pd(PPh$_3$)$_4$/Me$_3$SiCN*
Cyanoallenes from 2-acetylenecarbonic acid esters   C≡C-C(OCOOR) → C(CN)=C=C

347.

An equimolar mixture of startg. acetylenecarbonic acid ester and trimethylsilyl cyanide treated with 5 mol% Pd(PPh$_3$)$_4$ in THF under argon, and refluxed for 1 h → product. Y 91%. An excess of trimethylsilyl cyanide provides dicyanated products. Reaction is thought to involve initial oxidative addition of the carbonate to Pd(0) to produce an **allenylpalladium(II) alkoxide**, followed by ligand exchange with cyanide and reductive elimination of the metal. Other Pd-catalysts were ineffective. F.e.s. Y. Tsuji et al., Org. Lett. *2*, 2635-7 (2000).

Bis(allylpalladium chloride)/chiral 2-(o-phosphinoaryl)pyridines or 1,1'-bi-2-   ←
naphthol N-(2-pyridyl)phosphoromonoamidites or 1,4-diamino-2,3-bis(phosphinites) or
2-[2-(arylseleno)ferrocenyl]-Δ^2-oxazolines or N,O-bis(phosphino)-2-aminoalcohols or
N,N'-bis(2-phosphinoalkyl)dicarboxylic acid amides or 2-amino-2'-diarylphosphino-
5,5',6,6',7,7',8,8'-octahydro-1,1'-binaphthyls or phosphinobis(5,6-dihydro-4H-
1,3-oxazines) or o-aminosulfoxides or phenanthrolines or bicyclic 8-quinolyl
phosphorodiamidites or polymer-based 2-(2-pyridyl)-Δ^2-oxazolines or β-turn
peptidylphosphines/N,O-bis(trimethylsilyl)acetamide/potassium or lithium acetate
Asym. α-allylation   H → C-C≡C
update s. *48*, 772s59; with (R)-2-(2-diphenylphosphinophenyl)-7-isopropyl-6,7-dihydro-5*H*-[1]pyrindine as ligand s. K. Ito, T. Katsuki et al., Synlett *2001*, 284-6; with chiral 1,1'-bi-2-naphthol N-(2-pyridyl)phosphoromonoamidites s. C.G. Arena et al., Tetrahedron:Asym. *11*, 4753-9 (2000); with chiral 1,4-diamino-2,3-bis(phosphinites) s. A. Zhang, B. Jiang et al., ibid. 3123-30; with 2-[2-(arylseleno)ferrocenyl]-Δ^2-oxazolines s. S.-L. You, X.-L. Hou et al., ibid. 1495-500; with N,O-bis(phosphino)-2-aminoalcohols s. L. Gong, A. Mi et al., ibid. 4297-302; with N,N'-bis(2-phosphinoalkyl)dicarboxylic acid amides s. A. Saitoh, T. Morimoto et al., ibid. 4049-53; with chiral 2-amino-2'-diarylphosphino-5,5',6,6',7,7',8,8'-octahydro-1,1'-binaphthyls s. ibid. 4153-62; with chiral phosphinobis(5,6-dihydro-4*H*-1,3-oxazines) s. S. Lee et al., ibid. *10*, 1795-802 (1999); with chiral o-aminosulfoxides s. K. Hiroi et al., ibid. *9*, 3797-817 (1998); with chiral phenanthrolines s. G. Chelucci, A. Saba, ibid. *9*, 2575-8 (1998); asym. α-allylation of cyclic β-keto esters with a bicyclic 8-quinolyl phosphorodiamidite (QUIPHOS) for generating chiral quaternary hydrocarbon groups s. J.M. Brunel, A. Tenaglia et al., ibid. *11*, 3585-90 (2000); asym. α-allylation with chiral polymer-based 2-(2-pyridyl)-Δ^2-oxazolines s. K. Hallman, C. Moberg et al., ibid. *10*, 4037-46 (1999); with a library of polymer-based β-turn

peptidylphosphines (as a prelude to optimization of homogeneous counterparts) s. S.R. Gilbertson et al., J. Am. Chem. Soc. *122*, 6522-3 (2000).

Bis(η^3-allylpalladium chloride)/(1R,2R)-N,N'-bis[(S_p)-2-(diphenylphosphino)- ←
ferrocenoyl]cyclohexane-1,2-diamine dihydrate/lithium diisopropylamide
Asym. α-allylation of non-stabilized cyclic ketone enolates   H → C-C=C
with Trost's ligand cf. *59*, 354; improved enantioselectivity with (1R,2R)-N,N'-bis[(S_p)-2-(diphenylphosphino)ferrocenoyl]cyclohexane-1,2-diamine dihydrate as ligand s. S.-L. You, X.-L. Hou et al., Org. Lett. *3*, 149-51 (2001).

Via intermediates   *v.i.*
Synthesis of nitriles from aldehydes via cyanohydrins s. *31*, 49s*60*   CHO → CH_2CN

Nitrogen ↑   CC ↓↑ N

Without additional reagents   *w.a.r.*
β-*tert*-Amino-α-methyleneketones from methyl ketones CH_3 → C(=CH_2)CH_2N<
Double Mannich reaction with dimethyl(methylene)ammonium chloride s. *60*, 337

cis-2-Silyl-1-sulfonylaziridines from N-sulfonylimines

348.

2.5 eqs. Trimethylsilyldiazomethane (1.8 *M* in hexanes) added to a soln. of startg. N-tosylimine in dioxane, and the mixture stirred at 40° for 7 h → product. Y 72% (*cis*-selectivity 95%). These conditions apply only for trimethylsilyldiazomethane and not for phenyldiazomethane or ethyl diazoacetate. *cis*-Selectivity prevailed except in one example in which an alternate mechanism may be operating. The reaction seems to be general for a wide range of N-sulfonylimines, and tolerates sulfonyl, phenolether, ar. chloride, ar. nitro, ethylene, and carboxylic acid ester groups. F.e. and regiospecific ring opening with nucleophiles s. V.K. Aggarwal, M. Ferrara, Org. Lett. *2*, 4107-10 (2000); **with asym. induction** s. R. Hori, T. Shiori et al., Tetrahedron Lett. *41*, 9455-8 (2000).

Pyridine ring from enamines   ○
and diethyl ethoxymethyleneoxalacetate cf. *9*, 807; benz[*g*]isoquinoline-5,10-diones from 2-acyl-1,4-naphthoquinones s. K. Kobayashi et al., Tetrahedron Lett. *41*, 7657-60 (2000).

Irradiation
2-Azetidinones from α-diazoketones and azomethines
s. *13*, 681; from chiral N-(*p*-methoxyphenethyl)aldimines with asym. induction under irradiation, and oxidative removal of the protective group (*49*, 330s*60*) s. J. Podlech, S. Steurer, Synthesis 1999, 650-4.

Microwaves s. under Pyrrolidine   ←

Sodium hydride   *NaH*
(E)-α,β-Ethylene-α-sulfinylphosphonic acid esters from N-tosylimines CH_2 → C=C
Aza-Knoevenagel condensation with addition of one C-atom

349.   Ar = p-FC_6H_4

A stirred mixture of diethyl benzenesulfinylmethylphosphonate, 1 eq. startg. N-tosylimine, *0.2 eq.* 60% NaH and THF heated in a capped vessel under N_2 at 70° for 2 min (monitored by TLC) → diethyl (E)-2-(4-fluorophenyl)-1-(phenylsulfinyl)ethenylphosphonate. Y 87%. This contrasts with the Horner synthesis which eliminates the phosphoryl group on reaction with aldehydes. Furthermore only a catalytic amount of base is required whereas in the Horner route or other condensation methods stoichiometric amounts are necessary. The reaction is rapid and affords the (E)-isomers exclusively. F.e.s. Y. Shen, G.-F. Jiang, Synthesis *2000*, 99-102; (1E,3Z)-3-cyano-1,3-dienephosphonic acid esters from diethyl 3-cyanoallylphosphonate with DBU s. J. Chem. Soc. Perkin Trans. 1 *1999*, 495-8.

Sodium/alcohol   NaOR
α-Arylketones from aldehydes and ar. aldehyde tosylhydrazones   CHO → COCH$_2$Ar
via *in situ*-generated aryldiazomethanes

350.
$$\text{Ph}_2\text{CHCHO} + [\text{N}_2\text{=CHPh}] \xrightarrow{\text{TsNHN=CHPh}} \text{Ph}_2\text{CHCOCH}_2\text{Ph}$$

A 2 *M* soln. of freshly prepared NaOMe in *methanol* added to 1 eq. startg. tosylhydrazone in the same solvent, 0.5 eq. startg. aldehyde added with exclusion of light, and heated at 65° for 4 h → 1,1,3-triphenyl-2-propanone. Y 84%. The procedure is notably applicable to branched aldehydes (not simple aliphatic aldehydes!) and is thereby complementary with the Anselme method (cf. *7, 816s37*); more importantly, it avoids isolation and handling of potentially toxic and explosive aryldiazomethanes (especially with *electron-releasing* groups). Water and ethylene glycol were also suitable solvents. F.e.s. S.R. Angle, M.L. Neitzel, J. Org. Chem. *65*, 6458-61 (2000).

Potassium tert-*butoxide*   KOBu-t
Ring closures with vinylogous amidinium salts
review s. *25, 581s31*; pyridine N-oxides from oxo compds. and NH$_2$OH·HCl with KOBu-*t* s. I.W. Davies, J.-F. Marcoux et al., Org. Lett. *3*, 209-11 (2001).

Potassium tert-*butoxide/acetic acid/trifluoroacetic acid*   ←
Pyridines from vinylogous formamidinium salts and ketones
Regiospecific conversion

351.

3-Arylpyridines. 1.04 eqs. 20 wt% KOBu-*t* in THF added dropwise to a suspension of startg. ketone in dry THF at 0°, the yellow slurry stirred at room temp. for 45 min, 1.04 eqs. startg. [thermally stable] vinamidinium salt added in one portion, stirred at room temp. for 45 min, transferred dropwise via cannula under N$_2$ to a soln. of 7 eqs. acetic acid and 0.8 eq. trifluoroacetic acid in THF at 25-30°, stirred for 45 min, treated in one portion with 10 eqs. concd. NH$_4$OH, and heated at reflux for 5 h → 5-chloro-3-[4-(methanesulfonyl)phenyl]-2-(2-methyl-5-pyridyl)-pyridine. Y 94% (as a single regioisomer after chromatography on silica gel). An electron-withdrawing group at the α-site of the vinamidinium salt appears essential for high yields. Poor results were obtained with NaOBu-*t* or LDA as base. F.e.s. J.-F. Marcoux, E.G. Corley et al., Org. Lett. *2*, 2339-41 (2000).

n-*Butyllithium*   BuLi
Ring closures with functionalized benzotriazoles

352.

5-Aminothiazoles. 1 eq. *n*-BuLi (1.6 *M* in hexane) added to a soln. of startg. aldimine in THF at -78° under argon, stirred 5 min, then 1.1 eqs. phenyl isothiocyanate added, and quenched with 2 *M* NaOH at -78° when TLC indicated completion of reaction (or left overnight at room temp. if no precipitate had formed) → N-phenyl-2,4-diphenylthiazole-5-amine. Y 81%. No imidazole by-products were observed reflecting the nucleophilicity of sulfur. F.e.s. A.R. Katritzky et al., J. Org. Chem. *65*, 8077-9 (2000); **2-aminobenzothiazoles** from 2-amino-4-(benzotriazol-1-yl)methylthiazoles and enones, also pyrid[1,2-*a*]indoles, s. ibid. 8059-62; **1-acylaminoindenes** from N-[α-(acylamino)benzyl]benzotriazoles and aldehydes with ZnBr$_2$ s. ibid. 8066-8; imidazo-[1,2-*a*]pyridines from 2-amino-1-α-benzotriazol-1-ylpyridinium salts (with DBU) s. ibid. 9201-5; **pyrroles** from N-α-(thioacylamino)benzotriazoles and ethylene derivs. with KOBu-*t*/MeI s. ibid. 8819-21.

N-Protected 3-silyl-2-pyrrolidones from aziridines
and trimethylsilyldiazomethane s. *60*, 243

Lithium diisopropylamide   i-*Pr*$_2$*NLi*
(Z)-β-Amino-α,β-ethyleneketones   C(=N)CH → C(N<)=C-CO
from ketimines and 1-acylbenzotriazoles

353.

A soln. of startg. ketimine in THF added to a stirred soln. of 2 eqs. LDA in the same solvent at -10°, stirred at 0° for 30 min, cooled to -78°, a soln. of startg. 1-acylbenzotriazole in THF added dropwise, allowed to warm to room temp. while stirring overnight, and quenched by addition of satd. NH$_4$Cl → (Z)-3-(*tert*-butylamino)-1,3-diphenylprop-2-en-1-one. Y 85%. F.e.s. A.R. Katritzky et al., Synthesis *2000*, 2029-32.

Sodium acetate   NaOAc
Benzene ring from three 1,1-nitroethylene deriv. molecules

354.

Sym. 1,3,5-triarylbenzenes. Startg. nitrostyrene treated with 2 eqs. NaOAc in DMF at 80-90° for 20 h → 1,3,5-triphenylbenzene. Y 40%. The method is practical and based on readily available starting materials. It is also an improvement on the use of dimethylformamide dimethyl acetal (T.Y. Kim et al., Bull. Korean Chem. Soc. *20*, 1255 (1999)). Lewis basic DMF is believed to facilitate the elimination of nitrous acid in the final step. Other dipolar, aprotic solvents such as DMSO, NMP or HMPA may also be used. The reaction failed with β-subst. β-nitrostyrenes or with alkyl derivs. such as 1-nitro-3-methyl-1-butene or 1-nitro-1-cyclohexene. F.e. incl. tri-2-furyl- and tri-2-thienyl-derivs. s. T.Y. Kim, J.N. Kim et al., Bull. Korean Chem. Soc. *21*, 521-2 (2000).

Triethylamine Et_3N
**Asym. synthesis of cyclopropanecarboxylic acid esters
from electron-deficient ethylene derivs.
Chiral 2,2-dicyanocyclopropanecarboxylic acid esters**

355.

A mixture of 0.12 eq. startg. chiral pyridinium salt, 0.18 eq. Et_3N, and startg. alkylidenemalononitrile in dichloromethane stirred for 12 h at 0° → product. Y 99% (d.r. 89:14). For phenyl-subst. substrates, the benzene ring substituents did not influence the diastereoselectivity, and for alkyl-subst. substrates, the bulky *t*-butylmethylidenemalononitrile gave the highest d.r. Cyclohexylidenemalononitrile was found to be essentially non-selective (d.r. 55:45). This method is especially useful for preparing optically active, highly electron-deficient cyclopropanes. F.e.s. S. Kojima et al., Tetrahedron Lett. *41*, 9847-51 (2000).

Pyrrolidine/microwaves ←
Pyridine from 1,2,3-triazine ring ←
and α,β-ethylenehalides cf. *24*, 836; from enamines under microwave enhancement (or from ketones and pyrrolidine) s. A. Díaz-Ortiz, A. De la Hoz et al., Synlett *2001*, 236-7.

1,8-Diazabicyclo[5.4.0]undec-7-ene   DBU
(1E,3Z)-3-Cyano-1,3-dienephosphonic acid esters from N-tosylimines ←
Regiospecific synthesis with addition of four C-atoms s. *60*, 349

Imidazo[1,2-*a*]pyridines from 2-amino-1-α-benzotriazol-1-ylpyridinium salts s. *60*, 352 ○

Cuprous or cupric trifluoromethanesulfonate/chiral 2,2'-bipyridyls or ←
*bis(Δ²-oxazolines) or chiral clay-supported bis(Δ²-oxazolines) or
polymer-based azabis(Δ²-oxazolines)*
Asym. cyclopropanation with diazo compds. ▽
with chiral bis(Δ²-oxazolines) s. *23*, 819s59; with chiral 2,2'-bipyridyls s. H.L. Wong et al., Tetrahedron Lett. *41*, 7723-6 (2000); with pinene-derived 2,2'-bipyridyls s. G. Chelucci et al., Tetrahedron:Asym. *11*, 3419-26 (2000); s.a. D. Lötscher et al., ibid. 4341-57; with chiral 2,2':6',2''-terpyridyls s. H.-L. Kwong, W.-S. Lee, ibid. 2299-308; asym. cyclopropanation of α,β-ethylenefluorides s. O.G.J. Meyer, G. Haufe et al., Synthesis *2000*, 1479-90; with clay-supported chiral bis(Δ²-oxazolines) s. J.M. Fraile, J.A. Mayoral et al., Tetrahedron:Asym. *9*, 3997-4008 (1998); with chiral polymer-based azabis(Δ²-oxazolines) s. M. Glos, O. Reiser, Org. Lett. *2*, 2045-8 (2000).

Magnesium   Mg
N-Protected (E)-ethyleneamines from cyclic α-amino-N-tosylhydrazones C

356.

N-Protected (E)-4-ethyleneamines. 2.5 eqs. Phenylmagnesium bromide in anhydrous THF added dropwise to a cold soln. of startg. hydrazone in the same solvent under N_2, stirred at room temp. for 4 h, and quenched with satd. aq. NH_4Cl → product. Y 75%. A complex mixture was obtained if the ring nitrogen was unprotected. Reaction is applicable to aryl- and alkyl-magnesium bromides. F.e. and ring opening of piperidine analogs s. S. Chandrasekhar et al., Tetrahedron Lett. *41*, 10131-4 (2000).

Zinc bromide $ZnBr_2$
Indenes from 1-benzylbenzotriazoles and aldehydes
1-Acylaminoindenes s. *60*, 352

Triphenyl borate/(R)-2,2′-diphenyl-3,3′-bi-4-phenanthrol
Asym. synthesis of aziridine-2-carboxylic acid esters from aldimines

357.

A soln. of 2.5 mol% (R)-2,2′-diphenyl-3,3′-biphenanthryl-4,4′-diol in dichloromethane treated under argon with 7.5 mol% triphenyl borate, the mixture heated at 55° for 1 h, heating continued for 30 min at the same temp. under a vacuum of 0.5 mm Hg, diluted with dichloromethane, a soln. of the startg. aldimine in the same solvent added via syringe at 0°, the mixture stirred for 10 min, 1.1 eqs. ethyl diazoacetate added rapidly, and stirring continued for 6 h at 0° and for 14 h at 22° → product. Y 85% (e.e. 96%; *cis*-selectivity >98%). The procedure is more reliable and generally applicable than *58*, 344 (which largely depends on the purity of BH_3-THF), and affords higher yields and diastereoselectivity while maintaining high levels of asym. induction. Chiral BINOL was less effective. F.e.s. J.C. Antilla, W.D. Wulff, Angew. Chem. Int. Ed. Engl. *39*, 4518-21 (2000).

Boron fluoride BF_3
3-Hydroxytetrahydrofurans from diazo compds.
and pre-formed β-siloxyaldehydes cf. *59*, 418; 4-aryl-3-hydroxytetrahydrofuran-2-carboxylic acid esters from 1-aryl-3-siloxyepoxides via *in situ*-generated α-aryl-β-siloxyaldehydes, stereoselectivity and with asym. induction s. S.R. Angle, S.L. White, Tetrahedron Lett. *41*, 8059-62 (2000).

Indium(III) chloride $InCl_3$
***cis*-Aziridines from aldimines and diazo compds.**
with 10 mol% Nd(OTf)₃ in protic media cf. *58*, 342; *cis*-aziridine-2-carboxylic acid esters with 2 mol% $InCl_3$ in methylene chloride s. S. Sengupta, S. Mondal, Tetrahedron Lett. *41*, 6245-8 (2000).

Samarium s. under $TiCl_4$ Sm

Titanium tetrachloride/samarium $TiCl_4$/Sm
2,3-Dihydro-1*H*-1,5-benzodiazepines from *o*-nitroazides and ketones
o-Diamine N,N,N′,N′-tetraanions as intermediates

358.

under mild, neutral conditions. 2.2 eqs. $TiCl_4$ added dropwise via syringe to a stirred suspension of 3 eqs. Sm powder in freshly distilled, dry THF at room temp. under N_2, refluxed for 2 h, the resulting suspension of low-valent titanium cooled to room temp., a soln. of startg. *o*-nitroazide in anhydrous THF added, followed slowly, after 5-10 min, by 2.2 eqs. startg. ketone in the same solvent, stirred at 60° for 4 h, and quenched with distilled water → 2,4-dibutyl-2,3-dihydro-2-methyl-1*H*-1,5-benzodiazepine. Y 85%. The procedure is based on readily available substrates, is convenient in manipulation, and moderate-to-high yielding. The intermediate may be trapped with methanol as the corresponding *o*-diamine. F.e.s. W. Zhong, Y. Zhang et al., Tetrahedron Lett. **42**, 73-5 (2001).

Phosphorus oxide chloride $POCl_3$
Ketones from carboxylic acid amides H → COR
s. **9**, 872; regiospecific C-acylation of indoles with phthalimidoacetamides s. J.D. Kreisberg, P. Magnus et al., Tetrahedron Lett. **42**, 627-9 (2001).

Phosphorus oxide chloride/hexadecyltrimethylammonium bromide $POCl_3/[Me_3C_{16}H_{33}N]Br$
Ring closures with dimethylformamide
N-heterocyclic aldehydes s. **22**, 826, and 2-halogenoquinolines from acetanilides s. **34**, 756; 2-chloro-3-formylquinolines with added hexadecyltrimethylammonium bromide s. M.M. Ali, K.C. Rajanna et al., Synlett *2001*, 251-3.

Dichlorobis(triphenylphosphine)ruthenium(II)/stannous chloride $RuCl_2(PPh_3)_2/SnCl_2$
Quinolines from prim. ar. amines and quaternary N,N-diallylammonium salts

359.

A mixture of 6 eqs. aniline, 1 eq. N,N-diallyl-N,N-dipropylammonium chloride, 5 mol% RuCl₂(PPh₃)₂, and 1 eq. SnCl₂ in 9:1 dioxane/water flushed with argon in a pressure vessel, and heated at 180° for 20 h with stirring → 2-ethyl-3-methylquinoline. Y 57%. A dual catalytic cycle is invoked, wherein amine exchange takes place between aniline and the ammonium salt to form N-(prop-1-ylidene)aniline, which dimerizes prior to deaminative ring closure. Poor yields were obtained with N,N-dibutyl-N,N-dicrotylammonium chloride. F.e.s. C.S. Cho, S.C. Shim et al., Tetrahedron **56**, 7747-50 (2000).

Rhodium(II) acetate or Rhodium(II) acetate/ytterbium(III) triflate ←
Regio- and stereo-specific 1,3-dipolar cycloaddition with cyclic carbonyl ylids
s. **43**, 943s**55**; with indole as dipolarophile s. S. Muthusamy, S.A. Babu et al., Tetrahedron Lett. **42**, 523-6 (2001); with enones s. ibid. **41**, 8839-42 (2000); cycloaddition of *in situ*-generated 1-alkoxyisobenzopyrylium-4-olates with aldehydes (with added Yb(OTf)₃), also asym. variant, s. H. Suga, T. Ibata et al., Org. Lett. **2**, 3145-8 (2000).

Chlorotris(triphenylphosphine)rhodium(I)/2-amino-3-picoline ←
Regiospecific synthesis of *o*-subst. ketones from ar. aldehydes and ethylene derivs. ←
via chelation-assisted hydroacylation-*o*-alkylation

360.

A mixture of freshly purified benzaldehyde, 0.2 eq. 2-amino-3-picoline (as co-catalyst), 0.2 eq. benzylamine, 10 mol% RhCl(PPh₃), and 5 eqs. *tert*-butylethylene stirred at 170° for 12 h in a screw-capped pressure vial → 1-[2-(3,3-dimethylbutyl)phenyl]-4,4-dimethylpentan-1-one. Y 94%. Activation of the *ortho*-C-H bond towards oxidative addition of rhodium(I) is facilitated by coordination to intermediate imide nitrogen. F.e. and ***o*-alkylation of the intermediate ar. ketimines** s. C.-H. Jun et al., Angew. Chem. Int. Ed. Engl. **39**, 3440-2 (2000).

Halogen ↑   CC ↓↑ Hal

Irradiation (s.a. under KOBu-t, NaNH$_2$ and K$_2$CO$_3$)
Regiospecific [3+2]-cycloaddition with 2-ketoiodonium ylids

361.

2,3-Dihydrofuran ring from ethylene derivs. A soln. of startg. iodonium ylid and excess of styrene in acetonitrile irradiated with a 400 W medium-pressure Hg-lamp until reaction complete → product. Y 96%. With Rh$_2$(OAc)$_4$ the yield was 75%. High regio- and diastereo-selectivity were recorded, the O atom of the ylid adding exclusively at the *more* substituted carbon atom of the olefin. F.e., **also furan and oxazole ring from acetylene derivs. and nitriles**, respectively, with Rh$_2$(OAc)$_4$ s. E.P. Gogonas, L.P. Hadjiarapoglou, Tetrahedron Lett. *41*, 9299-303 (2000); 2-vinyl-2,3-dihydrofuran ring cf. *55*, 367s58; oxazole ring with Cu(acac)$_2$ cf. S. Spyroudis, P. Tarantili, J. Org. Chem. *58*, 4885-9 (1993).

Electrolysis/palladous acetate/potassium carbonate/tetra-n-butylammonium bromide   ←
Sym. diaryls from ar. halides   2 ArHal → Ar-Ar
with NiBr$_2$(bipy) as catalyst s. *35*, 549s55; with NiBr$_2$(dmbp) (dmbp = 6,6'-dimethyl-2,2'-bi-pyridyl) or with Pd(OAc)$_2$/K$_2$CO$_3$/Bu$_4$NBr in DMF/water/*i*-PrOH s. T.M. Cassol, F.W.J. Demnitz et al., Tetrahedron Lett. *41*, 8203-6 (2000).

Microwaves s. under K$_2$CO$_3$, In(OTf)$_3$, Pd(OAc)$_2$ and Pd(PPh$_3$)$_4$ ←

Sodium hydride   NaH
C-α-Alkylation of phosphonic acid esters   H → R
with *n*-BuLi cf. *30*, 43; of cyanomethylphosphonic acid esters with NaH, also conversion to α-cyano-α-isocyanophosphonic acid esters, s. J.R. Simon, R. Neidlein, Synthesis *2000*, 1101-8.

Sodium hydroxide   NaOH
Darzens condensation in an aq. suspension
Chalcone epoxides from ar. aldehydes

362.

A suspension of benzaldehyde, 1 eq. phenacyl chloride and 1.03 eqs. NaOH stirred in water at room temp. for 2 h, and the obtained crystalline powder collected by filtration, washed with water, then dried in a desiccator → 2,3-epoxy-1,3-diphenyl-1-propanone. Y 94%. Organic solvent is avoided completely in this method, which is also simple and relatively rapid, with easy work-up. KOH, LiOH or Ca(OH)$_2$ were also effective as base. F.e. and with N-benzyl-cinchonidinium chloride as phase transfer catalyst with high diastereoselectivity (but low enantioselectivity) s. K. Tanaka, R. Shiraishi, Green Chem. *2001*, 135-6.

Sodium hydroxide/(S)-2-amino-2'-hydroxy-1,1'-binaphthyl or   ←
polymer-based N-benzylcinchonidinium chloride
Asym. C-α-alkylation of α-(alkylideneamino)carboxylic acid esters   H → R
under phase transfer catalysis s. *54*, 394s58; with benzyl or allyl halides using polymer-based N-benzylcinchonidinium chloride as recyclable phase transfer catalyst s. R. Chinchilla, C. Najera et al., Tetrahedron:Asym. *11*, 3277-81 (2000); with (S)-2-amino-2'-hydroxy-1,1'-binaphthyl as phase transfer catalyst cf. Y.N. Belokon, H.B. Kagan et al., ibid. *10*, 1723-8 (1999).

Sodium hydroxide/tetra-n-hexylammonium chloride $NaOH/(n\text{-}C_6H_{13})_4NCl$
Δ^2-Pyrazolines from 1,1-halogenohydrazones and ethylene derivs.
1,3-Dipolar cycloaddition in aq. medium

363.

under micellar catalysis. A *heterogeneous* mixture of startg. halogenohydrazone, 4 eqs. ethyl acrylate, and 0.1 eq. tetra-*n*-hexylammonium chloride in 0.1 M aq. NaOH stirred at room temp. for 0.75 h → 1-(4-methoxyphenyl)-3-methoxycarbonyl-5-ethoxycarbonyl-4,5-dihydropyrazole. Y 100%. The procedure is inexpensive and environmentally favourable in that no (or little) organic solvent is required. Strong mechanical shaking enhanced reaction in certain instances, as also did the inclusion of an organic solvent under homogeneous conditions. Carboxylic acid esters, cyclic ethers, nitriles and alcohols remained unaffected. F.e.s. G. Molteni et al., J. Chem. Soc. Perkin Trans. 1 *2000*, 3742-5; with $KHCO_3$ and a little Et_3N in water, and further base combinations, s. G. Schlegel, German patent DE-19739489 (Hoechst-Schering-Agrevo Gmbh).

Sodium hydroxide/benzyltriethylammonium chloride $NaOH/BnEt_3NCl$
Synthesis of 1-chloro-3,3-difluorocyclopropenes from difluoromethylene compds.

364.

1-Aryl-2-chloro-3,3-difluorocyclopropenes. A mixture of 2′,2′-difluorostyrene, chloroform (1 part), 40% aq. NaOH (1 part), and a little benzyltriethylammonium chloride stirred at room temp. for 8 h, and the mixture poured into water → 1-phenyl-2-chloro-3,3-difluorocycloprop-1-ene. Y 55%. F.e.s. C.-C. Lee, S.-T. Lin, Synthesis *2000*, 496-8.

Potassium hydroxide/chiral spirocyclic quaternary ammonium salts   ←
Asym. C-α-alkylation under phase transfer catalysis   H → R
s. *58*, 353; rate enhancement *under ultrasonication* s. T. Ooi, K. Maruoka et al., Synlett *2000*, 1500-2.

Potassium tert-butoxide/irradiation   KOBu-t/*ℋℋ*
Indoles from *o*-halogenamines and ketones
s. *36*, 795s*40*; 2-[het]arylindoles, also with KOBu-*t*/FeBr$_2$, s. M.T. Baumgartner, R.A. Rossi et al., Synthesis *1999*, 2053-6.

Organolithium compd./cuprous cyanide   RLi/CuCN
Cross-coupling with lithium divinylcuprates   Hal → R
s. *33*, 806; S-chiral β,γ-ethylenesulfoximines s. S. Bosshammer, H.-J. Gais, Synlett *1998*, 99-100.

Hydrocarbon groups from halides
s. *23*, 831s*32*; chiral N-ethyl- from N-chloromethyl-5-oxazolidones with 1:1 MeLi/CuCN s. H. Schedel, K. Burger, Monatsh. Chem. *131*, 1011-8 (2000).

n-Butyllithium   BuLi
Asym. synthesis of α-aminocarboxylic acid esters
via 2-alkylation of 3,6-dialkoxy-2,5-dihydropyrazines
s. *38*, 802; chiral heteroarylalanines s. P. Dalla Croce et al., Tetrahedron:Asym. *11*, 2635-42 (2000); asym. 2-propargylation s. C. Ma, J.M. Cook et al., Tetrahedron Lett. *41*, 2781-5 (2000); benzocyclobutene analogs of phenylalanine s. S. Kotha et al., Bioorg. Med. Chem. Lett. *9*, 2565-8 (1999).

n-*Butyllithium/potassium* tert-*butoxide*   BuLi/KOBu-t
Styrenes from benzyl halides and 1-alkylbenzotriazoles s. *60*, 352   ArCH$_2$Hal → ArCH=C

n-*Butyllithium/zinc chloride*   BuLi/ZnCl$_2$
Functionalized α-halogenoboronic from boronic acid esters   CH(Hal)B(OR)$_2$
Asym. synthesis with insertion of one C-atom

365.

Chiral α-halogeno-β-siloxyboronic acid esters. A soln. of startg. chiral silyloxyboronic acid ester in THF added via cannula to a soln. of 1.16 eqs. (dichloromethyl)lithium (prepared from *n*-BuLi and dichloromethane) in the same solvent at -100° under argon, 2 eqs. anhydrous ZnCl$_2$ added, the mixture allowed to warm to 20-25°, and stirred for 24 h → product. Y 97%. Silyl ethers and azido groups serve as masked hydroxyl and amino groups, respectively, which normally interfere with this homologization as a result of binding with the metal cations. F.e. and chiral β-azido-α-bromoboronic acid esters s. R.P. Singh, D.S. Matteson, J. Org. Chem. *65*, 6650-3 (2000).

n-*Butyllithium/ytterbium(III) triflate*   BuLi/Yb(OTf)$_3$
1-Aryllactones from ketocarboxylic acid esters and ar. halides

366.

1 eq. *n*-BuLi (1.6 *M* in hexanes) added dropwise at -100° to a soln. of 1-bromo-2-(2-chloroethyl)-benzene in THF, the mixture stirred 10 min at the same temp. then added via a dry ice-cooled cannula to a soln. of 1 eq. anhydrous Yb(OTf)$_3$ in THF at -78°, stirred an additional 30 min at the same temp., a soln. of ethyl levulinate in THF added via cannula, the resulting mixture allowed to warm to room temp. over 3 h, stirred an additional 2 h, then quenched with satd. aq. NaHCO$_3$ → 5-[2-(2-chloroethyl)phenyl]-5-methyldihydrofuran-2-one. Y 70%. Reaction proceeds by selective metal-halogen exchange at the sp^2-hybridized bromide, followed by *in situ*-generation of an organoytterbium compd. which undergoes selective addition to the keto group. Yields were poor in the absence of Yb(OTf)$_3$ as deprotonation of the keto ester complicated the ring closure. Synthesis via organomagnesium compds. gave a very low yield, while organocerium derivs. were only moderately successful. F.e. and **1-vinyllactones**, also subsequent reductive ring closure to bi- and tri-cyclic lactols with SmI$_2$, s. G.A. Molander, C Köllner, J. Org. Chem. *65*, 8333-9 (2000).

n-*Butyllithium/hexamethylphosphoramide*   BuLi/HMPA
Metalation of allenes   ←
with *n*-BuLi s. *31*, 806; α-metalation of N-allenyl-2-oxazolidones and -2-imidazolones with added HMPA (1 eq.), also further bases, s. H. Xiong, R.P. Hsung et al., Org. Lett. *2*, 2869-71 (2000).

tert-*Butyllithium*   t-BuLi
Cyclopentadienones from 1,4-diiodo-1,3-dienes

367.

4 eqs. *t*-BuLi (1.6 *M* in pentane) added to a soln. of the startg. diiodo compd. in THF at -78° under N$_2$, the mixture stirred for 1 h at the same temp., CO$_2$ bubbled into the soln. for 1 min, stirring

continued for 5 min at -30° under a slight positive pressure of CO_2, and quenched with 3 N HCl → 2,5-bis(trimethylsilyl)-3,4-dibutylcyclopentadienone. Y 57%. This one-pot synthesis is fast and efficient with good yields. F.e.s. Z. Xi, Q. Song, J. Org. Chem. 65, 9157-9 (2000).

Lithium diisopropylamide   i-Pr_2NLi
Asym. C-α-alkylation of carboxylic acid esters   H → R
s. 39, 754; asym. synthesis of α-acylaminocarboxylic acid esters s. J.M. McIntosh, S. Peters, Synthesis 1999, 635-8.

2-Alkylation of succinic acid monoesters via O,C-dianions
with $LiNH_2$ in liq. NH_3 cf. 27, 840; with LDA, also conversion to 2-alkylated succinic acids and succinic acid anhydrides s. S.C. Bergmeier, K.A. Ismail, Synthesis 2000, 1369-71.

Asym. α-alkylation of hydrazones
with t-BuLi cf. 31, 812s48; aldehyde hydrazones with LDA s. Z. Yang, J. Meinwald et al., Synthesis 2000, 1936-43; stereospecific 4-alkylation of Δ^2-pyrazolines s. Y.R. Huang, J.A. Katzenellenbogen, Org. Lett. 2, 2833-6 (2000).

Asym. C-α-alkylation of 1-acyl-2-imidazolones
of restricted substrates cf. 44, 776s54; unusual enhancement of enantioselectivity, notably with bulky 1-acyl-3-arenesulfonyl-2-imidazolidones having a tert-butyl or adamantyl group at the 5-position, s. A.A.-M. Abdel-Aziz, T. Kunieda et al., Tetrahedron Lett. 41, 8533-7 (2000).

Asym. α-lateral alkylation of Δ^2-oxazolines
s. 32, 637s34 (review); 53, 399; of (1S)-1-amino-2-exo-hydroxyapocamphane-based Δ^2-oxazolines s. S. Chandrasekhar, A. Kausar, Tetrahedron:Asym. 11, 2249-53 (2000).

Asym. synthesis of β-aminocarboxylic acids
via 5-alkylation of 2,3,5,6-tetrahydro-4(1H)-pyrimidinones
s. 48, 802; chiral α,α-disubst. β-aminocarboxylic acids via sequential asym. dialkylation of 1-benzoyl-2(S)-tert-butyl-3-methyl-2,3,5,6-tetrahydro-4(1H)-pyrimidinone s. E. Juaristi et al., Tetrahedron:Asym. 9, 3881-8 (1998); asym. 2-alkylation of 2-benzyl-4-isopropyl-2,4-dihydro-1H-pyrazino[2,1-b]quinazoline-3,6-dione s. F.L. Buenadicha, C. Avendaño et al., ibid. 9, 4275-84 (1998).

α-prim-Aminocarboxylic acid esters from halides with two extra C-atoms
via alkylation of N,N-bis(trimethylsilyl)glycine ethyl ester cf. 24, 852; via alkylation of 2-α-carbalkoxy-1H-naphtho[1,8-de]-1,2,3-triazin-2-ium N-ylids (with LDA) s. R. Anilkumar, S. Chandrasekhar et al., Tetrahedron Lett. 41, 6665-8 (2000).

Lithium diisopropylamide/lithium chloride   i-$Pr_2NLi/LiCl$
Asym. Darzens-Claisen condensation
glycidic acid esters with KOBu-t cf. 52, 389; α,β-epoxyketones from aldehydes with asym. induction using a camphor auxiliary (with LDA/LiCl) s. C. Palomo et al., J. Org. Chem. 65, 9007-12 (2000).

Lithium diisopropylamide/hexamethylphosphoramide   HMPA/i-Pr_2NLi
α-Alkylation of α-halogenoketimines   H → R
s. 54, 96; with azidoiodides (with added HMPA) s. W. Aelterman, N. De Kimpe et al., J. Org. Chem. 66, 53-8 (2001).

Lithium N,N,N-trimethyl-2-aminoethylamide/potassium carbonate   ←
Benzo[b]thiophene-2-carbonyl compds. from ar. aldehydes   O

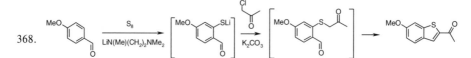

1.05 eqs. 2.5 M n-BuLi in hexanes added to a soln. of 1.1 eqs. N,N,N'-trimethylethylenediamine in anhydrous THF at -20° under argon, the mixture stirred for 15 min, startg. benzaldehyde added, stirring continued at the same temp. for 24 h, cooled to -40°, treated with 2.6 eqs. S_8, after stirring at -20° for 3 h the intermediate thiolate treated with 2 eqs. chloroacetone and K_2CO_3, and shaken for 24 h → product. Y 42% overall. The method is simple, general, direct, and rapid, and may be carried out in one pot. F.e. and isolation of the intermediate *o*-mercaptoaldehydes s. T. Gallagher, R.A. Porter et al., Tetrahedron Lett. *41,* 5415-8 (2000).

Lithium bis(trimethylsilyl)amide $LiN(SiMe_3)_2$
Asym. α-alkylation of conformationally rigid 3-acyl-2-thiazolidones   H → R

369.

1.05 eqs. LiHMDS (prepared *in situ* from hexamethyldisilazane and BuLi) added slowly to (-)-5-butyryl-1,7,8,9,10,10-hexamethyl-3-thia-5-azatricyclo[5.2.1.0^{2,6}]dec-8-en-4-one in THF at -50° under argon, allowed to stand for 1 h, treated with methyl iodide, and quenched with satd. aq. NH_4Cl → product. Y 91% (diastereoselectivity 118:1). High asym. induction was recorded, even with methyl iodide. F.e.s. S. Hoshimoto, T. Kunieda et al., Chem. Pharm. Bull. *48,* 1541-4 (2000).

Lithium bis(trimethylsilyl)amide/tris(dibenzylideneacetone)dipalladium/   ←
   1,1'-bis(diphenylphosphino)ferrocene
Synthesis of 1-vinylisochromenes from *o*-bromocinnamyloxysilanes and methyl ketones ○
via α-arylation-intramolecular O-allylation

370.

A soln. of 2 eqs. pinacolone in dioxane/THF added slowly to a soln. of 3 eqs. 1 *M* LiHMDS in THF at 5°, a soln. of 5 mol% Pd$_2$(dba)$_3$ and 10 mol% dppf in dioxane/THF added at room temp., followed by a soln. of *tert*-butyldimethyl[3-(2-bromophenyl)allyloxy]silane, the mixture heated to 100° overnight, then quenched with 1 *M* HCl at room temp. → 3-*tert*-butyl-1-vinyl-1*H*-isochromene. Y 71%. The choice of trialkylsilyl as O-protective group was crucial to the success of the tandem reaction, the corresponding acetate giving a complex mixture with pinacolone or the deacylated alcohol with acetophenone. A lithium base and a coordinating solvent are also required. F.e.s. R. Mutter et al., Chem. Commun. *2000,* 1675-6.

Sodium bis(trimethylsilyl)amide $NaN(SiMe_3)_2$
Asym. C-α-alkylation of 3-acyl-2-oxazolidones   H → R
with $LiN(SiMe_3)_2$/LiCl cf. *44,* 776s56; with SuperQuat, (S)-4-benzyl-5,5-dimethyl-2-oxazolidone, as auxiliary, also conversion to **chiral aldehydes**, s. S.D. Bull, S.G. Davies et al., Tetrahedron:Asym. *11,* 3475-9 (2000); asym. C-α-allylation of 3-(alkoxyacetyl)-2-oxazolidones with $NaN(SiMe_3)_2$, and conversion to **chiral 4-ene-1,2-diol 2-monoethers**, s. M.T. Crimmins et al., Org. Lett. *2,* 2165-7 (2000).

Potassium amide/liq. ammonia/irradiation $KNH_2/NH_3/\mathit{HH}$
Potassium carbonate/irradiation K_2CO_3/HH
S$_{RN}$1-Photoarylation   H → Ar
of ketones cf. *32,* 800; α- and *p*-arylation of arylacetic acids via O,C-dianions with KNH_2 in liq. NH_3 (cf. *15,* 629) s. G.C. Nwokogu, J.F. Wolfe et al., Org. Lett. *2,* 2643-6 (2000); photoarylation of 5-membered heteroarenes with K_2CO_3 in acetonitrile s. B. Guizzardi, A. Albini et al., Tetrahedron *56,* 9383-9 (2000).

Potassium carbonate/tetra-n-butylammonium bromide/microwaves ←
Benzofurans from *o*-hydroxyaldehydes under microwave irradiation without solvent ○
2-acylbenzofurans from α-tosyloxymethyl ketones cf. *56*, 334; benzofuran-2-carboxylic from chloroacetic acid esters under phase transfer catalysis with K_2CO_3/Bu_4NBr s. D. Bogdal, M. Warzala, Tetrahedron *56*, 8769-73 (2000).

Cesium carbonate or Potassium carbonate Cs_2CO_3 or K_2CO_3
3-Amino-1,2-diacylindoles from *o*-acylaminonitriles and α-bromoketones

371.

1.5 eqs. Startg. bromoketone in DMF and ca. 1.7 eqs. Cs_2CO_3 added to a soln. of startg. benzonitrile in DMF, and the mixture stirred at room temp. overnight → product. Y 63% (>99% purity). Tedious work-up was avoided by simple filtration to remove the base. The method is simple, general and amenable to building combinatorial libraries in the solution phase using parallel filtration and automated reverse phase HPLC (205 compds. being prepared in acceptable yield on the multi-milligram scale out of 280 reactions). F.e.s. M. Nettekoven, Tetrahedron Lett. *41*, 8251-4 (2000); N-carbalkoxy-derivs. with K_2CO_3 cf. S. Radl, P. Vachal et al., J. Heterocycl. Chem. *37*, 855-62 (2000).

Sodium acetate NaOAc
Pyrrole ring from α-halogenaldehydes

372.

Pyrrolo[2,3-*d*]pyrimidines. A mixture of startg. bromaldehyde, 1 eq. 2,4-diamino-6-hydroxy-pyrimidine, and 2 eqs. Na-acetate in 1:1 acetonitrile/water warmed to 40°, and the homogeneous soln. stirred for 1-2 h → product. Y 80%. The cyclocondensation was sensitive to the O-protecting group employed. F.e.s. C.J. Barnett, L.M. Grubb, Tetrahedron Lett. *41*, 9741-5 (2000); **from α-halogenoketones or enolethers** cf. European patent EP-89720 (Eli Lilly & Co.).

Lithium chloride s. under i-Pr_2NLi LiCl

Triethylamine Et_3N
2-Azetidinones from azomethines ○
s. *7*, 836s*55*, *56*; *trans*-4-acylamino-2-azetidinones from N-acylamidines s. K. Thiagarajan, B.M. Bhawal et al., Tetrahedron *56*, 7811-6 (2000); *trans*-4-aryl-3-oxy- or -3-phthalimido-2-azetidinones from polycyclic N-arylaldimines s. B.K. Banik, F.F. Becker, Tetrahedron Lett. *41*, 6551-4 (2000); functionalized α-(2-azetidinon-1-yl)carboxylic acid esters with asym. induction s. S. Mignani, L. Stella et al., Synth. Commun. *30*, 3685-91 (2000); chiral 3-acoxy-4-aryl-2-azetidinones from chiral $Cr(CO)_3$-complexed ar. aldimines s. P. Del Buttero et al., Tetrahedron:Asym. *11*, 1927-41 (2000).

Triethylamine/potassium hydrogen carbonate $Et_3N/KHCO_3$
Δ²-Pyrazolines from 1,1-halogenohydrazones and ethylene derivs. ○
in an aq.-organic 2-phase medium s. *60*, 363

Triethylenediamine
Baylis-Hillman-type reaction with β,γ-ethylenehalides *dabco*
C≡C-C-C≡C

373.
$$NC-C(=O)-H \;+\; Ph-CH(Br)-CH_2-CO_2Me \longrightarrow NC-C(Ph)(H)-C(=CH_2)-CO_2Me$$
(with second CN=C group)

2-Cyano-1,4-dienes. A soln. of startg. allyl halide (itself obtainable via the Baylis-Hillman reaction) and 2 eqs. dabco allowed to react in acrylonitrile at room temp. for 7 days → 4-cyano-2-methoxycarbonyl-3-phenylpenta-1,4-diene. Y 67%. This is the first example of the use of an allyl halide as electrophile under Baylis-Hillman reaction conditions. The procedure was also applicable to α-(chloromethyl)-α,β-ethyleneketones, but reaction with methyl (2Z)-2-(bromomethyl)hex-2-enoate failed. F.e.s. D. Basavaiah et al., Tetrahedron Lett. *42*, 85-7 (2001).

1,8-Diazabicyclo[5.4.0]undec-7-ene *DBU*
Asym. C-α-alkylation of α-(alkylideneamino)carboxylic acid derivs. H → R
with DBU/LiClO₄ cf. *27*, 843s*57*; of 1-(N-alkylideneglycyl)-2-imidazolidones with DBU or Schwesinger's base (BEMP) s. G. Guillena, C. Najera, Tetrahedron:Asym. *9*, 3935-8 (1998).

1,8-Diazabicyclo[5.4.0]undec-7-ene/cuprous iodide *DBU/CuI*
Furans from activated ketones via α-propargylation ○
One-pot conversion

374.
NC-CH₂-C(=O)-Ph + Br-CH₂-C≡CH → [NC-CH(C(=O)Ph)-CH₂-C≡CH → furanone intermediate] → 2-Ph-5-Me-3-cyanofuran

2,5-Disubst. 3-cyanofurans. A soln. of startg. ketone *in toluene* treated with 1 eq. propargyl bromide, 2 eqs. DBU, and 0.1 eq. CuI, and heated at 90° under N₂ for 3 h → product. Y 55%. This one-pot procedure is simpler and more efficient than the previous stepwise methods. Other catalysts (ZnCl₂, RuCl₃ and ZnCO₃) were less effective, and poor results were obtained in water or without solvent. F.e. incl. **3-acylfurans** and furan-3-carboxylic acid esters s. A. Arcadi et al., Tetrahedron Lett. *41*, 9195-8 (2000).

Quinidine/ethyldiisopropylamine ←
β-Lactones from aldehydes via asym. [2+2]-cycloaddition ⌐ₒ
with ketene using Me₃Al/(S)-3,3'-bis(triphenylsilyl)-1,1'-bi-2-naphthol cf. *50*, 397; chiral γ,γ-dichloro-β-lactones from acetyl chloride with quinidine/*i*-Pr₂NEt s. R. Tennyson, D. Romo, J. Org. Chem. *65*, 7248-52 (2000).

Tetrakis(acetonitrile)copper(I) fluoroborate s. under R₂Zn	[Cu(MeCN)₄]BF₄
Cuprous cyanide s. under RLi	CuCN
Cupric trifluoromethanesulfonate	Cu(OTf)₂
Friedel-Crafts acylation s. *60*, 327	H → COR
Cupric sulfate/alumina s. under PdCl₂(PPh₃)₂	CuSO₄/Al₂O₃
Cuprous bromide s. under Zn	CuBr
Cuprous iodide s. under DBU, Pd₂(dba)₃, and PdCl₂(PPh₃)₂	CuI
Magnesium	Mg

Asym. synthesis with α-halogenoboronic acid esters ←
review s. *52*, 390s*57*; chiral α-(fluoroaryl)boronic acid esters and conversion to the corresponding sec. fluorobenzylalcohols s. R.P. Singh, J.M. Shreeve et al., J. Org. Chem. *65*, 8123-5 (2000).

Bartoli indole synthesis from ar. nitro compds. ○
s. *44*, 786; scope and variations s. A.P. Dobbs et al., Synlett *1999*, 1594-6.

Magnesium/calcium bis(tetrahydridoborate) Mg/Ca(BH₄)₂
4-Ethylenealcohols from carboxylic acid esters COOR → CH(OH)CH₂CH₂CH=CH₂
Synthesis by condensing two vinylmagnesium bromide molecules s. *60*, 328

Magnesium/manganese(II) chloride
trans-1,2-Diarylbenzocyclobutenes from o-bis(dibromomethyl)arenes Mg/MnCl$_2$

A soln. of 1,2-bis(dibromomethyl)benzene in THF added to a soln. of Ph$_4$Mn(MgBr)$_2$ (generated from 2.2 eqs. MnCl$_2$ and 8.8 eqs. phenylmagnesium bromide) at 0° under argon, stirred for 10 min, and quenched with water → 1,2-diphenylbenzocyclobutene. Y 60% (93% *trans*-selectivity). With less manganate or MnCl$_2$ the yield decreased and the *trans:cis* ratio was inferior. Reaction is thought to involve sequential 1,2-Mn→C-aryl migration, followed by formation of a manganacyclic prior to reductive elimination of Mn(0). F.e. and with Ph$_3$MnMgBr, **also enesilanes** from dibromomethylsilanes, s. H. Kakiya, K. Oshima et al., Bull. Chem. Soc. Jpn. *73*, 2139-47 (2000).

Magnesium/dichlorobis(triphenylphosphine)nickel(II) Mg/NiCl$_2$(PPh$_3$)$_2$
N,N-Disubst. carboxylic acid amides from carbamyl chlorides ClCON< → RCON<

n-BuMgCl + ClCONEt$_2$ ⟶ n-BuCONEt$_2$

A soln. of N,N-diethylcarbamyl chloride and 5 mol% NiCl$_2$(PPh$_3$)$_2$ in THF stirred for a few min under N$_2$, 1.1 eqs. 2 M n-BuMgCl in the same solvent carefully added dropwise at room temp., and quenched after 2 h with satd. NH$_4$Cl soln. → product. Y 68%. The procedure is rapid, generally applicable to alkyl- and aryl-magnesium halides, adaptable on the large scale, and avoids the use of carbon monoxide. F.e. and with Bu$_3$P via [crystalline] **carbamylphosphonium salts** (notably for reaction with branched Grignards) s. L. Lemoucheux, J. Rouden et al., Tetrahedron Lett. *41*, 9997-10001 (2000).

Zinc (s.a. under NiCl$_2$(PPh$_3$)$_2$ and NiBr$_2$(PPh$_3$)$_2$) Zn
Sym. β-diketones by methylene insertion with bis(iodozincio)methane RCOCH$_2$COR
from acylcyanides cf. *55*, 471; from acyl chlorides in THF/tetrahydrothiophene s. S. Matsubara et al., Synlett *1999*, 1471-3.

Zinc/cuprous bromide Zn/CuBr
Synthesis of α,α-difluorophosphonic acid esters ←
s. *49*, 805s*53*; β,γ-acetylene-α,α-difluorophosphonic acid esters from α,β-acetylenebromides s. X. Zhang, D.J. Burton, Tetrahedron Lett. *41*, 7791-4 (2000).

Zinc/aluminum chloride Zn/AlCl$_3$
β,γ-Ethyleneketones from nitriles and β,γ-ethylenebromides RCN → RCOC-C=C
with Zn-Ag cf. *36*, 838; with Zn and 0.4 eq. AlCl$_3$ s. A.S.-Y. Lee, L.-S. Lin, Tetrahedron Lett. *41*, 8803-6 (2000).

Zinc/trimethylsilyl chloride/lead(II) chloride Zn/Me$_3$SiCl/PbCl$_2$
Syntheses via reductive ring opening of ω-iodo-ω-deoxyglycosides C
with subsequent Barbier-type allylation cf. *58*, 371; chiral dienepolyols by subsequent methylenation (with ICH$_2$ZnI) in the presence of Me$_3$SiCl/PbCl$_2$, or by vinylation with divinylzincs, s. L. Hyldtoft, R. Madsen, J. Am. Chem. Soc. *122*, 8444-52 (2000).

Zinc/bis(dibenzylideneacetone)palladium(0)/3-n-butyl-1-methyl- ←
2-(diphenylphosphino)imidazolium hexafluorophosphate/
1-n-butyl-2,3-dimethylimidazolium fluoroborate
Negishi diaryl synthesis Ar-Ar'
s. *38*, 836s*59*; **in ionic liquids** (e.g. 1-butyl-2,3-dimethylimidazolium fluoroborate) in the presence of 3-butyl-1-methyl-2-(diphenylphosphino)imidazolium hexafluorophosphate as ligand, also cross-coupling with benzylzinc halides s. J. Sirieix, P. Knochel et al., Synlett *2000*, 1613-5.

Ethylmagnesium bromide s. under Ti(OPr-i)$_4$ EtMgBr

Isopropylmagnesium bromide
Syntheses via *o*-(chloromethyl)arylmagnesium bromides

i-PrMgBr
←

377.

Phthalans from aldehydes. 1.1 eqs. *i*-PrMgBr in THF added dropwise to a stirred soln. of 2-chloromethyl-1-iodobenzene in the same solvent at -10° under argon, the mixture stirred for 1.5 h, 1.5 eqs. startg. aldehyde added, allowed to warm slowly to room temp., refluxed for 12 h, and quenched with brine → product. Y 91%. F.e., **also phthalimidines from isocyanates,** and cross-coupling with an allyl halide (followed by conversion to 2,3,4,5-tetrahydro-1*H*-2-benzazepines) s. T. Delacroix, P. Knochel et al., J. Org. Chem. *65*, 8108-10 (2000).

Lithium trialkylmagnesiates/titanium tetrachloride
Sym. diaryls from ar. halides
Oxidative dimerization under mild conditions

$R_3MgLi/TiCl_4$
2 ArHal → Ar-Ar

378.

1 eq. 1.6 M BuLi in hexane added at 0° to a soln. of 0.5 eq. 1 M BuMgBr *in THF*, stirred for 10 min, cooled to -40°, a soln. of 4-bromo-N,N-diethylbenzamide added dropwise, stirred for 0.5 h, 1.5 eqs. TiCl$_4$ added, warmed to 0° gradually, and poured into satd. aq. NH$_4$Cl → N,N,N',N'-tetraethylbiphenyl-4,4'-dicarboxamide. Y 65%. Reaction is thought to involve *in situ* formation of an aryltitanium species, and is applicable to both ar. bromides and iodides, leaving esters, amides and nitriles unaffected. F.e. and intramolecular coupling s. A. Inoue, K. Oshima et al., Tetrahedron *56*, 9601-5 (2000).

Dialkylzinc/tetrakis(acetonitrile)copper(I) fluoroborate/
(S)-3,3'-bis(methylthio)-1,1'-bi-2-naphthol
Regiospecific synthesis of α-methylenecarboxylic
from (Z)-α,β-ethylene-α-(halogenomethyl)carboxylic acid esters

←

$C=CCH_2Hal → C(R)C=CH_2$

379.

Asym. synthesis. 10 Mol% [Cu(MeCN)$_4$]BF$_4$ and 20 mol% (S)-3,3'-bis(methylthio)-BINOL in dry THF stirred at -20° in the presence of 0.19 eq. Et$_2$Zn, solns. of startg. allyl chloride in THF and 1.5 eqs. 1 M Et$_2$Zn in hexane added by syringe pump over 20 min, stirred for a further 20 min at -20°, and quenched with 2 M aq. HCl → (-)-ethyl 2-methylene-3-(4-nitrophenyl)pentanoate. Y 76% (e.e. 60%). Enantioselectivity was low (or zero) from the corresponding bromide or mesylate. F.e. and racemic products without the chiral auxiliary s. C. Börner, S. Woodward et al., Chem. Commun. *2000*, 2433-4.

Diethylzinc Et_2Zn
Simmons-Smith reaction with asym. induction
s. *41*, 797s56,59; chiral cyclopropylboronic acid esters s. J. Pietruszka, A. Witt, J. Chem. Soc. Perkin Trans. 1 *2000*, 4293-300.

Diethylzinc/2,4,6-trichlorophenol $Et_2Zn/ArOH$
Simmons-Smith-type reaction with iodomethylzinc phenoxides

380.

Readily prepared, stable iodomethylzinc phenoxides are useful, highly reactive alternatives to traditional zinc carbenoids for cyclopropanation, notably of *unfunctionalized* ethylene derivs. E: 2 eqs. Diethylzinc added to a soln. of 2 eqs. 2,4,6-trichlorophenol in dichloromethane at -40°, the mixture stirred for 15 min, 2 eqs. diiodomethane added, stirring continued for 15 min, the startg. alkene added, the cooling bath removed, and the mixture stirred for 12 h at room temp. → product. Y 98% (>95% conversion). These novel reagents self-destruct unless two *ortho*-substituents are present, and reactivity is enhanced with an electron-withdrawing *para*-substituent. The phenolic by-product is retrievable. F.e.s. A.B. Charette et al., Angew. Chem. Int. Ed. Engl. *39*, 4539-42 (2000).

Dilithium tetramethylzincate/ethyl dichlorovanadate $Me_4ZnLi_2/VO(OEt)Cl_2$
Methylarenes from ar. bromides via dilithium aryl(trimethyl)zincates   Br → Me

381.

1.5 eqs. Me_4ZnLi_2 added under argon to a stirred soln. of 2-bromobiphenyl in THF at room temp., stirring continued at 0° for 2 h, 3 eqs. $VO(OEt)Cl_2$ added, stirred for a further 15 h at room temp., then quenched with aq. 1.5 M HCl → 2-phenyltoluene. Y 99%. Methylation also takes place with *o*-subst. ar. bromides having methoxy, methylthio or cyano groups, as well as with 2-bromonaphthalene and 9-bromoanthracene. Reaction is thought to involve either single-electron oxidation between the arylzincate and the vanadium reagent or transmetalation. F.e. and sequential replacement of the bromine by methyl in 1,1-dibromo-2-phenylcyclopropane under the same conditions s. T. Takada, T. Hirao et al., J. Org. Chem. *66*, 300-2 (2001).

Zinc cyanide s. under Pd(PPh$_3$)$_4$ $Zn(CN)_2$
Zinc chloride s. under BuLi $ZnCl_2$

Indium   *In*
2-α-Hydroxy-1,3-dienes from aldehydes   CHO → $CH(OH)C(=CH_2)CH=CH_2$
Synthesis with addition of four C-atoms in aq. medium
from diisopropyl 2,3-butadien-2-ylboronate cf. *53*, 442; from 1,4-dibromo-2-butyne in water with In s. W. Lu, T.H. Chan et al., Org. Lett. *2*, 3469-71 (2000).

Triethylborane Et_3B
β-Alkoxylamino-γ-lactones from α-acryloyloxyalkoximes
Stereospecific radical 1,4-addition-ring closure

382.

5 eqs. Triethylborane in hexane added to a refluxing soln. of startg. ethyleneoxime and 6 eqs. ethyl iodide in toluene, and the mixture stirred at the same temp. for 15 min → product. Y 64% (12:1 mixture of diastereomers). Cleavage of the lactone ring affords the corresponding **chiral β-aminocarboxylic acid derivs.** so that the overall conversion is effectively an alternative to Mannich-type reactions. F.e.s. H. Miyabe, T. Naito et al., Org. Lett. 2, 4071-4 (2000).

Pinacolborane s. under $Pd(OAc)_2$
Triisopropyl borate s. under $Pd(PPh_3)_4$ $(i\text{-}PrO)_3B$

Indium(III) triflate/microwaves
N-(α,β-Ethyleneacyl)ation-intramolecular Diels-Alder reaction
with furfurylamines cf. *59*, 335; rapid N-allylation-intramolecular Diels-Alder reaction with 10 mol% recoverable, water-tolerant $In(OTf)_3$ under microwave irradiation without solvent s. D. Prajapati et al., Tetrahedron Lett. *41*, 8639-43 (2000).

Aluminum chloride s. under Zn $AlCl_3$

Ytterbium(III) triflate (s.a. under BuLi) $Yb(OTf)_3$
Ytterbium(III)-catalyzed ar. thioalkylation $H \to CH(SR)COOR'$

383.

under mild conditions. A mixture of 5 mol% $Yb(OTf)_3$, *o*-allyloxychlorobenzene, and ethyl α-chloro-α-(ethylthio)acetate in distilled nitromethane stirred at room temp. for 5 h, then filtered through Celite and subjected to chromatographic separation → product. Y 85%. In this milder version of the Tamura synthesis (*30*, 599s*36*), leading to arylacetic acid esters, a variety of functional groups (e.g. amides, alkoxy, allyloxy, ester, halide and phenols) are tolerated. F.e. incl. alkoxyaryl and furyl derivs., also desulfurization to **arylacetic acid esters**, s. S. Sinha, S. Chandrasekaran et al., Tetrahedron Lett. *41*, 9109-12 (2000).

1-Butyl-3-methylimidazolium hexafluorophosphate s. under $NiCl_2(PPh_3)_2$

Tris(trimethylsilyl)silane/azodiisobutyronitrile $(Me_3Si)_3SiH/AIBN$
Radical arylation $H \to Ar$
under irradiation cf. *37*, 765s*40*; also hetarylation with $(Me_3Si)_3SiH/AIBN$ s. V. Martinez-Barrasa, C. Burgos et al., Org. Lett. 2, 3933-5 (2000).

1,1,3,3-Tetramethyldisilazane s. under Platinum complexes $(Me_2SiH)_2NH$
Titanium tetraisopropoxide s.a. under $Pd_2(dba)_3$ $Ti(OPr\text{-}i)_4$

Titanium tetraisopropoxide/ethylmagnesium bromide $Ti(OPr\text{-}i)_4/EtMgBr$
1-ω-Hydroxycyclopropanols from lactones via modified Kulinkovich reaction

384.

2.25 eqs. EtMgBr added dropwise over 2 h to a soln. of the startg. lactone and 0.2 eq. Ti(OPr-*i*)$_4$ in dry THF under N$_2$ while keeping the temp. below 20°, the mixture stirred for 2 h, then quenched with satd. aq. NH$_4$Cl → 1-(2-hydroxypropyl)cyclopropanol. Y 70%. The reaction was extended to γ- and δ-lactones. F.e. and by classical Kulinkovich reaction with oxy-functionalized esters s. A. Esposito, M. Taddei, J. Org. Chem. *65*, 9245-8 (2000).

Trimethylsilyl chloride s. under Zn Me_3SiCl

Titanium tetrachloride (s.a. under R$_3$MgLi) $TiCl_4$
Ar. thioalkylation   ArH → ArC-SR
with SnCl$_4$ cf. *30*, 599s*36*; **with asym. induction** using a chiral chloro(phenylthio)acetic acid ester (with TiCl$_4$) s. S. Madan, S.S. Bari et al., Tetrahedron:Asym. *11*, 2267-70 (2000).

Bis(tri-n-butyltin) oxide $(Bu_3Sn)_2O$
Generation of nitrile oxides from hydroximinohalides under neutral conditions   ←

385.

1,3-Dipolar cycloaddition. A mixture of 2 eqs. phenyl vinyl sulfide and 0.5 eq. bis(tri-*n*-butyl)-tin oxide in benzene added to a soln. of startg. chiral hydroximinochloride in the same solvent, and the mixture stirred for 2 h at room temp. → product. Y 59% (1:1 mixture of diastereo-isomers). Yields were generally higher than by the traditional route with Et$_3$N. F.e. and oxazoles from acetylene derivs. s. V.P. Sandanayaka, Y. Yang, Org. Lett. *2*, 3087-90 (2000).

Lead(II) chloride s. under Zn $PbCl_2$

tert-Butyl nitrite   *t-BuONO*
Radical synthesis of allylarenes from ar. amines   NH$_2$ → CH$_2$CH=CH$_2$

386.

3,5-Dinitroaniline added portionwise to a soln. of 1.7 eqs. *tert*-butyl nitrite and 15 eqs. allyl bromide in acetonitrile at 11-15° (a further 0.44 eq. *tert*-butyl nitrite being added with the second half of the aniline), and the mixture stirred at room temp. for 1 h → product. Y 93%. The process is rapid and applicable for large-scale manufacture. F.e.s. F. Ek, L.G. Wistrand, European patent EP-1013636 (Nycomed Imaging AS).

Tert. phosphines and di(phosphines) s. under Palladium complexes   ←
Ethyl dichlorovandate s. under Me$_4$ZnLi$_2$ $VO(OEt)Cl_2$

Perfluorobutanesulfonic acid $C_4F_9SO_3H$
Ketones from carboxylic acid chlorides   COCl → COR
aryl ketones from arenes with CF$_3$SO$_3$H cf. *28*, 827; predominant *o*-benzoylation of alkylbenzenes with perfluorobutanesulfonic acid s. F. Effenberger et al., Synthesis *2000*, 1427-30l.

Manganese(II) chloride s. under Mg $MnCl_2$

Cobalt carbonyl/potassium carbonate $Co_2(CO)_8/K_2CO_3$
**3-Azabicyclo[3.3.0]oct-1(8)-ene-2,7-diones
from α,β-acetylenechlorides and 2-ethyleneamines
via (α,β-acetylenecarboxylic acid N-allylamide)dicobalt hexacarbonyl complexes
Aminocarbonylation-intramolecular Pauson-Khand reaction**

387.

A soln. of startg. cobalt complex (prepared along with $Co_4(CO)_{10}Cl_2$-complexed 1,4-diphenyl-1,3-butadiyne by treatment of chloroethynylbenzene with $Co_2(CO)_8$ in toluene at 0-25°) mixed with a slight excess (1.1-1.3 eqs.) of startg. amine and 2.2 eqs. K_2CO_3, and the intermediate amide (isolable in 89% yield) heated at 45° for 4 h → product. Y 55% (d.r. 1.2:1). F.e. incl. an angular triquinane system, and intermolecular Pauson-Khand reaction s. J. Balsells, M.A. Pericàs et al., Org. Lett. *1*, 1981-4 (1999).

Dichlorobis(triphenylphosphine)nickel(II) s.a. under Mg $NiCl_2(PPh_3)_2$
Dichlorobis(triphenylphosphine)nickel(II)/triphenylphosphine/zinc/   ←
1-butyl-3-methylimidazolium hexafluorophosphate
Sym. diaryls from ar. halides   2 ArHal → Ar-Ar
with $Ni(CO)_2(PPh_3)_2$ cf. *27*, 870s57; with $NiCl_2(PPh_3)_2$/Zn/Ph_3P **in ionic liquids** (e.g. 1-butyl-3-methylimidazolium hexafluorophosphate) for efficient solvent and catalyst recycling s. J. Howarth et al., Tetrahedron Lett. *41*, 10319-21 (2000).

Dibromobis(triphenylphosphine)nickel(II)/triphenylphosphine/zinc/potassium tert-*butoxide*   ←
Nickel-catalyzed C-arylation of malononitriles   H → Ar

388.

Startg. aryl chloride and 1.1 eqs. potassiomalononitrile (prepared from 1.1 eqs. malononitrile and 2.2 eqs. *t*-BuOK) added to a soln. of $Ni(PPh_3)_3$ (generated *in situ* from 20 mol% $NiBr_2(PPh_3)_2$, 40 mol% PPh_3 and 60 mol% Zn in dry THF under N_2), the mixture heated with stirring to 60° for 48 h, then quenched with dil. HCl → α-(*p*-trifluoromethylphenyl)malononitrile. Y 85% by GC (selectivity 93%). The method is simple, efficient, inexpensive, highly selective, generally applicable to ar. chlorides, bromides and iodides, and does not require an electron-rich or bulky phosphine (as in palladium-catalyzed arylation). F.e.s. H.J. Cristau, M. Taillefer et al., Tetrahedron Lett. *41*, 8457-60 (2000).

Rhodium(II) acetate $Rh_2(OAc)_4$
Furan ring from 2-ketoiodonium ylids and acetylene derivs. s. *60*, 361

Chloro(1,5-cyclooctadiene)rhodium(I) dimer/potassium iodide/formic acid   ←
Rhodium-catalyzed carbonylation   Hal → COO(H,R)
arylacetic acid esters from formic acid esters under 1 atm. CO s. *41*, 819; also arylacetic acids in formic acid s. A. Giroux, Y. Han et al., Tetrahedron Lett. *41*, 7601-4 (2000).

Chloro(1,5-cyclooctadiene)rhodium(I) dimer/tri-n-propylphosphine/potassium carbonate ←
Aryl ketones from N-(2-pyrazinyl)aldimines and ar. iodides H → Ar
via rhodium-catalyzed Heck-type arylation

389.

Diaryl ketones. A mixture of 4-iodoanisole, 1.5 eqs. startg. aldimine, 2.5 mol% [RhCl(COD)]$_2$, 5 mol% n-Pr$_3$P, and 1.5 eqs. K$_2$CO$_3$ in diglyme stirred under N$_2$ at 160° for 50 h → product. Y quantitative. The yield of the corresponding diaryl ketone (obtained by hydrolysis with 6 N HCl at room temp. over 16 h) was 99%. Reaction is thought to involve 1,2-addition of an aryl rhodium(III) species to the imine (facilitated by coordination of pyrazinyl nitrogen to the metal) prior to β-hydride elimination and reductive elimination of hydrogen halide to regenerate the catalyst. Nickel and palladium catalysts were ineffective. Ar. bromides reacted more slowly, while ar. chlorides were unreactive. F.e. and with NaOBu-t in m-xylene s. T. Ishiyama, J. Hartwig, J. Am. Chem. Soc. *122*, 12043-4 (2000).

Palladous acetate s.a. under ↯ Pd(OAc)$_2$
Palladous acetate/pyridine/tetra-n-butylammonium chloride ←
Coumarins from o-iodophenols and acetylene derivs. O
Cyclocarbonylation

390.

Startg. iodophenol, 5 eqs. oct-4-yne, 2 eqs. pyridine, 1 eq. Bu$_4$NCl and 5 mol% Pd(OAc)$_2$ in DMF stirred under a CO balloon at 120° for 12 h → product. Y 66%. This method is mild, efficient, and based on readily available starting materials, tolerating carboxylic acid esters, ethers, ketones and silyl groups. Interestingly, insertion of the alkyne into the initially generated arylpalladium iodide takes place in preference to insertion of CO. Terminal alkynes failed to react. F.e.s. D.V. Kadnikov, R.C. Larock, Org. Lett. *2*, 3643-6 (2000).

Palladous acetate/pinacolborane/triethylamine ←
***in situ*-Suzuki coupling** Ar-Ar'
from two different aryl triflates and a tetraalkoxydiborane cf. *53*, 481; from two different o-subst. ar. halides with pinacolborane/Pd(OAc)$_2$/Et$_3$N s. O. Baudoin et al., J. Org. Chem. *65*, 9268-71 (2000).

Palladous acetate/triphenylphosphine/cesium carbonate Pd(OAc)$_2$/Ph$_3$P/Cs$_2$CO$_3$
Regiospecific Heck-type arylation of azoles H → Ar
s. *57*, 376; polymer-based synthesis with supported ar. halides s. Y. Kondo et al., Org. Lett. *2*, 3111-3 (2000).

Palladous acetate/triphenylphosphine/sodium acetate/tetra-n-*butylammonium chloride*
9-Alkylidenefluorenes from arylacetylenes and ar. iodides

391.

A mixture of 5 mol% Pd(OAc)$_2$, 10 mol% Ph$_3$P, 2 eqs. NaOAc, 1 eq. Bu$_4$NCl, 1 eq. 2-(trifluoromethyl)iodobenzene, and diphenylacetylene in DMF heated in an oil bath at 100° for 48 h → product. Y 75%. An unprecedented palladium migration from ethylene to arene is invoked as the key intermediate stage. The effects of varying bases were also studied. F.e. and regio- and stereoselectivity, also aza-analogs from 3-iodopyridine, and coupling with an eneiodide (1,2,2-triphenyl-1-iodoethylene) s. Q.P. Tian, R.C. Larock, Org. Lett. *2*, 3329-32 (2000).

Palladous acetate/triphenylphosphine/triethylamine   *Pd(OAc)$_2$/Ph$_3$P/Et$_3$N*
α,β-Ethylenehydroxamic acid esters from α,β-ethyleneiodides   I → CONHOR
by carbonylation s. *43*, 808s60

Palladous acetate/triphenylphosphine/ethyldiisopropylamine/
tetra-n-*butylammonium chloride*
Synthesis of 2-alkylidenecyclopentanones
from 1-(alk-1-ynyl)cyclobutanols and unsatd. halides

392.

2-Benzylidenecyclopentanones. A mixture of 10 mol% Pd(OAc)$_2$, 0.2 eq. PPh$_3$, 2 eqs. 2-iodoanisole, 2 eqs. *i*-Pr$_2$NEt, 2 eqs. Bu$_4$NCl, and startg. alkynylcyclobutanol in DMF heated in an oil bath under argon at 80° for 12 h → product. Y 74%. Different combinations of catalyst and base were also studied. F.e. and regio- and stereo-selectivity, also coupling with α,β-ethyleneiodides and enol triflates s. R.C. Larock, C.K. Reddy, Org. Lett. *2*, 3325-7 (2000).

Palladous acetate/tri-2-furylphosphine/ethyldiisopropylamine   *Pd(OAc)$_2$/Fu$_3$P/i-Pr$_2$NEt*
Intramolecular acylpalladation-carbonylation-allene insertion-anion capture
s. *53*, 429; polymer-based synthesis of α-aminomethyl-α,β-ethylene-α'-(oxindol-3-yl)ketones with Pd(OAc)$_2$/tri-2-furylphosphine/*i*-Pr$_2$NEt s. R. Grigg et al., Chem. Commun. *2000*, 2241-2.

Palladous acetate/1,3-bis(diphenylphosphino)propane/triethylamine/
1-butyl-3-methylimidazolium fluoroborate
Acetophenones from ar. halides   H → Ar
via regiospecific Heck arylation of vinyl ethers in ionic liquids s. *60*, 405

Palladous acetate/4-(diphenylphosphinyl)phenylphosphonic acid/
tetra-n-*butylammonium hydrogen sulfate/potassium carbonate/microwaves*
Heck arylation in aq. media
with a water-soluble Pd-phosphine complex cf. *27*, 871s48; rapid procedure in 1:1 acetonitrile/water with Pd(OAc)$_2$/4-(diphenylphosphinyl)phenylphosphonic acid in the presence of K$_2$CO$_3$/Bu$_4$NHSO$_4$ under microwave irradiation s. D. Villemin, B. Nechab, J. Chem. Res., Synop *2000*, 429-31; in aq. micelles s. G. Oehme et al., German patent DE-19809166.

Bis(dibenzylideneacetone)palladium(0) s.a. under Platinum complexes   *Pd(dba)$_2$*

Bis(dibenzylideneacetone)palladium(0)/triethylenediamine/tetra-n-butylammonium chloride ←
Heck arylation H → Ar
update s. *27,* 871s59; terminal arylation of allylsilanes using Pd(dba)$_2$/dabco/Bu$_4$NCl, also allylarenes by desilylative Heck arylation with allyl shift using Ph$_3$P/Bu$_4$NOAc s. T. Jeffery, Tetrahedron Lett. *41,* 8445-9 (2000).

Bis(dibenzylideneacetone)palladium(0)/di-1-adamantyl-n-butylphosphine/ ←
  potassium phosphate
Heck arylation with ar. chlorides
using *t*-Bu$_3$P/Cs$_2$CO$_3$ cf. *57,* 380; of styrenes with non-activated or deactivated ar. chlorides using di-1-adamantyl-*n*-butylphosphine/K$_3$PO$_4$ s. A. Ehrentraut, M. Beller et al., Synlett *2000,* 1589-92.

Tris(dibenzylideneacetone)dipalladium s.a. under LiN(SiMe$_3$)$_2$ Pd$_2$(dba)$_3$

Tris(dibenzylideneacetone)dipalladium/tri-2-furylphosphine/triethylamine ←
Regio- and stereo-specific Heck arylation of enesilanes
2-Pyridyldimethylsilyl as directing group and phase tag

393.

1.1 eqs. Startg. ar. iodide, 2-pyridyldimethyl(vinyl)silane, and 1.2 eqs. Et$_3$N added to a soln. of 0.5 mol% Pd$_2$(dba)$_3$·CHCl$_3$ and 2 mol% tri-2-furylphosphine in THF at room temp. under argon, and the mixture stirred at 50° for 24 h → product. Y 95% (>99% E). Reaction is applicable to the coupling of both unsubst. and β-subst. enesilanes with a wide array of electronically and structurally diverse aryl, heteroaryl, and alkenyl iodides. The 2-pyridyldimethylsilyl group fulfils two roles: to facilitate reaction by coordination of pyridyl nitrogen with palladium; and to simplify purification by acid-base extraction. The latter involves work-up with 1 *N* aq. HCl which takes the product and Et$_3$N into the aq. phase (where the product is extracted following neutralization), while the excess of halide and catalyst are left in the organic phase (from which the catalyst can be retrieved and reused up to three times). The choice of *mono*phosphine ligand is critical. F.e.s. K Itami, J. Yoshida et al., J. Am. Chem. Soc. *122,* 12013-4 (2000).

Tris(dibenzylideneacetone)dipalladium/tri-2-furylphosphine or ←
  tris(p-methoxyphenyl)phosphine/cuprous iodide/triethylamine
1,3-Diynes from dibromomethylene compds.   C≡C-C≡C

394.

Sym. 1,3-diynes. A mixture of startg. dibromomethylene compd., 0.025 eqs. Pd$_2$(dba)$_3$, 0.15 eqs. tri-2-furylphosphine, 0.2 eq. CuI, and 3 eqs. Et$_3$N in DMF flushed with N$_2$, and heated at 80° for 4 h → product. Y 71%. Moderate to good yields of the homocoupling products were obtained with both 2-aryl- and 2-alkyl-1,1-dibromoethenes, but *ortho*-substituents may interfere with reaction. The electronic nature of the phosphine ligand has a determining effect: with the weak tri-2-furylphosphine, homocoupling prevails; however, with the electron-rich tris(*p*-methoxyphenyl)phosphine, Sonogashira coupling with a terminal (aryl or alkyl) alkyne is possible to give **unsym. 1,3-diynes.** Reaction is thought to involve intermediate formation of a 1-acetylenepalladium(II) bromide (in equilibrium with a 1-acetylenecopper(II) iodide). F.e.s. W. Shen, S.A. Thomas, Org. Lett. *2,* 2857-60 (2000).

Tris(dibenzylideneacetone)dipalladium/2-(dicyclohexylphosphino)- ←
2'-dimethylaminobiphenyl/cesium carbonate
Palladium-catalyzed N-arylation-intramolecular Heck arylation   ○

395.

Indoles from *o*-dibromides. 5 Mol% $Pd_2(dba)_3$ added to a mixture of 1.5 eqs. *o*-dibromobenzene, startg. enaminone, 10 mol% 2-(dicyclohexylphosphino)-2'-dimethylaminobiphenyl, and 1.6 eqs. Cs_2CO_3 in THF, stirred at 80° under N_2 for 12 h, a further portion of the Pd catalyst and ligand added, and refluxing continued for a further 24 h → product. Y 61%. F.e. and with intramolecular C→N-aryl migration, also quinoline ring from an *o*-bromaldehyde, s. S.D. Edmondson et al., Org. Lett. *2*, 1109-12 (2000).

Tris(dibenzylideneacetone)dipalladium/1,1'-bis(diphenylphosphino)ferrocene/ ←
titanium tetraisopropoxide/lithium/trimethylsilyl chloride
N-Unsubst. carboxylic acid amides from ar. bromides   Hal → $CONH_2$
via palladium-catalyzed carbonylation-nitrogenation

396. $ArBr \xrightarrow[CO]{Pd(0)} [ArCOPdBr + LnTiClN(SiMe_3)_2] \xrightarrow{\substack{N_2 \\ Ti(OPr-i)_4 \downarrow Li/Me_3SiCl}} ArCOPdNSiMe_3 \xrightarrow{\substack{TiX \\ |}} \xrightarrow{H_2O} ArCONH_2$

Ar = p-$EtO_2CC_6H_4$

A soln. of 2 eqs. $Ti(OPr-i)_4$, 20 eqs. Li, and 20 eqs. trimethylsilyl chloride in THF stirred under N_2 (1 atm.) at room temp. overnight, diluted with DMF, excess of trimethylsilyl chloride removed by heating with THF, 2.5 mol% $Pd_2(dba)_3 \cdot CHCl_3$ and 10 mol% dppf added, followed by a soln. of the startg. ar. halide in DMF, the N_2 atmosphere replaced by carbon monoxide (1 atm.), and the soln. heated to 90° for 14 h → ethyl *p*-carbamylbenzoate. Y 75%. The corresponding benzonitriles were obtained as by-products (8-27%). Reaction is presumed to involve generation of an aroylpalladium bromide, followed by conversion to an amide-titanium complex prior to hydrolysis to the amide (possibly via condensation with DMF). The ar. bromide may possess electron-withdrawing or -donating groups. F.e.s. K. Ueda, M. Mori et al., J. Am. Chem. Soc. *122*, 10722-3 (2000).

Bis(tricyclohexylphosphine)palladium(0)/cesium pivalate $Pd(PCy_3)_2$/$CsOCOBu$-t
9-Fluorenones from o-halogenodiaryls by cyclocarbonylation   ○

397.

A mixture of startg. *o*-halogenodiaryl, 5 mol% $Pd(PCy_3)_2$ and 2 eqs. anhydrous Cs-pivalate in DMF stirred under argon at room temp. for 5 min, flushed with 1 atm. CO, and heated to 110° for 7 h → product. Y 100%. Reaction is thought to involve intermediate formation of an aroylpalladium halide, which undergoes cyclodehydrohalogenation to a palladacyclic prior to reductive elimination of the ketone with regeneration of Pd(0). The choice of base and catalyst is critical. 3'-Subst. 2-iodobiphenyls reacted with high regioselectivity, while phenolethers and aldehydes remained unaffected. Both electron-withdrawing and -donating substituents were tolerated. F.e. incl. polycyclic and heterocyclic fluorenones s. M.A. Campo, R.C. Larock, Org. Lett. *2*, 3675-7 (2000).

Tetrakis(triphenylphosphine)palladium(0)/n-butyllithium/triisopropyl borate
Arylacetylenes from ar. bromides via lithium alk-1-ynyl(trialkoxy)borates Hal → C≡C
Suzuki-Miyaura cross-coupling

398.
$n\text{-}C_6H_{13}C{\equiv}CH \xrightarrow[B(OPr\text{-}i)_3]{n\text{-}BuLi} [n\text{-}C_6H_{13}C{\equiv}C-B(OPr\text{-}i)_2] \xrightarrow{Br-C_6H_4-Me} n\text{-}C_6H_{13}C{\equiv}C-C_6H_4-Me$

A soln. of 1.4 eqs. *n*-BuLi in hexane slowly added to a cooled (-78°) soln. of 1.36 eqs. 1-octyne in DME under argon, after 1 h 1.35 eqs. triisopropyl borate (freshly distilled) slowly added, the mixture stirred for 2 h at the same temp., warmed to 20° over 30 min, THF added (to solubilize the ate complex), treated with 0.9 mol% Pd(PPh$_3$)$_4$ and 1 eq. *p*-bromotoluene in DME (via cannula), the mixture refluxed at 80° for 5 h, cooled to room temp., and quenched with water → product. Y 98%. This *in situ*-Suzuki coupling is notably applicable to *o*-subst. and unactivated aryl bromides, and thereby is complementary with Sonogashira coupling which normally fails with these substrates. F.e. and coupling with enehalides s. A.-S. Castanet, F. Colobert et al., Org. Lett. **2**, 3559-61 (2000).

Tetrakis(triphenylphosphine)palladium(0)/potassium carbonate Pd(PPh$_3$)$_4$/K$_2$CO$_3$
Carbonylation of diaryliodonium salts ←
arylcarboxylic acid esters cf. *55*, 324; arylthiolic acid esters with Pd(PPh$_3$)$_4$/K$_2$CO$_3$ s. L. Wang, Z.C. Chen, J. Chem. Res., Synop **2000**, 372-3.

Tetrakis(triphenylphosphine)palladium(0)/potassium carbonate/ ←
tetra-n-*butylammonium bromide*
Synthesis of vinylcyclopropanes from allenes ▽
2-cyclopropyl-1,3-dienes from vinyl halides cf. *43*, 806; α-styrylcyclopropanes from ar. iodides with Pd(PPh$_3$)$_4$/K$_2$CO$_3$/Bu$_4$NBr s. S. Ma, S. Zhao, Org. Lett. **2**, 2495-7 (2000).

Tetrakis(triphenylphosphine)palladium(0)/zinc cyanide/microwaves ←
Ar. nitriles from bromides Hal → CN
with Pd$_2$(dba)$_3$/dppf/Zn(CN)$_2$ cf. *29*, 845s58; rapid procedure under microwave irradiation without solvent using Pd(PPh$_3$)$_4$ as catalyst s. M. Alterman, A. Hallberg, J. Org. Chem. **65**, 7984-9 (2000).

Soluble polymer-based palladium phosphine complex/triethylamine ←
Soluble polymer-based palladium-catalyzed syntheses ←
in a 2-phase liq.-liq. medium under thermomorphic conditions

399.
$C_6H_5I + CH_2{=}CH{-}CO_2Bu\text{-}t \xrightarrow{\text{polymer-Pd(0) catalyst}} C_6H_5{-}CH{=}CH{-}CO_2Bu\text{-}t$

Simplification of catalyst and product isolation in palladium-mediated syntheses (e.g. Heck arylation, Suzuki coupling) can be effected in a liq.-liq. 2-phase medium by using a *soluble polymer-based Pd complex* which is preferentially soluble at room temp. in one of the phases while the product is preferentially soluble in the other. Reaction takes place on heating in a homogeneous phase, although complete miscibility of the two phases may not be essential (and in some instances may be detrimental). E: A soln. of iodobenzene, 1.2 eqs. *tert*-butyl acrylate, and 1.5 eqs. Et$_3$N in *heptane* added under N$_2$ via forced syphon through a cannula to the palladium complex (prepared from 2 mol% Pd(PPh$_3$)$_4$ and poly(N-isopropylacrylamide)-based phosphine ligand) in the same volume of 90% *ethanol*, heated to 75° with stirring to induce miscibilization, cooled (for phase separation), and worked up with GC monitoring (ca. 48 h reaction time) → product. Y >99% (after three cycles). The product was isolated from the heptane layer, while the catalyst remained in the ethanolic phase (for separation or reuse in a second or third cyclic). Air-stable tridentate SCS-Pd(II) catalysts bound to the same ligand or poly(ethylene glycol) were also described, the former being particularly advantageous in that no precautions against adventitious oxidation of the catalyst are required. F.e. and reactions s. D.E. Bergbreiter et al., J. Am. Chem. Soc. *122*, 9058-64 (2000).

Palladous chloride $PdCl_2$
3-Allyl-2,5-dihydrofurans from 2-allenealcohols and β,γ-ethylenebromides ○

400.

A mixture of startg. 2-allenealcohol, 5 eqs. allyl bromide, and 5 mol% $PdCl_2$ stirred in dimethyl-acetamide at room temp. for 18.5 h → 4-allyl-3-(*n*-butyl)-2,5-dihydro-2-methylfuran. Y 76%. The procedure is mild and based on readily available substrates and an air-stable catalyst. However, the yields were low with allyl chlorides. With α-subst. allyl bromides, reaction proceeded regiospecifically with double bond shift. Significantly, there was no formation of vinyloxiranes as occurs under Pd(0) catalysis (cf. *52*, 415s55). F.e. and Pd catalysts s. S. Ma, W. Gao, Tetrahedron Lett. *41*, 8933-6 (2000).

Palladous chloride/potassium carbonate $PdCl_2/K_2CO_3$
1,2,4-Oxadiazoles from amidoximes and unsatd. iodonium salts via carbonylation ○

401.

3,5-Diaryl-derivs. *p*-Chloro-N-hydroxybenzamidine added under 1 atm. CO at 95° in N-methyl-2-pyrrolidone or N-methyl-2-pyrrolidone/toluene to a mixture of 1 eq. diphenyliodonium tetra-fluoroborate, 5-10 mol% $PdCl_2$, and 2 eqs. K_2CO_3, and stirred for 7 h → 3-(*p*-chlorophenyl)-5-phenyl-1,2,4-oxadiazole. Y 77%. This method complements Young's method from ar. iodides (*55*, 415), which was restricted to reaction of *acet*amidoxime. F.e. incl. 3-aryl-5-vinyl- and 5-aryl-3-vinyl-derivs. s. H-C. Ryu, Y-T. Hong, S-K. Kang, Heterocycles *54*, 985-8 (2001).

Palladous chloride/1,1'-bis(diphenylphosphino)ferrocene/cesium carbonate ←
Ar. amidines from ar. bromides, amines and isonitriles ArBr → ArC(N<)=NR
s. *60*, 158

Bis(benzonitrile)dichloropalladium(II)/cuprous iodide/piperidine ←
1,3-Enynes from terminal acetylene derivs. C≡C-C≡C
(Z)-2-(Alkylseleno)-1,3-enynes s. *60*, 174; (Z)-2-alkoxy-1,3-enynes s. *60*, 175

Dichlorobis(triphenylphosphine)palladium(II)/n-butyllithium $PdCl_2(PPh_3)_2/BuLi$
Regio- and stereo-specific 3-component synthesis of 3-alkylidenetetrahydrofurans ○
from ethynylcarbinols, electron-deficient ethylene derivs. and unsatd. halides
via Michael addition-intramolecular carbopalladation

402.

3-Arylidenetetrahydrofurans. 1 eq. *n*-BuLi (2 *M* in hexanes) added dropwise to an ice-cooled soln. of propargyl alcohol in THF under N_2, allowed to warm to room temp. (15 min), diethyl 2-benzylidenemalonate and iodobenzene added sequentially, followed by the palladium catalyst [prepared by adding 10 mol% *n*-BuLi (2 *M* in hexanes) at room temp. under N_2 to a suspension of 5 mol% PdCl$_2$(PPh$_3$)$_2$ in DMSO], the mixture stirred 15 min, and quenched with satd. aq. NH$_4$Cl → product. Y 89%. This is a versatile and practical route with potential for combinatorial library synthesis not requiring slow addition techniques. *n*-Butyllithium was the preferred base for generating Pd(0) *in situ*. Carboxylic acid esters, ar. halides, cyclic acetals and phenolethers remain unaffected. F.e. and **3-allylidenetetrahydrofurans** from α,β-ethylenehalides or enol triflates s. M. Bottex, G. Balme et al., J. Org. Chem. *66*, 175-9 (2001).

Dichlorobis(triphenylphosphine)palladium(II)/cuprous iodide/triethylamine ←
1,3-Enynes from terminal acetylene derivs. C≡C-C≡C
with Pd(PPh$_3$)$_4$ cf. *27*, 851s*46*; (E)-1,3-enynes from α,β-ethyleneiodides with PdCl$_2$(PPh$_3$)$_2$/CuI/ Et$_3$N s. M.P. Jennings, P.V. Ramachandran et al., J. Org. Chem. *65*, 8763-6 (2000); (Z)-2-(alkylseleno)-1,3-enynes with PdCl$_2$(PhCN)$_2$/CuI/piperidine s. *60*, 174; (Z)-2-alkoxy-1,3-enynes s. *60*, 175.

2-Aryl-2,3-dihydro-1,5-benzothiazepines
from *o*-aminomercaptans, ethynylcarbinols and ar. halides
via isomerizing Sonogashira coupling

403.

A soln. of 1.05 eqs. 1-phenylpropyn-1-ol in THF added dropwise over 30 min at room temp. under N$_2$ to a mixture of 4-bromobenzonitrile, 2 mol% Pd(PPh$_3$)$_2$Cl$_2$ and 1 mol% CuI in degassed 12:7 THF/Et$_3$N, the mixture refluxed for 16 h, cooled to room temp., 1.1 eqs. 2-aminothiophenol added, and refluxed for 8 h → product. Y 67%. The intermediate chalcone-forming step is limited to electron-poor (hetero)aryl halides. F.e., **also 2-aryl-2,3-dihydro-1,5-benzoxazepines and -1*H*-1,5-benzodiazepines** from *o*-aminophenols and *o*-diamines, respectively, s. R.U. Braun, T.J.J. Muller et al., Org. Lett. *2*, 4181-4 (2000).

Dichlorobis(triphenylphosphine)palladium(II)/cuprous iodide/tetramethylguanidine ←
Benzofurans from *o*-iodophenols and acetylene derivs.
via cuprous acetylides cf. *32*, 820; *44*, 781; chiral 2-α-aminobenzofurans with PdCl$_2$(PPh$_3$)$_2$/CuI/ tetramethylguanidine, **also chiral 2-α-amino-N-sulfonylindoles** (cf. *44*, 815) s. F. Messina, F. Corelli et al., Tetrahedron:Asym. *11*, 1681-5 (2000); chiral tryptophan analogs with Pd(OAc)$_2$/ Na$_2$CO$_3$/LiCl s. C. Ma, J.M. Cook et al., Tetrahedron Lett. *41*, 2781-5 (2000).

Dichlorobis(triphenylphosphine)palladium(II)/triphenylphosphine/cupric sulfate-alumina/ ←
triethylamine
Aceanthrylenes from 9-bromoanthracenes and terminal acetylene derivs.

404.

A suspension of 9-bromoanthracene, 4.5 mol% PdCl$_2$(PPh$_3$)$_2$, 0.23 eq. Ph$_3$P and 0.23 eq. CuSO$_4$ (50 wt% on alumina) in deoxygenated benzene stirred under argon, 4 eqs. Et$_3$N and 4 eqs. startg. alkyne added, the mixture refluxed for 16 h, then quenched with satd. aq. NH$_4$Cl → product. Y 73%. This unprecedented transformation is thought to involve regiospecific insertion of the Pd-coordinated alkyne residue into the adjacent *ortho*-C-H bond, followed by intramolecular carbopalladation and rearomatization. Bulky groups on the acetylene led to an increase in 9-alkynylanthracene by-product formation when the reaction was conducted in a sealed tube. F.e.s. H. Dang, M.A. Garcia-Garibay, J. Am. Chem. Soc. *123*, 355-6 (2001).

Dichloro[1,1'-bis(diphenylphosphino)ferrocene]palladium(II)/potassium acetate ←
Carbonylation of ar. halides Hal → COOH
arylcarboxylic acids with PdCl$_2$/Ph$_3$P/Ca(OCHO)$_2$ cf. *43*, 808; arylcarboxylic acids-^{14}C with ^{14}CO using Pd(dppf)Cl$_2$/KOAc, also esters in methanol, s. C.S. Elmore et al., J Labelled Compd. Radiopharm. *43*, 1135-44 (2000); 3-chromene-4-carboxamides with Pd(OAc)$_2$/Ph$_3$P/KI or KBr s. N. Taka et al., Synth. Commun. *30*, 4263-9 (2000); steroidal α,β-**ethylenehydroxamic acid esters** from α,β-ethyleneiodides with Pd(OAc)$_2$/Ph$_3$P/Et$_3$N s. Z. Szarka, R. Skoda-Földes et al., Synth. Commun. *30*, 1945-53 (2000).

Bis(3-methylbenzothiazolin-2-ylidene)palladium(II) diiodide/sodium formate/ ←
sodium hydrogen carbonate/tetra-n-butylammonium bromide
Heck arylation with ar. bromides in ionic liquids H → Ar

405.

Tetra-*n*-butylammonium bromide heated to a melt at 130°, 0.01 eq. bis(3-methylbenzothiazolin-2-ylidene)palladium(II) diiodide added, followed by 0.02 eq. Na-formate, 1 eq. 4-chloronitrobenzene, ca. 2 eqs. NaHCO$_3$ and 1 eq. *n*-butyl acrylate, and the mixture stirred for 1 h → product. Y 95%. High reaction rates are achieved under these conditions, even with less-reactive bromides. The product was readily extracted in a Soxhlet apparatus with cyclohexane (1 h), the catalyst being retrievable from the ionic liquid for recycling. Reaction was even faster with Na$_2$CO$_3$ as base, but side reactions were more prominent. F.e.s. V. Calò et al., Tetrahedron Lett. *41*, 8973-6 (2000); **acetophenones from ar. halides via Heck arylation of vinyl ethers** in 1-butyl-3-methylimidazolium fluoroborate with Pd(OAc)$_2$/dppp/Et$_3$N s. L. Xu, J. Xiao et al., Org. Lett. *3*, 295-7 (2001).

Silica-supported (2-pyridylaldimine)palladium(II) complex/triethylamine ←
Heterogeneous Heck arylation
with a glass bead-supported palladium phosphinoguanidinium complex cf. *46*, 799s59; with a silica-supported (2-pyridylaldimine)palladium(II) complex s. J.H. Clark et al., Green Chem. *2*, 53-5 (2000).

Cyclopalladated complexes/potassium carbonate ←
Heck arylation with highly active cyclopalladated complexes
with *peri*-cyclopalladated tri(1-naphthyl)phosphines s. *51*, 416s55; and with pincer-type complexes cf. *55*, 412s58; with a thermally stable, readily prepared, non-phosphine complex, di-μ-chlorobis(benzophenone oxime-6-C,N)dipalladium(II) (or the benzaldehyde oxime analog) s. S. Iyer, C. Ramesh, Tetrahedron Lett. *41*, 8981-4 (2000).

(1,3-Divinyl-1,1,3,3-tetramethyldisiloxane)platinum(0)/bis(dibenzylideneacetone)- ←
palladium(0)/1,1,3,3-tetramethyldisilazane/tetra-n-butylammonium fluoride
Synthesis of (E)-3-ethylenealcohols CH=C(R)C-C(OH)
from 3-acetylenealcohols and unsatd. halides
via intramolecular hydrosilylation-cross-coupling

406.

One-pot procedure under dual catalysis. 1 eq. Tetramethyldisilazane added dropwise to 1.3 eqs. 3-octyn-1-ol at room temp., the mixture stirred for 1 h, evacuated at room temp. for 10 min to remove excess of the disilazane, dry THF added followed by a little (1,3-divinyl-1,1,3,3-tetramethyldisiloxane)platinum(0) complex (in xylenes), the mixture stirred at room temp. for 1 h, a soln. of 2.2 eqs. Bu$_4$NF in THF added dropwise, followed successively by 1 eq. iodobenzene and 5 mol% Pd(dba)$_2$ in one portion, and stirred for 1.5 h at room temp. → product. Y 85% (E/Z 98.3:1.7). The silyl group assumes a triple role: as temporary protective group, as a temporary tether to fix the geometry of the enesilane, and as a displacable function for the cross-coupling. The procedure is mild and environmentally friendly, and affords trisubst. **(E)-3-aryl-3-ethylenealcohols** with high stereospecificity and in high yield, leaving a wide range of functionality unaffected. F.e. and cross-coupling with hetaryl iodides and enebromides s. S.E. Denmark, W. Pan, Org. Lett. *3*, 61-4 (2001).

Sulfur ↑ CC ↕ S

Irradiation s. *under Bu$_3$SnH*

Sodium hydride/scandium(III) triflate NaH/(Sc(OTf)$_3$
Cyclopentanones from cyclobutanones
Regiospecific ring expansion with retention of configuration

407.

3-Alkoxycyclopentanones. A suspension of 1.2 eqs. NaH (60 wt%) and 1.3 eqs. trimethylsulfoxonium iodide in DMF stirred at room temp. for 30 min, cooled to 0°, a soln. of startg. (±)-cyclobutanone in DMF added slowly over 5-30 min, the ice-bath removed, stirring continued for a further 15 min, treated with 0.25 eqs. Sc(OTf)$_3$ for 10 min, then heated to 50° for 4-5 h → product. Y 79%. Small amounts (9-25%) of 2-cyclopentenones were also formed as by-products on two occasions. The method is simple and general, leaving phenolethers, benzyl ethers, cyclic ethers, and cyclic urethans unaffected. Use of Et$_3$Al in place of Sc(OTf)$_3$ gave poor yields (the cyclopentenone being the major product). F.e. and application to the synthesis of (+)-carbovir and (+)-aristeromycin s. B. Brown, L.S. Hegedus, J. Org. Chem. *65*, 1865-72 (2000).

Sodium hydroxide/tetra-n-butylammonium iodide NaOH/Bu$_4$NI
Pyrrolo[1,2-c]pyrimidines from isonitriles
and 2-formylpyrroles cf. *32*, 739a; 3-carbalkoxypyrrolo[1,2-c]pyrimidine from 2-bromomethyl-1-carbalkoxypyrrole ring and tosylmethyl isocyanide via N→C-carbalkoxy group migration with NaOH/Bu$_4$NI s. J. Mendiola, J.M. Minguez et al., Org. Lett. *2*, 3253-6 (2000).

Potassium tert-*butoxide/methyl iodide* KOBu-t/MeI
Pyrroles from N-α-(thioacylamino)benzotriazoles and ethylene derivs. s. *60*, 352

n-*Butyllithium* BuLi
Ar. aldehydes from sulfoxides s. *60*, 334 S(O)R → CHO

Lithium diisopropylamide
Asym. Mannich-type reaction with N-protected α-aminosulfones
Synthesis with addition of two C-atoms

i-Pr$_2$NLi
SO$_2$Ar → R

408.

A soln. of startg. (1R)-camphor-based methyl α-siloxyketone in THF treated with 3.5 eqs. LDA at -78° for 1 h, 2 eqs. startg. N-protected aminosulfone in THF added, and worked up after 15 min at the same temp. → product. Y 94% (d.r. 98:2). This is part of a multistep route to **chiral N-protected β-aminocarboxylic acids** from inexpensive acetylene and an accessible chiral auxiliary (which may be recovered in 75-80% yield). F.e. and with (R)-camphenilone as auxiliary, also chiral β-amino-γ-benzyloxy derivs. with double asym. induction, s. C. Palomo et al., Angew. Chem. Int. Ed. Engl. *39*, 1063-5 (2000).

Sodium hydrogen carbonate
1-Arylsulfonyl-4-arylthio-5,6-dihydro-2-pyridones
from 3-arylthio-2,5-dihydrothiophene 1,1-dioxides and N-(arylsulfonyl)isocyanates via regiospecific hetero-Diels-Alder reaction

NaHCO$_3$

409.

A mixture of startg. 3-sulfolene, 3 eqs. N-tosylisocyanate, a catalytic amount of hydroquinone, and 1 eq. anhydrous NaHCO$_3$ in toluene heated at 110° under N$_2$ for 4.5 h → product. Y 71%. NaHCO$_3$ was added to remove the generated sulfur dioxide, and the hydroquinone was used to prevent polymerization of the intermediate diene. Polar solvents were not used because they reacted with the isocyanates. F.e.s. S.-S.P. Chou, C.-C. Hung, Tetrahedron Lett. *41*, 8323-6 (2000).

Sodium hydrogen carbonate/tetra-n-butylammonium bromide
trans-**Cyclopropanecarbonyl**
from α-(benzothiazol-2-ylthio)carbonyl compds. and electron-deficient ethylene derivs.

NaHCO$_3$/Bu$_4$NBr

410.

in ionic liquids. Tetra-*n*-butylammonium bromide heated to a melt at 110°, startg. lactone added followed by 3 eqs. *n*-butyl acrylate and 2 eqs. NaHCO$_3$, and the mixture stirred for 2.3 h → product. Y 70%. *trans*-Cyclopropanes were formed with a high degree of stereoselectivity. The products were readily isolated by vacuum distillation or Soxhlet extraction (with hexane), thereby allowing recycling of the ionic liquid. There was no Michael addition, as takes place in toxic DMF. F.e.s. V. Calò, A. Nacci et al., Tetrahedron Lett. *41*, 8977-80 (2000).

1,8-Diazabicyclo[5.4.0]undec-7-ene
Asym. cyclopropanation of ethylene derivs. with sulfonium ylids

DBU

s. *44*, 819s52; chiral 1,2,3-trisubst. cyclopropanes from ethyl dimethylsulfonioacetate and Garner's [oxazolidine-type] aldehyde with DBU s. D. Ma, Y. Jiang, Tetrahedron:Asym. *11*, 3727-36 (2000); diastereoselectivity under Lewis acid catalysis with Yb(OTf)$_3$ or BF$_3$-etherate s. A. Mamai, J.S. Madalengoitia, Tetrahedron Lett. *41*, 9009-14 (2000).

Cuprous bromide s. under Mg   CuBr

Magnesium (s.a. under Cp_2TiCl_2)   Mg
Synthesis of hydrocarbon groups from sulfones $SO_2R \rightarrow R'$
s. *11*, 905; chiral 5-subst. 4-alkoxy-2-pyrrolidones s. P.Q. Huang et al., Synth. Commun. *30*, 2259-68 (2000).

Magnesium/cuprous bromide/diethylzinc   ←
Synthesis of allenes from α,β-acetylenesulfoxides   C(R)=C=C

411.

1,1-Disubst. allenes. A soln. of startg. α,β-acetylenesulfoxide in THF added to a suspension of startg. organocopper compd. (freshly prepared from 1 eq. phenylmagnesium bromide and 1 eq. CuBr) in THF at -70°, kept at the same temp. for 30 min then warmed to 20° over 30 min, a soln. of bis(iodomethyl)zinc (prepared from 3 eqs. diethylzinc and 6 eqs. diiodomethane in THF) at -50° added via cannula over 5 min while maintaining the temp. at 20°, stirred for 10 min at 25°, and quenched with 2:1 aq. NH_4Cl/NH_4OH → product. Y 95%. Arylmagnesium bromides as well as prim., sec. and even *tert*-alkylmagnesium bromides participated in the reaction, as did subst. zinc carbenoids (generated from 1,1-diiodides with $Bu_2Zn/2LiBr$) to give **trisubst. allenes**. Reaction takes place by initial **regiospecific carbocupration** of the alkyne, followed by cross-coupling with the Zn-carbenoid and β-elimination. F.e.s. J.P. Varghese, P. Knochel, I. Marek, Org. Lett. *2*, 2849-52 (2000).

Zinc/tetrakis(triphenylphosphine)palladium(0) $Zn/Pd(PPh_3)_4$
Palladium-catalyzed cross-coupling of alkylthio-N-heterocyclics with benzylzinc bromides   SR → R'

412.

1.9 eqs. 0.82 M Benzylzinc bromide in THF added to a mixture of 2-(methylthio)benzothiazole and 1 mol% $Pd(PPh_3)_4$ under N_2, and the mixture heated at 55-60° for 2.5 h under N_2 → product. Y 89%. Reaction is applicable to methylthio-pyridines, -pyrazines, and -benzimidazoles, and 2-(methylthio)pyrimidines (the latter being particularly reactive); furthermore, with 2,4-bis(methylthio)pyrimidines, the 4-methylthio group was unaffected (this regioselectivity complementing that of 2,4-dichloropyrimidines). F.e.s. M.E. Angiolelli, T.P. Selby et al., Synlett *2000*, 905-7.

Ethylmagnesium bromide   EtMgBr
Ketones from thiolic acid esters   COSR → COR
from 2-pyridyl thiolates and Grignard reagents cf. *29*, 853; regiospecific monoacylation of di(2-pyrryl)methanes with EtMgBr, also sequential acylation to provide a rational synthesis of porphyrins having up to four different *meso* substituents, s. P.D. Rao, J.S. Lindsey et al., J. Org. Chem. *65*, 7323-44 (2000).

Diethylzinc s. under Mg Et_2Zn
Boron fluoride s. under Me_3SiCN BF_3
Scandium(III) triflate s. under NaH $Sc(OTf)_3$

Ytterbium(III) triflate $Yb(OTf)_3$
Asym. cyclopropanation of ethylene derivs. with sulfonium ylids s. *44*, 819s60 ▽

Samarium diiodide/nickel iodide SmI_2/NiI_2
Stereospecific synthesis of C-(α-hydroxyalkyl) glycosides from glycosyl sulfones and oxo compds. $SO_2R \rightarrow C(OH)R$

with SmI_2 cf. *50*, 531; improved yields of 1,2-*trans*-C-glycosides with added NiI_2 as catalyst (without the need for bulky protective groups at C_2) s. N. Miquel, J.-M. Beau et al., Angew. Chem. Int. Ed. Engl. *39*, 4111-4 (2000).

Trimethylsilyl cyanide/boron fluoride Me_3SiCN/BF_3
Nitriles from monothiolphosphoric acid esters $SPO(OR)_2 \rightarrow CN$

413.

α-Glycosyl cyanides. A soln. of startg. glycosyl monothiolphosphate and trimethylsilyl cyanide in acetonitrile treated at room temp. with BF_3-etherate and 4 Å molecular sieves for 3 h → product. Y 73%. F.e.s. W. Kudelska, Z. Naturforsch. *53B*, 1277-80 (1998).

Titanocene dichloride/magnesium/triethyl phosphite $Cp_2TiCl_2/Mg/(EtO)_3P$
Enolethers from mercaptals $C(SR)_2 \rightarrow C=C(OR')$
On-resin Takeda synthesis s. *60*, 113

Syntheses with 1,3-bis(arylthio)ethylenes via (E)-vinyltitanocene chlorides   ←
s. *51*, 423; (E,E)-1,3-dienes from acetylene derivs., also **acoxy-2-ethylenes from aldehydes**, s. T. Takeda et al., Chem. Lett. *2000*, 1198-9; 5-silyl-1,3-dienes from 2,4-bis(arylthio)-3-ethylenesilanes and ketones, also 5-silyl-(1,3-dien)olethers (from carboxylic acid esters) and their hydrolysis to β,γ-ethylene-δ-silylketones, and 3-cyclopropyl-2-ethylenesilanes from ethylene derivs., s. Tetrahedron Lett. *41*, 8377-81.

Tri-n-butyltin hydride/irradiation $Bu_3SnH/⫙$
Syntheses with sulfonic acid esters via 1,5-hydrogen atom transfer   ←
ar. 4-hydroxythioethers with $(Bu_3Sn)_2$ under irradiation cf. *54*, 212; ζ-hydroxycarbonyl from α,β-ethylenecarbonyl compds. via radical 1,4-addition (with Bu_3SnH), as part of a cyclohexane ring synthesis, s. G. Petrovic, Z. Cekovic Org. Lett. *2*, 3769-72 (2000).

4-Ethyl-1,1,1,3,3-pentakis(dimethylamino)diphosphazene   ←
Asym. synthesis of epoxides from oxo compds. and sulfonium salts
from ketones cf. *55*, 421; *trans*-3-ethyleneepoxides with 4-ethyl-1,1,1,3,3-pentakis(dimethylamino)diphosphazene s. A. Solladie-Cavallo et al., Tetrahedron Lett. *41*, 7309-12 (2000).

Nickel(II) iodide s. under SmI_2 NiI_2
Tetrakis(triphenylphosphine)palladium(0) s. under Zn $Pd(PPh_3)_4$

Remaining Elements ↑   CC ↓↑ Rem

Without additional reagents   *w.a.r.*
Wittig synthesis $CO \rightarrow C=C$
s. *10*, 633; of 2-alkoxy-1,3-dienes, e.g. (E)-3-alkoxy-2,4-dienecarboxylic acid esters, and hydrolysis to γ,δ-ethylene-β-ketocarboxylic acid esters, s. D. Chapdelaine, P. Deslongchamps et al., Synlett *2000*, 1819-21.

α-(Benzotriazol-1-yl)ketones from carboxylic acid chlorides $COCl \rightarrow COCH_2Bt$
Synthesis with addition of one C-atom

414.

1 eq. 1-[(Trimethylsilyl)methyl]benzotriazole added to the startg. carboxylic acid chloride in THF, and the mixture refluxed for 24 h → product. Y 95%. This is the key step in a convenient and safe **alternative to the Arndt-Eistert homologation**, being applicable to both aliphatic and aromatic carboxylic acids. F.e., also conversion to **carboxylic acid esters and acids** via 2-(benzotriazol-1-yl)enol triflates and 1-(alk-1-ynyl)benzotriazoles, s. A.R. Katritzky et al., Org. Lett. *2*, 3789-91 (2000).

Petasis (boron-Mannich) reaction $\quad\quad\quad\quad\quad\quad\quad\quad\quad\quad\quad\quad\quad\quad\quad\quad$ CHO → CH(R)N<
s. *48*, 856s*59*; *anti*-1-trifluoromethyl-2-aminoalcohols from vinyl- or aryl-boronic acids s. G.K.S. Prakash, N.A. Petasis, G.A. Olah et al., Org. Lett. *2*, 3173-6 (2000); further polymer-based syntheses of tert. amines s. N. Schlienger, T.K. Hansen et al., Tetrahedron *56*, 10023-30 (2000).

Arylglycines from arylboronic acid esters $\quad\quad\quad\quad\quad\quad\quad\quad\quad$ B(OH)$_2$ → CH(N<)COOH
Petasis synthesis with addition of two C-atoms via resin-to-resin transfer

415.

ca. 1.1 eqs. Glyoxylic acid monohydrate in dry THF added to startg. piperazinyltrityl resin, the suspension allowed to mix at room temp. under N_2 for 2 h, an excess (4 eqs.) of N,N-diethanol-aminomethylated polystyrene-based boronic acid ester added, followed by 8:3 THF/ethanol, the suspension mixed at 65° for 48 h under N_2, cooled to room temp., the resin mixture filtered off, washed, and cleaved with 5% trifluoroacetic acid/methylene chloride in the same vessel at room temp. for 1 h, filtered, and the filtrate (and washings) worked up → product. Y >95% crude (90% conversion). Transesterification of the boronate resin with ethanol affords a soluble diethyl boronate, which adds to the generated immonium ion in the key product-forming step. Yields were higher with electron-rich arylboronic acid derivs., and the procedure was adapted with vinylboronate analogs. F.e.s. K.A. Thompson, D.G. Hall, Chem. Commun. *2000*, 2379-80.

Synthesis of piperazinones from 1,2-diamines and boronic acids

416.

An equimolar mixture of startg. diamine, and glyoxylic acid monohydrate in acetonitrile treated with startg. styrylboronic acid, and refluxed for 2 h → product. Y 76%. This one-step, 3-component method is simple and efficient for sec. 1,2-diamines, whereas prim. 1,2-diamines failed to react. F.e. incl. 3-subst. **3,4-dihydroquinoxalin-2(1*H*)-ones**, also coupling with [het]aryl-boronic acids s. N.A. Petasis, Z.D. Patel, Tetrahedron Lett. *41*, 9607-11 (2000).

Dötz reaction
s. *43*, 860s*59*; benzo-condensed O-heterocyclics, e.g. 7-alkoxy-4-hydroxy-2,3-dihydrobenzo-furans, from cyclic chromium α-alkoxy-β-chloro-α,β-ethylene(alkoxy)carbene complexes s. S.A. Eastham, P. Quayle et al., Synlett *1998*, 61-3; preparation of chromium alkoxycarbene complexes s. H. Matsuyama, M. Iyoda et al., J. Org. Chem. *65*, 4796-803 (2000).

α-(2-Alkoxyvinyl)lactones
from acetylenealcohols and chromium alkoxycarbene complexes
Interrupted Dötz reaction

417.

α-(2-Alkoxy-2-arylvinyl)lactones. A soln. of pentacarbonyl[methoxy(2,6-dimethoxyphenyl)-methylene]chromium(0) and 2 eqs. 4-pentyn-1-ol in dry THF heated to reflux for 4.75 h → 3-

[2-methoxy-2-(2,6-dimethoxyphenyl)ethenyl]tetrahydro-2(2H)-pyranone. Y 74% (after flash chromatographic purification). Yields were lower under ultrasonication (65% for this example). The procedure provides rapid and direct access to functionalized γ-, δ- and ε-lactones possessing an enolether residue (as masked carbonyl group). It is thought that the o-methoxy groups retard the competing Dötz benzannelation, thereby directing reaction to lactonization. F.e. and with Ac_2O/Et_3N s. G.M. Good, W.J. Kerr et al., Tetrahedron Lett. *41*, 9323-6 (2000); β-lactone analogs s. J.P.A. Harrity et al., Synlett *1996*, 1184.

Benzofurans from 1,3-dien-5-ynes and chromium alkoxycarbene complexes

418.

via Dötz reaction. A soln. of 1.2 eqs. startg. chromium methylcarbene complex and startg. dienyne in dioxane added dropwise to refluxing dioxane over 2 h, refluxed for a further 24 h, allowed to cool to room temp., concentrated *in vacuo*, worked up, the crude mixture of arenechromium complexes dissolved in chloroform, treated with 1.3 eqs. I_2, stirred for 12-24 h, and poured into aq. $Na_2S_2O_3$ → product. Y 86%. The procedure is very clean and tolerant of a variety of substitution patterns within the terminating alkene, yields being higher from *trans*-isomers. F.e., also from **3-ene-1,5-diynes** in the presence of a H-atom donor (optimally 0.2 M cyclohexadiene in dioxane), s. J.W. Herndon et al., Tetrahedron Lett. *41*, 8687-90 (2000).

Bicyclo[3.n.1]alk-2-en-(n+6)-ones
from chromium or tungsten α,β-ethylene(alkoxy)carbene complexes
and isocyclic enamines

419.

A soln. of startg. chromium (or tungsten) carbene complex and 1-(cyclopent-1-enyl)pyrrolidine *in hexane* allowed to react at 25°, and the precipitated complex (Y 94%) treated in THF with trifluoroacetic acid at 0°, followed by warming in the presence of water → product. Y 89%. The intermediate σ-chromium complexes were converted to the corresponding **1-alkoxy-9-aminotricyclo[3.3.0.0^{2,8}]octanes** (semibullvalene derivs.) on heating in THF, or obtained directly from the startg. carbene complex and enamine under the same conditions, thereby securing three new carbon-carbon bonds and five stereogenic centres in one operation. F.e. and with the corresponding cyclohexene complexes, **also with asym. induction,** s. J. Barluenga et al., J. Am. Chem. Soc. *122*, 12874-5 (2000).

Hetero-Diels-Alder reaction-allylboration with asym. induction

420.

Chiral 1-amino-6-α-hydroxy-1,2,3,6-tetrahydropyridines. A soln. of startg. L-proline-derived chiral boronylazadiene in toluene added to 1 eq. of startg. aldehyde under N_2 at 0°, the mixture allowed to reach room temp., 2 eqs. startg. maleimide added, and the soln. heated at 80° for 3 days → product. Y 55% (d.e. >95%). F.e. (without asym. induction) s. J. Tailor, D.G. Hall, Org. Lett. 2, 3715-8 (2000).

Irradiation (s.a. under Bu_3SnH)
Radical C-alkylation of *p*-quinones   H → R
with trialkylboranes cf. 27, 907s57; with tellurides under irradiation s. S. Yamago, J. Yoshida et al., Chem. Lett. 2000, 1234-5.

4-Alkoxy-2,4,6-cyclooctatrienones
from chromium 2,4-diene(alkoxy)carbene complexes and terminal acetylene derivs.

421.

Dötz-type reaction. A soln. of the startg. chromium dienylcarbene complex and 3 eqs. phenylacetylene in THF refluxed under N_2 until TLC indicated complete consumption of carbene, hexane added, and the mixture exposed to sunlight and air (for demetalation) → product. Y 71% (1.2:1 mixture of diastereomers). The reaction proceeds via insertion of CO and alkyne to give a trienylketene complex (in Dötz fashion), followed by regiospecific ring closure to the 8-membered ring rather than the expected 6-electron cyclization. β-Styrylcarbene complexes, however, underwent the expected benzannulation. Comparable results were obtained with a photochemically driven process. F.e.s. J. Barluenga et al., Angew. Chem. Int. Ed. Engl. 39, 4346-8 (2000).

Electrolysis
Cathodic generation and reactions of alkoxycarbenium ions
from 1,1-alkoxysilanes   ←

422.

Alkoxy-3-ethylenes. A soln. of startg. 1,1-alkoxysilane (0.4 mmol scale) in deuteriated dichloromethane containing Bu_4NBF_4 as electrolyte electrolyzed in the presence of CF_3SO_3H *at -72°* in an H-type divided cell fitted with a carbon felt anode and a Pt plate cathode until 2.5 F/mol had been consumed, 2 eqs. allyltrimethylsilane added, and worked up *after a few min* → product. Y 80%. This is the first report of the generation of simple [cyclo]aliphatic alkoxycarbenium ions, which are stable below -50°, characterizable by ^{1}H NMR, and react smoothly with a range of nucleophiles. The process is rapid and has the advantage over conventional routes because nucleophiles that might otherwise be oxidized during an *in situ* procedure can be used without difficulty. However, diastereoselectivity is lower than in Lewis acid-catalyzed reactions. F.e. and nucleophiles (allylsilanes, enoxysilanes, enolesters, and even weakly nucleophilic β-diketones) s. S. Suga, J. Yoshida et al., J. Am. Chem. Soc. *122*, 10244-5 (2000).

Microwaves (s.a. under Pd)
Wittig synthesis with lactones $CO \rightarrow C=C$ ←
s. *29*, 861; C-glycosides under microwave enhancement with cyanomethylene(triphenyl)-phosphorane s. Y. Lakhrissi, Y. Chapleur et al., Tetrahedron:Asym. *11*, 417-21 (2000).

Sodium hydride NaH
Horner synthesis of α,β-ethylenecarboxylic acid esters
with two extra C-atoms s. *23*, 879; (Z)-isomers with the BINOLate ester of carbomethoxymethyl-phosphonic acid s. K. Ohmori, S. Yamamura et al., Angew. Chem. Int. Ed. Engl. *39*, 2290-4 (2000); (E)-2,4-dienones s. *60*, 1.

Potassium hydroxide KOH
Cyclopropanes from ethylene derivs. and telluronium salts ▽
Dispiro[2.1.2.n]alkan-4-ones and bicyclo[4.1.0]heptan-2-ones s. *60*, 426

Potassium tert-butoxide/18-crown-6 polyether KOBu-t/crown
(E)-α-Bromo-α,β-ethylenecarboxylic acid esters from aldehydes $CHO \rightarrow CH=C(Br)COOR$
Horner synthesis with addition of two C-atoms

423.

A soln. of 1.1 eqs. methyl bis(2,2,2-trifluoroethoxy)bromophosphonoacetate in THF treated with 1.05 eqs. KOBu-t and 1.3 eqs. 18-crown-6 at -78° for 30 min, startg. aldehyde added, and worked up after 20 min at the same temp. → product. Y 94% (E/Z 30/1). The yield was 65% (E/Z 10/1) after 2 h without the crown ether. Reaction was fast with ar. aldehydes, but less rapid with conjugated and branched aliphatic aldehydes. The yield was also low from the corresponding phosphonyl diethyl ester. F.e.s. K. Tago, H. Kogen, Org. Lett. *2*, 1975-8 (2000); details s. Tetrahedron *56*, 8825-31 (2000).

Organolithium compds./tetra-n-butylammonium fluoride RLi/Bu_4NF
Angular quinanes from 3-siloxy-1-vinylbicyclo[3.2.0]hept-2-en-7-ones ←
with vinyllithium s. *58*, 302; with 2-lithiothiophene, also linear quinanes, s. S.K. Verma, H.W. Moore et al., J. Org. Chem. *65*, 8564-73 (2000).

n-Butyllithium BuLi
Wittig synthesis with reactive aldehydes $CHO \rightarrow CH=C$
s. *18*, 912; iterative synthesis of β-(1→6)-ethylene-bridged C-pentagalactosides s. A. Dondoni et al., Tetrahedron Lett. *41*, 6657-60 (2000).

Stereospecific synthesis of ethylene derivs. $CO \rightarrow C=C$
from oxo compds. and phosphine N-carbalkoxyimines

424.

A soln. of carbomethoxymethyl(diphenyl)phosphine N-carbomethoxyimine in dry THF treated in a Schlenk tube with 1.2 eqs. 1.6 M n-BuLi in hexane at -35°, cooled to -70° after 30 min, 1 eq. benzaldehyde added, the mixture stirred for 6 h, allowed to reach room temp., and quenched with ether or water → methyl cinnamate. Y 81% (E/Z 99:1). F.e.s. E. Peralta-Pérez, F. López-Ortiz, Chem. Commun. *2000*, 2029-30.

n-*Butyllithium/zinc chloride*   BuLi/ZnCl$_2$
α-Alkoxyboronic acid esters from boronic acid esters   CO → C=C
Asym. synthesis with insertion of one C-atom

425.

1.2 eqs. *n*-BuLi (1.6 *M* in hexane) added dropwise to 1 eq. diethoxymethyltributyltin in THF at -100° (95% EtOH/liquid nitrogen bath), stirred at the same temp. for 15-20 min, the startg. (+)-pinanediol boronate (1 eq.) added slowly followed by a soln. of 3 eqs. ZnCl$_2$ in THF, warmed to room temp. slowly, then stirred overnight at the same temp. before quenching with satd. aq. NH$_4$Cl → (S)-pinanediol (1S)-[1-(ethoxy)-2-methylpropyl]boronate. Y 66% (d.e. >98%). The procedure is simple, straightforward, and based on easily prepared stannylacetals. F.e.s. L. Carmès, B. Carboni, F. Carreaux, J. Org. Chem. *65*, 5403-8 (2000).

Lithium diisopropylamide/hexamethylphosphoramide   i-Pr$_2$NLi/HMPA
Cyclopropanes from electron-deficient ethylene derivs. and telluronium salts   ▽

426.

***cis*-2-Vinylcyclopropanecarboxylic acid derivs.** 1.5 eqs. LDA in THF soln. added to 1.5 eqs. startg. telluronium salt at -78°, stirred for 5 min, 4.5 eqs. of *HMPA* and the startg. α,β-ethylenic ester added, the mixture stirred at the same temp. for 4 h, then warmed to room temp. → *trans*-2-(4-fluorophenyl)-*cis*-3-[(trimethylsilyl)vinyl]-1-(methoxycarbonyl)cyclopropane. Y 78% (93:7 mixture of diastereomers). The 1,2-*trans*-isomers were obtained in the absence of HMPA (or in the presence of 12-crown-4). F.e. incl. amide derivs. s. S. Ye, Y. Tang et al., J. Org. Chem. *65*, 6257-60 (2000); bicyclo[4.1.0]heptan-2-ones with KOH s. X. Guo, Q. Zhong et al., Synth. Commun. *30*, 3275-9 (2000); dispiro[2.2.n]alkan-4-ones, s. ibid. 3363-7.

Potassium bis(trimethylsilyl)amide   KN(SiMe$_3$)$_2$
Peterson olefination   CO → C=C
α,β-ethylenenitriles from aldehydes with two extra C-atoms (with LDA/MgI$_2$) cf. *28*, 856s37; (Z)-α,β-ethylenecarboxylic acid amides from N,N-dibenzyl-α-triphenylsilylacetamide (with KN(SiMe$_3$)$_2$) s. S. Kojima et al., Chem. Commun. *2000*, 1795-6.

Potassium carbonate   K$_2$CO$_3$
Horner synthesis
s. *41*, 862; *43*, 841; of 2-alkylidenecyclopentanone-3-carboxylic acid esters and α-alkylidene-γ-lactones in aq. THF s. F. Beji, H. Amri et al., Synth. Commun. *30*, 3947-54 (2000).

Triethylamine/magnesium chloride   Et$_3$N/MgCl$_2$
Wittig synthesis with lactols
ethylenealcohols s. *28*, 852; 2,4-dienecarboxylic acids from 5-hydroxy-2(5H)-furanones with Et$_3$N/MgCl$_2$, also 13-*cis*-retinoic acids (isoretinoins), s. X.C. Wang et al., World Intellectual Property Organization patent WO-9948866 (Abbott Lab).

Ethanolamine   H$_2$NCH$_2$CH$_2$OH
Asym. allylboration   CO → C(OH)C-C=C
s. *33*, 865s50; chiral 1-perfluoroalkyl- or 1-perfluoroaryl-3-ethylenealcohols s. D.J.S. Kumar et al., Tetrahedron:Asym. *11*, 4629-32 (2000); chiral (1,4,6-trien-3-ol)iron tricarbonyl complexes s. V. Prahlad, W.A. Donaldson et al., ibid. 3091-102; asym. allylboration of α-, β-, δ- or ε-aldehydocarboxylic acid esters, also conversion to chiral lactones, s. P.V. Ramachandran, H.C. Brown et al., ibid. *10*, 11-15 (1999).

Poly[4-(N-benzylaminomethyl)styrene]
3-Chromenes from *o*-hydroxyaldehydes and α,β-ethyleneboronic acids

427.

Equimolar amounts of salicylaldehyde and (E)-hex-1-enylboronic acid added to 80 mol% poly[4-(N-benzylaminomethyl)styrene] (pre-swollen in dioxane at room temp. for 15 min), and the mixture stirred at 90° for 12 h → product. Y 99%. The cyclization proceeds via an iminium ion interaction with phenol coordinated to the boronic acid. *o*-Hydroxyketones were unreactive under these conditions. F.e.s. Q. Wang, M.G. Finn, Org. Lett. **2**, 4063-5 (2000).

1,8-Diazabicyclo[5.4.0]undec-7-ene   DBU
(*all-E*)-2,4,6-Trienals from aldehydes   CH≡CHCH≡CHCH≡CHCHO
Synthesis with addition of six C-atoms

428.

One-pot procedure. A mixture of 2-[1-(1,3-diox-4-en-6-yl)-2-propenyl]-4,4,5,5-tetramethyl-1,3,2-dioxaborolane and 1 eq. startg. aldehyde in anhydrous toluene stirred at room temp. for ca. 2 days (or at 50°) until TLC indicated completion of reaction, kept at 110° for 3 h, cooled to room temp., 1 eq. DBU added, stirred for 12 h at room temp., and worked up → product. Y 60%. A 2-step conversion was also devised (in good yield) with isolation of the intermediate **(E,E)-7-hydroxy-2,4-dienals**, which were converted to the trienals by treatment with Burgess' reagent or (MeSO$_2$)$_2$O/*i*-Pr$_2$NEt. Where the intermediate was formed as a mixture of geoisomers, the latter were treated with a little I$_2$ to effect Z→E rearrangement. F.e. and preparation of the auxiliary s. R.W. Hoffmann et al., Synthesis *2000*, 2060-8.

1,5,7-Triazabicyclo[4.4.0]dec-5-ene   ←
Stereospecific Horner synthesis under mild conditions   CO → C≡C
with DBU cf. **39**, 854s48; of racemic 4-boronophenylalanine s. K.C. Park et al., Synthesis *1999*, 2041-4; **also Wittig synthesis** of (E)-ethylene derivs. from aldehydes with 1,5,7-triazabicyclo-[4.4.0]dec-5-ene, its 1-methyl deriv., or tetramethylguanidine, s. D. Simoni et al., Org. Lett. **2**, 3765-8 (2000).

Pyridine   C$_5$H$_5$N
C-Arylation with aryllead triacetates   H → Ar
of β-diketones cf. **36**, 877; 4-arylation of 3-benzopyranone-4-carboxylic acid esters and conversion to 4-aryl-3-chromenes (neoflavenes) s. D.M.X. Donnelly, J.-P. Finet et al., Tetrahedron **57**, 413-23 (2001); *o*-**arylation** of anilines with *t*-BuMgCl s. S. Saito, H. Yamamoto et al., Synlett *2000*, 1676-8.

Silver(I) oxide s. under Pd(PPh$_3$)$_4$ Ag$_2$O
Cuprous thiophene-2-carboxylate s. under Pd$_2$(dba)$_3$ ←
Cuprous iodide s. under Pd(PPh$_3$)$_4$ CuI
Magnesium s. under Et$_3$N   Mg

tert-*Butylmagnesium chloride*   t-BuMgCl
o-**Arylation of anilines with aryllead triacetates** s. **36**, 877s60   H → Ar

Zinc perchlorate/chiral bis(Δ^2-oxazolines)
1-Acyl-Δ^2-pyrazoline-5-carbonyl from α,β-ethylenecarbonyl compds.
Lewis acid-catalyzed asym. 1,3-dipolar cycloaddition

429.

0.1 eq. Zn(ClO$_4$)$_2$ added to a soln. of 0.1 eq. 2,2'-(4,6-dibenzofurandiyl)bis(4(R)-phenyl-Δ^2-oxazoline) in dichloromethane, the mixture stirred vigorously at room temp. under N$_2$ for 3 h, startg. N-acyl-2-oxazolidone added at -40° followed by 1.1 eqs. acetic anhydride, molecular sieves (4 Å) and 1.2 eqs. trimethylsilyldiazomethane (2 M in hexane), stirring continued at the same temp. for 72 h, then quenched with water → product. Y 87% (e.e. 99%). The enantioselectivity was reversed with the corresponding 4,4-dimethyl-2-oxazolidone in the presence of magnesium cation. F.e.s. S. Kanemasa, T. Kanai, J. Am. Chem. Soc. *122*, 10710-1 (2000).

Zinc triflate $Zn(OTf)_2$
β-Aminoketones from aldimines and enoxysilanes C=NR → C(NHR)C-CO
β'-Amino-α,β-ethyleneketones s. *50*, 551s60

Zinc triflate/pyridine $Zn(OTf)_2/C_5H_5N$
Synthesis of 3-ethylenealcohols from oxo compds. CO → C(OH)C-C=C
and 2-ethylenestannanes with Zn(OTf)$_2$/bis(Δ^2-oxazolines) cf. *51*, 433s52; and tetraallylstannane with Zn(OTf)$_2$/pyridine (10 mol% each) for reaction with ring-subst. aryl ketones s. R. Hamasaki, Y. Yamamoto et al., Tetrahedron Lett. *41*, 9883-7 (2000).

Zinc chloride (s.a. under BuLi) $ZnCl_2$
β-Aminocarboxylic acid esters C=NR → C(NHR)C-COOR'
from aldimines and O-silyl O-alkyl keteneacetals with asym. induction
from N-glycosylimines s. *45*, 528; s.a. P. Allef, H. Kunz, Tetrahedron:Asym. *11*, 375-8 (2000).

Chiral 3-sulfonyl-1,3,2-oxazaborolidin-5-ones ←
Asym. aldol-type condensation CHO → CH(OH)C-COOR
s. *47*, 885s48; effect of Lewis basicity of aldehydes on enantioselectivity s. I. Simpura, V. Nevalainen, Tetrahedron:Asym. *10*, 7-9 (1999); with chiral N-protected α-aminoaldehydes s. S. Kiyooka et al., Tetrahedron Lett. *41*, 6599-603 (2000).

Boron fluoride BF_3
N-Protected 3-ethyleneamines ←
from (E)-2-ethylenesilanes and *in situ* generated N-protected aldimines

430.

with asym. induction. 2 eqs. BF$_3$-Etherate added via microsyringe to a soln. of 2 eqs. benzaldehyde dimethyl acetal and 2.1 eqs. *p*-toluenesulfonamide in dichloromethane at -78° under

N_2, the mixture stirred for 30 min at the same temp., the startg. silane in dichloromethane added, stirred at -35° for 48 h, then quenched with satd. $NaHCO_3$ soln. → (5R,6R,3E)-methyl 5-methyl-6-tosylamino-6-phenylhexenoate. Y 80% (d.r. 20:1). Selectivity is dependent on the bulk of the acetal. The diastereoselectivity is interpreted in terms of *re*-face addition to a synclinal transition state. F.e. and N-carbomethoxy derivs. with $MeOCONH_2$ (at -78 to -20°) or **chiral N-carbalkoxy-3-silylpyrrolidines** (at -100°), also with aldehydes in place of acetals, s. J.V. Schaus, J.S. Panek et al., Tetrahedron *56*, 10263-74 (2000).

Chemoselective Lewis acid-catalyzed syntheses with 3-ethyleneepoxides as β,γ-ethylenealdehyde or (1,3-dienol)ate equivalents   C̣

431.

The amphoteric nature of 3-ethyleneepoxides has been noted for the first time under Lewis acid catalysis. **E: 1,6-Dien-4-ols from potassium 2-ethylene(trifluoro)borates.** 1.5 eqs. K-allyltrifluoroborate and 10 mol% BF_3-etherate *as nucleophile* added successively to a soln. of 2-(2(E)-phenylvinyl)oxirane in anhydrous THF at 0°, the suspension stirred for 30 min at the same temp., and quenched with satd. aq. NH_4Cl → (E)-1-phenyl-1,6-heptadien-4-ol. Y 88%. The dienolate activity was demonstrated with $Sc(OTf)_3$ as Lewis acid and aldehydes *as electrophile* in what is effectively **equivalent to a vinylogous aldol-type** (Mukaiyama) **condensation** with dienoxysilanes. **E:** A soln. of 15 mol% $Sc(OTf)_3$ and 4-nitrobenzaldehyde in anhydrous THF stirred at -15° under argon, 1.3 eqs. 2-methyl-2-vinyloxirane in the same solvent *added slowly* [to prevent homocondensation] via syringe pump over 70 min, stirred for a further 2 h at -15°, and quenched with satd. aq. $NaHCO_3$ → (E)-5-hydroxy-2-methyl-5-(4-nitrophenyl)-2-penten-1-al. Y 88%. A variety of aldehydes (ar., heteroar., enals, and enolizable aliphatic aldehydes) participated in the latter reaction. F.e.s. M. Lautens et al., Angew. Chem. Int. Ed. Engl. *39*, 4079-82 (2000).

1,2-*trans*-C-Allyl glycosides from carbohydrate 1,2-O-isopropylidene derivs.

432.

The first instance of anomeric activation by an O,O-isopropylidene group is reported. **E:** 4 eqs. allyltrimethylsilane added to a soln. of startg. 1,2-O-isopropylidene glycofuranoside in dry methylene chloride under N_2 at 0°, stirred for 10 min, 0.2 eq. BF_3-etherate added, stirred at 0° for 15 min, warmed to room temp., stirred again for 3 h, and quenched with satd. aq. $NaHCO_3$ → product. Y 68%. Reaction is presumed to proceed via a cyclic oxocarbenium species, the nucleophile adding to the *exo*-face of the bicyclic acetal to give the 1,2-*trans*-configuration exclusively. The stereochemistry depends on neither the size nor the configuration of the substituents on the ring, but the yield decreases with the bulk of the C_4-alkyl group. Conditions are mild, leaving acid-sensitive functions unaffected. F.e. and methods (e.g. with $TiCl_4$), s. F. García-Tellado, J.J. Marrero-Tellado et al., Angew. Chem. Int. Ed. Engl. *39*, 2727-9 (2000); **N-protected *trans*-2-allylpyrrolidin-3-ols** from pyrrolidine analogs s. Org. Lett. *2*, 3513-5 (2000).

4-Methylenetetrahydropyrans from aldehydes   ○
and 2-methylene-4-siloxysilanes with $EtOSiMe_3$ cf. *47*, 897; from 3-ethylene-3-(silylmethyl)-alcohols with BF_3 s. I.E. Markó, B. Leroy, Tetrahedron Lett. *41*, 7225-30 (2000).

5-Isoxazolidones from nitrones with asym. induction

433.

Chiral *anti*-3-α-alkoxy-5-isoxazolidones. A soln. of startg. chiral nitrone and 3 eqs. O-methyl O-*tert*-butyldimethylsilyl ketene acetal in methylene chloride treated at -80° with 2 eqs. BF$_3$-etherate for 30 min → product. Y 90% (100% purity; *syn:anti* <5:95). The proportion of the Lewis acid is critical to the yield and chemoselectivity. *tert*-Butyldimethylsilyl triflate gave predominantly *syn*-β-siloxylaminocarboxylic acid esters, while Et$_2$AlCl gave a mixture of acyclic products. F.e. and Vorbrüggen synthesis of **isoxazolidine nucleosides** s. P. Merino et al., Tetrahedron Lett. *41*, 9239-43 (2000).

Gallium trichloride   *GaCl$_3$*
3-Ethylenealcohols from aldehydes   CHO → CH(OH)C-C≡C
and allyltri-*n*-butylstannane - Enhancement by bidentate chelation with *o*-alkynyl groups s. *60*, 27

**3-Alkoxy-Δ^3-Pyrrolines
from chromium α,β-ethylene(alkoxy)carbene complexes and aldimines
Regio- and stereo-specific [3+2]-cycloaddition**

434.

A soln. of 0.2 eq. GaCl$_3$ (1 M in methylcyclohexane) added to a soln. of 1.2 eqs. of the startg. aldimine and carbene complex in 1,2-dichloroethane, the mixture refluxed under N$_2$ for 1 h, cooled to room temp., and quenched with satd. aq. NaHCO$_3$ → product. Y 86% (*trans*-selectivity 83%). Traditional Lewis acids were less effective. Reaction is thought to involve initial stereoselective [2+2]-cycloaddition between the imine and carbene residue, followed by ring expansion to a 6-membered chromacyclic prior to reductive elimination. F.e. **and 3-component synthesis** with *in situ*-generated aldimines s. H. Kagoshima, T. Akiyama, J. Am. Chem. Soc. *122*, 11741-2 (2000).

Indium(III) chloride   *InCl$_3$*
**Stereospecific synthesis of 5-subst. 2-pyrrolidones
from 5-acoxy-2-pyrrolidones and Si-nucleophiles**

435.

A soln. of the startg. pyrrolidone triacetate in dichloromethane added in one portion under argon to a suspension of 60 mol% dry InCl$_3$ in the same solvent at 0°, the mixture stirred for 10 min, 1.5 eqs. startg. silyl enolate added dropwise over 3 min, stirring continued for 3 h at the same temp., and quenched with satd. aq. NaHCO$_3$ → product. Y 85% (4,5-*trans*-selectivity 98%). An increase in the steric bulk of the nucleophilic carbon favours approach of the nucleophile from the opposite face of the 4-acetoxy group, thus leading to preferential 4,5-*trans*-selectivity. F.e. and with allylsilanes s. D. Russowsky et al., Tetrahedron Lett. *41*, 9939-42 (2000).

Ammonium ceric nitrate $(NH_4)_2Ce(NO_3)_6$
Maleic anhydrides from acetylenedicobalt hexacarbonyl complexes ○

436.

A soln. of startg. complex in aq. acetone treated with an excess of CAN at 0° until reaction complete → product. Y 83%. The reaction clearly involves incorporation of two carbonyl groups but the mechanism is unclear. F.e.s. K. Tanino, I. Kuwajima et al., J. Am. Chem. Soc. *122*, 6116-7 (2000).

Scandium(III) triflate $Sc(OTf)_3$
β-Aminocarboxylic acid esters C=NR → C(NHR)C-COOR'
from aldimines and O-silyl O-alkyl keteneacetals
s. *50*, 551s*51*; polymer-based synthesis s. S. Kobayashi et al., J. Comb. Chem. *2*, 438-40 (2000); **β-aminoketones** from enoxysilanes in methylene chloride containing 2 eqs. water, also with Zn(OTf)$_2$ for β'-amino-α,β-ethyleneketones, s. K. Ishimaru, T. Kojima, J. Org. Chem. *65*, 8395-8 (2000).

Ytterbium(III) triflate/2,6-bis(4(S)-isopropyl-Δ²-oxazolin-2-yl)pyridine/ ←
trifluoroacetic acid
2,3-Dihydro-4-pyrones from aldehydes via asym. hetero-Diels-Alder reaction ○
with Me$_3$Al/chiral 3,3'-bis(triarylsilyl)-BINOLs cf. *43*, 851; chiral 2-carbalkoxy-2,3-dihydro-4-pyrones from glyoxylic acid esters with Yb(OTf)$_3$ and chiral bis(Δ²-oxazolines) s. C. Qian, L. Wang, Tetrahedron Lett. *41*, 2203-6 (2000).

Samarium diiodide/hexamethylphosphoramide/methanol SmI$_2$/HMPA/MeOH
δ-Alkoxy-γ-ketocarbonyl compds. CH(OR)CO-C-C-COOR'
from chromium alkoxycarbene complexes and α,β-ethylenecarbonyl compds.

437.

13 eqs. HMPA added to a soln. of 4.5 eqs. SmI$_2$ (0.1 *M*) in THF under argon at room temp., the mixture added to a soln. of the startg. chromium carbene complex and 9 eqs. methanol in the same solvent at -78°, stirred for 20 min, 5.8 eqs. ethyl acrylate added, warmed to room temp., stirred for a further 1 h, then quenched with phosphate buffer (pH 7) → product. Y 85%. Reaction also proceeds with acrylonitrile, 2-cyclopentenone and β-substituted ethyl acrylates. F.e., **also γ-alkoxy-γ-arylcarboxylic acid esters** from tungsten aryl(alkoxy)carbene complexes, s. K. Fuchibe, N. Iwasawa, Org. Lett. *2*, 3297-9 (2000).

Chiral diols/dimethylformamide ←
Asym. synthesis of 3-ethylenealcohols CHO → CH(OH)C-C=C
from aldehydes and 2-ethylenesilanes
with a chiral Ti(IV)-aroxide cf. *49*, 898; from 2-ethylene(trichloro)silanes with a chiral diol (e.g. diisopropyl (2R,3R)-tartrate) and a base (Et$_3$N, DMF, HMPA) via chiral pentacoordinate allyl-silicates, also asym. allylation of glyoxylates with allyl(trialkoxy)silanes using TiCl$_4$, s. D. Wang et al., Tetrahedron:Asym. *10*, 327-38 (1999).

1-Butyl-3-methylimidazolium fluoroborate ←
Wittig synthesis in ionic liquids CO → C=C

438.

Recyclable, non-volatile ionic liquids are useful alternative solvents for the Wittig synthesis, permitting, in particular, a facile separation of the product from triphenylphosphine oxide by 2-

fold extractive work-up. **E:** A soln. of benzaldehyde and 1.08 eqs. startg. stabilized P-ylid in 1-butyl-3-methylimidazolium fluoroborate heated at 60° with stirring for 2.5 h, the mixture extracted first with *tert*-butyl methyl ether (three times) and then with toluene (three times), the extracts evaporated *in vacuo*, and the ethereal extract filtered on a short pad of SiO_2 → benzylideneacetone (Y 87%) and triphenylphosphine oxide (Y 11%). The toluene extract removed 78% of the phosphine oxide from the ionic liquid. Following extraction, the resulting ionic liquid was shown to contain only trace amounts of the startg. m., product, and by-product, and was reusable up to six times (for the same or a different synthesis) with the same high yields. Reaction is generally applicable to ar. and aliphatic aldehydes, as well as enals, and (E)-selectivity is the same or better than in conventional solvents. The *p*-nitro-analog was isolated in lower yield, however, due to poor solubility in the ether. F.e. incl. enoates and α,β-ethylenealdehydes s. V. Le Boulaire, R. Grée, Chem. Commun. *2000*, 2195-6.

1,1,3,3-Tetramethylguanidine $\qquad (Me_2N)_2C{=}NH$
C-Arylation with arylbismuth(V) compds. ←

s. *36*, 877s42; with aryl(biphenyl-2,2′-ylene)bismuth diacetates using $(Me_2N)_2C{=}NH$ as base, also N- and O-arylation with $Cu(OAc)_2$, s. A.Y. Fedorov, J.-P. Finet, J. Chem. Soc. Perkin Trans. 1 *2000*, 3775-8.

Amberlyst 15 ←
β-Aminocarboxylic acid esters $\qquad C{=}NR \rightarrow C(NHR)C\text{-}COOR$
from aldimines and O-silyl O-alkyl keteneacetals
with $TiCl_4$ cf. *33*, 879; also with readily retrievable Amberlyst 15 for preferential addition to N-anisylimines in the presence of N-tosyl- or N-carbethoxy-imines, and with retention of enoxysilanes, s. M. Shimizu, S. Itohara, Synlett *2000*, 1828-30.

Acetyl chloride $\qquad AcCl$
Hetero-Diels-Alder reaction with 3-siloxy-1,3-dienamines under mild conditions ○

439.

2,3-Dihydro-4-pyrones. 1.5 eqs. 3-Cyclohexenecarboxaldehyde added dropwise under N_2 to the startg. diene in chloroform, stirred at room temp. for 6 h, diluted with dichloromethane, the mixture cooled to -78°, 2 eqs. acetyl chloride added, stirred for 30 min, then quenched with satd. $NaHCO_3$ → product. Y 92%. Acetyl chloride facilitates both desilylation and dehydroamination without requiring a Lewis acid. F.e., **also 2,3-dihydro-4-pyridones** from aldimines, s. Y. Huang, V.H. Rawal, Org. Lett. *2*, 3321-3 (2000).

Titanium tetraisopropoxide/(R)-1,1′-bi-2-naphthol ←
Catalytic asym. synthesis of 3-ethylenealcohols $\qquad CHO \rightarrow CH(OH)C\text{-}C{=}C$
from aldehydes and 2-ethylenestannanes
with $Cl_2Ti(OPr\text{-}i)_2/(S)$-1,1′-bi-2-naphthol cf. *49*, 898; from tetraallylstannane with $Ti(OPr\text{-}i)_4/$ (R)-1,1′-bi-2-naphthol in toluene or pentane at room temp. (*without* molecular sieves) s. H. Doucet, M. Santelli, Tetrahedron:Asym. *11*, 4163-9 (2000).

Titanium tetraisopropoxide/(R)- or (S)-1,1′-bi-2-naphthol ←
Vinylogous asym. aldol-type condensation ←
with $CuF_2/(S)$-Tol-BINAP cf. *55*, 434; with $Ti(OPr\text{-}i)_4/(R)$-BINOL s. M. De Rosa, A. Scettri et al., Tetrahedron:Asym. *11*, 3187-95 (2000); chiral δ-siloxy-β-keto esters with (S)-BINOL s. ibid. 2255-8.

Tetra-n-butylammonium difluorotriphenylsilicate $\qquad [Bu_4N][Ph_3SiF_2]$
Trifluoromethyltrimethylsilane as trifluoromethyl carbanion equivalent ←
N-(perfluoro-*tert*-butyl)anilines with CsF cf. *44*, 577s59; 2-sulfinylamino-1,1,1-trifluorides from N-sulfinylimines with asym. induction using $[Bu_4N][Ph_3SiF_2]$ s. G.K.S. Prakash, G.A. Olah et al., Angew. Chem. Int. Ed. Engl. *40*, 589-90 (2001); α-hydroxy-α-trifluoromethyl- from α-keto-

carboxamides (with Bu₄NF) s. R.P. Singh, J.M. Schreeve et al., J. Org. Chem. *64*, 2579-81 (1999).

Titanium tetrachloride $TiCl_4$
N'-Acyl-3-ethylenehydrazines from acylhydrazones with asym. induction ←
by addition of 2-ethylene(trichloro)silanes cf. *58*, 406; chiral 1-acyl-3-allylpyrazolidines from 1-acyl-δ^2-pyrazolines and allylstannanes with TiCl₄, also addition of Me₃SiCN and silyl ketene acetals, s. F.M. Guerra, E.M. Carreira et al., Org. Lett. *2*, 4265-7 (2000).

Regiospecific synthesis of β-(*o*-carboxyaryl)carboxylic acid esters from phthalides and O-silyl O-alkyl keteneacetals

440.

β-**Aryl-β-(*o*-carboxyaryl)carboxylic acid esters.** A soln. of TiCl₄ in dichloromethane added dropwise to a stirred soln. of startg. phthalide and 3 eqs. 1-methoxy-1-(trimethylsilyloxy)ethene in the same solvent at 0°, the mixture stirred at room temp. overnight, treated with 5% KHSO₄, and worked-up after 1 h → product. Y 64%. The method is simple, efficient and general. The reaction appears to be sensitive to the steric effect of an *ortho*-substituent on the aromatic ring of the phthalide, but curiously not to the *ortho*-substituent on the 3-aryl group. F.e. and preparation of the startg. phthalides s. J. Epsztajn, A. Bieniek et al., Synthesis *2000*, 1603-7.

(E)-2-Acylenetetrahydrofurans from 1,3-disiloxy-1,3-dienes and epoxides ◯

441.

A soln. of 1.3 eqs. TiCl₄ in dichloromethane added to a soln. of startg. epoxide (1:1 mixture of diastereomers) and 1,3-bis(trimethylsilyloxy)-1-methoxy-2-methyl-1,3-butadiene in the same solvent at -78°, the mixture stirred at the same temp. for 5 h then at 20° for 12 h, and quenched with satd. aq. NaCl → product. Y 58% (E:Z >98:2). Reaction appears to involve regioselective, Ti(IV)-mediated nucleophilic attack of the terminal (C₄) carbon atom of the 1,3-diene on the epoxide, followed by Ti(IV)-promoted intramolecular Michael addition and (E)-selective elimination of silanolate. The method affords good yields with high chemo-, regio- and (E)-selectivities. Ester groups on the epoxide were unaffected as well as halogen in epihalohydrins. F.e. s. P. Langer, T. Eckardt, Angew. Chem. Int. Ed. Engl. *39*, 4343-6 (2000).

Tri-n-butyltin hydride/triethylborane/irradiation ←
Radical synthesis of C-α-glycosides from α-selenoglycosides C=C → CHC(R)

442.

N-Protected 2-amino-2-deoxy-C-α-glycosides. 2.2 eqs. Methyl acrylate added to startg. selenide in toluene under N₂ at room temp., 3.7 eqs. tributyltin hydride and 1 eq. triethylborane added, the mixture sonicated (in an ultrasonic bath), and irradiated (200 W lamp) for 1.5 h → product. Y 93%. Mild reaction conditions suppress reduction of the intermediate radical. A range of N-protecting groups (N-acetyl, N-trifluoroacetyl, N-carbo-*tert*-butoxy) were tolerated. F.e.s. R. SanMartin, T. Gallagher et al., Org. Lett. *2*, 4051-4 (2000).

Tri-n-butyltin hydride/azodiisobutyronitrile Bu$_3$SnH/AIBN
Radical addition of selenides to ethylene derivs. C=C → CHC(R)
s. *38*, 904; regiospecific synthesis of α,α-difluorophosphonic acid esters with Bu$_3$SnH/AIBN, also addition of the S-analogs, s. T. Lequeux, S.R. Piettre et al., Org. Lett. *3*, 185-8 (2001).

Stannous chloride SnCl$_2$
β-Hydroxy- from α-stannyl-carboxylic acid esters CHO → CH(OH)C-COOR
and aldehydes with TiCl$_4$ cf. *41*, 888s*43*; highly diastereoselective synthesis of γ-alkoxy-β-hydroxy- and β,γ-dihydroxy-esters with SnCl$_2$ under chelation control s. M. Yasuda, A. Baba, Chem. Commun. *2001*, 157-8.

Stannic chloride SnCl$_4$
Replacement of sulfonyl group by Si-nucleophiles SO$_2$R → R'
with retention of configuration
N-Acylimmonium ions as intermediates

Chiral β-(2-imidazolidon-1-yl)ketones. 1.25 eqs. SnCl$_4$ added dropwise over 10 min to a soln. of startg. chiral sulfone in dichloromethane at -78°, the mixture stirred at the same temp. for 30 min, then a soln. of 1.5 eqs. startg. silyl enol ether in the same solvent added dropwise, stirring continued for 1 h at the same temp., the mixture allowed to warm to -40°, stirred for 30 min, and quenched with brine → product. Y 98% (diastereoselectivity >99%). The stereochemistry is explained by attack of the nucleophile from the least hindered side of the intermediate. F.e. and with allyltrimethylsilanes (poor results being obtained with O-silyl O-alkyl keteneacetals) s. A. Giardina, M. Petrini et al., J. Org. Chem. *65*, 8277-82 (2000).

2,3-Dihydro-4-pyridones from aldimines ○
with ZnCl$_2$ cf. *38*, 887; with SnCl$_4$, asym. induction with chiral Cr(CO)$_3$-complexed ar. aldimines, and conversion to enantiopure hydro[iso]quinolines, s. H. Ratni, E.P. Kündig et al., Synlett *1999*, 626-8.

Stannic chloride/n-butyllithium/2,2,2-trifluoroethanol ←
***o*-Vinylation of phenols** H → C=CH
with terminal alkynes and SnCl$_4$/Bu$_3$N cf. *50*, 410; with trimethylsilylacetylene and 10 mol% SnCl$_4$/20 mol% BuLi/10 mol% 2,2,2-trifluoroethanol s. K. Kobayashi, M. Yamaguchi, Org. Lett. *3*, 241-2 (2001).

Lead(II) triflate/chiral 1,1'-binaphthyl crown ether ←
Asym. aldol-type condensation in aq. medium CHO → CH(OH)C-CO

A mixture of startg. aldehyde, 1.5 eqs. enoxysilane and 20 mol% chiral Pb(OTf)$_2$-1,1'-binaphthyl crown ether complex in water/isopropanol (1:4.5) allowed to react at 0° for 24 h → product. Y 99% (*syn:anti* 94:6; e.e. 87%). This represents the first successful use of a chiral crown polyether-based Lewis acid in an asym. aldol-type condensation. The reaction provides good to high yields of product with high levels of diastereo- and enantio-selectivity as compared to prior art. Such selectivity is explained by displacement of water molecules coordinated to lead(II) by the aldehyde (aliphatic or aromatic), followed by face-selective condensation with the enoxysilane in the environment of the multi-coordinated crown complex. Both Pb(II) and crown ether were readily retrievable. F.e., also with an O-silylketene-O,S-acetal, s. S. Nagayama, S. Kobayashi, J. Am. Chem. Soc. *122*, 11531-2 (2000).

Trimethylsilyl triflate Me_3SiOTf
Syntheses with 2-siloxyfurans with asym. induction
chiral 5-α-amino-2(5H)-furanones from aldimines with *t*-BuMe$_2$SiOTf cf. *53*, 453; chiral N-carbalkoxy derivs. from cyclic 1,1-alkoxyurethans via *si*-face addition to the intermediate N-acyliminium ions (with Me$_3$SiOTf) s. M.G.M. D'Oca, R.A. Pilli et al., Tetrahedron Lett. *41*, 9709-12 (2000).

5-(Carbalkoxyene)-3-hydroxy-2(5H)-furanones from 1-alkoxy-1,3-disiloxy-1,3-dienes

445.

0.09 eq. Trimethylsilyl triflate in dichloromethane added to a soln. of 1.2 eqs. oxalyl chloride and startg. 1,3-bis(trimethylsilyloxy)-1,3-diene in the same solvent at -78°, the mixture allowed to warm to 20° over 12 h, then stirred for 2 h at the same temp. → product. Y 84%. Regioselectivity is determined by the greater nucleophilicity of the terminal diene carbon rather than the central carbon. F.e.s. P. Langer, N.N.R. Saleh, Org. Lett. *2*, 3333-6 (2000).

Trimethylsilyl triflate/pyridine/benzyltriethylammonium chloride/acetic acid/triethylamine ←
1,3-Nitroethers from N,N-disiloxyenamines and acetals $C(OR)_2 \rightarrow C(OR)C\text{-}CH(NO_2)$

446.

A soln. of 1.06 eqs. startg. acetal in dichloromethane added to a soln. of 1.2 eqs. Me$_3$SiOTf and 0.05 eq. pyridine in the same solvent at -78°, followed by a soln. of startg. bis(silyloxy)enamine in the same solvent, the mixture stirred at the same temp. for 6 h, treated with a soln. of 1.4 eqs. benzyltriethylammonium chloride in the same solvent, stirred for 5 min, 2 eqs. acetic acid and 2.2 eqs. Et$_3$N in the same solvent added, followed by an additional 2 eqs. acetic acid in methanol, allowed to warm to room temp., and stirred for 1 h → (1-methoxy-3-nitrobutyl)benzene. Y 77% (1:1 mixture of diastereomers). The relative reactivities of aliphatic and ar. acetals suggest that Lewis acidic TMSOTf abstracts alkoxide to generate a carbonium ion which undergoes nucleophilic attack by the bis(silyloxy)enamine. The intermediate may then be trapped by a second electrophile (other than proton) so that sec. nitro compds. (precursors of the N,N-disiloxyenamines) can be extended at both β- and β'-sites by suitable combination of electrophiles. This is in contrast to the electrophilic behavior of N,N-disiloxyenamines in reaction with stabilized carbanions. F.e.s. A.D. Dilman, S.L. Ioffe et al., J. Org. Chem. *65*, 8826-9 (2000).

Trifluoromethanesulfonic acid CF_3SO_3H
3-Ethylenealcohols from aldehydes in aq. acidic media $CHO \rightarrow CH(OH)C\text{-}C\!\equiv\!C$
with tetraallyltin in aq. HCl cf. *49*, 909; with allyl(tributyl)stannane in aq. media with CF_3SO_3H, steroidal derivs., s. T.-P. Loh et al., Tetrahedron:Asym. *11*, 1565-9 (2000).

Chiral chromium(III) salen complex
Asym. Diels-Alder reaction of 1-amino-3-siloxy-1,3-dienes

447.

with α,β-ethylenealdehydes. 2 eqs. Methacrolein added to a stirred mixture of 5 mol% chiral Cr(III) salen complex and dry, powdered 4 Å molecular sieves in methylene chloride at -40°, startg. diene added, stirring continued at -40° for 2 days, solids removed by filtration, and the filtrate worked up → product. Y 93% (e.e. 97%). *endo*-Adducts were obtained exclusively with high enantioselectivity, which was largely dependent on the nature of the alkyl group on nitrogen. Chromium(III) is presumed to activate the enal by coordination with the carbonyl oxygen lone pair *anti* to the substituent at the α-position of the enal (the diene approaching the dienophile from the more open surface). F.e. and with α,β-ethylene-α-siloxyaldehydes s. Y. Huang, V.H. Rawal et al., J. Am. Chem. Soc. *122*, 7843-4 (2000).

Tetra-n-butylammonium fluoride (s.a. under RLi and Pd-catalysts) Bu_4NF
syn-1,3-Diols from enoxy(hydrido)silanes and aldehydes $CHO \rightarrow CH(OH)C\text{-}CH(OH)$
Reductive aldol-type condensation

448.

A soln. of startg. aldehyde and 1.2 eqs. startg. dimethylsilyl enolate (>98% Z) in THF treated with 6 mol% Bu_4NF at -78° for 3.5 h → product. Y 65% (*syn*/*syn*:*syn*/*anti* 99:1). The enolate may be derived from ar. or aliphatic ketones; however, reaction with aliphatic aldehydes was less efficient. The stereochemistry is attributed to a *syn*-selective aldol reaction and the subsequent 1,2-*syn*-selective intramolecular reduction (the fluoride ion facilitating both steps). F.e.s. K. Miura, A. Hosomi et al., Chem. Lett. *2000*, 150-1.

Tetra-n-butylammonium bromide Bu_4NBr
Catalytic Michael-type addition $C=C \rightarrow CHC(R)$
with enoxystannanes under mild conditions

449.

to 1,1-nitroethylene derivs. β-Nitrostyrene added to a mixture of 3 eqs. startg. tin enolate and 0.1 eq. Bu_4NBr in dry THF under N_2 at -45°, the mixture stirred at the same temp. for 4 h, and worked up by stirring vigorously after adding ether and aq. NH_4F for 15 min → product. Y 94%. Bu_4NBr greatly enhanced the reaction rate, yield and *syn*-selectivity with trisubst. enoxystannanes. F.e. and addition to α,β-ethylenenitriles (with Bu_4NCl), s. M. Yasuda, A. Baba et al., Chem. Lett. *2000*, 1266-7.

Dichloro[1,3-bis(diphenylphosphino)propane]nickel(II))/n-butyllithium/
  potassium phosphate
Nickel-catalyzed Suzuki diaryl coupling with ar. chlorides Ar-Ar'
with $NiCl_2$(dppf) cf. *51*, 453; with $NiCl_2$(dppp) using an automated, bench-top MEDLEY synthesizer under inert atmosphere, also suitable for reactions with air-sensitive organolithium compds. and Grignards over a wide temperature range, s. A. Orita, J. Otera et al., Org. Process Res. Dev. *4*, 337-41 (2000); apparatus s. ibid. 333-6.

Bis(1,5-cyclooctadiene)rhodium(I) fluoroborate/sodium fluoride [Rh(cod)$_2$]BF$_4$/NaF
Synthesis of benzylalcohols from aldehydes and arylstannanes in water CHO → CH(OH)Ar

450. NC-C$_6$H$_4$-CHO + Me$_3$SnPh ⟶ NC-C$_6$H$_4$-CH(OH)Ph

An eco-friendly alternative to 53, 470, not requiring an organic solvent or an inert atmosphere, is reported. E: A mixture of startg. aldehyde, 1 eq. phenyltrimethylstannane, 5 eqs. NaF and 5% mol [Rh(cod)$_2$]BF$_4$ in deionized water stirred in a capped vessel overnight at 110° → product. Y 92%. Organometallic additions to carbonyl groups generally require anhydrous conditions. F.e.s. C.J. Li, Y. Meng, J. Am. Chem. Soc. *122*, 9538-9 (2000).

Acetoacetonato(dicarbonyl)rhodium(I)/hydrogen peroxide/sodium hydrogen carbonate ←
7-Ethylene-1,3,5-triols from (diallylhydridosiloxy)-3-ethylenes ←
via regio- and stereo-specific intramolecular silylformylation-allylsilylation

451.

A mixture of startg. siloxyethylene and 3 mol% Rh(acac)(CO)$_2$ in benzene heated in a stainless-steel pressure reactor at 60° under 1000 psi CO until reaction complete, the solvent evaporated, and the residue treated with H$_2$O$_2$ and NaHCO$_3$ in THF/methanol at reflux under standard Tamao conditions → *syn*,*syn*-triol. Y 59% (77% diastereoselectivity). Silylformylation was slow at <45° or <800 psi CO. F.e. and with retention of *tert*-butyldimethylsilyl ethers s. M.J. Zacuto, J.L. Leighton, J. Am. Chem. Soc. *122*, 8587-8 (2000).

Acetoacetonatobis(ethylene)rhodium(I)/(S)-2,2'-bis(diphenylphosphino)- ←
1,1'-binaphthyl
Asym. 1,4-addition of arylboronic acids C=C → CHC(Ar)
to enones cf. 55, 452; to enoates s. S. Sakuma, N. Miyaura et al., J. Org. Chem. *65*, 5951-5 (2000); also asym. 1,4-addition with lithium aryl(trimethoxy)borates s. Y. Takaya, T. Hayashi et al., Tetrahedron:Asym. *10*, 4047-56 (1999).

Asym. 1,4-addition of boronic acids to 1,1-nitroethylene derivs. C=C → CHC(R)

452. 1-nitrocyclohexene + Ph-B(OH)$_2$ ⟶ 2-phenyl-nitrocyclohexane

Dioxane added to a mixture of 0.03 eq. Rh(acac)(CH$_2$=CH$_2$)$_2$, 0.033 eq. (S)-BINAP, and 5 eqs. phenylboronic acid, stirred for 3 min at room temp. under N$_2$, 1-nitrocyclohexene added along with 10 vol% water, and stirring continued for 3 h at 100° → product. Y 79% (*cis*-selectivity 87%; e.e. 98.3%). Stereoselectivity was even higher in aq. DMA. The thermodynamically less stable *cis*-isomer is produced with high enantioselectivity. F.e. and *cis*→*trans*-isomerization (with NaHCO$_3$ in refluxing ethanol), also with α,β-ethyleneboronic acids, s. T. Hayashi et al., J. Am. Chem. Soc. *122*, 10716-7 (2000).

(1,5-Cyclooctadiene)bis(methyldiphenylphosphine)rhodium(I) triflate [Rh(cod)(MePh$_2$P)$_2$]TfO
γ,δ-Ethyleneketones from 2-ethylenecarbonic acid esters and enoxysilanes ←

453.

(56 : 44)

under neutral conditions. A mixture of startg. allylic carbonate, 4 eqs. enoxysilane and 1 mol% (Rh(cod)(MePh$_2$P)$_2$)OTf in dichloromethane heated in a sealed tube at 80° for 19 h → product.

Y 92% (56:44 mixture of regioisomers). Reaction is initiated by oxidative addition of Rh(I) to the O-Si bond, and may subsequently involve an η^3-allylrhodium complex. Other Rh catalysts were less effective. Amides, ketones and esters remained intact. F.e.s. T. Muraoka, I. Matsuda et al., Tetrahedron Lett. *41*, 8807-11 (2000).

Palladium/potassium fluoride-alumina/microwaves ←
Palladium nanoparticles/poly(N-vinyl-2-pyrrolidone)/sodium phosphate ←
Heterogeneous Suzuki diaryl coupling Ar-Ar'
with polysiloxane-supported Pd-complexes cf. *37*, 902s*57*; with 5% Pd powder in 40% KF/Al$_2$O$_3$ under microwave irradiation without solvent s. G.W. Kabalka et al., Green Chem. *2*, 120-2 (2000); from electron-deficient ar. chlorides and chloropyridines *in a 3-phase medium* with a polymer-based dichloropalladium(II) phosphine complex cf. K. Inada, N. Miyaura, Tetrahedron *56*, 8661-4 (2000); with palladium nanoparticles stabilized with poly(N-vinyl-2-pyrrolidone) s. Y. Li, M.A. Elsayed et al., Org. Lett. *2*, 2385-8 (2000).

Palladous acetate/1,3-bis(2,6-diisopropylphenyl)imidazol-2-ylidene hydrochloride/ ←
tetra-n-butylammonium fluoride
Diaryls from aryl(trialkoxy)silanes and ar. halides
with Pd(OAc)$_2$/Ph$_3$P or Pd$_2$(dba)$_3$/2-(dicyclohexylphosphino)biphenyl cf. *57*, 439s*58*; from ar. bromides and electron-deficient ar. chlorides, as well as heteroaryl halides (e.g. 2-bromopyridine), with Pd(OAc)$_2$/1,3-bis(2,6-diisopropylphenyl)imidazol-2-ylidene hydrochloride/Bu$_4$NF, s. H.M. Lee, S.P. Nolan, Org. Lett. *2*, 2053-5 (2000).

Palladous acetate/triphenylphosphine/sodium tetraarylborates ←
Synthesis of (Z)-α-alkylidene-γ-lactones from 3-acetylenechloroformic acid esters ○
via intramolecular acylpalladation-cross-coupling

454.

Equimolar amounts of startg. chloroformate and NaBPh$_4$ in THF at oil bath temp. of 65-70° treated with 10 mol% Pd(OAc)$_2$ and 20 mol% PPh$_3$ → product. Y 73%. Internal acetylenes yielded significant amounts of acoxy-3-acetylenes as by-product. The (Z)-isomer produced can be converted to the (E)-isomer in quantitative yield by heating with an excess of sec. amine. F.e. incl. an example of ring closure-coupling with thienyltributylstannane s. R. Grigg, V. Savic, Chem. Commun. *2000*, 2381-2.

Palladous acetate/tri-o-tolylphosphine/potassium carbonate ←
Double Suzuki coupling
in aq. organic medium - 2,3-Diarylindoles s. *53*, 473s*60*

Palladous acetate/tri-2-furylphosphine Pd(OAc)$_2$/Fu$_3$P
Synthesis of lactams from unsatd. carbamyl chlorides ○
via intramolecular acylpalladation-cross-coupling

455.

3-Alkylideneoxindoles. A soln. of startg. carbamyl chloride and 1.1 eqs. tributyl(2-thienyl)tin in toluene heated at 50° for 5 min with 10 mol% Pd(OAc)$_2$ and 20 mol% tri-2-furylphosphine → product. Y 88% (91% after 30 min with just *1 mol%* Pd(OAc)$_2$ and *2 mol%* ligand). F.e. incl. isoquinoline and pyrrole analogs, also from ethylenic carbamyl chlorides, s. M.R. Fielding, R. Grigg et al., Chem. Commun. *2000*, 2239-40.

Palladous acetate/diadamantyl-n-butylphosphine/potassium phosphate
Suzuki diaryl coupling with non- or de-activated ar. chlorides   ← Ar-Ar'

456.

A suspension of 2-chlorotoluene, 1.5 eqs. phenylboronic acid, 2 eqs. K-phosphate, 0.01 mol% diadamantyl-*n*-butylphosphine and *0.005 mol%* Pd(OAc)$_2$ in dry toluene sealed under argon in an ACE pressure tube, and heated in an oil bath at 100° for 20 h → product. Y 87% (TON *17,400*). The catalyst/ligand combination is the most efficient system reported thus far for this coupling in terms of productivity and activity (TON values of 10,000-20,000 being achieved even for deactivated ar. chlorides, *independent* of steric and electronic effects). The industrial potential is underscored. F.e. incl. coupling with 3-chloropyridine s. A. Zapf, M. Beller et al., Angew. Chem. Int. Ed. Engl. *39*, 4153-5 (2000).

Palladous acetate/tetra-n-butylammonium bromide/potassium phosphate   ←
Suzuki diaryl coupling with a highly active catalyst s. *56*, 445s60

Bis(dibenzylideneacetone)palladium(0)/triphenylphosphine/tetra-n-butylammonium acetate   ←
Allylarenes from ar. iodides and 2-ethylenesilanes with allyl shift   I → C-C=C
s. *27*, 871s60

Tris(dibenzylideneacetone)dipalladium/tri-2-furylphosphine/   ←
cuprous thiophene-2-carboxylate
Ketones from thiolic acid esters and boronic acids   COSR → COR'

457.

α,β-Ethyleneketones. A mixture of startg. thiolic acid ester, 1.1 eqs. β-styrylboronic acid, 1.6 eqs. Cu(I) thiophene-2-carboxylate, 1% Pd$_2$dba$_3$·CHCl$_3$, and 3 mol% tri-2-furylphosphine in THF stirred under argon at 50° for 18 h *without base* → product. Y 81%. Reaction is presumed to involve initial oxidative addition of Pd(0) to the copper(I) *carboxylate*-bound thiolate, followed by Suzuki-type cross-coupling with the boronic acid (possibly via transmetalation with copper). The method is mild and efficient, and based on readily available reactants. Highly functionalized and base-sensitive residues were tolerated. F.e. incl. trifluoromethyl, heterocyclic and **aryl ketones**, also coupling with S-aryl thiolates, s. L.S. Liebeskind, J. Srogl, J. Am. Chem. Soc. *122*, 11260-1 (2000); aryl ketones with *trans*-di(μ-acetato)bis[*o*-(di-*o*-tolylphosphino)benzyl]dipalladium(II) in the presence of K$_2$CO$_3$ and NaI cf. Org. Lett. *2*, 3229-31 (2000).

Tris(dibenzylideneacetone)dipalladium/(S)-2-dimethylamino-2'-dicyclohexyl-   ←
phosphino-1,1'-binaphthyl/potassium phosphate/sodium iodide
Asym. Suzuki coupling   Ar-Ar'
chiral 1,1'-binaphthyls with a PdCl$_2$/chiral ferrocenyl aminophosphine complex cf. *59*, 447; chiral diaryls with Pd$_2$(dba)$_3$/(S)-2-dimethylamino-2'-dicyclohexylphosphino-1,1'-binaphthyl s. J. Yin, S.L. Buchwald, J. Am. Chem. Soc. *122*, 12051-2 (2000).

Palladium phosphine complex   ←
Suzuki coupling in aq. organic media   B(OH)$_2$ → R
in an aq. 2-phase medium with a water-soluble triarylphosphine as ligand cf. *53*, 473; in aq. THF with PdCl$_2$(PPh$_3$)$_2$/Na$_2$CO$_3$ for the synthesis of N-protected lactams s. E.G. Occhiato, A. Guarna et al., Org. Lett. *2*, 1241-2 (2000); heteroarylbenzoic acids in aq. acetonitrile with Pd(PPh$_3$)$_4$/Na$_2$CO$_3$ s. Y. Gong, H.W. Pauls, Synlett *2000*, 829-31; 2,3-diarylindoles from 2,3-dihalogenoindoles by double coupling with Pd(OAc)$_2$/tri-*o*-tolylphosphine s. Y. Liu, G.W. Gribble, Tetrahedron

Lett. *41*, 8717-21 (2000); β-perfluoroalkylstyrenes with PdCl$_2$(PPh$_3$)$_2$ s. M.P. Jennings et al., J. Org. Chem. *65*, 8763-6 (2000); removal of Pd(II) from aq. and organic solutions with polystyrene-bound trimercaptotriazine s. K. Ishihara et al., Chem. Lett. *2000*, 1218-9.

Polymer-based palladium phosphine complex
Suzuki diaryl coupling in a 2-phase medium ← Ar-Ar′
under thermomorphic conditions s. *60*, 399

Polymer-based palladium phosphine complex/potassium phosphate ←
Suzuki diaryl coupling with ar. chlorides in a 3-phase medium s. *37*, 902s60

Tetrakis(triphenylphosphine)palladium(0) Pd(PPh$_3$)$_4$
Stille diaryl coupling
s. *34*, 862s57; with chemically tagged bromobenzoic acids for reaction monitoring s. X. Wang, J.A. Porco et al., Org. Lett. *2*, 3509-12 (2000).

Tetrakis(triphenylphosphine)palladium(0)/potassium hydroxide Pd(PPh$_3$)$_4$/KOH
δ,ε-Acetylenefluorides from 1,3-fluoroiodides and acetylenestannanes Sn≤ → R

458. Ph—≡—SnBu$_3$ + I—CH$_2$—CHF—C$_7$H$_{15}$-n ⟶ Ph—C≡C—CH$_2$—CHF—C$_7$H$_{15}$-n

262 ml (Phenylethynyl)tributyltin added dropwise under argon to 71.5 mg 3-fluoro-1-iododecane and 28.9 mg Pd(PPh$_3$)$_4$ in dry benzene contained in a pressure tube fitted with a septum seal, stirred for 16 h at 120°, and mixed with aq. KOH for 1 day at room temp. → 5-fluoro-1-phenyl-1-dodecyne. Y undisclosed. The procedure is simple and efficient, and is suitable for the conversion of substrates which are unstable in strongly basic media. F.e.s. Japanese patent JP-11246448 (Sagami Chem. Res. Centre).

Tetrakis(triphenylphosphine)palladium(0)/sodium carbonate Pd(PPh$_3$)$_4$/Na$_2$CO$_3$
Intramolecular carbopalladation-Suzuki coupling ○
with Pd(OAc)$_2$/Ph$_3$P/Na$_2$CO$_3$/Et$_4$NCl cf. *55*, 454; 4-benzyl-3-methylene-1-tosylpyrrolidines with Pd(PPh$_3$)$_4$/Na$_2$CO$_3$ s. C.-W. Lee, K.H. Ahn et al., Org. Lett. *2*, 1213-6 (2000).

Tetrakis(triphenylphosphine)palladium(0)/potassium carbonate Pd(PPh$_3$)$_4$/K$_2$CO$_3$
Methylarenes from ar. halides Hal → Me
and 10-methyl-9-oxa-10-borabicyclo[3.3.2]decane cf. *46*, 899; with trimethylboroxine s. M. Gray et al., Tetrahedron Lett. *41*, 6237-40 (2000).

Tetrakis(triphenylphosphine)palladium(0)/sodium iodide Pd(PPh$_3$)$_4$/NaI
Arylferrocenes from ar. iodides via palladium-catalyzed cross-coupling ←

459. 2 Br—C$_6$H$_4$—I + Fc—Hg—Fc ⟶ 2 Br—C$_6$H$_4$—Fc

A mixture of bis(ferrocenyl)mercury, 2.3 eqs. *p*-bromoiodobenzene, 4 eqs. NaI and 2 mol% Pd(PPh$_3$)$_4$ in 3:2 freshly distilled THF/dry acetone refluxed under argon for 1.5 h → 1-(*p*-bromophenyl)ferrocene. Y 88%. The method is simple, efficient, general and chemoselective for ar. iodides in the presence of functional groups such as aromatic bromides, fluorides, nitro compds., and esters as well as phenolethers and trifluoromethyl groups. Substrates with electron-withdrawing substituents led to better yields than those with electron-donating ones, whereas substrates with *ortho*-substituents, predictably, required longer reaction times and gave lower yields. *p*-Iodophenol also gave a poor yield. F.e. incl. heteroarylferrocenes (e.g. 2-thienyl- or 3-pyridyl-analogs) and with PdCl$_2$(PPh$_3$)$_2$ (PdCl$_2$(dppf) and PdCl$_2$(MeCN)$_2$ giving poor results) s. A.V. Tsvetkov, I.P. Beletskaya et al., Tetrahedron Lett. *41*, 3987-90 (2000).

Tetrakis(triphenylphosphine)palladium(0)/silver(I) oxide $Pd(PPh_3)_4/Ag_2O$
Cross-coupling of unsatd. silanols with ar. halides ←
s. *45*, 555s*58*; with unsatd. silane-diols or -triols, or with dichlorosilanes in aniline/water, s. E. Hirabayashi, A. Mori et al., J. Org. Chem. *65*, 5342-9 (2000); with unsatd. 3,3,3-trifluoropropylsilanols cf. Bull. Chem. Soc. Jpn. *73*, 749-50 (2000); cross-coupling of 1-alkoxyenesilanols or 1-alkoxyvinylsilicon hydrides with ar. iodides using (η^3-allylpalladium chloride)$_2$/Bu$_4$NF cf. S.E. Denmark, L. Neuville, Org. Lett. *2*, 3221-4 (2000).

Tetrakis(triphenylphosphine)palladium(0)/cuprous iodide $Pd(PPh_3)_4/CuI$
Stille coupling with enestannanes   Sn⇐ → R
s. *39*, 887s*58*; α-trifluoromethylstilbenes s. I.H. Jeong, B.T. Kim et al., Tetrahedron Lett. *41*, 8917-21 (2000); (E)-β-germylstyrenes (without CuI) s. F. David-Quillot, A. Duchêne et al., ibid. 9981-4; silylated β-aryl-β,γ-ethylenecarboxylic acids s. ibid. 8893-6.

Tetrakis(triphenylphosphine)palladium(0)/cuprous iodide/lithium bromide ←
Palladium-catalyzed cross-coupling with lithium alk-1-ynyl(trialkoxy)borates ←

460.

1,3-Enynes. A mixture of 2 eqs. lithium *tert*-butylethynyl(triisopropoxy)borate, 5 mol% Pd(PPh$_3$)$_4$, and 5 mol% CuI in dry DMF purged with argon, a soln. of startg. vinyl bromide in DMF added via cannula, and stirred at 60° for 36 h → product. Y 98%. The method is free of coupling by-products and serves as an alternative to Stille, Sonogashira or Suzuki coupling. F.e. **and arylacetylenes** from electronically and sterically diverse ar. bromides or [preferably] iodides s. C.H. Oh, S.H. Jung, Tetrahedron Lett. *41*, 8513-6 (2000).

trans-Di(μ-acetato)bis[o-(di-o-tolylphosphino)benzyl]dipalladium(II)/
potassium carbonate/sodium iodide ←
Aryl ketones from arylboronic acids and thiolic acid esters s. *60*, 457   ArB(OH)$_2$ → ArCOR

Palladous chloride $PdCl_2$
Coupling of triarylantimony diacetates with enoxysilanes
under mild conditions ←

461.

5-Aryl-2(3H)-furanones. 1 eq. (Furan-2-yloxy)trimethylsilane added to a mixture of triphenylantimony diacetate and 5 mol% PdCl$_2$ in DME/acetonitrile (1:1), and stirred at room temp. for 8 h under N$_2$ → 5-phenylfuran-2(3H)-one. Y 82%. Conditions are *mild* and reaction can be carried out even in the presence of water (10% v/v). F.e. and with PdCl$_2$(MeCN)$_2$, **also α-arylketones** and α'-aryl-α,β-ethyleneketones, and **cinnamic acid esters from 1-alkoxy-1-siloxycyclopropanes**, s. S.-K. Kang et al., J. Chem. Soc. Perkin Trans. 1 *2000*, 3350-1.

Palladous chloride/lithium chloride/acetic acid ←
β-Arylketones from α,β-ethyleneketones and arylstannanes   C=C → CHC(Ar)
with [Rh(cod)(MeCN)$_2$]BF$_4$ cf. *54*, 466; from aryl(chloro)stannanes or tetraarylstannanes with Pd(II)-catalysts (e.g. PdCl$_2$/LiCl/AcOH) s. T. Ohe, S. Uemura, Bull. Chem. Soc. Jpn. *73*, 2149-55 (2000).

Bis(η^3-allylpalladium chloride)/tetra-n-butylammonium fluoride ←
Cross-coupling of 1-alkoxyenesilanols with ar. iodides s. *45*, 555s*60* ←

Dichlorobis(diethyl sulfide)palladium(II)/potassium phosphate $PdCl_2(SEt_2)_2/K_3PO_4$
Suzuki diaryl coupling with a highly active catalyst s. *56*, 445s*60*   Ar-Ar'

Dichlorobis(triphenylphosphine)palladium(II) $PdCl_2(PPh_3)_2$
2-Alkoxy-1,3-dienes by Stille coupling s. *60*, 175   Sn≤ → R

Fluorous dichlorobis(triarylphosphine)palladium(II)/lithium chloride   ←
Stille coupling in a fluorous 2-phase medium

462.

A soln. of methyl 4-bromobenzoate, 1.2 eqs. 2-(tributylstannyl)furan, and 1 eq. LiCl *in DMF* added to a suspension of 1.5 mol% fluorous dichlorobis(triarylphosphine)palladium(II) *in perfluoromethylcyclohexane* under argon, heated at 80° for 3 h, cooled to room temp., and the two phases worked up → methyl 4-(2-furyl)benzoate. Y 91%. The product was isolated from the DMF layer (and washings), while the fluorous phase, containing the catalyst, was set aside for reuse (up to 6 times with only a slight reduction in the yield). F.e. and comparison of ligands s. S. Schneider, W. Bannwarth, Angew. Chem. Int. Ed. Engl. *39*, 4142-5 (2000).

[2,6-Bis(diphenylphosphinoxy)-4-methylphenyl]palladium(II) trifluoroacetate   ←
Suzuki diaryl coupling with highly active catalysts   Ar-Ar'
with σ-palladated phosphite complexes cf. *56*, 445; with *0.001 mol%* pincer-type [2,6-bis(diphenylphosphinoxy)-4-methylphenyl]palladium(II) trifluoroacetate (TON up to 190,000) s. R.B. Bedford, S.L. Welch et al., New J. Chem. *24*, 745-7 (2000); with 0.2 mol% *phosphine-free* palladacyclic 2-[1-(*tert*-butylthio)ethyl]phenylpalladium(II) chloride dimer cf. D. Zim, J. Dupont et al., Org. Lett. *2*, 2881-4 (2000); with 0.2 mol% $PdCl_2(SEt_2)_2/K_3PO_4$ or $Pd(OAc)_2/Bu_4NBr/K_2CO_3$ s. Tetrahedron Lett. *41*, 8199-202 (2000).

Dichlorobis[tris[4-[2-(perfluorohexyl)ethyl]phenyl]phosphine]platinum(II) $PtCl_2(PAr_3)_2$
3-Ethylenealcohols from aldehydes $CHO → CH(OH)CH_2CH=CH_2$
Synthesis with addition of three C-atoms
using a fluorous allylstannane under thermal conditions cf. *54*, 431; under milder conditions in benzotrifluoride with a readily recyclable fluorous Pt-complex, dichlorobis[tris[4-[2-(perfluorohexyl)ethyl]phenyl]phosphine]platinum(II), s. Q. Zhang, D.P. Curran et al., J. Org. Chem. *65*, 8866-73 (2000).

Carbon ↑   CC ↓↑ C

Without additional reagents   w.a.r.
2,4-Diaminoquinolines from 1-acylbenzotriazoles and three isocyanate molecules   ○

463.

A mixture of 1-butanoylbenzotriazole and 3 eqs. phenyl isocyanate heated in a sealed tube at 210° for 24 h → 2,4-dianilino-3-ethylquinoline. Y 86%. Reaction is limited to higher 1-alkanoylbenzotriazoles. 1-Acetylbenzotriazole gave 5-aminopyrimido[5,4-*c*]quinoline-2,4(1*H*,3*H*)-diones, while 1-aroylbenzotriazoles gave 6-aminophenanthridines. F.e. and reactions with 1-acetoacetyl- and 1-cinnamoyl-benzotriazoles, s. A.R. Katritzky, P.J. Steel et al., J. Org. Chem. *65*, 8069-73 (2000).

Electrolysis   ⚡
Kolbe synthesis   2 RCOOH → R-R
bis(phosphine oxides) s. *1*, 715s59; P-chiral bis(phosphine oxides) s. M. Sugiya, H. Nohira, Bull. Chem. Soc. Jpn. *73*, 705-12 (2000).

Microwaves s. under Piperidine ←

Potassium hydroxide   KOH
Benzene ring from 2-pyrones and ketones ←

464.

Functionalized diaryls. A mixture of startg. 6-aryl-3-cyano-4-methylthio-2-pyrone, 1.5 eqs. acetone, and 1.5 eqs. KOH in dry DMF stirred at room temp. under N_2 for 30 h, poured into ice-water, stirred vigorously for 30 min, and acidified with 10% HCl → 4-(4-fluorophenyl)-2-methyl-6-(methylthio)benzonitrile. Y 65%. The procedure is versatile, mild, and cost-effective, the work-up is easy, and no catalyst is required. A carbanionic mechanism is invoked. F.e. and 3-amino-4-cyanobiphenyls s. V.J. Ram et al., J. Chem. Soc. Perkin Trans 1 *2000*, 3719-23.

n-Butyllithium/lithium chloride   BuLi/LiCl
Asym. synthesis of 5-alkylidenecyclopent-2-en-2-olones   ○
from α-alkoxy-α-allenecarboxylic acid amides cf. *58*, 465; from α,β-ethylenecarboxylic acid amides and 1-glycosyloxyallenes with *n*-BuLi/LiCl, followed by quenching with aq. HCl or [preferably] 1,1,1,3,3,3-hexafluoroisopropanol s. P.E. Harrington, M.A. Tius, Org. Lett. 2, 2447-50 (2000).

Sodium carbonate Na_2CO_3
Furans from oxazoles and acetylene derivs. ←
s. *28*, 869; 2-arylfuran-3,4-dicarboxylic acid esters with Na_2CO_3 in the absence of solvent, s. W. Pei, J. Pei et al., Synthesis *2000*, 2069-77.

Sodium hydrogen carbonate $NaHCO_3$
C-β-(β-Ketoalkyl) glycosides from unprotected aldoses   OH → C-C(O)
via Knoevenagel condensation in aq. medium

465.

1.5 eqs. $NaHCO_3$ and 1.2 eqs. acetylacetone added to a soln. of D-glucose in water, stirred at 90° for 6 h, washed with methylene chloride, and treated with Dowex resin 50X8-200 (H⁺ form) → product. Y 96% (100% β). This provides an efficient and convenient one-step synthesis of pure C-β-glycosidic ketones directly from unprotected sugars under thermodynamic control. F.e. and furans by Yb(OTf)$_3$-catalyzed reaction s. F. Rodrigues, A. Lubineau et al., Chem. Commun. *2000*, 2049-50.

Cesium fluoride   CsF
2-Siloxy-1,1,1-trifluorides from oxo compds. s. *60*, 224   CO → C(OSi≤)CF$_3$

Lithium chloride s. under BuLi   LiCl

Piperidine/microwaves ←
Doebner synthesis of α,β-ethylenecarboxylic acids   CHO → CH=CHCOOH
under microwave irradiation
with NH$_4$OAc without solvent cf. *1*, 569s55; with piperidine s. R.F. Pellón et al., Synth. Commun. 30, 3769-74 (2000).

1,8-Diazabicyclo[5.4.0]undec-7-ene **DBU**
(all-E)-2,4,6-Trienals from aldehydes CH=CHCH=CHCH=CHCHO
Synthesis with addition of six C-atoms s. *60*, 428

4-Pyrrolidinopyridine ←
α-Siloxycarboxylic acid amides from oxo compds. and amines CHO → CH(OSi≤)CON≤
3-Component synthesis by insertion of a zwitterionic carbonyl equivalent

466.

with asym. induction. 1 eq. 4-pyrrolidinopyridine added to a mixture of startg. aldehyde, 1.2 eqs. L-leucine benzyl ester, and 2 eqs. *tert*-butyldimethylsiloxymalononitrile in ether at 0°, and stirred for 5 h at the same temp. → product. Y 80% (diastereoselectivity 79%). With other prim. or sec. amine components, no additional base was usually needed. F.e.s. H. Nemoto et al., Org. Lett. *2*, 4245-7 (2000).

Magnesium/silver fluoroborate **Mg/AgBF$_4$**
2-Ethylene-*tert*-amines from α-*tert*-aminonitriles CN → C=C
Modified Bruylants reaction with vinylmagnesium bromides

467.

A soln. of startg. α-aminonitrile in THF treated at room temp. with 1.6 eqs. AgBF$_4$ (as azomethinium ion promoter) for 10 min, cooled to -78°, 4 eqs. vinylmagnesium bromide added, and allowed to react at -78 to 0° for 1 h → product. Y 98% (32% without AgBF$_4$). Other vinylic organometallics were less effective or inactive. 1-Alkynylmagnesium halides reacted similarly to give the corresponding **2-acetylene-*tert*-amines**, preferably *without* AgBF$_4$. F.e. and diastereoselectivity s. C. Agami, G. Evano et al., Org. Lett. *2*, 2085-8 (2000).

Indium *In*
Sym. α-diketones from acylcyanides 2 COCN → COCO

468.

under mild, neutral conditions. A mixture of benzoyl cyanide and 1.5 eqs. In powder (100 mesh) in DMF stirred at room temp. under sonication for 8 h, then quenched with a few drops of water → benzil. Y 62%. The method is simple and convenient but limited to aroyl cyanides, aliphatic acyl cyanides being unreactive. Phenolethers, aromatic chlorides and bromides were unaffected. DMF is the most suitable solvent and reaction failed in THF, THF/water, DMF/water, acetonitrile or water. Acyl chlorides gave a complex product mixture. F.e.s. H.S. Baek, S.H. Kim et al., Tetrahedron Lett. *41*, 8097-99 (2000).

Trimethylaluminum/1,1'-bi-2-naphthol **Me$_3$Al/BINOL**
Catalytic aldol transfer condensation CHO → CH(OH)C-CO

469.

via aluminum enolates. A soln. of 0.1 eq. trimethylaluminum (2 *M* in toluene/heptane) added to a degassed soln. of 5 mol% 1,1'-bi-2-naphthol in dry dichloromethane at room temp. under argon, stirred 1 h, cooled to 0°, 1 eq. pivaldehyde and 1 eq. diacetone alcohol added simultaneously when gas evolution had stopped, allowed to warm to room temp., stirred for 40 h, and poured into 0.5 M HCl → product. Y 72%. Diacetone alcohol, the aldol self-condensation product of acetone, undergoes retro-aldol reaction to liberate acetone, which condenses with the startg. aldehyde. Reaction failed with β-hydroxy-esters or -aldehydes, as did condensation with ketones. F.e. and aluminum chelate catalysts s. I. Simpura, V. Nevalainen, Angew. Chem. Int. Ed. Engl. *39*, 3422-4 (2000).

Montmorillonite s. under TsOH ←

Boron fluoride BF_3
1,2-*trans*-C-Allyl glycosides from carbohydrate 1,2-O-isopropylidene derivs. s. *60*, 432 C

Silver fluoroborate s. under Mg $AgBF_4$

Aluminum chloride $AlCl_3$
Friedel-Crafts α-*prim*-aminoacylation with 2,5-oxazolidiones

470.

with retention of chirality. 3 eqs. $AlCl_3$ added portionwise over 25 min with vigorous stirring to a cooled soln. of methyl L-aspartate N-carboxyanhydride in benzene, stirred at room temp. for 8 h, and the product recrystallized from ethanol → ethyl (S)-3-benzoyl-3-aminopropionate hydrochloride. Y 43% (e.e. 94%). F.e.s. O. Itoh, A. Amano, Synthesis *1999*, 423-8.

Trifluoroacetic acid CF_3COOH
Decarboxylative Biginelli synthesis of 3,4-dihydro-2(1*H*)-pyrimidinones O

471.

Trifluoroacetic acid added with stirring to a mixture of 1.36 eqs. urea and 1 eq. 3,5-dimethylbenz-aldehyde in anhydrous dichloroethane, 0.91 eq. oxalacetic acid added, and the mixture refluxed 12 h → 6-(3,5-dimethylphenyl)-2-oxo-1,2,3,6-tetrahydropyrimidine-4-carboxylic acid. Y 99%. Yields are high, and the method is clean and generally applicable to electron-rich or -deficient aldehydes. F.e. and method, also conversion to the corresponding esters or 6-acyl-derivs. s. J.C. Bussolari, P.A. McDonnell, J. Org. Chem. *65*, 6777-9 (2000).

Titanium tetrachloride $TiCl_4$
4(1*H*)-Quinolones from 3,1-benzoxazine-2,4-diones ←
s. *28*, 418s33; quinoline-2,4-diones from O-silyl O-alkyl keteneacetals, also 3,3-disubst. 1,8-naphthyridine-2,4-diones, s. A.L. Zografos et al., Org. Lett. *1*, 1953-5 (1999).

Stannic chloride/pyridine/1,8-diazabicyclo[5.4.0]undec-7-ene ←
2-α-Hydroxy-1-monothiolglutaric acid esters C
from 2,2-dialkoxy-6-alkylthio-3,4-dihydro-2*H*-pyrans and aldehydes
threo-adducts cf. *54*, 477; *erythro*-adducts via epimerization of the intermediate 2,2-dialkoxytetra-hydropyran with added DBU/pyridine prior to hydrolysis s. C.-M. Yu et al., J. Chem. Soc. Perkin Trans. 1 *2000*, 3622-6.

Phosphorus oxide chloride $POCl_3$
Condensation of carboxylic acids with two dimethylformamide molecules ←
β-amino-α,β-ethylenealdehydes s. *30*, 634; vinylogous amidinium salts, also from carboxylic

acid chlorides (notably for preparing α-chloro-derivs.), s. I.W. Davies et al., J. Org. Chem. 65, 4571-4 (2000).

p-Toluenesulfonic acid/montmorillonite ←
Porphyrin synthesis ○
s. *21*, 948s*37, 59*; with TsOH-on-montmorillonite K10, also pyrromethanes from two 5-(acetoxymethyl)pyrrole molecules, s. B.A. Freeman, K.M. Smith, Synth. Commun. 29, 1843-55 (1999).

Tungsten pentacarbonyl/p-toluenesulfonic acid $W(CO)_5/TsOH$
Naphthalenes from *o*-acetyleneketones and enolethers or enamines ○
via Diels-Alder reaction with tungsten 2-benzopyran-3-ylidene complexes

472.

One-pot procedure. A soln. of *o*-ethynylbenzophenone in THF treated with 3 eqs. $W(CO)_5$·THF at room temp. for 1 day in the presence of a little TsOH, 4 eqs. butyl vinyl ether added to the intermediate benzopyranylidene complex, and worked up when reaction complete (<10 h) → 1-phenylnaphthalene. Y 95%. The intermediate complexes are dark-blue, air-stable, crystalline solids, isolable in high yield. Diels-Alder reaction with enamines was instantaneous. F.e.s. N. Iwasawa et al., J. Am. Chem. Soc. *122*, 10226-7 (2000).

Hydrogen chloride HCl
3-Acyl-2-azetidinone ring from 5-acyl-1,3-dioxane-4,6-diones with asym. induction CO

473.

Chiral 6-acylpenams. HCl gas bubbled through a soln. of startg. thiazoline (obtained from commercially available L-cysteine methyl ester hydrochloride in 2 steps) and 1.66 eqs. startg. Meldrum's acid deriv. in dry benzene at 5° for 10 min, heated at 79° for 1.5 h, and cooled to room temp. → product. Y 93%. F.e., also selective reduction of the ester function to the aldehyde (with DIBAL-H), s. H. Emtenäs, F. Almqvist et al., Org. Lett. *2*, 2065-7 (2000).

Manganese dioxide/triethylamine MnO_2/Et_3N
Indolizines from pyridinioacetic acids and electron-deficient ethylene derivs. ○
via oxidative 1,3-dipolar cycloaddition with pyridinium methylids

474.

3-Unsubst. indolizines. A suspension of startg. N-(carboxymethyl)pyridinium chloride, 5 eqs. startg. alkene, 8 eqs. MnO_2, and Et_3N (0.15 ml per mmol substrate) in toluene stirred at 90° for 2 h → product. Y 67%. The procedure is simple and convenient, and not susceptible to base. F.e. and [less conveniently] from acetylene derivs., also with pyridine, morpholine or K_2CO_3 as base, s. L. Zhang, Y. Hu et al., Synthesis *2000*, 1733-7.

Ruthenium carbene complex
Ring opening metathesis-cross metathesis-ring closing metathesis
Regio- and stereo-specific conversion
8-Vinyl-2,7-dioxabicyclo[4.3.0]non-4-enes

475.

8 Mol% $(Cy_3P)_2Cl_2Ru{=}CHPh$ in methylene chloride added to a soln. of startg. oxanorbornene and 1 eq. allyl acetate in the same anhydrous solvent, and stirred at room temp. for 3 h → product. Y 80%. F.e. and from propargyloxy analogs s. O. Arjona, J. Plumet et al., Tetrahedron Lett. *41*, 9777-9 (2000).

(Imidazolidin-2-ylidene)ruthenium carbene complex
Cross-metathesis of ethylene derivs.

s. *49*, 932s59; improved conversion with a *phosphine-free* (imidazolidin-2-ylidene)ruthenium carbene complex s. S. Gessler, S. Blechert et al., Tetrahedron Lett. *41*, 9973-6 (2000).

Dodecacarbonyltetrarhodium $Rh_4(CO)_{12}$
Pyrroles from β-diketones under mild neutral conditions

476.

Pyrrole-2-carboxylic acid esters. A mixture of ethyl isocyanoacetate, 0.75 mol% $Rh_4(CO)_{12}$, and 2 eqs. acetylacetone in dry toluene stirred under argon at 25° for 4 h → product. Y 84%. Cyclocondensation with unsym. 1,3-dicarbonyl compds. gives pyrroles regiospecifically on the basis of either steric or electronic effects. Mechanistically, reaction is presumed to involve oxidative addition of Rh(0) to the C-α-hydrogen bond of the isonitrile, followed by addition to a keto group prior to decarbonylation and cyclohydration. Simple ketones afforded **α,β-ethylene-α-formamidocarboxylic acid esters**. F.e. incl. a 4-fluoro deriv. s. H. Takaya, S. Murahashi et al., Org. Lett. *3*, 421-4 (2001).

Elimination

Hydrogen ↑

Lithium diisopropylamide/N-(tert-butylimino)benzenesulfinyl chloride
(E)-α,β-Ethyleneketones from ketones under mild conditions

477.

via lithium enolates. A soln. of cyclohexanone in THF added under argon to a freshly-prepared soln. of 1.23 eqs. LDA in the same solvent *at -78°*, stirred for 10 min at the same temp., treated with 1.2 eqs. N-(*tert*-butylimino)benzenesulfinyl chloride in THF, stirred for 30 min, and quenched with 1% HCl → product. Y 93% (by GC). Reaction is mild and generally applicable to acyclic and cyclic ketones, the former yielding (E)-isomers exclusively. With unsym. ketones, the *less*-substituted position is selectively dehydrogenated. F.e. and with added 15-crown-5 s. T. Mukaiyama et al., Chem. Lett. *2000*, 1250-1.

Boron fluoride s. under Polymer-based phenyl iodosotrifluoroacetate BF_3

Ammonium ceric nitrate $(NH_4)_2Ce(NO_3)_6$
4(3H)-Pyrimidinones from 5,6-dihydro-4(3H)-pyrimidinones s. *40*, 227s*58*   CHCH → C=C

Dimethyl sulfoxide   *DMSO*
Imidazoles from Δ²-imidazolines
with Pd-C cf. *4*, 810s*46*; 2-aryl derivs. with DMSO s. M. Anastassiadou, M. Payard et al., Synthesis *2000*, 1814-6.

Polymer-based phenyl iodosotrifluoroacetate/boron fluoride   ←
Intramolecular oxidative non-phenolic coupling   O
with $PhI(OCOCF_3)_2/BF_3$ cf. *29*, 910s*51*; of phenolether derivs. with polymer-based phenyl iodosotrifluoroacetate s. H. Tohma, Y. Kita et al., Tetrahedron *57*, 345-52 (2001).

N-(tert-Butylimino)benzenesulfinyl chloride s. under i-Pr₂NLi $PhS(Cl)=NBu\text{-}t$

Sulfuryl chloride/pyridine SO_2Cl_2/C_5H_5N
Dehydrogenation   CHCH → C=C
thioenethers with Et_3N as base cf. *31*, 905s*53*; $\Delta^{7,9(11)}$-steroids from Δ^7-steroids with pyridine as base s. N. Aksara et al., J. Chem. Res., Synop *2000*, 454.

Manganese dioxide or Nickel peroxide MnO_2 or NiO_2
Aromatization by dehydrogenation
thiazoles s. *25*, 649s*54*; chiral N-protected thiazol-2-yl α-aminoketones, also with Ni-peroxide, s. M. Groarke, M.A. McKervey et al., Tetrahedron Lett. *41*, 1279-82 (2000); 4-alkyl-1,3,5-triarylpyrazoles from Δ^2-pyrazoline derivs. s. Y.R. Huang, J.A. Katzenellenbogen, Org. Lett. *2*, 2833-6 (2000).

Ferric chloride $FeCl_3$
Aromatization by cyclodehydrogenation   O
extended polyarenes with $CuCl_2/AlCl_3$ cf. *35*, 655s*55*; also with $Cu(OTf)_2/AlCl_3$ or $FeCl_3$ (for alkyl-subst. polycyclic arenes) s. F. Dötz, K. Müllen et al., J. Am. Chem. Soc. *122*, 7707-17 (2000).

Bis(acetonitrile)dichloropalladium(II)/p-benzoquinone   ←
2-Methylenecyclopentanones from 1-vinylcyclobutanols   O
with $Pd(OAc)_2/O_2$ cf. *57*, 456; 3-alkoxy-2-methylenecyclopentanones with $Pd(MeCN)_2Cl_2/p$-benzoquinone s. L.S. Hegedus, P.B. Ranslow, Synthesis *2000*, 953-8.

Oxygen ↑   CC ↑ O

Without additional reagents   *w.a.r.*
Jacobs-Gould 4(1H)-pyridone ring closure   O
in boiling Dowtherm cf. *5*, 355, 614; energy efficient process without solvent by high throughput passage through a stainless steel coil at 380° s. T. Cablewski, C.R. Strauss et al., Green Chem. *2*, 25-7 (2000).

Potassium tert-butoxide   *KOBu-t*
Dieckmann cyclization
s. *33*, 918s*58*; chiral 3-pyrrolidone-4-carboxylic acid esters, chemoselectivity at -78°, s. A.C. Pinto, P.R.R. Costa et al., Tetrahedron:Asym. *11*, 4239-43 (2000).

Lithium bromide/Amberlite 15   ←
Ethylene derivs. from epoxides   ▽ → C=C
(E)-α,β-ethyleneketones cf. *36*, 932s*58*; (E,E)-2,4-dienecarboxylic acid esters s. R. Antonioletti, G. Righi et al., Tetrahedron Lett. *41*, 9315-8 (2000).

Sodium iodide s. under CeCl₃   *NaI*

1,8-Diazabicyclo[5.4.0]undec-7-ene/magnesium chloride   *DBU/MgCl₂*
Quinoline ring from ar. nitro compds.

and β,γ-ethylenesulfones by intermolecular coupling cf. *53*, 371; 1*H*-isothiazolo[5,4,3-*d,e*]-quinoline 2,2-dioxides from β,γ-ethylenesulfonic acid N-(*m*-nitroaryl)amides via double ring closure with added MgCl₂ s. Z. Wróbel, Tetrahedron Lett. *41*, 7365-6 (2000).

Methylmagnesium chloride   *MeMgCl*
α,β-Ethylenenitriles from β-hydroxynitriles   C(OH)CH → C=C

478.

(E)-α,β-Ethylenenitriles. A soln. of 2.1 eqs. MeMgCl in THF added at room temp. to a soln. of startg. hydroxynitrile in the same solvent, stirred for 10 h, and quenched with satd. aq. NH₄Cl → product. Y 81%. Reaction is applicable to nitriles possessing sec., tert. or allylic hydroxyl groups, and proceeds **via magnesium O,C-dianions** which *eliminate MgO* as the driving force of the conversion. Highly substituted substrates reacted stereospecifically, while sterically unbiased compds. exhibited stereoselectivity comparable with that of Wittig synthesis. Alkyl chlorides were unaffected. F.e.s. F.F. Fleming, B.C. Shook, Tetrahedron Lett. *41*, 8847-51 (2000).

Magnesium chloride s. under DBU   *MgCl₂*

Cerium(III) chloride/sodium iodide   *CeCl₃/NaI*
(E)-α,β-Ethylenecarbonyl from β-hydroxycarbonyl compds.

479.

(E)-α,β-Ethyleneketones. 1.5 eqs. NaI added to a stirred suspension of 4-hydroxy-4-phenylbutan-2-one and 1.5 eqs. CeCl₃·7H₂O in acetonitrile, and stirring continued for 10 h at reflux → product. Y 89% (100% E). The procedure is mild, inexpensive, and highly efficient, and can be undertaken with exposure to atmospheric oxygen and moisture. There was no furan ring opening under these conditions. F.e. incl. hindered and unhindered **(E)-α,β-ethylenecarboxylic acid esters** s. G. Bartoli, E. Marcantoni et al., Org. Lett. *2*, 1791-3 (2000).

Samarium diiodide/diacetatotetrakis(tri-n-butylphosphine)palladium(II)   ←
Cyclic 2-vinylalcohols from ethylenelactolides   ○
chiral polyoxy-2-vinylcyclopentanols s. *55*, 478; also from ε,ζ-ethylene-δ-lactol esters, and 2-vinylcyclobutanol and 2-alkynylcyclopentanol analogs, s. J.M. Aurrecoechea et al., J. Org. Chem. *65*, 6493-501 (2000).

Amberlite 15 s. under LiBr   ←

Bis(4-perfluorohexylphenyl) diselenide/sodium tetrahydridoborate   *ArSeSeAr/NaBH₄*
Ethylene derivs. from 1,2-disulfonates   C(OSO₂R)C(OSO₂R) → C=C
with NaHTe/C₅H₅N cf. *49*, 938s*52*; from 1,2-dimesylates with bis(4-perfluorohexylphenyl) diselenide/NaBH₄ (with simple recovery of the 'minimally' fluorous by-product by continuous extraction with FC 72) s. D. Crich et al., Org. Lett. *2*, 4029-31 (2000).

Phosphazene base (P₄-t-Bu)   ←
Benzofurans from *o*-alkoxyoxo compds.   ○
with KF/Al₂O₃ cf. *23*, 927s*51*; 2-arylbenzofurans with the hindered phosphazene base, N,N',N''-tris[tris(dimethylamino)phosphoranylidene]phosphoric triamide *tert*-butylimine, under mild conditions s. G.A. Kraus et al., Org. Lett. *2*, 2409-10 (2000).

Methanesulfonic acid   *MeSO₃H*
Cyclodehydration
with TsOH cf. *19*, 954; 2,3,4,5-tetrahydro-1*H*-3-benzazepines with MeSO₃H s. S.W. Gerritz et al., Org. Lett. *2*, 4099-102 (2000).

Sulfuric acid H_2SO_4
Cyclobutanones from 1-α-alkoxycyclopropanols

480.

Startg. cyclopropanol in wet ether treated with 1 drop of concd. H_2SO_4, the mixture refluxed for 12-24 h, and quenched with satd. aq. $NaHCO_3$ → product. Y 87%. Rearrangement of unsubst. 1-alkoxymethylcyclopropanols failed, as did reaction with BF_3-etherate or $HClO_4$. F.e.s. T.A. Shevchuk, O.G. Kulinkovich, Russ. J. Org. Chem. *36*, 491-5 (2000).

Hydrogen chloride HCl
Ring closure with acetals
4-hydroxy-1,2,3,4-tetrahydroisoquinolines s. *23*, 928s25; chiral products s. A. Gluszynska, M.D. Rozwadowska, Tetrahedron:Asym. *11*, 2359-66 (2000).

Diacetatotetrakis(tri-n-butylphosphine)palladium(II) s. under SmI_2 $Pd(OAc)_2(PBu_3)_4$

Nitrogen ↑ CC ↑ N

Rhodium(II) acetate $Rh_2(OAc)_4$
3-Hydroxy-5,6-dihydro-4-pyridones from diazomethyl β-acylaminoketones

481.

A little $Rh_2(OAc)_4$ added to a soln. of 4-(benzoylpropylamino)-1-diazobutan-2-one in methylene chloride under dry argon, and stirred at room temp. for 2 h → 2,3-dihydro-5-hydroxy-1-propyl-6-phenyl-4(1*H*)-pyridone. Y 82%. Reaction is also effective with γ- and δ-lactam analogs to give the corresponding indolizinones and quinolizinones. In certain instances, it is possible to trap the intermediate cyclic carbonyl ylid with a dipolarophile. F.e.s. A. Padwa et al., J. Org. Chem. *65*, 7124-33 (2000).

Rhodium(II) triphenylacetate $Rh_2(OCOCPh_3)_4$
Intramolecular carbene insertion into carbon-hydrogen bonds
with $Rh_2(OAc)_4$ cf. *38*, 954s59; with $Rh_2(OCOCPh_3)_4$ for enhanced stereoselectivity s. D.F. Taber et al., J. Org. Chem. *65*, 5436-9 (2000).

Halogen ↑ CC ↑ Hal

Without additional reagents w.a.r.
2-Ethylenealcohols from 2,3-epoxyhalides
with Zn cf. *29*, 968; in water at 38°, chiral products, s. Z. Liu, Y. Li et al., Tetrahedron:Asym. *9*, 3755-62 (1998).

n-Butyllithium BuLi
1,3,5-Triynes from 3-(dibromomethylene)-1,4-diynes C≡C-C≡C-C≡C
Fritsch-Buttenberg-Wiechell rearrangement with alkynyl migration

482.

The startg. dibromomethylene compd. added to a stirred soln. of 1.22 eqs. *n*-BuLi (2.5 *M* in hexane) in dry *hexane* at -78° under N_2, warmed to -40° over 1 h, and quenched with NH_4Cl →

product. Y 80%. This is the first example of an alkynyl shift in this established rearrangement. Reaction failed in THF. Silyl, aryl and alkyl groups are tolerated, and the method allows *multiple rearrangements* within the same compd. F.e. incl. unsym. triynes s. S. Eisler, R.R. Tykwinski, J. Am. Chem. Soc. *122*, 10736-7 (2000).

n-Butyllithium/(-)-sparteine
1,5-Cyclononadien-3-ol from 9-chloro-2,7-dienol carbamates
Asym. intramolecular α,α'-diallyl coupling

483.

2 eqs. 1.6 M *n*-BuLi in hexanes slowly injected under argon into a soln. of startg. (2Z,7Z)-9-chloro-2,7-nonadienyl carbamate and 2 eqs. (-)-sparteine in toluene at -88°, stirred for 2 h, quenched with methanol and satd. aq. NH$_4$Cl, and allowed to warm to room temp. → (1R,2Z,7Z)-2,7-cyclononadienyl carbamate. Y 73% (e.e. 88%). Reaction proceeds with inversion of configuration at the metal-bearing C-atom. The corresponding (2E,7E)-substrate, however, underwent intramolecular lithium-ene reaction. F.e.s. A. Deiters, R. Fröhlich, D. Hoppe, Angew. Chem. Int. Ed. Engl. *39*, 2105-7 (2000).

tert-Butyllithium/diethylaluminum chloride   t-BuLi/Et$_2$AlCl
1,2,3,4-Tetrahydroisoquinoline ring closure
by intramolecular ring opening of tetrahydro-1,3-oxazines with asym. induction

484.

A soln. of startg. perhydrobenzoxazine in anhydrous ether cooled to -90° under argon, 2.2 eqs. *t*-BuLi (1.5 M in pentane) added over 10 min, ca. 2 eqs. Et$_2$AlCl (1 M in hexane) added, the cooling bath removed, the mixture warmed to room temp., stirred overnight, then quenched with 2 M NaOH → (1R)-N-(8-menthyl)-1-(3',4'-dimethoxyphenylmethyl)-6,7-dimethoxy-1,2,3,4-tetrahydroisoquinoline. Y 79%. The reaction proceeds through bromine-lithium exchange, Li→Al-transmetalation, N,O-ring opening and finally ring closure, permitting stereospecific introduction of an alkyl, aryl, benzyl or homobenzyl substituent at C-1 of the isoquinoline ring. The use of other Lewis acids resulted in debromination as a side reaction. F.e.s. R. Pedrosa et al., J. Org. Chem. *66*, 243-50 (2001).

Lithium bis(trimethylsilyl)amide   LiN(SiMe$_3$)$_2$
Stereospecific intramolecular α-alkylation of sulfones with α-bromacetals

485.

β-Sulfonyllactolides. 1.1 eqs. LiN(SiMe$_3$)$_2$ added to a soln. of startg. bromacetal in THF at -78°, and the mixture warmed to room temp. → product. Y 98%. F.e. incl. tetrahydrofuran analogs, also intramolecular Michael addition of α,β-ethylene(hydroxy)carboxylic acid esters, s. C. Jin, A.S. Gopalan et al., Tetrahedron Lett. *41*, 9753-7 (2000).

1,8-Diazabicyclo[5.4.0]undec-7-ene s. under Bu$_3$SnH *DBU*

Silver carbonate *Ag$_2$CO$_3$*
Intramolecular 1,3-dipolar cycloaddition with nitrilimines ○
pyrazole ring s. *17*, 871s*46*; **with asym. induction** via dehydrochlorination of 1,1-chlorohydrazones with Ag$_2$CO$_3$ s. G. Molteni, T. Pilati, Tetrahedron:Asym. *10*, 3873-6 (1999).

9-Borabicyclo[3.3.1]nonane s. under Pd(PPh$_3$)$_4$ *9-BBN*
Thallous ethoxide s. under Pd(PPh$_3$)$_4$ *TlOEt*
Diethylaluminum chloride s. under t-BuLi *Et$_2$AlCl*

1,1,2,2-Tetraphenyldisilane/triethylborane *Ph$_2$Si(H)Si(H)Ph$_2$/Et$_3$B*
Radical ring closure of ethylenebromides
with Ph$_2$SiH$_2$/(PhCOO)$_2$ cf. *45*, 577; carbohydrate-condensed tetrahydrofuran ring from *vic*-allyl oxybromides, regio- and stereo-selectivity, with Ph$_2$Si(H)Si(H)Ph$_2$/Et$_3$B, also ring closure of ethyleneselenides, s. O. Yamazaki, H. Togo et al., J. Org. Chem. *65*, 5440-2 (2000).

Tris(trimethylsilyl)silane/triethylborane/oxygen ←
Stereospecific intramolecular radical 1,4-addition to 1,3-oxazine-2,6-diones

486.

2 eqs. 1 *M* Et$_3$B in hexane added via syringe with stirring to a 0.02 *M* soln. of startg. bromide in anhydrous benzene at 0°, followed by 2 eqs. of tris(trimethylsilyl)silane, allowed to warm to room temp., and stirred under an atmosphere of dry air until reaction complete → product. Y 97% (≥ 19:1 mixture of diastereoisomers). Non-bonded interaction between the C$_2$ carbonyl group and the incipient α-amino stereogenic centre is thought to be responsible for the high diastereoselectivity. The thermally and solvolytically unstable products are readily converted to the corresponding ***trans*-1,1'-disubst. cyclic β-aminocarboxylic acid esters.** F.e. incl. 5,6-, 6,6- and 7,6-azabicyclics, also intramolecular radical addition of selenolic acid esters s. P.A. Evans et al., J. Am. Chem. Soc. *122*, 11009-10 (2000).

Tris(trimethylsilyl)silane/azodiisobutyronitrile *(Me$_3$Si)$_3$SiH/AIBN*
Serial radical ring closure
with Bu$_3$SnH/AIBN s. *41*, 955; *46*, 971s*54*; regio- and stereo-specific double ring closure of n-iodo-2,4,n-trienecarboxylic acid esters with (Me$_3$Si)$_3$SiH/AIBN s. K. Takasu, M. Ihara et al., Org. Lett. *2*, 3579-81 (2000).

Titanocene dichloride/manganese/trimethylsilyl chloride ←
Glycals from O^2-acylglycosyl bromides C(OAc)C(Br) → C=C
with ca. 2 eqs. Cp$_2$TiCl$_2$ cf. *57*, 479; with added Me$_3$SiCl (1.3 eqs.) and *30 mol%* Cp$_2$TiCl$_2$ s. T. Hansen, T. Skrydstrup et al., Tetrahedron Lett. *41*, 8645-9 (2000).

Tri-n-butyltin hydride/sodium trihydridocyanoborate/azodiisobutyronitrile ←
Radical ring closure of acetylenehalides ○
3-methylenetetrahydrofuran ring s. *38*, 965s*42*; 4-alkylidene-2-oxa-8-azabicyclo[3.3.0]octanes s. M.R.P.N. Matos, R.A. Batey et al., Tetrahedron Lett. *40*, 9189-93 (1999).

Tri-n-butyltin hydride/triethylborane *Bu$_3$SnH/Et$_3$B*
Transannular radical ring closure
review s. *29*, 970s*53*; *cis*-fused bicyclic acetals with Bu$_3$SnH/Et$_3$B s. H. Nagano et al., Synlett *2000*, 1193-5.

Tri-n-butyltin hydride/triethylborane/methylaluminum bis(2,6-di-tert-butyl- ←
4-methylphenoxide)
Lewis acid-catalyzed asym. intramolecular radical 1,4-addition of vinyl radicals ○
s. *50*, 579; with chiral *trans*-2-[N-(arylsulfonyl)-N-benzylamino]cyclohexanols as auxiliary s. A. Nishida et al., Tetrahedron:Asym. *11*, 3789-805 (2000).

Tri-n-butyltin hydride/azodiisobutyronitrile $Bu_3SnH/AIBN$
Radical ring closure of ethylenehalides
s. *29*, 970s*58*; bridgehead quaternary ammonium salts s. E.W. Della, P.A. Smith, J. Org. Chem. *65*, 6627-33 (2000); 3,3-difluorotetrahydrofurans from 2-allyloxy-1,1,1-chlorodifluorides with asym. induction s. Japanese patent JP-11255760 (Sumitomo Chem. Co. Ltd.)

Regio- and stereo-specific radical ring closure of 2-iodo-1,n-enynes

Exocyclic 1,3-dienes. 1.12 eqs. Bu₃SnH and a little AIBN in dry benzene added slowly via syringe pump during 4 h to a soln. of startg. vinyl iodide in the same solvent under reflux, refluxing continued for 2 h, cooled to room temp., solvent removed, and the residue worked up with 2:1 ether/satd. aq. KF by stirring at room temp. for 2 h → product. Y 82% (E/Z 2.4:1). The procedure is generally applicable to both 5- and the less common 6-(π-*exo*)-*exo-dig* cyclization to give isolated or condensed systems. Yields were highest where the alkyne terminus was substituted by an aryl or isopropenyl group. F.e. incl. 3,4-bis(alkylidene)tetrahydro-furans and thiophenes, also pyrrolidine and cyclopentane or cyclohexane analogs s. C.-K. Sha, F.-S. Wang et al., Org. Lett. *2*, 2011-3 (2000).

Ring closures via aryl radicals
s. *43*, 957s*59*; 2,3-dihydrospiro[pyrrole-3,3'-oxindoles] s. C. Escolano, K. Jones, Tetrahedron Lett. *41*, 8951-5 (2000); 8-oxoberbine, protoberberine and pavine alkaloids s. K. Orito, M. Tokuda et al., Org. Lett. *2*, 2535-7 (2000).

Tri-n-butyltin hydride/azodiisobutyronitrile/1,8-diazabicyclo[5.4.0]undec-7-ene ←
Radical ring closure via 1,5-hydrogen atom transfer
s. *43*, 958s*58*; of cyclic (α,β-ethyleneiodo)ketones via intramolecular radical 1,4-addition with added DBU s. C.-K. Sha et al., Tetrahedron Lett. *41*, 9865-9 (2000); 3-spiro-2-pyrrolidones s. J.M.D. Storey, ibid. 8173-6.

Hexamethyldistannane s. under PdCl₂(PPh₃)₂ $(Me_3Sn)_2$

Stannic chloride $SnCl_4$
Cyclic ketones from carboxylic acid chlorides
s. *10*, 688; homochiral isoflavanones s. J.L. Vicario, D. Badía et al., Tetrahedron Lett. *41*, 8297-300 (2000).

Dichlorobis(triphenylphosphine)nickel(II)/triphenylphosphine/n-butyllithium ←
Intramolecular reductive diaryl coupling
with Ni(PPh₃)₄ cf. *31*, 964; dibenz[*c,e*]oxepin-5-ones with NiCl₂(PPh₃)₂/Ph₃P/*n*-BuLi s. G. Bringmann, E.-M. Peters et al., Synlett *2000*, 1822-4.

Tetrakis(triphenylphosphine)palladium(0)/potassium tert-butoxide $Pd(PPh_3)_4/KOBu$-t
Intramolecular coupling of ketones with unsatd. halides
intramolecular α-arylation with Pd(dba)₂/BINAP/NaOBu-*t* cf. *54*, 489s*59*; intramolecular α-vinylation with Pd(PPh₃)₄/KOBu-*t* s. D. Sole, J. Bonjoch et al., Org. Lett. *2*, 2225-8 (2000).

Tetrakis(triphenylphosphine)palladium(0)/triphenylarsine/9-borabicyclo[3.3.1]nonane/ ←
thallous ethoxide
Hydroboration-Suzuki coupling in one-pot ←
intermolecular process cf. *49*, 836; transannular macrocyclization by *intramolecular* conversion with 9-BBN/TlOEt and Pd(PPh₃)₄/Ph₃As as catalyst s. S.R. Chemler, S.J. Danishefsky, Org. Lett. *2*, 2695-8 (2000); with Pd(dppf)Cl₂/Ph₃As/Cs₂CO₃ s. N.C. Kallan, R.L. Halcomb, ibid. 2687-90.

Dichlorobis(triphenylphosphine)palladium(II)/triethylamine $PdCl_2(PPh_3)_2/Et_3N$
Intramolecular double Heck arylation with asym. induction

488.

Chiral 3,3'-bi(oxindoles). A soln. of startg. chiral C_2-symmetric diiodide in N,N-dimethylacetamide treated with 10% $PdCl_2(PPh_3)_2$ and excess of Et_3N at 100° → product. Y 90%. The procedure is convenient for generating **chiral vicinal quaternary hydrocarbon groups.** There was no trace of the *meso*-compd. This is part of a total synthesis of (-)-chimonanthine and (+)-calycanthine. F.e.s. L.E. Overman et al., J. Am. Chem. Soc. *121*, 7702-3 (1999).

Dichlorobis(triphenylphosphine)palladium(II)/hexamethyldistannane $PdCl_2(PPh_3)_2/(Me_3Sn)_2$
Intramolecular reductive diaryl coupling
via *in situ*-Stille coupling s. *45*, 583; phenanthro[9,10-*d*]pyrazoles with $PdCl_2(PPh_3)_2$/$Me_3SnSnMe_3$, also **phenanthro[9,10-*d*]isoxazoles**, s. R. Olivera, E. Domínguez et al., Synlett *2000*, 1028-30.

Sulfur ↑   CC ↑ S

Without additional reagents w.a.r.
1,3-Dienes from 2,5-dihydrothiophene 1,1-dioxides C
s. *17*, 968; dendralenes s. S. Fielder, M.S. Sherburn et al., Angew. Chem. Int. Ed. Engl. *39*, 4331-3 (2000).

***o*-Ethyleneamines from Δ³-isothiazoline 2,2-dioxide ring**
3-Amino-2-vinylpyridines from 1,3-dihydroisothiazolo[4,3-*b*]pyridine 2,2-dioxides

489.

via 2-alkylidene-3-imino-2,3-dihydropyridines. A soln. of 1,3,3-trimethyl-1,3-dihydroisothiazolo[4,3-*b*]pyridine 2,2-dioxide in trichlorobenzene refluxed (215°) for 15 min → N-(2-isopropenyl-3-pyridyl)-N-methylamine. Y 90%. This provides an easy route to these products, which may be used for azaindole synthesis. F.e. and 2-(1,3-dien-2-yl) derivs. s. S. Kosinski, K. Wojciechowski, Eur. J. Org. Chem. *2000*, 1263-70.

Potassium tert-*butoxide* KOBu-t
1-Acylpyrroles from 1-acyl-3-sulfonyl-Δ³-pyrrolines ←

490.

1 eq. 1 *M* KOBu-*t* in *tert*-butanol added to a 0.075 *M* soln. of startg. sulfone in dry THF, and refluxed for 2 h → product. Y 93%. Reaction is presumed to involve deconjugation of the vinyl sulfone, followed by elimination of benzenesulfinic acid. This is part of a 3-step conversion **from α-acylaminoketones** and 1-(phenylthio)vinyl(triphenyl)phosphonium iodide, and is versatile

in terms of the substituents on C_2 and C_3. However the method failed with an α-acylaminoaldehyde. F.e.s. I. Burley, A.T. Hewson et al., Tetrahedron Lett. *41*, 8969-72 (2000).

Ramberg-Bäcklund rearrangement ◯
s. *20*, 685; medium-ring and macrocyclic N-protected ethyleneamines s. D.I. Magge, E.J. Beck, J. Org. Chem. *65*, 8367-71 (2000).

Dimethyl sulfate/1,8-diazabicyclo[5.4.0]undec-7-ene Me_2SO_4/DBU
2,3-Dihydro-4-pyridone ring from N-(γ-keto)thiolactams ◯

491.

1.7 eqs. Freshly distilled dimethyl sulfate added dropwise via syringe to a stirred suspension of the startg. thiolactam in anhydrous toluene under N_2 at room temp., the mixture heated in an oil bath at 140-150°, refluxed for 15 min, 1.7 eqs. DBU added dropwise via syringe, and reflux continued for an additional 20 min → 4-methyl-8-chloro-2,3,5,6-tetrahydro-(1*H*)-benzo[*c*]quinolizin-3-one. Y 46%. The procedure is fast and suitable for large-scale synthesis, being based on inexpensive starting materials and reagents. F.e.s. A. Guarna et al., J. Org. Chem. *65*, 8093-5 (2000).

Remaining Elements ↑   CC ↑ Rem

Irradiation s. under CF_3COOH ∭
N-Methyl-1,4-dihydropyridine ←
4,5-Condensed 2(5*H*)-furanone ring from chromium alkynyloxycarbene complexes ◯

492.

A soln. of 3 eqs. N-methyl-1,4-dihydropyridine in dichloromethane added dropwise to a soln. of the startg. carbene complex in the same solvent at -10° under argon, the ice-bath removed after 15 min, and the mixture stirred at room temp. for 24 h → product. Y 46% (single isomer). Reaction is thought to involve initial hydride addition to the carbene, followed by double ring closure with insertion of two CO molecules (without requiring a CO atmosphere). F.e. and 7-membered analogs, **also condensed 5-α-alkoxy-2(5*H*)-furanones** from chromium alkynyl-(alkoxy)carbene complexes, s. H. Rudler et al., Angew. Chem. Int. Ed. Engl. *39*, 3417-9 (2000).

Silver fluoride AgF
1,3-Dipolar cycloaddition with non-stabilized azomethinium ylids
from 1,1'-disilylamines intermolecularly cf. *49*, 388s*50*; *intramolecular* conversion s. G. Pandey et al., Org. Lett. *2*, 2299-301 (2000).

Trifluoroacetic acid/irradiation CF_3COOH/∭
4-Alkylidene-2-cyclopentenones from 3-silyl-2,5-cyclohexadienones ◯
Photolytic rearrangement under acidic conditions

493.

A 0.06-0.08 *M* soln. of 4-acetoxymethyl-4-(3'-butenyl)-3-(trimethylsilyl)-2,5-cyclohexadien-1-one and 2 eqs. trifluoroacetic acid in benzene degassed by bubbling argon for 15 min, then irradiated through uranyl glass (366 nm) for 3 h → product. Y 75% (E:Z 1:1). Irradiation in the absence of trifluoroacetic acid yielded diastereomeric mixtures of bicyclo[3.1.0]hex-3-en-2-ones. The conversion can also be conducted stepwise by first irradiating under neutral conditions, followed by treatment with trifluoroacetic acid. F.e.s. A.G. Schultz, L.O. Lockwood, Jr., J. Org. Chem. 65, 6354-61 (2000).

Thiophenol/azodiisobutyronitrile/potassium fluoride   PhSH/AIBN/KF
anti-2-Ethylene-2'-hydroxyhydrazines from α-vinylsiloxyhydrazones ←
via thiyl-mediated intramolecular vinyl group transfer

with asym. induction. 10 Mol% AIBN added to a soln. of the startg. hydrazone and 1.2 eqs. thiophenol in deoxygenated cyclohexane under N_2, the mixture refluxed 2-3 h, and the crude product treated at room temp. with KF/methanol → product. Y 89% (*anti*-selectivity >98%). The vinyl group is transferred by a radical mechanism from the silyloxy tether, thus controlling the stereochemistry of the developing amine group. F.e.s. G.K. Friestad, S.E. Massari, Org. Lett. 2, 4237-40 (2000).

Tris(trimethylsilyl)silane/triethylborane $(Me_3Si)_3SiH/Et_3B$
Stereospecific intramolecular radical 1,4-addition to 1,3-oxazine-2,6-diones s. *60*, 486   ○

Tris(trimethylsilyl)silane/azodiisobutyronitrile $(Me_3Si)_3SiH/AIBN$
2-Oxabicyclo[3.2.1]oct-6-en-3-ones ←
from 8-arylseleno-2-oxabicyclo[2.2.2]oct-5-en-3-ones
Radical skeletal rearrangement

The startg. selenide (readily obtained by Diels-Alder reaction with 3-carbomethoxy-2-pyrone), 1.5 eqs. tris(trimethylsilyl)silane, and 0.15 eq. AIBN in benzene heated to reflux for 45 min → product. Y 98%. The stereochemical and optical integrity of the starting material are maintained throughout the process. F.e., also transformation of the bridged lactones into the corresponding *cis*-fused **2-oxabicyclo[3.3.0]oct-7-en-3-ones** (with silica gel or TsOH), or their direct formation with $Ph_3SnH/AIBN$ s. I.E. Markó et al., Org. Lett. 2, 3123-5 (2000).

Tri-n-butyltin hydride/azodiisobutyronitrile/methyllithium ←
Synthesis via radical 1,5-Si→C-aryl migration ←
3-arylalcohols s. *56*, 490; *o*-α-hydroxydiaryls from *o*-α-(diarylsilyloxy)bromides s. A. Studer, M. Bossart et al., Org. Lett. 2, 985-8 (2000).

Sodium periodate $NaIO_4$
Ethylene derivs. from selenides $C(SeR)CH → C≡C$
s. *29*, 912s*32*; cysteine-containing dehydropeptides s. N.M. Okeley, W.A. van der Donk et al., Org. Lett. 2, 3603-6 (2000).

Tetra-n-butylammonium fluoride Bu_4NF
Ethylene derivs. from 1,1-iodosilanes $CHC(I)Si \leqslant \rightarrow C{=}CH$
Polyfluoroethylene derivs.

496.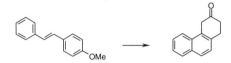

A soln. of startg. fluorinated 1,1-iodosilane and ca. 2.9 eqs. Bu$_4$NF in THF heated to 65° for 5 min, diluted with water, extracted with perfluoromethylcyclohexane, and the fluorous extract distilled after washing with water → product. Y 87%. This is part of a 2-step synthesis **from polyfluoroalkyl iodides** and trimethyl(vinyl)silane. F.e. incl. functionalized polyfluoroalkenes and in DMSO s. Z. Szlávik, J. Rábai et al., *Org. Lett.* **2**, 2347-9 (2000).

Hydrogen chloride/irradiation   *HCl/∭*
Photochemical 2-tetralone ring closure   O

497.

1,2-Dihydro-3(4*H*)-phenanthrenones. A soln. of (E)-*p*-methoxystilbene and 0.5 *M* aq. HCl in degassed acetonitrile irradiated at 350 nm with a Rayonet apparatus for 43 h → product. Y 96% (52% conversion). In the absence of the acid simple *trans→cis*-isomerization was observed. The reaction is thought to proceed via a [1,9]-hydrogen shift (in fact a substrate with a substituent at the 9-position failed to react). F.e., also 7,8-hetaryl (furan and thiophene)-fused 2-tetralones from β-hetaryl-*p*-alkoxystyrenes, s. T.-I. Ho, J.-H. Ho, J.-Y. Wu, J. Am. Chem. Soc. *122*, 8575-6 (2000).

Tris(dibenzylideneacetone)dipalladium/triphenylarsine/ethyldiisopropylamine   ←
Intramolecular Stille coupling with enestannanes   O
s. **42**, 980s*50*; macrocyclic 1,3,9-trienes with Ph$_3$As as ligand for improved efficiency s. E. Marsault, P. Deslongchamps, *Org. Lett.* **2**, 3317-20 (2000).

Carbon ↑   CC ↑ C

Without additional reagents   *w.a.r.*
Retro-Diels-Alder reaction   C
s. *17*, 198s*59*; polymer-based conversion using a traceless linker strategy, e.g. for generating thioenolethers, s. L. Blanco, S. Deloisy et al., *Tetrahedron Lett.* **41**, 7875-8 (2000).

α-Methylenealdehydes from 1,3-dioxins s. **60**, 254

4(1*H*)-Quinolones from 5-(α-arylaminoalkylidene)-1,3-dioxane-4,6-diones   C O
s. **42**, 919; 2-cyano-4(1*H*)-quinolones s. M.-K. Jeon, K. Kim, *Tetrahedron Lett.* **41**, 1943-5 (2000).

Irradiation s. *under HCl*   ∭
Microwaves s. *under Ru-carbene complex*   ←

Cesium carbonate Cs_2CO_3
2-β-Carboxypyridines from 6-cyano-4-oxa-5-azabicyclo[4.4.0]dec-8-en-3-ones   C
s. **60**, 314

Chiral molybdenum carbene complex   ←
Asym. ring-closing metathesis   O
with chiral BINOL-based complexes cf. *48*, 988s*58*; with a modular chiral 2,2'-dihydroxybiphenyl-based complex for preparing chiral cyclic 2-ethyleneethers and α,β-unsatd. lactolides s. G.S. Weatherhead, A.H. Hoveyda et al., *Tetrahedron Lett.* **41**, 9553-9 (2000); 4a-carba-D-arabinofuranosides with Schrock's catalyst s. C.S. Callam, T.L. Lowary, *Org. Lett.* **2**, 167-9 (2000).

Four-coordinate ruthenium carbene complex
Ring-closing metathesis
with Grubbs' complex, update, s. *49,* 985s59; improved conversion with 4-coordinate benzylidene-(tricyclohexylphosphine)bis(1,1,1,3,3,3-hexafluoro-2-methylpropyl-2-oxy)ruthenium(II), preferably in the presence of HCl, s. M.S. Sanford, R.H. Grubbs et al., Angew. Chem. Int. Ed. Engl. *39,* 3451-3 (2000).

Ruthenium carbene complex/microwaves
Polymer-based ring-closing metathesis
s. *49,* 985s55; *52,* 494s56; using a *soluble*, PEG-based substrate for preparing unsatd. N-protected cyclic α-aminocarboxylic acid esters, also enhancement under microwave irradiation and in the presence of 2 eqs. 1-octene, s. S. Varray, F. Lamaty et al., J. Org. Chem. *65,* 6787-90 (2000).

Soluble polymer-based ruthenium carbene complex
Ring-closing metathesis using a soluble polymer-based catalyst

498.

Startg. α,ω-diene (0.1 *M*) in dry dichloromethane treated with 2.5 mol% freshly prepared PEG-based Grubbs' complex at room temp. for 12 h → product. Conversion >99%. This is the first example of a ring-closing metathesis with a *soluble* polymer-based complex. The latter is remarkably stable, similar in activity to homogeneous, non-polymer-based catalysts, and more readily recoverable (by precipitation with ether) and reusable than insoluble polymer-based complexes (cf. *58,* 497s60). Benzyl and silyl ethers remain unaffected. F.e.s. Q. Yao, Angew. Chem. Int. Ed. Engl. *39,* 3896-8 (2000).

Polymer-based (imidazolidin-2-ylidene)ruthenium carbene complex
Ring-closing metathesis
with a polymer-based dichloro(Δ⁴-imidazolin-2-ylidene)ruthenium carbene complex cf. *58,* 497s59; with a 'permanently' immobilized, readily retrievable, polymer-based (imidazolidin-2-ylidene)ruthenium complex s. S.C. Schürer, S. Blechert et al., Angew. Chem. Int. Ed. Engl. *39,* 3898-901 (2000); with a macroporous analog s. L. Jafarpour, S.P. Nolan, Org. Lett. *2,* 4075-8 (2000).

Formation of Electron Pair on Nitrogen

Elimination

Oxygen ↑   EIN ⇑ O

Methyl[2-(thiolatomethyl)phenylthio](triphenylphosphine)rhenium(V) oxide/ triphenylphosphine
Deoxygenation of cyclic N-oxides under mild conditions   ⩾NO → ⩾N

499.

Startg. N-oxide treated with 1 eq. Ph₃P and 0.1 mol% methyl[2-(thiolatomethyl)phenylthio]-(triphenylphosphine)rhenium(V) oxide in benzene for 0.3 h → product. Y 100%. Deoxygenation of N-oxides of pyridines, quinolines and trimethylamine was effected. Nitro groups and halides were unaffected. Suitable solvents included benzene, toluene and THF. F.e. and acceleration with Bu₄NBr s. Y. Wang, J.H. Espenson, Org. Lett. 2, 3525-6 (2000).

Nickel(II) chloride/lithium/4,4′-di-tert-butylbiphenyl
Deoxygenation   ⩾NO → ⩾N
of cyclic N-oxides with Ni cf. *17*, 983; with Ni(0) generated from NiCl₂/Li with 4,4′-*tert*-butylbiphenyl or 4-vinylbiphenyldivinylbenzene copolymer as electron transfer agent, also deoxygenation of tert. amine oxides and azoxy compds., s. F. Alonso, M. Yus et al., Tetrahedron *56*, 8673-8 (2000).

Nitrogen ↑   EIN ⇑ N

Carbon disulfide CS_2
2H-1,2,3-Triazoles from 2H-1,2,3-triazolium-1-imides s. *60*, 160   ←

Formation of Electron Pair on Sulfur

Elimination   ⇑

Oxygen ↑   EIS ⇑ O

2-Chloro-1,3-dimethyl-Δ²-imidazolinium chloride/triethylamine   ←
Thioethers from sulfoxides   ⩾SO → ⩾S
with dimethyl(benzylthio)formiminium chloride cf. *34*, 999s*41*; with 2-chloro-1,3-dimethyl-Δ²-imidazolinium chloride or the analogous tetrahydro-2-bromopyrimidinium salt s. T. Isobe, T. Ishikawa, J. Org. Chem. *64*, 5832-5 (1999).

Formation of Electron Pair on Remaining Elements

Elimination   ⇑

Sulfur ↑   ElRem ⇑ S

Tris(trimethylsilyl)silane/azodiisobutyronitrile $(Me_3Si)_3SiH/AIBN$
Phosphines from phosphine sulfides   ⩾PS → ⩾P
with Si₂Cl₆ cf. *25*, 701; with (Me₃Si)₃SiH/AIBN, also from phosphine selenides, s. R. Romeo, C. Chatgilialoglu et al., Tetrahedron Lett. *41*, 9899-902 (2000).

Remaining Elements ↑   ElRem ⇑ Rem

Tris(trimethylsilyl)silane/azodiisobutyronitrile $(Me_3Si)_3SiH/AIBN$
Phosphines from phosphine selenides s. *25*, 701s*60*   ⩾PSe → ⩾P

Resolutions

(s.a. Subject Index under Resolution)

Electrolysis/(S)-3,5-dihydro-3,3,5,5-tetramethyl-4H-dinaphth[2,1-c:1',2'-e]azepine-N-oxyl
Ketones from sec. alcohols with kinetic resolution
with a little chiral N-oxyl and NaOCl as reoxidant cf. *51*, 500; electrocatalytic procedure s. M. Kuroboshi, H. Tanaka et al., Tetrahedron Lett. *41*, 8131-5 (2000).

Chromatography
Determination of enantiomeric purity
update s. *5, 666s59*; review of NMR methods s. R. Rothchild, Enantiomer *5*, 457-71 (2000).

Atropisomeric 3-aryl-4-dimethylaminopyridines
Kinetic resolution of sec. alcohols by catalytic asym. O-acylation
with planar-chiral 4-dimethylaminopyridine-fused ferrocenes cf. *52*, 497s55; with inexpensive atropisomeric 3-aryl-4-dimethylaminopyridines s. A.C. Spivey et al., J. Org. Chem. *65*, 3154-9 (2000).

Planar-chiral 4-dimethylaminopyridine-fused ferrocenes
Kinetic resolution of amines via asym. N-acylation
with 5-(carbalkoxyoxy)oxazoles s. *60*, 138

Boric acid H_3BO_3
Resolution of 1,1'-bi-2-naphthols
via separation of diastereoisomeric cyclic 1,1'-bi-2-naphthol boric acid esters

Racemic 1,1'-bi-2-naphthol refluxed with 1.13 eqs. boric acid in toluene → 1,1'-bi-2-naphthol boric anhydride (Y 98%), mixed with 2 eqs. (S)-proline in dry THF, refluxed for 3 h, cooled to room temp., and the precipitate removed, washed and dried → 1,1'-bi-2-naphthol boric acid (S)-proline deriv. (Y 96%), treated successively with 2 N NaOH and 2 N HCl at room temp. with stirring, followed by addition of ether, and the organic phase worked up → (S)-1,1'-binaphthol (Y 85.3%: e.e. 94%). The mother liquor from the (S)-proline treatment, on evaporation to dryness and work up in the same way, gave (R)-1,1'-binaphthol (Y 74%; e.e. 100%). The method is considered generally applicable. Details and one-pot procedure (with slightly lower e.e.) s. Z. Shan et al., Tetrahedron:Asym. *9*, 3985-9 (1998).

(R,R,R,R)-2,3,10,11-Tetrakis(hydroxydiphenylmethyl)-1,4,9,12-tetraoxa-dispiro[4.2.4.2]tetradecane
Optical resolution via host-guest complexation
of alcohols with O,O'-dibenzoyl-(2R,3R)-tartaric acid cf. *5, 666s59*; of low molecular weight [volatile] alcohols in the solid state with (R,R,R,R)-2,3,10,11-tetrakis(hydroxydiphenylmethyl)-1,4,9,12-tetraoxadispiro[4.2.4.2]tetradecane s. K. Tanaka et al., Eur. J. Org. Chem. *2000*, 3171-6.

Octapeptides
Kinetic resolution of alcohols by catalytic asym. O-acylation
with tripeptides cf. *52*, 497s*58*; of 2-acylaminoalcohols with octapeptides s. E.R. Jarvo, S.J. Miller, Tetrahedron *56*, 9773-9 (2000).

Lipase
Kinetic resolution of alcohols and carboxylic acids by asym. hydrolysis s. *28*, 13s*60*

Kinetic resolution of alcohols by asym. O-acylation with enolesters
Also double kinetic resolution, and in ionic liquids s. *44*, 214s*60*

Kinetic resolution of amines by asym. N-acylation s. *44*, 314s*60*

Parallel kinetic resolution of sec. alcohols and prim. amines
by asym. aminolysis s. *60*, 9

Lipase or Microorganisms
Kinetic resolution of diols
by asym. alcoholysis of cyclic carbonic acid esters s. *53*, 4s*60*

Lipase/hydridoruthenium(II) chloride complex/triethylamine
Deracemization of sec. alcohols
by enzymatic asym. O-acylation-catalytic racemization
with a hydridoruthenium carbonyl complex cf. *53*, 500s*59*; deracemization of 2-ethylenealcohols with a hydridoruthenium(II) chloride complex s. D. Lee, M.J. Kim et al., Org. Lett. *2*, 2377-9 (2000).

Immobilized lipase
α-Acylaminocarboxylic acid esters from Δ^2-5-oxazolones
with dynamic kinetic resolution s. *50*, 54s*60*

Hydrolase or Yeast
Glycols from epoxides with kinetic resolution s. *48*, 108s*60*

Phosphotriesterase
Kinetic resolution of phosphoric acid esters by asym. hydrolysis s. *60*, 10

Other Reactions   Oth

Acetic acid/irradiation   AcOH/*☇*
Demetalation of dihydro-η^3-pyranylmolybdenum complexes

Reviews

This is a collection of reviews in the field of synthetic organic chemistry published up to August 2001. The layout is to aid access via the Supplementary Reference Index, each entry being entered in the Subject Index, e.g.

Scavengers, polymer-based
-, review **8,** 927s**60**

5, 666   **NMR** methods for determination of **enantiomeric excess,** R. Rothchild, Enantiomer *5,* 457-71 (2000).

7, 281   **Calixarenes** bearing azaaromatic moieties, W. Sliwa, Heterocycles *55,* 181-99 (2001).

7, 823   **Pyrrolidinetrione derivs.**: synthesis and applications in heterocyclic chemistry, B. Zaleska, S. Lis, Synthesis *2001,* 811-27.

8, 927   **Polymeric scavenger reagents** in organic synthesis, J. Eames, M. Watkinson, Eur. J. Org. Chem. *2001,* 1213-24.

11, 744   **Intramolecular carbocyclization** of 1,5-diketones and oxo-1,5-diketones, V.G. Kharchenko, N.V. Pchelintseva, L.I. Markova, Russ. J. Org. Chem. *36,* 919-42 (2000).

12, 653   **Nucleophilic addition of hydrogen sulfide and thiols** to diacetylene alcohols and diols, A.N. Volkov, B.A. Trofimov et al., Sulfur Reports *22,* 195-214 (2000).

14, 711 α,β-**Ethylenephosphonic acid ester** α-**carbanions** in synthesis, T. Minami et al., Synthesis *2001,* 349-57.

16, 698   New developments in the chemistry of **organoantimony and bismuth rings,** H.J. Breunig, R. Rosler, Chem. Soc. Rev. *29,* 403-10 (2000).

16, 820   New concepts in **tetrathiafulvalene chemistry,** J.L. Segura, N. Martín, Angew. Chem. Int. Ed. Engl. *40,* 1372-409 (2001).

16, 888   Regio- and stereo-chemistry of **1,3-dipolar cycloaddition of nitrile oxides to alkenes,** R.P. Litvinovskaya, V.A. Khripach, Russ. Chem. Rev. *70,* 464-85 (2001); 1,3-dipolar cycloaddition with chiral allyl ethers, L. Raimondi, M. Benaglia, Eur. J. Org. Chem. *2001,* 1033-43.

17, 169   Biocatalytic selective **modifications of conventional nucleosides,** carbocyclic nucleosides, and C-nucleosides, M. Ferrero, V. Gotor, Chem. Rev. *100,* 4319-47 (2000); **synthesis of glycopeptides** containing carbohydrate and peptide recognition motifs, H. Herzner, H. Kunze et al., ibid. 4495-537; synthesis of **complex carbohydrates and glycoconjugates:** enzyme-based and programmable one-pot strategies, K.M. Koeller, C.-H. Wong, ibid. 4465-93; methods for **anomeric carbon-linked and fused sugar amino acid synthesis:** the gateway to artificial glycopeptides, A. Dondoni, A. Marra, ibid. 4395-421; **peptide nucleic acids:** analogs and derivs., K.N. Ganesh, P.E. Nielsen, Curr. Org. Chem. *4,* 931-43 (2000).

19, 33   Contemporary methods for **peptide and protein synthesis,** S. Aimoto, Curr. Org. Chem. *5,* 45-87 (2001); synthesis of peptides by solution methods, Y. Okada, ibid. 1-43; efficient syntheses of biologically active peptides of aquatic origin involving unusual α-amino acids, T. Shioiri, Y. Hamada, Synlett *2001,* 184-201; development and application of expressed **protein ligation,** T.W. Muir, ibid. 733-40; combinatorial methods for the discovery and **optimisation of homogeneous catalysts,** S. Dahmen, S. Bräse, Synthesis *2001,* 1431-49.

19, 764   Photoinduced **ortho [2+2]-cycloaddition** of double bonds to triplet benzenes, P.J. Wagner, Angew. Chem. Int. Ed. Engl. *34,* 1-8 (2001).

20, 685   The electrophilic and **radical behaviour of α-halogenosulfonyl derivs.,** L.A. Paquette, Synlett *2001,* 1-12.

23, 819 The use of **polypyrazolylborate copper(I) complexes as catalysts** in the conversion of olefins into cyclopropanes, aziridines and epoxides, and alkynes into cyclopropenes, M. Mar Diaz-Requejo, P.J. Perez, J. Organometal. Chem. *617-8*, 110-8 (2001).

26, 817 **Synthesis of 4-(phenylamino)pyrimidine derivs.** as ATP-competitive protein kinase inhibitors with potential for cancer chemotherapy, G.W. Rewcastle, H.D. Showalter et al., Curr. Org. Chem. *4*, 679-706 (2000).

27, 435 The modular approach to **acetylenic phthalocyanines and phthalocyanine analogues**, R. Faust, Eur. J. Org. Chem. *2001*, 2797-803.

27, 675 **Annulation reactions of azoles and azolines with heterocumulenes**, M.C. Elliott et al., Tetrahedron *57*, 6651-77 (2001); **formation of cumulenes**, triple-bonded, and related compounds by flash vacuum thermolysis of five-membered heterocycles, G.I. Yranzo, C. Wentrup et al., Eur. J. Org. Chem. *2001*, 2209-20.

28, 13 **Preparative biotransformations**, S.M. Roberts, J. Chem. Soc. Perkin Trans. 1 *2001*, 1465-99.

28, 584 Synthesis of **1-hydroxy-1,1-bis(phosphonates)**, M. Lecouvey, Y. Leroux, Heteroatom Chem. *11*, 556-61 (2000).

29, 391 Synthesis and reactions of **2-hetero-4H-3,1-benzoxazin-4-ones**, G.M. Coppola, J. Heterocyc. Chem. *37*, 1369-88 (2000).

29, 591 The latest advances in chemistry of **thiophene 1-oxides and selenophene 1-oxides**, J. Nakayama, Sufur Reports *22*, 123-49 (2000).

29, 665 Annulation by **double Michael addition of tethered diacids to α,β-acetyleneketones**: history and scope, R.B. Grossman, Synlett *2001*, 13-21.

29, 932 Methods of synthesis of substituted **cyclopentadienes and indenes**, N.B. Ivchenko, I.E. Nifant'ev et al., Russ. J. Org. Chem. *36*, 609-37 (2000).

31, 592 The catalytic synthesis of **thia-crowns** from thietanes by metal carbonyl complexes, R.D. Adams, Aldrichimica Acta *33*, 39-48 (2000).

32, 460 **Polyfluoroalkoxysulfur(IV) fluorides**, V.E. Pashinnik, Russ. J. Org. Chem. *36*, 350-8 (2000).

32, 928 **Heavy allenes and cumulenes** E:C:E' and E:C:C:E' (E = P, As, Si, Ge, Sn, E' = C, N, P, As, O, S), J. Escudie, L. Rigon et al., Chem. Rev. *100*, 3639-96 (2000).

33, 325 **2H-Azirines** as synthetic tools in organic chemistry, F. Palacios et al., Eur. J. Org. Chem. *2001*, 2401-14.

33, 466 Chemistry of **xenon derivs., synthesis and chemical properties**, V.K. Brel, N.S. Zefirov et al., Russ. Chem. Rev. *70*, 231-64 (2001).

33, 627 Recent developments in **imino-Diels-Alder reactions**, P. Buonora, T. Oh et al., Tetrahedron *57*, 6099-138 (2001).

34, 681 Methods of synthesis of **conjugated ω-amino ketones**, Y.V. Smirnova, Z.A. Krasnaya, Russ. Chem. Rev. *69*, 1021-36 (2000).

36, 36 Progress of **enantio-differentiating hydrogenation of prochiral ketones** over asymmetrically modified nickel catalysts and a newly proposed enantio-differentiation model, T. Osawa, T. Harada, O. Takayasu, Topics Catalysis *13*, 155-68 (2000).

36, 470 **Fluorination of alkenes** by iodoarene difluorides, M. Sawaguchi, N. Yoneda et al., J. Fluorine Chem. *105*, 313-7 (2000).

36, 667 **Diels-Alder reactions** on solid supports, J. Yli-Kauhaluoma, Tetrahedron *57*, 7053-71 (2001).

37, 807 Preparations and reactions of **β-fluoro-subst. vinamidinium salts**, H. Yamanaka, T. Ishihara, J. Fluorine Chem. *105*, 295-303 (2000).

38, 363 Multiple isotope effects on the **acyl group transfer reactions** of amides and esters, J.F. Marlier, Acc. Chem. Res. *34*, 283-90 (2001).

39, 83 **Vanadium-catalyzed enantioselective sulfoxidations:** rational design of biocatalytic and biomimetic systems, F. Van de Velde, R.A. Sheldon et al., Topics Catalysis, *13*, 259-65 (2000).

39, 450 **Tartaric acid and tartrates** in the synthesis of bioactive molecules, A.K. Ghosh et al., Synthesis *2001*, 1281-301.

39, 646 Factors controlling the **addition of carbon-centred radicals to alkenes** - an experimental and theoretical perspective, H. Fischer, L. Radom, Angew. Chem. Int. Ed. Engl. *40*, 1340-71 (2001).

40, 235 **Selective reactions of** reactive amino groups in **polyamines** by metal-chelated or -mediated methods, S.H. Lee, C.S. Cheong, Tetrahedron *57*, 4801-15 (2001).

41, 556 An overview of recent advances on the synthesis and biological activity of α-aminophosphonic acid derivs., J. Huang, R. Chen, Heteroatom Chem. *11*, 480-92 (2000); syntheses, characterization, stereochemistry and complexing properties of acyclic and macrocyclic compounds possessing **α-amino- or α-hydroxy-phosphonate units**, S. Failla, G.A. Consiglio et al., Heteroatom Chem. *11*, 493-504 (2000).

41, 694 **Nucleophilic aromatic substitution of hydrogen** as a tool for the synthesis of indole and quinoline derivs., M. Makosza, K. Wojciechowski, Heterocycles *54*, 445-74 (2001).

41, 737 Combinatorial and evolution-based methods in the **creation of enantioselective catalysts**, M.T. Reetz, Angew. Chem. Int. Ed. Engl. *40*, 284-310 (2001).

42, 91 Amides of trivalent phosphorus acids as **phosphorylating reagents** for proton-donating nucleophiles, E.E. Nifantiev, S.Y. Burmistrov et al., Chem. Rev. *100*, 3755-99 (2000).

42, 462 **Electrochemical fluorination** as a locomotive for the development of fluorine chemistry, T. Abe, J. Fluorine Chem. *105*, 181-3 (2000).

42, 597 Advances in the directed **metalation of azines and diazines** (pyridines, pyrimidines, pyrazines, pyridazines, quinolines, benzodiazines and carbolines). Part 1: Metalation of pyridines, quinolines and carbolines, F. Mongin, G. Quéguiner, Tetrahedron *57*, 4059-90 (2001); Part 2: Metalation of pyrimidines, pyrazines, pyridazines and benzodiazines, A. Turck, G. Quéguiner et al., ibid. 4489-505.

42, 979 Formation of **acetylenes** by ring opening of 1,1,2-trihalocyclopropanes, L.K. Sydnes, Eur. J. Org. Chem. *2000*, 3511-8.

42, 993 Perspectives in **reductive lanthanide chemistry**, W.J. Evans, Coord. Chem. Rev. *206-7*, 263-83 (2000).

43, 219 **Tetra-*n*-propylammonium perruthenate** (TPAP) - an efficient and selective reagent for oxidation reactions in solution and on the solid phase, P. Langer, J. Prakt. Chem. *342*, 728-30 (2000).

43, 628 Chemistry of **allenesulfones and α,β-acetylenesulfones,** T.G. Back, Tetrahedron *57*, 5263-301 (2001).

43, 700 **Multicomponent reactions** with isocyanides, A. Domling, I. Ugi, *39*, 3168-210 (2000).

43, 943 Rhodium(II)-mediated **cyclizations of diazo alkynyl ketones**, A. Padwa, J. Organometal. Chem. *617-8*, 3-16 (2000).

43, 944 **The transannular Diels-Alder strategy:** applications to total synthesis, E. Marsault, P. Deslongchamps et al., Tetrahedron *57*, 4243-60 (2001).

44, 241 Recent advances in the synthesis of **heterocycles from oximes**, E. Abele, E. Lukevics, Heterocycles *53*, 2285-336 (2000).

44, 646 Chemistry of **hetaryladamantanes**. Part 3: 6-, 7-, and 8-membered hetaryladamantanes, V.P. Litvinov, M.G.A. Shvekhgeimer, Russ. J. Org. Chem. *36*, 299-335 (2000).

44, 782 **Metal glycosylidenes:** novel organometallic tools for C-glycosidation. Part 19. Organotransition metal-modified sugars, K.H. Dötz et al., J. Organometal. Chem. *617-8*, 119-32 (2001).

44, 972 **Radical cascade processes,** A.J. McCarroll, J.C. Walton, Angew. Chem. Int. Ed. Engl. *40,* 2224-48 (2001).

44, 976 Synthesis of **2,2-dimethyl-4-chromanones,** T. Timar, P. Sebok et al., J. Heterocyc. Chem. *37,* 1389-417 (2000).

45, 56 The concept of docking and **protecting groups in biohydroxylation,** A. De Raadt, H. Weber et al., Chem. A Eur. J. *7,* 27-31 (2001).

45, 542 **Arylation with organo-lead and -bismuth compds.,** G.I. Elliott, J.P. Konopelski, Tetrahedron *57,* 5683-705 (2001).

45, 579 Toward a carbohydrate-based chemistry: progress in the development of general-purpose **chiral synthons from carbohydrates,** R.I. Hollingsworth, G. Wang, Chem. Rev. *100,* 4267-82 (2000).

46, 55 Transition metal-mediated **kinetic resolution,** G.R. Cook, Curr. Org. Chem. *4,* 869-85 (2000); practical considerations in kinetic resolution reactions, J.M. Keith, E.N. Jacobsen et al., J. Prakt. Chem. *343,* 5-26 (2001).

46, 641 Decoding the "black box" reactivity that is **organocuprate conjugate addition** chemistry, S. Woodward, Chem. Soc. Rev. *29,* 393-401 (2000).

46, 667 **Catalysis of Michael addition** and the vinylogous Michael addition **by ferric chloride hexahydrate,** J. Christoffers, Synlett *2001,* 723-32.

46, 767 **Asym. synthesis of** novel sterically **constrained amino acids,** Tetrahedron Symposium-in-Print No. 88 (31 papers), Tetrahedron *57(30),* 6329-641 (2001).

47, 114 *Cinchona* **alkaloids** and their derivs.: versatile catalysts and ligands in asym. synthesis, K. Kacprzak, J. Gawronski, Synthesis *2001,* 961-98.

47, 262 Phenyl iodosotrifluoroacetate, G. Pohnert, J. Prakt. Chem. *342,* 731-34 (2000); synthetic uses of **hypervalent organoiodine compds.** through radical pathways, H. Togo, M. Katohgi, Synlett *2001,* 565-81.

47, 569 "Phospha-variations" on the themes of Staudinger and Wittig: **phosphorus analogs of Wittig reagents,** S. Shah, J.D. Protasiewicz, Coord. Chem. Rev. *210,* 181-201 (2000).

47, 625 Substituent effects on the reactivity of the **silicon-carbon double bond,** T.L. Morkin, W.J. Leigh, Acc. Chem. Res. *34,* 129-36 (2001).

47, 639 Metal-mediated **carbometalation of alkynes and alkenes** containing adjacent heteroatoms, A.G. Fallis, P. Forgione, Tetrahedron *57,* 5899-913 (2001).

47, 646 **Nonlinear effects in asym. catalysis:** a personal account, H.B. Kagan, Synlett *2001,* 888-99.

47, 715 Some recent applications of **α-aminonitrile chemistry,** D. Enders, J.P. Shilvock, Chem. Soc. Rev. *29,* 359-73 (2000).

47, 863 **Reactions of α,β-ethylene-α-halogenocarboxylic acid esters** with nucleophilic reagents, D. Caine, Tetrahedron *57,* 2643-84 (2001).

47, 933 **Cyclic diaryliodonium ions:** old mysteries solved and new applications envisaged, V.V. Grushin, Chem. Soc. Rev. *29,* 315-24 (2000).

47, 955 Applications of **carbene complexes** toward organic synthesis, J.W. Herndon, Coord. Chem. Rev. *206-7,* 237-62 (2000); recent progress in **asym. intermolecular C-H activation by rhodium carbenoid intermediates,** H.M.L. Davies, E.G. Antoulinakis, J. Organomet. Chem. *617-8,* 47-55 (2001).

49, 419 Progress in the preparation of **organofluorine compds.** using HF or HF-base molten salts, N. Yoneda, J. Fluorine Chem. *105,* 205-7 (2000).

49, 985 Synthesis and transformations of **8- to 10-membered heterocycles bearing internal (E)-olefin groups,** U. Nubbemeyer, Eur. J. Org. Chem. *2001,* 1801-16.

50, 382 **Addition of carbon-centred radicals to imines** and related compds., G.K. Friestad, Tetrahedron *57,* 5461-96 (2001).

50, 514 Preparation of **highly reactive metals** and the development of novel organometallic reagents, R.D. Rieke, Aldrichimica Acta *33*, 52-60 (2000).

50, 555 Dendrimers and hyperbranched polymers as **high-loading supports for organic synthesis,** R. Haag, Chem. A Eur. J. *7*, 327-35 (2001); the **'resin-capture-release' hybrid technique:** a merger between solid- and solution-phase synthesis, A. Kirschning et al., ibid. 4445-50; **traceless linkers-only disappearing link** in solid-phase organic synthesis, S. Bräse, S. Dahmen, ibid. *6*, 1899-905 (2000); solid-phase organic synthesis: **a critical understanding of the resin,** A.R. Vaino, K.D. Janda, J. Comb. Chem. *2*, 579-96 (2000).

51, 429 Alkoxide-mediated preparation of **enolates from silyl enol ethers and enol acetates** - from discovery to synthetic applications, D. Cahard, P. Duhamel, Eur. J. Org. Chem. *2001*, 1023-31.

52, 75 **Polymer-supported catalysis** in synthetic organic chemistry, B. Clapham, K.D. Janda et al., Tetrahedron *57*, 4637-62 (2001).

52, 125 **Stereoselective construction of the tetrahydrofuran nucleus** by alkoxyl radical cyclizations, J. Hartung, Eur. J. Org. Chem. *2001*, 619-32.

52, 170 **Axially chiral bidentate ligands** in asym. synthesis, M. McCarthy, P.J. Guiry, Tetrahedron *57*, 3809-44 (2001).

52, 221 **Novel aromatic compds.,** Tetrahedron Symposium-in-Print No. 87 (28 papers), Tetrahedron *57(17),* 3507-799 (2001).

52, 274 Synthesis of **heterocycles via intramolecular annulation of nitrene intermediates,** B.C.G. Soderberg, Curr. Org. Chem. *4*, 727-64 (2000).

52, 497 **Asym. catalysis of acyl transfer** by Lewis acids and nucleophiles, A.C. Spivey, A.J. Redgrave et al., Org. Prep. Proced. Int. *32*, 331-65 (2000).

53, 59 **Transition metal-functionalized dendrimers as catalyst,** G.E. Oosterom, J.N.H. Reek, P.W.N.M. van Leeuwen et al., Angew. Chem. Int. Ed. Engl. *40*, 1828-49 (2001).

53, 326 New and selective **transition metal-catalyzed reactions of allenes,** A.S.K. Hashmi, Angew. Chem. Int. Ed. Engl. *39*, 3590-3 (2000).

55, 35 Stoichiometric applications of **acyclic π-organoiron complexes** to organic synthesis, W.A. Donaldson, Curr. Org. Chem. *4*, 837-68 (2000).

55, 297 30 Years of **chiral ligand exchange,** V.A. Davankov, Enantiomer, *5*, 209-23 (2000).

56, 495 Development of **olefin metathesis catalyst precursors** bearing nucleophilic carbene ligands, L. Jafarpour, S.P. Nolan, J. Organometal. Chem. *617-8,* 17-27 (2001); synthesis of **chiral triazolinylidene and imidazolinylidene transition metal complexes** and first application in asym. catalysis, D. Enders, H. Gielen, ibid. 70-80.

57, 48 **Isotopic desymmetrization** as a stereochemical probe, G.C. Lloyd-Jones, Synlett *2001,* 161-83.

57, 99 **Automated solid-phase synthesis of oligosaccharides,** O.J. Plante, P.H. Seeberger et al., Science *291,* 1523-7 (2001); solid-phase oligosaccharide synthesis and combinatorial carbohydrate libraries, P.H. Seeberger, W.-C. Haase, Chem. Rev. *100*, 4349-93 (2000); **combinatorial chemistry** toward understanding the functions **of carbohydrates and their conjugates,** A. Barkley, P. Arya, Chem. A Eur. J. *7*, 555-63 (2001); adventures in carbohydrate chemistry: new technologies, synthesis, molecular design and chemical biology, K.C. Nicolaou, H.J. Mitchell, Angew. Chem. Int. Ed. Engl. *40*, 1576-624 (2001).

57, 106 Use of **supercritical fluids** in synthesis, R.S. Oakes et al., J. Chem. Soc. Perkin Trans. 1 *2001,* 917-41.

58, 49 **Biocatalytic transformation of racemates** into chiral building blocks **in 100% chemical yield and 100% e.e.,** U.T. Strauss, K. Faber et al., Tetrahedron:Asym. *10*, 107-17 (1999).

58, 499 **Aldolase antibody 38C2:** a biocatalyst expanding the scope of enzymatic transformations, C. Hertweck, J. Prakt. Chem. *342*, 832-5 (2000); recent progress in the design and synthesis of **artificial enzymes,** W.B. Motherwell et al., Tetrahedron *57*, 4663-86 (2001).

59, 124 **Intramolecular O-glycosidation,** K.-H. Jung, R.R. Schmidt et al., Chem. Rev. *100*, 4423-42 (2000); stereocontrolled **glycosyl transfer reactions** with unprotected glycosyl donors, S. Hanessian, B. Lou, ibid. 4443-63; recent advances in **O-sialylation,** G.-J. Boons, A.V. Demchenko, ibid. 4539-65.

59, 159 Ligating ability of **1,1'-bis(diphenylphosphino)ferrocene:** a structural survey (1994-8), G. Bandoli, A. Dolmella, Coord. Chem. Rev. *209*, 161-96 (2000).

59, 174 **Arginine mimetics,** L. Peterlin-Masic, D. Kikelj, Tetrahedron *57*, 7073-105 (2001).

59, 206 Recent advances in **catalytic asym. Michael addition,** N. Krause, A. Hoffmann-Roder, Synthesis *2001*, 171-96.

59, 407 **Tandem Michael reactions,** E.V. Gorobets, F.A. Valeev et al., Russ. Chem. Rev. *69*, 1001-19 (2000).

60, 233 **Bis(iodozincio)methane** - preparation, structure and reactions, S. Matsubara, K. Utimoto et al., J. Organometal. Chem. *617-8*, 39-46 (2001).

60, 395 **Axially chiral aminophosphine ligands** in asym. catalysis, P.J. Guiry et al., Curr. Org. Chem. *4*, 821-36 (2000).

60, 452 Rhodium-catalyzed **asym. 1,4-addition of boronic acids** and derivs. to electron-deficient olefins, T. Hayashi, Synlett *2001*, 879-87.

Index to Volume 60

As in previous volumes, reactions are indexed from both the starting material and product aspects, e.g. '**Azides** startg. m.f. amines' and '**Amines** from azides'. Nomenclature for complex functions can be located under the 'special s.' sub-entry, e.g. '**Carboxylic acids** special s. aminocarboxylic acids' or by consulting the Formula Index of Complex Functional Groups (Volume *48*, p. 471).

Hydrogenated and functionalized ring systems are indexed by the conventional reversal, e.g. '**Pyridines, aryl-**', the only important exception to the rule being alkylideneisocyclics which are indexed as such, e.g. '**Alkylidenecyclopentanes**'.

As from Volume *51*, '**Epoxides**' has been used in place of 'Oxido compds.'; '**Thiiranes**' in place of 'Sulfido compds.'; '**Diels-Alder reaction**' in place of 'Diene synthesis'; and '**Benzo[*b*]thiophenes**' in place of 'Thianaphthenes'.

References to abstracts in this volume are in the format **60**, 234. An entry such as '**Peptidyl aldehydes, C-terminal, 19**, 33s60' refers to the indexing of a supplementary reference, which must be followed up via the Supplementary References Index (p. 313).

Aceanthrylenes
- from
  anthracenes, 9-halogeno- and
  acetylene derivs., terminal **60**, 404
Acetaldehyde equivalent 43, 607s60
Acetals (s.a. Ketals)
–, cleavage in aq. organic media **60**, 11
–, – under neutral conditions **60**, 12
– special s.
  ethyleneacetals
  halogenacetals
  hydroxyacetals
  stannylacetals
  sulfonylacetals
– startg. m. f.
  carboxylic acid esters **60**, 98
  syn-α,β-dialkoxyketones, with 2 extra
    C-atoms **60**, 271
  1,3-nitroethers **60**, 446
  pyrrolidines, 3-silyl-, synthesis, asym.
    60, 330
–, cyclic
–, radical 1,4-addition, regiospecific,
    oxidative with – **60**, 282
– special s.
  carbohydrate 1,2-O-isopropylidene
    derivs.
  (1,3-diene)acetals, cyclic
– startg. m. f.
  diol monoethers, reduction,
    regiospecific **60**, 26
  β-hydroxyacetals, cyclic **60**, 282
Acetic acid
– as reactant **60**, 316
Acetic acid esters, chiral
– as reactant **60**, 244
Acetic anhydride
– as reagent **60**, 76, 105
Acetoacetic acid amides
– startg. m. f.
  oxamic acid esters **60**, 100
Acetone
– as reagent **60**, 110
Acetone dimethyl acetal
– as reactant **60**, 341
Acetoxy... s. Acoxy...
Acetyl s.a. Acyl...
Acetylation (s.a. Acylation)
Acetylation, heterogeneous 35, 105s60
Acetyl chloride
– as reagent **60**, 439
Acetylene
– as reactant **60**, 408
Acetylenealcohols
– startg. m. f.
  α-(2-alkoxyvinyl)lactones **60**, 417
2-Acetylenealcohols
– special s.
  2-acetylene-1,4-diol...
  4-acetylene-1,3-diol 1-(monobenzyl
    ethers)
– startg. m. f.
  (Z)-α,β-ethylene-β'-hydroxyketones
    60, 277
–, cyclic
– special s.
  cyclobutanols, 1-(alk-1-ynyl)-
–, terminal
– special s.
  ethynylcarbinols
3-Acetylenealcohols
– startg. m. f.

(E)-3-ethylenealcohols, synthesis
    60, 406
  2H-pyrans, 3,4-dihydro-, 4-alkylidene-
    60, 297
4-Acetylenealcohols
– from
  4-acetylene-1,3-diol 1-(monobenzyl
    ethers), asym. conversion **60**, 46
o-**Acetylenealcohols**
–, reactions, Lewis acid-catalyzed under
    bidentate chelation **60**, 27
α,β-**Acetylenealdehydes**
– startg. m. f.
  β-allenecarboxylic acids, synthesis
    with asym. induction **60**, 228
Acetylene ate complexes
– special s.
  lithium alkynyl(trialkoxy)borates
2-Acetylenecarbonic acid esters
– startg. m. f.
  cyanoallenes **60**, 347
α,β-**Acetylenecarbonyl compds.**
–, Baylis-Hillman reaction, chalcogen-
    mediated with – **60**, 275
– startg. m. f.
  α,β-ethylene-β-halogeno-β'-
    hydroxycarbonyl compds. **60**, 275
α,β-**Acetylenecarboxylic acid esters**
– startg. m. f.
  cis-β-hetaryl-α,β-ethylenecarboxylic
    acid esters, 1,4-addition **60**, 295
Acetylene derivs.
–, [2+2]-cycloaddition, regiospecific, Ru-
    catalyzed with – **60**, 288
– from
  cyclopropanes, 1,1,2-trihalogeno-,
    review **42**, 979s60
– special s.
  alkynyl...
  arylacetylenes
  diynes
  enynes
  triynes
  phthalocyanines, acetylenic
    silylacetylenes
  thiazoles, alkynyl-
– startg. m. f.
  amines, prim. **58**, 126s60
  anthracene ring, 9,10-dihydro- **60**, 274
  coumarins, carbonylation **60**, 390
  1,4-cycloheptadienes **60**, 292
  (E,E)-1,3-dienes **51**, 423s60
  enamines, polymer-based synthesis
    60, 124
  2-ethylenealcohols, regiospecific
    conversion **60**, 284
  3-ethylenesilanes with 2 extra C-atoms
    60, 270
  2(5H)-furanones, 5-acoxy- **60**, 316
  furan ring **60**, 361
  ketones **60**, 66
  – via thioenolethers **7**, 217s60
  2-pyridones, 3,4-dihydro-, polymer-
    based synthesis **60**, 124
  succinimides, biscarbonylation
    54, 312s60
– –, O- or N-functionalized
– startg. m. f.
  ethylene derivs., O- or
    N-functionalized, synthesis,
    regiostereospecific **60**, 258
– –, heteroatom-functionalized

–, carbometalation, review **47**, 639s60
Acetylene derivs., terminal
– special s.
  di(propargyl)arenes
– startg. m. f.
  aceanthrylenes **60**, 404
  acetylenestannanes **60**, 219
  arylacetylenes **60**, 398
  2,4,6-cyclooctatrienones, 4-alkoxy-
    60, 421
  1,3-diynes **60**, 394
  2H-pyrans, 3,4-dihydro-, 4-alkylidene-
    60, 297
  silylacetylenes **60**, 219
Acetylene-α-diazoketones
–, ring closure, Rh(II)-mediated, review
    43, 943s60
**Acetylenedicobalt hexacarbonyl
complexes**
– startg. m. f.
  maleic anhydrides **60**, 436
**3-Acetylene-2,2-difluoroalcohols,
C-protected 40**, 567s60
β,γ-**Acetylene-α,α-difluorophosphonic
acid esters 49**, 805s60
**Acetylene-α,β-dihydroxycarboxylic
acid esters, chiral 47**, 114s60
**4-Acetylene-1,3-diol 1-(monobenzyl
ethers)**
  4-acetylenealcohols, asym. conversion
    60, 46
2-Acetylene-1,4-diol monocarbonates
– startg. m. f.
  1,3-dioxolan-2-ones, 4-(1-aryloxy-
    vinyl)- **60**, 84
2-Acetylene-1,4-diols, chiral 58, 236s60
α,β-**Acetylenehalides**
– startg. m. f.
  3-azabicyclo[3.3.0]oct-1(8)-ene-2,7-
    diones **60**, 387
3-Acetylenehalogenoformic acid esters
– startg. m. f.
  (Z)-α-alkylidene-γ-lactones, synthesis
    60, 454
2-Acetylenehydroxylamines
– startg. m. f.
  Δ⁴-isoxazolines **60**, 75
**3-Acetylene-3'-hydroxythioethers
32**, 231s60
α,β-**Acetyleneketones**
–, Michael addition, double with –, review
    29, 665s60
β,γ-**Acetyleneketones**
– startg. m. f.
  furans **60**, 74
o-**Acetyleneketones**
– startg. m. f.
  naphthalenes **60**, 472
γ,δ-**Acetylenemalonic acid esters**
– startg. m. f.
  bicyclo[3.3.0]oct-1-en-3-one-7,7-
    dicarboxylic acid esters **60**, 345
1-Acetylenephosphonic acid esters
– from
  dihalogenomethylene compds. **60**, 214
1-Acetylene-1-selenides
– startg. m. f.
  (Z)-2-alkylseleno-1,3-enynes **60**, 174
  α,β-ethylene-β-halogeno-α-
    (organoseleno)carboxylic
    esters, carbonylation, stereospecific
    60, 301

Ace – Add

(1-Acetylene-1-selenides
- startg. m. f.)
  (Z)-1-halogeneneselenides **60**, 174

Acetylenestannanes
- from
  acetylene derivs., terminal **60**, 219
- startg. m. f.
  δ,ε-ethylenefluorides **60**, 458

α,β-Acetylenesulfoxides
-, carbocupration-cross-coupling,
  regiospecific **60**, 411
-, hydrozirconation **48**, 48s60
- startg. m. f.
  allenes, synthesis **60**, 411

1-Acetylene-1-thioethers
- startg. m. f.
  thiophenes, 2,3-dihydro- **60**, 305

2-Acoxyacroleins 60, 254

α-Acoxyalkoximes
- special s.
  α-acryloyloxyalkoximes

(3E)-5-Acoxy-2-alkoxy-1,3-dienes
52, 77s60

Acoxy compds. (s.a. Carboxylic acid esters)
- from
  enolesters, asym. reduction **60**, 123
  ketones, – – **60**, 23
  tetrahydropyran-2-yl ethers **60**, 102
- special s.
  alkoxyacoxy compds.
  glycol esters

2-α-Acoxy-1,3-dienes 54, 320s60

5-Acoxy-3-en-1-ynes
- startg. m. f.
  1,2,4-trienes, synthesis with remote
  asym. induction **60**, 326

α-Acoxy-α,β-ethylenealdehydes
- special s.
  2-acoxyacroleins

Acoxy-2-ethylenes
- from
  aldehydes **51**, 423s60
-, isomerization **60**, 76
- special s.
  5-acoxy-3-en-1-ynes
  (siloxy)acoxy-2-ethylenes
- startg. m. f.
  bicyclo[3.3.0]oct-1-en-3-one-7,7-
  dicarboxylic acid esters **60**, 345

α-Acoxynitriles s. Cyanohydrin acetates

o-Acoxynitro compds., polymer-based
-, N-acylation with – **60**, 159

α-Acryloyloxyalkoximes
- startg. m. f.
  β-alkoxylamino-γ-lactones, synthesis
  60, 382

Activation
- of borane-amine complexes **60**, 15

Activation-deactivation, asym.
- of racemic complexes with 2 chiral
  auxiliaries **60**, 30

N-Acylalleneamines cyclic 21, 749s60

Acylamines (s.a. Carboxylic acid amides)
-, N-acylation, asym. with – **60**, 148
- from
  carbobenzoxyamines **60**, 165
  oximes **60**, 74
- special s.
  alkoxyacylamines
  chloroacetylamines

Acylamines, ar.

- startg. m. f.
  acylamino-*p*-quinones **60**, 77
  o-quinone mono-N-acylimines **60**, 77

ω-Acylaminoaldehydes
- from
  N-acyllactams **60**, 251

α-Acylaminocarboxylic acid esters
- from
  Δ²-5-oxazolones, asym. conversion
  50, 54s60
- special s.
  α-formamidocarboxylic acid esters
– – –, chiral **39**, 754s60

α-Acylaminocarboxylic acids
- startg. m. f.
  2-acylamino-1,2-iodohydrins, cyclic
  60, 187
– –, chiral **39**, 612s60

α-Acylamino-α,β-ethylenecarboxylic acid esters
- special s.
  α,β-ethylene-α-formamidocarboxylic
  acid esters

ω-Acylamino-α′-hydroxyketones
60, 251

2-Acylamino-1,2-iodohydrins, cyclic
- from
  α-acylaminocarboxylic acids **60**, 187

α-Acylaminoketones
- startg. m. f.
  pyrroles, 1-acyl- **60**, 490

β-Acylaminoketones
- special s.
  diazomethyl β-acylaminoketones

ω-Acylaminoketones
- from
  N-acyllactams **60**, 251
  oxo compds. **60**, 251

o-Acylaminonitriles
- startg. m. f.
  indoles, 3-amino-1,2-diacyl- **60**, 371

Acylamino-*p*-quinones
- from
  acylamines, ar. **60**, 77

α-Acylaminosulfones
- startg. m. f.
  N-acylimmonium ions, with stereo-
  retention **60**, 443

Acylation (s.a. Acyl group transfer,
  Hydroacylation...)
- special s.
  acetylation

**Acylation, asym., Lewis acid- or
  nucleophile-catalyzed**
-, review **52**, 497s60

C-Acylation (s.a. Friedel-Crafts...)

C-Acylation, Rh-catalyzed
- of amines, cyclic (α-acylation) **60**, 293

C-γ-Acylation
- of β,γ-ethylenephosphonic acid esters
  60, 1

N-Acylation
- special s.
  N-trifluoroacetylation
- with benzotriazoles, 1-acyl- **53**, 152s60
- with resolution, kinetic **60**, 138, 148

N-Acylation, asym.
-, resolution, kinetic of amines by –
  60, 138, 148
- with oxazoles, 5-carbalkoxyoxy-
  60, 138

N-Acylation, intramolecular (s.a.

Michael addition-intramolecular
  N-acylation)
-, polymer-based **60**, 159
- with N-hydroxysuccinimide esters,
  polymer-based **32**, 317s60

O-Acylation
- special s.
  O-benzoylation
- with simultaneous O-tritylation and
  O-silylation **60**, 90

O-Acylation, partial
- of glycols **55**, 81s60
-, reductive, asym. **60**, 23

N-Acylazomethinium salts (s.a. N-Acylimmonium ions)

Acyl carbanion equivalents
-, 1-lithioenolethers as – **60**, 175

Acylcyanides
- startg. m. f.
  α-diketones, sym. **60**, 468

1,5-O→N-Acyl group migration
-, N-hydroxypeptides by – **14**, 400s60

Acyl group transfer
- from amides or esters, isotope effects,
  review **38**, 363s60

Acyl halides s. Carboxylic acid halides

Acylhydrazines (s.a. Carboxylic acid
  hydrazides)

N-Acylimmonium ions
- from
  α-acylaminosulfones, with stereo-
  retention **60**, 443

N-Acylimmonium ions, cyclic
- as intermediates **60**, 187, 435

N-Acyllactams
-, ring opening, reductive, regiospecific
  60, 251
- startg. m. f.
  ω-acylaminoketones **60**, 251

Acylophenones (s.a. Aryl ketones)
- from
  halides, ar., with 2 extra C-atoms
  60, 405

O-Acyloximes
- special s.
  cyclobutanone O-acyloximes
- startg. m. f.
  α-siloxymalononitriles **60**, 333

α-(O-Acylimino)nitriles
-, Diels-Alder reactions, intramolecular
  with – **60**, 314

**Acylpalladation, intramolecular-cross-
  coupling 60**, 454, 455

Acylstannanes
- startg. m. f.
  β,γ-ethylene-β(δ)-stannylketones
  60, 283

1,4-Acylstannylation, stereospecific
60, 283

N-Acylsultams
- startg. m. f.
  N-[(β-phosphinylamino)acyl]sultams
  60, 245

α-(Acylthio)nitriles, chiral 11, 666s60

**Adamantanes, hetaryl-, 6- to 8-
  membered**
-, review **44**, 646s60

Adenines
- from
  hypoxanthines **60**, 141

1,4-Addition (s.a. CC⇆CC, Aminopalladation, intramolecular-1,4-addition,

Michael addition, Radical 1,4-addition)
- with
  arylstannanes **54**, 466s**60**
  organocuprates, review **46**, 641s**60**
1,4-Addition, asym.
- of azide ion **60**, 127
- to 1,1-nitroethylene derivs. **60**, 259
- with
  arylboronic acids (to enoates) **55**, 452s**60**
  lithium aryl(trimethoxy)borates **55**, 452s**60**
-, -, **regiospecific**
- of N-(2-ethylene)urethans **60**, 259
-, -, **Rh-catalyzed**
- of boronic acids **60**, 452 (s.a. Review section)
- to 1,1-nitroethylene derivs. **60**, 452
-, **regiostereospecific, Pd-catalyzed**
- with hetarenes **60**, 295
1,4-Addition-alkylation
- of α,β-ethyleneketones, cyclic **60**, 263
- with lithium trialkylzincates, unsym. **60**, 263
1,6-Addition
- to arylcarboxylic acid halides **60**, 257
Alcohols (s.a. Hydroxylation and under Replacement)
-, O-acylation s. under O-Acylation
-, O-benzylation **60**, 79
-, deoxygenation via trifluoroacetates **60**, 39
- from
  oxo compds., reduction (s.a. HC⇃OC)
  – – (prim. alcohols), – in ionic liquids **60**, 21
  – – (sec. alcohols), –, asym., enzymatic **60**, 24
  – – (– –), –, –, with chirally modified Ni catalysts, review **36**, 36s**60**
  – – (– –), synthesis, asym. **42**, 616s**60** (update); **44**, 565s**60** (update); **60**, 229 (solvent-less), 230
-, O-silylation s. under O-Silylation
- special s.
  acetylenealcohols
  allenealcohols
  aminoalcohols
  azidoalcohols
  benzylalcohols
  deuterioalcohols
  diol...
  ethylenealcohols
  halogenhydrins
  hydroxylaminoalcohols
  nitroalcohols
  phthalimidoalcohols
  polyols
  trihalogenalcohols
  triols
- startg. m. f.
  carbonic acid esters, sym. **60**, 135
  iodides **60**, 180
  mercaptans, with inversion **60**, 196
  oxo compds. (s.a. OC⇃H) **60**, 107, 114
  – – (aldehydes) **60**, 105
  – – (ketones) **60**, 110
  sulfamides **60**, 163
  sulfuric acid amide esters **60**, 51
  urethans, oxidative carbonylation **60**, 132
Alcohols, prim.

- from
  amines, prim. **60**, 85
  carboxylic acid esters **60**, 38
  carboxylic acids **60**, 38
- startg. m. f.
  halides, prim. **60**, 182
Alcohols, sec.
- from
  carboxylic acid esters, synthesis **60**, 328
-, resolution, kinetic, parallel (with prim. amines) **60**, 9
- special s.
  benzhydrols
Alcohols, tert.
- startg. m. f.
  amines, prim. **60**, 146
Alcohols, tert., cyclic
- startg. m. f.
  halogenoketones **60**, 176
Aldehyde equivalents
- special s.
  acetaldehyde equivalent
Aldehydes (s.a. Carbonyl compds., Hydroformylation, Oxo compds.)
- from
  alcohols s. OC⇃H and under Oxo compds.
  carboxylic acid amides, N,N-disubst. **60**, 41
  2-ethylenealcohols, asym. isomerization **60**, 73
  ethylene derivs., terminal **46**, 116s**60**
  methyl groups **60**, 109
  2-oxazolidones, 3-acyl-, asym. synthesis **44**, 776s**60**
- special s.
  acetylenealdehydes
  acylaminoaldehydes
  epoxyaldehydes
  ethylenealdehydes
  halogenaldehydes
  hydroxyaldehydes
  ketoaldehydes
  mercaptoaldehydes
- startg. m. f.
  acoxy-2-ethylenes **51**, 423s**60**
  alcohols, prim. s. HC⇃OC and under Oxo compds.
  anti-α-alkoxy-β-hydroxyketones, with 2 extra C-atoms **60**, 271
  β-aminoketones, N-protected, asym. conversion **60**, 336
  α-arylketones **60**, 350
  benzimidazoles, polymer-based synthesis **60**, 139
  benzylalcohols, sec., synthesis **60**, 450
  1-enolcarbonates, prim., reduction, asym. **60**, 28
  syn-1,3-diols **60**, 448
  2-ethylenealcohols, synthesis, regiostereospecific **60**, 284
  3-ethylenealcohols, –, – **60**, 302
  (E)-α,β-ethylene-α-halogenocarboxylic acid esters, with 2 extra C-atoms **60**, 423
  β,γ-ethyleneketones, synthesis **60**, 329
  anti-3-ethylene-2-silylalcohols, asym. conversion **60**, 221
  glycol esters, sym. **60**, 240
  – ethers, sym. **60**, 239
  3-hydroxysilanes, with 2 extra C-atoms **60**, 270

imidazole ring **46**, 321s**59**
indenes **60**, 352
mercaptals **60**, 194
nitriles **60**, 143, 144
(E)-α-nitro-β,γ-ethylenesulfones, with 1 extra C-atom **60**, 321
Δ²-oxazolines, under microwaves **60**, 319
phthalans **60**, 377
pyridines **60**, 335
4-pyrones, 2,3-dihydro- **60**, 439
Δ¹-pyrrolines **60**, 434
(*all*-E)-2,4,6-trienals, with 6 extra C-atoms **60**, 428
Aldehydes, ar.
- startg. m. f.
  benzo[*b*]thiophene-2-carbonyl compds. **60**, 368
  ketones, *o*-subst. **60**, 360
Aldehyde tosylhydrazones
- startg. m. f.
  α-arylketones **60**, 350
Aldimines (s.a. Azomethines)
-, Heck-type arylation, Rh-catalyzed **60**, 389
- special s.
  N-α-(benzotriazol-1-yl)aldimines
  N-(2-pyrazinyl)aldimines
- startg. m. f.
  anti-α-alkoxy-β-aminocarboxylic acid esters, with 2 extra C-atoms **60**, 252
  β-aminocarboxylic acid esters, synthesis, asym. with 2 extra C-atoms **60**, 244
  2-azetidinones, polymer-based synthesis **60**, 322
  aziridine-2-carboxylic acid esters, asym. conversion **60**, 357
  α-(borylamino)boronic acid esters **60**, 207
  4-pyridines, 2,3-dihydro- **60**, 439
  Δ¹-pyrrolines **60**, 434
Aldimines, N-protected, *in situ* generated
- startg. m. f.
  3-ethyleneamines, N-protected, synthesis, asym. **60**, 430
  pyrrolidines, 3-silyl-, –, –, – **60**, 330
Aldolase antibody 38C2
-, review **58**, 499s**60**
Aldol condensation, asym.
- with 1,4-dioxan-2-ones **60**, 235
- –, –, **Ti(IV)-catalyzed**
- of ketones with aldehydes **60**, 238
- –, –, **Zn-catalyzed**
- of ketones with aldehydes **60**, 231
- **condensation-Michael addition 60**, 335
Aldol-Tishchenko reaction, asym., catalytic 60, 236
Aldol transfer reaction, catalytic
- via aluminum enolates **60**, 469
Aldol-type condensation
- via vanadyl allenolates **60**, 277
- –, **asym., Pb(II)-catalyzed**
- in aq. medium **60**, 444
- –, **reductive 60**, 448
- –, **vinylogous**
-, alternative **60**, 431
Aldoses
-, Knoevenagel condensation in aq. medium with – **60**, 465
-, O²-phosphorylation **60**, 52

Aldoses
- special s.
  deoxyaldoses
- startg. m. f.
  glycosyl halides **60**, 181
  – phosphates **60**, 54
  C-β-(β-ketoalkyl)glycosides **60**, 405
Aldoximes
- startg. m. f.
  nitriles **60**, 169
Alkali metal enolates
- from
  enolesters or enoxysilanes, review
  51, 429s60
Alkanes s. Hydrocarbons
Alkenes s. Ethylene derivs.
Alkoximes
- special s.
  acoxyalkoximes
  hydroxyalkoximes
(E)-Alkoximes
- from
  oxo compds. **60**, 151
Alkoxyacetylenes
- startg. m. f.
  1-halogenenolethers, stereospecific
  conversion **60**, 175
1-Alkoxy-1-acoxy-3-allenes 40, 81s60
1,1-Alkoxyacoxy compds.
- from
  carboxylic acid esters, asym. reduction
  60, 20
1,1-Alkoxyacylamines
- special s.
  1,1-alkoxyaroylamines
Alkoxyallenes
- special s.
  1-alkoxy-1-acoxy-3-allenes
  methoxyallene
**1,2-Alkoxyallylation-transetherification
60**, 346
anti-α-**Alkoxy-β-aminocarboxylic acid
esters**
- from
  aldimines, with 2 extra C-atoms **60**, 252
1,1-Alkoxyaroylamines
- from
  o-aminocarboxylic acid N-*tert*-alkyl-
  amides **60**, 86
γ-**Alkoxy-γ-arylcarboxylic acid esters**
- from
  tungsten aryl(alkoxy)carbene
  complexes **60**, 437
α-**Alkoxyboronic acid esters**
- from
  boronic acid esters, asym. synthesis
  with 1 extra C-atom **60**, 425
Alkoxycarbenium ions
- from
  1,1-alkoxysilanes, cathodic generation
  60, 422
Alkoxycarbenium ions, bicyclic
– as intermediates **21**, 177s60
α-**Alkoxycarboxylic acid esters**
- special s.
  methyl methoxyacetate
γ-**Alkoxycarboxylic acid esters**
- special s.
  γ-alkoxy-γ-arylcarboxylic acid esters
o-**Alkoxycarboxylic acid thioamides**
- startg. m. f.
  2(3*H*)-benzofuranones **60**, 113

Alkoxydiboranes
- special s.
  tetraalkoxydiboranes
1-Alkoxy-1,3-dienes s. (1,3-Dien)olethers
2-Alkoxy-1,3-dienes 54, 374s60
– by Stille coupling, with inversion
  60, 175
- special s.
  acoxy-2-alkoxy-1,3-dienes
1-Alkoxy-1,3-disiloxy-1,3-dienes
- startg. m. f.
  2(5*H*)-furanones, 5-(carbalkoxyene)-
  3-hydroxy- **60**, 445
1-Alkoxyenesilanols
–, cross-coupling **45**, 555s60
1-Alkoxyenesilicon hydrides
–, cross-coupling **45**, 555s60
α-**Alkoxy-α,β-ethylenecarboxylic acid
esters**
- special s.
  α-allyloxy-α,β-ethylenecarboxylic
  acid esters
Alkoxy-2-ethylenes (s.a. 2-Ethylene-
  ethers)
–, cycloaddition, 1,3-dipolar with asym.
  induction, review **16**, 888s60
–, oxymercuration **16**, 681s60
- special s.
  3-ene-1,2-diol 2-monoethers
  4-oxa-1,7(8)-dienes
Alkoxy-3-ethylenes
- from
  1,1-alkoxysilanes **60**, 422
  2-ethylenesilanes **60**, 422
- special s.
  4-oxa-1,7-dienes
Alkoxy-3-ethylenes, exocyclic 60, 304
–, isocyclic
- from
  (1,3-diene)acetals, cyclic **57**, 321s60
Alkoxy-4-ethylenes
- special s.
  4-oxa-1,8-dienes
γ-**Alkoxy-β-hydroxycarboxylic acid
esters 41**, 888s60
anti-α-**Alkoxy-β-hydroxyketones**
- from
  aldehydes, with 2 extra C-atoms
  60, 271
δ-**Alkoxy-γ-ketocarbonyl compds.**
- from
  α,β-ethylenecarbonyl compds. and
  chromium alkoxycarbene complexes
  60, 437
α-**Alkoxyketones**
- special s.
  α-alkoxy(hydroxy)ketones
  α,β-dialkoxyketones
Alkoxylamines
- special s.
  N,O-dimethylhydroxylamine
β-**Alkoxylamino-γ-lactones**
- from
  α-acryloyloxyalkoximes, synthesis
  60, 382
Alkoxyl radicals, unsatd.
- startg. m. f.
  furans, tetrahydro-, review **52**, 125s60
α-**Alkoxyoximes 43**, 60s60
Alkoxysilanes (s.a. O-Desilylation,
  Siloxy..., O-Silylation, Silyl ethers)
- special s.

cinnamyloxysilanes
tetraalkoxysilanes
1,1-Alkoxysilanes
- startg. m. f.
  alkoxycarbenium ions, cathodic
  generation **60**, 422
  alkoxy-3-ethylenes **60**, 422
p-**Alkoxystilbenes**
- startg. m. f.
  3(4*H*)-phenanthrenones, 1,2-dihydro-
  60, 497
α-**(2-Alkoxyvinyl)lactones**
- from
  acetylenealcohols and chromium
  alkoxycarbene complexes **60**, 417
β-**(2-Alkoxyvinyl)lactones 43**, 616s60
C-α-Alkylation
- special s.
  radical C-α-alkylation
C-α-Alkylation, asym.
- of
  1*H*-naphtho[1,8-*de*]-1,2,3-triazin-2-
  ium N-ylids, 2-α-carbalkoxy **24**,
  852s60
  2-thiazolidones, 3-acyl-, conformationally rigid **60**, 369
–, –, **sequential 48**, 802s60
–, intramolecular, stereospecific
– with α-halogenacetals **60**, 485
Alkylation, ar. (s.a. Friedel-Crafts...)
- special s.
  thioalkylation, ar.
o-**Alkylation** (s.a. Hydroacylation-*o*-
  alkylation)
N-Alkylation
– of purines, 2-amino- (N⁹-alkylation)
  60, 154
- special s.
  mono-N-alkylation
N-Alkylation, reductive 60, 32
O-Alkylation
- special s.
  O-benzylation
  O-tritylation
α-**(Alkylideneamino)carboxylic acid
esters**
- startg. m. f.
  oxazoles, 5-alkoxy- **60**, 106
Alkylidenecyclobutanes
- special s.
  dialkylidenecyclobutanes
2-Alkylidenecyclopentanones
- from
  cyclobutanols, 1-(alk-1-ynyl)-,
  synthesis **60**, 392
Alkylidenecyclopropanes
–, silaboration, regiospecific **60**, 211
- startg. m. f.
  α/β-alkylidene-γ-silylboronic acid
  esters **60**, 211
β-**Alkylidene-γ-lactolides 53**, 291s60
α-**Alkylidene-γ-lactones 41**, 862s60
(Z)-α-**Alkylidene-γ-lactones**
- from
  3-acetylenehalogenoformic acid esters,
  synthesis **60**, 454
β-**Alkylidenenitriles, isocyclic 51**,
  337s60
α/β-**Alkylidene-γ-silylboronic acid
esters**
- from
  alkylidenecyclopropanes **60**, 211

O-Alkyllactims
- from
  lactams and diazo compds. **60,** 89
(Z)-2-Alkylseleno-1,3-enynes 27, 851s60
- from
  1-acetylene-1-selenides **60,** 174
  (Z)-1-halogeneneselenides **60,** 174
Alkylthio-N-heterocyclics
-, cross-coupling with benzylzinc halides **60,** 412
N-(Alkylthio)phthalimides
- startg. m. f.
  β-glycosides **39,** 189s60
Alkynes s. Acetylene derivs.
β-Alk-1-ynyl-β-lactones
- startg. m. f.
  β-allenecarboxylic acids, synthesis with asym. induction **60,** 228
2-Allenealcohols
- startg. m. f.
  furans 2,5-dihydro-, 3-allyl- **60,** 400
Alleneamines
- special s.
  N-acylalleneamines
ω-Alleneamines
- startg. m. f.
  2-ethyleneamines, cyclic **60,** 287
α-Allenecarboxylic acid amides
- startg. m. f.
  Δ³-2-pyrrolones, 4-halogeno-5-hydroxy- **60,** 177
α-Allenecarboxylic acid esters
- startg. m. f.
  (E)-β,γ-ethylenecarboxylic acid esters **60,** 325
β-Allenecarboxylic acids
- from
  α,β-acetylenealdehydes, synthesis with asym. induction **60,** 228
  β-alk-1-ynyl-β-lactones, – – – – **60,** 228
α-Allene-δ-hydroxycarboxylic acid esters
- startg. m. f.
  2-pyrones, 3,6-dihydro- via 1,4-addition **60,** 325
α-Alleneketones
- startg. m. f.
  furans **60,** 74
Allenes
- from
  α,β-acetylenesulfoxides, synthesis **60,** 411
-, reactions, transition metal-catalyzed, review **53,** 326s60
- special s.
  alkoxyallenes
  bis(allenes)
  cyanoallenes
  glycosyloxyallenes
  1,2,4-trienes
- startg. m. f.
  2-aryl-3-ethyleneazides **60,** 298
  1,2-dialkylidenecyclobutanes, dimerization, regiospecific **60,** 285
  3-ethylenealcohols, synthesis, regiostereospecific **60,** 302
  β,γ-ethylene-β-stannylketones **60,** 283
Allenes, 1,1-disubst. 60, 411
Allenes, trisubst. 60, 411
α-Allenesulfones
-, Diels-Alder reaction, intramolecular with – **38,** 723s60

Allenolates
- special s.
  vanadyl allenolates
Allenylpalladium(II) alkoxides
- as intermediates **60,** 347
Allyl alcohol
- as reactant **60,** 346
Allyl alcohols s. 2-Ethylenealcohols
Allylamines s. 2-Ethyleneamines
Allylarenes
- by Heck arylation, desilylative **27,** 871s60
- from
  amines, ar., prim. **60,** 386
  β,γ-ethylenehalides **60,** 386
C-Allylation (s.a. 1,2-Oxyallylation...)
C-Allylation, intramolecular
- special s.
  α,α'-diallyl coupling, intramolecular
C-α-Allylation, asym., Pd-catalyzed
-, update **48,** 772s60
C-Allylation-intramolecular Pauson-Khand reaction, dual catalytic 60, 345
N-Allylation, Pd-catalyzed, accelerated 60, 145
-, intramolecular, regiostereospecific **60,** 130
N-Allylation-intramolecular Diels-Alder reaction 59, 335s60
O-Allylation, intramolecular (s.a. C-α-Arylation-intramolecular O-allylation)
Allylboration (s.a. Hetero-Diels-Alder reaction-allylboration)
Allylboration, asym.
-, *anti*-3-ethylene-2-silylalcohols via – **60,** 221
Allyl bromide
- as reactant **60,** 249
- as reagent **60,** 11
Allyl ethyl carbonate
- as reactant **60,** 346
1,2-*trans*-C-Allyl glycosides
- from
  carbohydrate 1,2-O-isopropylidene derivs. **60,** 432
O→C-Allyl group migration 59, 322
Allyl halides s. β,γ-Ethylenehalides
Allyl homoallyl ethers s. 4-Oxa-1,7-dienes
Allylindium(III) compds.
- as intermediates **43,** 699s60
α-Allyloxy-α,β-ethylenecarboxylic acid esters
-,Claisen rearrangement, Yb(III)-catalyzed **60,** 311
η³-Allylrhodium complexes
- as intermediates **60,** 453
Allylsilanes s. 2-Ethylenesilanes, Allyltrimethylsilane
Allylsilylation, intramolecular (s.a. Silylformylation-allylsilylation, intramolecular)
Allylstannanes s. 2-Ethylenestannanes
Allyltrimethylsilane
- as reactant **60,** 432
Allylzincation, asym.
- under high pressure **43,** 770s60
Aluminum alkoxides, halogeno-, chiral
  chloroaluminum bis(phosphinyl-alkoxides), chiral **57,** 238s60
- aroxides
- tris(2,6-diphenylphenoxide) **60,** 257

Aluminum aroxides, organo-trimethylaluminum/1,1'-bi-2-naphthol 60, 469
- bromide **60,** 315
- chloride **12,** 686s60; **36,** 754s60; **60,** 116, 470
- as Lewis acid in water **60,** 119
- cyanide, diethyl- **60,** 250
- enolates
-, aldol transfer reaction, catalytic via – **60,** 469
- halides organo-
  diethylaluminum chloride **47,** 651s60; **60,** 433, 484
  dimethylaluminum – **37,** 657s60
- hydrides, organo-
  diisobutylaluminum hydride **60,** 7
- oxide **19,** 419s60; **60,** 48
Amberlite IRA 400 dithiocarbamate 38, 506s60
- – – hydrogen sulfide **60,** 196
- – 900 hydroxide **52,** 358s60
Amberlyst 15 33, 879s60
Amberlyst 15W (Er(III) or Yb(III) form) 28, 141s60
Amides s. Acylamin..., Carboxylic acid amides
Amidines, ar.
- from
  amines **60,** 158
  halides, ar. **60,** 158
Amidinium salts, vinylogous
- from
  carboxylic acids **30,** 634s60
- special s.
  formamidinium salts, vinylogous
- startg. m. f.
  pyridine N-oxides **25,** 581s60
Amidinotriazenes, polymer-based
- as intermediates **60,** 161
Amidoximes
- special s.
  sulfonylamidoximes
- startg. m. f.
  cyanamides, N,N-disubst. **60,** 131
  1,2,4-oxadiazoles, 5-aryl-, carbonylation **60,** 401
Amine N-oxides
-, deoxygenation **60,** 499
Amines (s.a. Hydroamination)
- from
  carboxylic acid amides **60,** 38, 40
  enamines **60,** 32
-, resolution, kinetic **60,** 138, 148
-, scavengers **1,** 319s60; **60,** 159
- special s.
  alleneamines
  benzylamines
  diamines
  enamines
  ethyleneamines
  nitramines
  siloxyamines
  ynamines
- startg. m. f.
  amidines, ar. **60,** 158
  guanidines **60,** 158
- (from 2 different molecules) **60,** 147
- (– 2 – –), polymer-based synthesis **60,** 147
  hydroxylamines **49,** 85s60
  sulfamides **60,** 163

Ami 274

(Amines
- startg. m. f.)
urethans, oxidative carbonylation
60, 132
Amines, ar.
- from
chlorides, ar. 60, 156
nitro compds., ar. 60, 15
- special s.
triarylamines
- startg. m. f.
benzylamines, ar., asym. hydroamination 60, 128
Amines, ar., prim.
- startg. m. f.
allylarenes 60, 386
fluorides, ar. 60, 183
nitro compds., ar. 60, 47
quinolines 60, 359
triarylamines 60, 157
-, N-trifluoroacetylation, preferential
60, 137
Amines, cyclic
-, C-acylation, Rh-catalyzed (α-
acylation) 60, 293
Amines, prim.
- from
acetylene derivs. 58, 126s60
alcohols, tert. 60, 146
azo compds. 20, 356s60
N-sulfonylimines, synthesis, asym.
60, 247
-, resolution, kinetic, parallel (with sec.
alcohols) 60, 9
- startg. m. f.
alcohols, prim. 60, 85
nitriles, oxidation, aerobic 60, 167
ureas, sym. 58, 133s60
Amines, N-protected
- startg. m. f.
β-aminoketones, N-protected, asym.
conversion 60, 336
Amines, sec.
- from
azomethines, reduction 60, 32, 38
-, synthesis, asym. 60, 248
-, N-nitrosation 60, 116
- startg. m. f.
α-tert-aminooximes 60, 140
Amines, sec., cyclic
- from
ethyleneazides 60, 172
Amines, silica-supported
- as scavenger 60, 154
Amines, tert., sym.
- from
halides 60, 150
Aminoalcohols
- special s.
diaminoalcohols
Aminoalcohols, chiral
- as reagent 42, 616s60 (update)
2-Aminoalcohols
- from
α-aminoketones, reduction, asym.
59, 33s60
- special s.
α-amino-β-hydroxy...
2-(arylamino)alcohols
trifluoromethyl-2-aminoalcohols
- startg. m. f.

2-oxazolidones, 3-carbo-tert-butoxy-
60, 135
2-Aminoalcohols, chiral
- as reagent 60, 20
- special s.
(R,R)-2,6-bis[2-(diphenylhydroxymethyl)-1-pyrrolidinylmethyl]-p-
cresol
(1S,2R)-N,N-dibutylnorephedrine
-, dendritic, chiral
- as reagent 48, 625s60
p-Aminoarylacetylenes, N-protected
- from
2-amino-1,3-enynes 54, 325s60
α-Amino-δ-arylcarboxylic acid esters
49, 836s60
α-Amino-β-arylcarboxylic acids
- from
cinnamic acids, asym. conversion
60, 126
2-Aminoazides
- from
aziridines, asym. induction 60, 123
Aminocarbonylation-intramolecular
Pauson-Khand reaction 60, 387
o-Aminocarboxylic acid N-tert-alkyl-
amides
- startg. m. f.
1,1-alkoxyaroylamines 60, 86
β-Aminocarboxylic acid derivs. 60, 382
Aminocarboxylic acid esters
- special s.
diaminocarboxylic acid esters
α-Aminocarboxylic acid esters
- special s.
glycinates
- - -, cyclic, N-protected 49, 985s60
β-Aminocarboxylic acid esters
- from
aldimines, polymer-based synthesis
50, 551s60
-, synthesis, asym. with 2 extra C-atoms
60, 244
- special s.
alkoxy-β-aminocarboxylic acid esters
β-arylaminocarboxylic - -
trans-β-Aminocarboxylic - -, cyclic
60, 486
α-Aminocarboxylic acids
-, N-arylation 60, 153
- from
α-ketocarboxylic acids 22, 421s60
- special s.
α-amino-β-arylcarboxylic acids
arginine...
C-(glycosyl)-α-aminocarboxylic -
α-Aminocarboxylic acids, fluorescent
- as reactant 19, 33s60
- -, sterically constrained
-, synthesis, asym., symposium reports
46, 767s60
β-Aminocarboxylic acids, N-protected,
chiral 60, 127, 408
2-Amino-2-deoxyglycosides 60, 17
2-Amino-2-deoxy-C-α-glycosides,
N-protected 60, 442
3-Amino-1,2-diols, chiral 53, 59s60
2-Aminodisulfides, chiral
- as reagent 42, 616s60
2-Amino-1,3-enynes
- startg. m. f.

p-aminoarylacetylenes, N-protected
54, 325s60
2-Aminoethers
- special s.
(1R,2R)-1-dimethylamino-2-(2-
methoxyphenoxy)-1,2-diphenyl-
ethane
α-Amino-β,γ-ethylenecarboxylic acids
- special s.
α-amino-α-vinylcarboxylic acids
(Z)-β-Amino-α,β-ethyleneketones
- from
benzotriazoles, 1-acyl- 60, 353
ketimines 60, 353
α-Aminohydroxamic acids,
N-protected, chiral
- as reagent 60, 65
(S)-2-Amino-2'-hydroxy-1,1'-binaphthyl
- as catalyst 54, 394s60
β-Amino-α-hydroxycarboxylic acid
amides, N-protected, chiral
20, 511s60
α-Amino-β-hydroxycarboxylic acids,
chiral 60, 250
β-Amino-α-hydroxyketones,
N-protected, chiral 60, 336
β-Amino-α-hydroxyphosphonic acid
esters 14, 380s60
α-Aminoketones (s.a. Friedel-Crafts
α-aminoacylation)
α-Aminoketones, cyclic
- from
ethylene derivs., carbonylation 60, 293
β-Aminoketones
- from
enoxysilanes 50, 551s60
β-Aminoketones, N-protected
- from
ketones, aldehydes and amines,
N-protected, asym. conversion
60, 336
ω-Aminoketones, conjugated
-, synthesis, review 34, 681s60
α-Amino-γ-lactones
- from
ethylene derivs. 60, 332
Aminolysis, asym. 60, 9
o-Aminomercaptans
- startg. m. f.
1,5-benzothiazepines, 2,3-dihydro-,
2-aryl- 60, 403
β-tert-Amino-α-methyleneketones
- from
methyl ketones 60, 337
δ-Amino-γ-methyleneketones, cyclic
60, 287
Aminonitriles
- special s.
diaminonitriles
α-Aminonitriles
-, chemistry (recent), review 47, 715s60
α-tert-Aminonitriles
- startg. m. f.
2-ethylene-tert-amines 60, 467
α-tert-Aminooximes
- from
1,1-nitroethylene derivs. and amines,
sec. 60, 140
Aminopalladation, intramolecular-1,4-
addition 60, 296
Aminophosphines
- special s.

2-(dicyclohexylphosphino)-2'-
  dimethylaminobiphenyl
α-Aminophosphonic acid derivs.
–, synthesis, review 41, 556s60
α-Aminophosphonic acid esters
– from
  azomethines 60, 226
– – –, acyclic and macroheterocyclic
–, synthesis and properties, review 41,
  556s62
– – monoesters 41, 556s60
1,1-Aminosilanes, protected
– from
  enamines, protected 29, 603s60
1-Amino-3-siloxy-1,3-dienes
–, Diels-Alder reaction, asym., Cr(III)-
  catalyzed with – 60, 447
α-Aminosulfones, N-protected
–, Mannich-type reaction, asym., with 2
  extra C-atoms 60, 408
2-Amino-3-sulfonyl-2,3-dideoxy-
  glycosides 17, 653s60
α-Amino-N-sulfonylhydrazones, cyclic,
  N-protected
– startg. m. f.
  (E)-ethyleneamines, N-protected,
  synthesis 60, 356
o-Aminosulfoxides, chiral
– as reagent 48, 772s60
Aminosulfur trifluorides
– special s.
  [bis(2-methoxyethyl)amino]sulfur
  trifluoride
α-Amino-α-vinylcarboxylic acids,
  chiral 60, 216
Ammonia
– as reactant 60, 126
Ammonium cerium(IV) nitrate 28,
  141s60; 43, 456s60; 60, 100, 436
– – –/montmorillonite 27, 162s60
– diacetoxybromate, quaternary,
  polymer-based 45, 120s60
– fluorides, quaternary, azabicyclic,
  chiral
– as reagent 48, 600s60
– hydrogen carbonate
– as reactant 60, 144
– hydroxides, quaternary, polymer-
  based, amphiphilic
– as reagent 52, 247s60
– molybdate 60, 58
– nitrate 60, 136
– salts, quaternary
– special s.
  bridgehead ammonium salts,
  quaternary
  N,N-diallylammonium –, –
Ammonolysis, asym. 44, 314s60
Anilines s. Amines, ar.
Anilinium hypophosphite
– as reactant 60, 215
Anthracene ring, 9,10-dihydro-
– from
  acetylene derivs. 60, 274
  o-di(propargyl)arenes 60, 274
Anthracenes, 9-α-alkoxy-
–, Diels-Alder reaction, asym., polymer-
  based with – 27, 694s60
Anthracenes, 9-halogeno-
– startg. m. f.
  aceanthrylenes 60, 404
Antibody catalysis

– with aldolase antibody 38C2, review
  58, 499s60
Anti-Michael addition, asym.,
  enzymatic
– of ammonia 60, 126
Antimony compds., organo-
– special s.
  Sb-heterocyclics
  triarylantimony diacetates
Arenedicarbonylchromium(0)
  complexes, polymer-based 46, 631s60
Arenes (s.a. Benzene ring)
– from
  thiolcarbamic acid aryl esters 3, 73s60
– special s.
  allylarenes
  di(propargyl)arenes
  methylarenes
  polyarenes
– startg. m. f.
  α-aryl-α-hydroxycarboxylic acid
  esters, asym. conversion 60, 227
Arenes, deactivated
–, iodination, ar. 60, 178
Arenes, electron-rich
– startg. m. f.
  arylglycinates, N-protected, synthesis,
  asym., regiospecific 60, 246
Arenes, novel
–, symposium reports 52, 221s60
Arginine mimetics
–, review 59, 174s60
Aromatization, on-resin 8, 823s60
Arsine oxides
– special s.
  triphenylarsine oxide
Arylacetic... s.a. α-Arylcarboxylic...
Arylacetic acid esters 60, 383
Arylacetic acids
–, photoarylation 32, 800s60
Arylacetylenes (s.a. under
  Sonogashira...)
– from
  acetylene derivs., terminal 60, 398
  halides, ar. 60, 398, 460
  lithium alk-1-ynyl(trialkoxy)borates
  60, 460
– special s.
  aminoarylacetylenes
  hydroxyarylacetylenes
– startg. m. f.
  fluorenes, 9-alkylidene- 60, 391
2-(Arylamino)alcohols
– from
  epoxides 60, 120, 122
β-Arylaminocarboxylic acid esters,
  cyclic
– from
  α,β-ethylene(nitro)carboxylic acid
  esters 60, 170
Arylation
– with aryl(biphenyl-2,2'-ylene)bismuth
  diacetates 36, 877s60
C-Arylation s.a. Heck arylation
C-Arylation, Ni-catalyzed
– of malononitriles 60, 388
o-Arylation
– of anilines 36, 877s60
C-α-Arylation-intramolecular O-
  allylation 60, 370
N-Arylation, heterogeneous, Ni-
  catalyzed 60, 156

N-Arylation, polymer-based 8, 563s60
N-Arylation-intramolecular Heck
  arylation 60, 395
Aryl(biphenyl-2,2'-ylene)bismuth
  diacetates
–, arylation with – 36, 877s60
Arylboronic acid esters
– startg. m. f.
  arylglycines, with 2 extra C-atoms
  60, 415
Arylboronic acids
–, 1,4-addition, asym. to enoates with –
  55, 452s60
– from
  diazonium fluoroborates 60, 218
– startg. m. f.
  thioethers, ar. (with mercaptans)
  60, 201
α-Arylcarbonyl compds.
– from
  enoxysilanes and triarylantimony
  diacetates 60, 461
Arylcarboxylic acid amides
–, p-metalation 60, 223
– – –, N-unsubst.
– from
  halides, ar., carbonylation-
  nitrogenation 60, 396
β-Arylcarboxylic acid esters
– special s.
  β-(carboxyaryl)carboxylic acid esters
γ-Arylcarboxylic acid esters
– special s.
  alkoxy-γ-arylcarboxylic acid esters
δ-Arylcarboxylic acid esters
– special s.
  amino-δ-arylcarboxylic acid esters
Arylcarboxylic acid halides
–, 1,4(6)-addition to – 60, 257
Arylcarboxylic acids (s.a. Carboxyaryl...)
– from
  benzyl halides 60, 91
Arylcarboxylic acids, ^{14}C-labelled
  43, 808s60
β-Arylcarboxylic acids
– special s.
  amino-β-arylcarboxylic acids
Aryldiazomethanes
–, generation in situ 60, 350
(E)-3-Aryl-3-ethylenealcohols 60, 406
2-Aryl-2-ethylene-prim-amines 60, 298
2-Aryl-3-ethyleneazides
– from
  allenes 60, 298
  halides, ar. 60, 298
α-Aryl-α,β-ethylenecarboxylic acid
  esters
– from
  styrenes, carbonylation 60, 317
(E)-γ-Aryl-β,γ-ethylene-α-keto-
  carboxylic acid esters 29, 752s60
Arylglycinates, N-protected
– from
  arenes, electron-rich, synthesis, asym.,
  regiospecific 60, 246
Arylglycines
– from
  arylboronic acid esters, with 2 extra
  C-atoms 60, 415
Arylglyoxylic acid esters 51, 76s60

α-Aryl-α-hydroxycarboxylic acid
esters (s.a. Mandelic acid esters)
- from
  arenes, asym. conversion **60**, 227
  α-oxocarboxylic acid esters, – –
    60, 227
γ-Aryl-α-ketocarboxylic acid esters,
  chiral **60**, 260
Aryl ketones (s.a. Friedel-Crafts...)
- from
  arylboronic acids **60**, 457
  halides, ar. **60**, 389
  N-(2-pyrazinyl)aldimines **60**, 389
- special s.
  acylophenones
  diaryl ketones
- startg. m. f.
  arylboronic acids **60**, 457
α-Arylketones
- from
  aldehydes **60**, 350
  aldehyde tosylhydrazones **60**, 350
Aryl lactolides
-, [1,2]-Wittig rearrangement, asym.
  53, 335s60
1-Aryllactones
- from ketocarboxylic acid esters and
  halides ar. **60**, 366
Arylmagnesium halides
-, preparation **60**, 208
- special s.
  (halogenomethyl)arylmagnesium
  bromides
Aryl mercaptans
-, Michael addition, asym. with – **60**, 191
1,4-Sn→C-Aryl migration
-, 3-arylstannanes via – **56**, 490s60
Aryloxo compds.
- from
  benzylalcohols (s.a. OCflH) **60**, 109
Aryloxysilanes
- from
  halides, ar. **53**, 86s60
α-Arylphosphonic...
  s.a. Benzylphosphonic...
Arylphosphonous acids
- from
  halides, ar. **60**, 215
Arylsilanes
- from
  halides, ar. **31**, 592s60
β-Arylsilanes **36**, 825s60
Arylstannanes
-, 1,4-addition with – **54**, 466s60
- startg. m. f.
  benzylalcohols, sec., synthesis **60**, 450
3-Arylstannanes
- via 1,4-Sn→C-aryl migration
  56, 490s60
N-(Arylsulfonyl)isocyanates
- startg. m. f.
  2-pyridones, 5,6-dihydro-, 1-
    arylsulfonyl-4-arylthio- **60**, 409
(Z)-β-Aryltelluro-α,β-ethylenesulfones
  50, 343s60
Arylthio-2-acetylenes
- startg. m. f.
  diaryl sulfides **60**, 204
3-Arylthio-1,5-dienes **42**, 733s60
Arylthiolic acid esters
- by carbonylation **55**, 324s60
Asym. synthesis s. Synthesis, asym.

Automated synthesis
- for air-sensitive compds. **51**, 453s60
Automultiplication, asym. **60**, 230
4-Aza-1-azoniabicyclo[2.2.2]octane
  tetrahydridoborate, 1-benzyl-
- as reagent **60**, 32
3-Azabicyclo[3.3.0]oct-1(8)-ene-2,7-
  diones
- from
  α,β-acetylenehalides and 2-ethylene-
    amines **60**, 387
Azabis(Δ²-oxazolines), polymer-based,
  chiral
- as reagent **23**, 819s60
3-Aza-1,5-dien-4-ones
- startg. m. f.
  2-piperidone ring **60**, 276
Aza-Horner synthesis
- with phosphine N-carbalkoxyimines
  60, 424
Aza-Knoevenagel condensation **60**, 349
Azetidines, 3-bromo-1-tosyl- **48**, 434s60
Azetidines, chiral
- as reagent **42**, 616s60
2-Azetidinone ring, 3-acyl-
- from
  1,3-dioxane-4,6-diones, 5-acyl-, asym.
    induction **60**, 473
2-Azetidinones
- from
  aldimines, polymer-based synthesis
    60, 322
  carboxylic acid esters, – – **60**, 322
  α-diazoketones, asym. induction
    13, 681s60
2-Azetidinones, *trans*-3-acylamino-
  60, 322
-, *trans*-4-acylamino- **7**, 836s60
-, 4-aryl-, 3-oxy-, chiral **7**, 836s60
-, 3-hydroxy-, chiral **1**, 775s60
Azides
- special s.
  aminoazides
  ethyleneazides
  nitroazides
- startg. m. f.
  hydrazones, dimethyl- **60**, 155
2-Azidoalcohols
- from
  1,3-dioxolane-2-thiones **52**, 175s60
- special s.
  β-azido-α-hydroxy...
β-Azidocarboxylic acid amides
- from
  α,β-ethylenecarboxylic acid amides,
    asym. conversion **60**, 127
Azidocarboxylic acid esters
- startg. m. f.
  lactams **52**, 13s60
β-Azido-α-halogenoboronic acid esters,
  chiral **60**, 365
β-Azido-α-hydroxycarboxylic acids
  10, 262s60
- from
  glycidic acids **60**, 119
α-Azidoketones **30**, 232s60
Aziridine-2-carboxylic acid esters
- from
  aldimines, asym. conversion **60**, 357
Aziridines
- from

ethylene derivs., Cu(I)-catalyzed,
  review **23**, 819s60
- startg. m. f.
  2-aminoazides, asym. induction **60**,123
  2-pyrrolidones, 3-silyl- (N-protected
    derivs.) **60**, 243
Aziridines, *cis*-2-silyl-1-sulfonyl-
- from
  N-sulfonylimines, asym. induction
    60, 348
-, 1-sulfonyl-
-, cycloaddition, 1,3-dipolar with –
    60, 265
- startg. m. f.
  pyrrolidines, 1-sulfonyl- **60**, 265
Δ¹-Azirines
- in synthesis, review **33**, 325s60
Azo compds.
- startg. m. f.
  amines, prim. **20**, 356s60
Azomethines (s.a. Imines)
- special s.
  aldimines
  alkylideneamino...
  ketimines
- startg. m. f.
  amines, sec., reduction **60**, 32, 38
  –, –, synthesis, asym. **60**, 248
  α-aminophosphonic acid esters **60**, 227
Azomethinium salts
- special s.
  N-acylazomethinium salts
Azomethinium salts, cyclic
  s. Cyclimmonium salts
Azomethinium ylids, non-stabilized
-, cycloaddition, 1,3-dipolar,
  intramolecular with – **49**, 388s60

Balz-Schiemann reaction
- in ionic liquids **60**, 183
Barbier fragmentation
  s. Retro-Barbier fragmentation
Bases, super s. Superbases
Baylis-Hillman reaction
-, alternative **60**, 277
- –, accelerated
- via β'-halogeno-β-hydroxyketones
  60, 275
- –, chalcogen-mediated
- with α,β-acetylenecarbonyl compds.
  60, 275
- –, polymer-based, soluble **39**, 593s60
Baylis-Hillman-type reaction
- with β,γ-ethylenehalides **60**, 373
Beckmann rearrangement (s.a. Solid-
  state Beckmann rearrangement)
- –, In(III)-catalyzed **60**, 169
1*H*-2-Benzazepines, 2,3,4,5-tetrahydro-
  60, 377
1*H*-3-Benzazepines, – **19**, 954s60
Benzene ring (s.a. Arenes,
  Aromatization)
-, [2+2]-cycloaddition, photochemical at
  the *ortho*-site, review **19**, 764s60
- from
  ketones **60**, 464

1,1-nitroethylene derivs. (3 molecules) **60**, 354
2-pyrones **60**, 464
Benzenes, 1,3,5-triaryl-, sym. 60, 354
Benzhydrols
– startg. m. f.
diarylmethanes **60**, 45
Benzils
– from
benzoins **37**, 234s60
Benzimidazoles
– from
aldehydes, polymer-based synthesis **60**, 139
o-nitramines, – – **60**, 139
Benzimidazoles, 2-alkylthio- 53, 228s60
1,2-Benziodazol-3-one 1-oxides, N-condensed
– as reagent **50**, 123s60
1,2-Benziodoxol-3-one, 1,1,1-triacetoxy-
– as reagent **60**, 62, 77, 78
1,2-Benziodoxol-3-one 1-oxides
– special s.
tetrabutylammonium 1-oxido-1,2-benziodoxol-3-one
1,2-Benziodoxol-3-ones, 1-aryloxy-
– as reagent **55**, 183s60
1,2-Benzisothiazoline 1,1-dioxides, 2,3-oxido-, chiral
– as reagent **35**, 57s60
1,2-Benzisoxazol-3(2H)-ones
– from
o-hydroxyhydroxamic acids **60**, 48
Benzocyclobutenes, *trans*-1,2-diaryl-
– from
o-bis(dihalogenomethyl)arenes **60**, 375
1H-1,5-Benzodiazepines, 2,3-dihydro-
– from
o-nitroazides and ketones **60**, 358
Benzofuran-2-carboxylic acid esters 56, 334s60
Benzofuran-3-carboxylic – –
– from
o-hydroxyarylacetylenes, carbonylation **60**, 318
Benzofuran-2,3-dihydro-
–, closure with ring expansion of bridged N,S-heterocycles **60**, 206
Benzofurans
– from
chromium alkoxycarbene complexes **60**, 418
1,3-dien-5-ynes **60**, 418
3-ene-1,5-diynes **60**, 418
o-hydroxy-β-styryl ethers **60**, 113
–, 2-α-amino-, chiral **32**, 820s60
–, 2,3-dihydro-, 7-alkoxy-4-hydroxy- **43**, 860s60
–, –, 5-subst.
– from
p-hydroxysulfoxides and ethylene derivs. **60**, 334
Benzoins
– startg. m. f.
benzils **37**, 234s60
Benzophenones s. Diaryl ketones
2-Benzopyran-3-ylidene complexes
– special s.
tungsten 2-benzopyran-3-ylidene complexes
1H-2,1,3-Benzothiadiazine 2,2-dioxides, 3,4-dihydro- 59, 250s60

1,5-Benzothiazepines, 2,3-dihydro-8, 657s60
–, –, 2-aryl-
– from
o-aminomercaptans, ethynylcarbinols, and halides, ar. **60**, 403
2,1-Benzothiazine 2,2-dioxides 48, 434s60
Benzothiazoles 8, 657s60
– from
thiazoles, 4-[(benzotriazol-1-yl)methyl]- **60**, 352
–, 2-alkylthio- **29**, 547s60
–, 2-amino- **60**, 352
α-(Benzothiazol-2-ylthio)carbonyl compds.
– startg. m. f.
trans-cyclopropanecarbonyl compds. **60**, 410
Benzo[b]thiophene-2-carbonyl compds.
– from
aldehydes, ar. **60**, 368
α-halogenocarbonyl compds. **60**, 368
Benzotriazole, 1-trimethylsilylmethyl-
– as reactant **60**, 414
Benzotriazoles
– special s.
thiazoles, 4-[(benzotriazol-1-yl)-methyl]-
–, 1-acyl-
–, N-acylation with – **53**, 152s60
– startg. m. f.
(Z)-β-amino-α,β-ethyleneketones **60**, 353
quinolines, 2,4-diamino- **60**, 463
–, 1-amidino-
– from
amines **60**, 147
– startg. m. f.
guanidines **60**, 147
–, 1-benzyl-
– startg. m. f.
indenes **60**, 352
–, N-α-thioacylamino-
– startg. m. f.
pyrroles **60**, 352
N-α-(Benzotriazol-1-yl)aldimines
– startg. m. f.
thiazoles, 5-amino- **60**, 352
α-(Benzotriazol-1-yl)ketones
– from
carboxylic acid halides, with 1 extra C-atom **60**, 414
Benzotriazol-1-ylmethanimines
– special s.
bis(benzotriazol-1-yl)methanimine
1-(Benzotriazol-1-yl)-1,1-silylthioethers
– startg. m. f.
thioacylsilanes **43**, 597s60
2H-1,4-Benzoxazines 13, 539s60
–, lactam-condensed
– from
ethylenecarboxylic acid anilides **60**, 78
4H-3,1-Benzoxazin-4-ones, 2-heterocyclyl-
–, review **29**, 391s60
2H-1,4-Benzoxazin-3-ylacetic acid esters, 3,4-dihydro- 60, 170
Benzoxazoles, 3-alkylthio- 29, 547s60
O-Benzoylation, partial and preferential 40, 99s60
Benzylalcohols

– special s.
deuteriobenzylalcohols
p-methoxybenzylalcohols
sulfamidobenzylalcohols
– startg. m. f.
aryloxo compds. **60**, 109
Benzylalcohols, sec.
–, determination of abs. configuration **24**, 245s60
– from
aldehydes, synthesis **60**, 450
arylstannanes, – **60**, 450
Benzylamines, ar.
– from
styrenes and amines, ar., asym. hydroamination **60**, 128
Benzylamines, sec., chiral 60, 248
O-Benzylation
– with benzyl mesylate **60**, 79
Benzyl carbamate
– as reactant **51**, 132s60
Benzyl halides
– special s.
o-(halogenomethyl)...
– startg. m. f.
arylcarboxylic acids **60**, 91
N-Benzylhydroxylamine
– as reactant **60**, 149
Benzyl mesylate
–, O-benzylation with – **60**, 79
N-Benzyl-N-(6-methyl-2-pyridyl)-carbamyl chloride
– as reagent **40**, 99s60
Benzylphosphonic acid derivs.
–, α-fluorination, asym. **39**, 458s60
Benzyltriethylammonium chloride
– as reagent **60**, 131
– **permanganate 60**, 162
– **tetrathiomolybdate**
– as reagent **60**, 199
Benzyltriphenylphosphonium chlorochromate 26, 235s60
– **dichromate 41**, 241s60
– **peroxymonosulfate 46**, 237s60
Benzyl trityl ether
–, O-tritylation, preferential with – **60**, 101
Benzylzinc halides
–, cross-coupling with alkylthio-N-heterocyclics **60**, 412
Biaryls s. Diaryl...
Bicyclo[3.n.1]alk-2-en-(n+6)-ones
– from
chromium α,β-ethylene(alkoxy)-carbene complexes and enamines, isocyclic, asym. induction **60**, 419
Bicyclo[3.3.0]octan-2-ols, 5,6-disiloxy-, chiral 60, 309
Bicyclo[5.1.0]oct-5-en-2-ols, chiral 60, 309
Bicyclo[3.3.0]oct-1-en-3-one-7,7-dicarboxylic acid esters
– from
γ,δ-acetylenemalonic acid esters and acoxy-2-ethylenes **60**, 345
Biginelli synthesis, decarboxylative 60, 471
BINAP s. 2,2'-Bis(diphenylphosphino)-1,1'-binaphthyl
1,1'-Bi-2-naphthol
s.a. under Aluminum aroxides

(R)-1,1'-Bi-2-naphthol, 3,3'-diphenyl-
– as reagent **54**, 296s60
–, polymeric, bridged
– as reagent **44**, 565s60
–, 5,6,7,8-tetrahydro-
– as reagent **44**, 565s60
(S)-1,1'-Bi-2-naphthol, 3,3'-bis(methylthio)-
– as reagent **60**, 379
–, 4,4',6,6'-tetrakis(perfluorooctyl)-
– as reagent **58**, 235s60
1,1'-Bi-2-naphthol bis(phosphites), cyclic, chiral
– as reagent **52**, 297s60
1,1'-Binaphthol crown ethers, chiral
– as reagent **60**, 444
(S)-1,1'-Bi-2-naphthol-3,3'-dicarboxamides, polymer-based
– as reagent **44**, 565s60
1,1'-Bi-2-naphthol phosphoromonoamidites, cyclic, chiral
– as reagent **52**, 297s60
– N-(2-pyridyl)phosphoromonoamidites, chiral
– as reagent **48**, 772s60
1,1'-Bi-2-naphthols
–, resolution **60**, 500
– special s.
di(phosphinyl)-1,1'-bi-2-naphthols
1,1'-Binaphthyls
– special s.
2-amino-2'-hydroxy-1,1'-binaphthyl
2,2'-bis[bis[3,5-bis(trifluoromethyl)phenyl]phosphino]-1,1'-binaphthyl
2,2'-bis(diphenylphosphino)-1,1'-binaphthyl
Biotransformations, preparative
–, review **2**, 13s60
3,3'-Bi(oxindoles), chiral 60, 488
3,3'-Bi-4-phenanthrols, chiral
– as reagent **60**, 357
Biphenyls s. Diaryl...
2,2'-Bipyridyls, chiral
– as reagent **23**, 819s60
Bis(allenes)
–, carbopalladation, intramolecular-1,4-distannylation and 1,4-silylstannylation, regiostereospecific **60**, 300
– startg. m. f.
1,3-dienes, exocyclic **60**, 300
Bis(benzotriazol-1-yl)methanimine
– as reagent **60**, 147
(R)-2,2'-Bis[bis[3,5-bis(trifluoromethyl)phenyl]phosphino]-1,1'-binaphthyl
– as reagent **47**, 542s60
2,2'-Bis[bis(3,5-dimethylphenyl)phosphino]biphenyl
– as ligand **54**, 30s60
Bis(catecholato)diborane
– as reactant **60**, 207
Bis(collidine)bromine(I) hexafluorophosphate
– as reagent **60**, 114
Bis(2,4-dichlorophenyl)phosphoromonochloridate
– as reagent **49**, 312s60
o-Bis(dihalogenomethyl)arenes
– startg. m. f.
benzocyclobutenes, *trans*-1,2-diaryl- **60**, 375

1,4-Bis(9-O-dihydroquinine)phthalazine
– as reagent **60**, 166
1,4-Bis(9-O-dihydroquin[id]ine)-phthalazine, polymer-based
– as reagent **56**, 131s60
(R,R)-2,6-Bis[2-(diphenylhydroxymethyl)-1-pyrrolidinylmethyl]-*p*-cresol
– as reagent **60**, 231
3,3'-Bis(diphenylphosphino)-2,2'-biindoles, chiral
– as ligand **43**, 51s60
(R or S)-2,2'-Bis(diphenylphosphino)-1,1'-binaphthyl
– as reagent **60**, 171, 452
2,2'-Bis(diphenylphosphino)-1,1'-binaphthyl, 7,7'-dimethoxy-, chiral
– as reagent **46**, 738s60
1,4-Bis(diphenylphosphino)butane
– as reagent **60**, 345
(S)-2,2'-Bis(diphenylphosphino)-3,3'-dibenzo[*b*]thiophene
– as reagent **46**, 738s60
1,2-Bis(diphenylphosphino)ethane
– as reagent **60**, 84
1,1'-Bis(diphenylphosphino)ferrocene
– as ligand, review **59**, 159s60
– as reagent **60**, 156, 158, 214, 344, 370, 396
(1R,2R)-N,N'-Bis[(S)-2-(diphenylphosphino)ferrocenyl]cyclohexane-1,2-diamine
– as reagent **59**, 354s60
1,3-Bis(diphenylphosphino)propane
– as reagent **60**, 405
(S)-2,2'-Bis(di-*o*-tolylphosphino)-1,1'-binaphthyl
– as reagent **60**, 303
(R)-2,2'-Bis(di-*p*-tolylphosphino)-1,1'-binaphthyl
– as reagent **60**, 246
Bis(ferrocenyl)mercury
– as reactant **60**, 459
Bis(iodomethyl)zinc
– as reactant **60**, 411
Bis(iodozincio)methane
– as reactant **60**, 233; s.a. Reviews section
[Bis(2-methoxyethyl)amino]sulfur trifluoride
– as reagent **60**, 142
(R)-2,3:2',3'-Bis(methylenedioxy)-6,6'-bis(diphenylphosphino)biphenyl
– as reagent **49**, 699s60
Bismuth(III) chloride 60, 12
Bismuth compds., organo-
– special s.
aryl(biphenyl-2,2'-ylene)bismuth diacetates
Bi-heterocyclics
Bismuth(III) nitrate 60, 12
Bis(4H-1,3-oxazines, 5,6-dihydro-), phosphino-
– as reagent **48**, 772s60
Bis(Δ²-oxazolines)
– special s.
azabis(Δ²-oxazolines)
Bis(Δ²-oxazolines), chiral
– as reagent **43**, 851s60; **54**, 261s60; **60**, 149, 227, 260, 261, 429
–, clay-supported, chiral
– as reagent **23**, 819s60

Bis(4-perfluorohexylphenyl) diselenide
– as reagent **49**, 938s60
Bis(phosphine oxides), P-chiral 1, 715s60
Bis(pinacolato)diborane
– as reactant **60**, 218
Bis(tri-*n*-butyltin) oxide
– as reagent **50**, 71s60; **60**, 385
Bis(trichloromethyl) carbonate
– as reagent **16**, 493s60; **60**, 181
N,O-Bis(trimethylsilyl)acetamide
– as reagent **60**, 345
Bis(trimethylsilyl) peroxide
– as reagent **60**, 272
Borane (s.a. Diborane)
Borane-amine complexes
–, activation, Pd-catalyzed **60**, 15
Borane-*tert*-benzylamine complex
–, prepn. **15**, 39s60
Borane-N,N-diethyl-1,1,3,3-tetramethylbutylamine
– as reagent **21**, 174s60
Borane-dimethyl sulfide 60, 19
Borane-tetrahydrofuran 60, 264
Borane-trimethylamine 60, 15
Boranes
– special s.
ethyleneboranes
silylboranes
trialkylboranes
Borate, tetrahydro- s. 4-Aza-1-azoniabicyclo[2.2.2]octane tetrahydridoborate, 1-benzyl-, Calcium bis(–), Sodium –, Zirconium(IV) –
Borates, organo-
– special s.
lithium alkynyl(trialkoxy)borates
lithium aryl(trimethoxy)borates
potassium 2-ethylene(trifluoro)borates
Boration (s.a. Silaboration)
Boric acid esters
– special s.
triisopropyl borate
triphenyl borate
Borinic acid esters
– special s.
diisopinocampheyl(methoxy)borane
Borinyl triflates
– special s.
dicyclohexylborinyl triflate
Boron enolates
– as intermediates **60**, 235
Boron chloride 51, 3s60; **60**, 113
Boron fluoride 7, 761s60; **44**, 819s60; **47**, 897s60; **60**, 46, 98, 180, 222, 265, 309, 330, 331, 413, 430
Boronic acid amide esters
s.a. Borylamino...
Boronic acid esters (s.a. Suzuki coupling)
– special s.
alkoxyboronic acid esters
arylboronic –
α-(borylamino)boronic –
cyclopropaneboronic – –
di(boronic – –)
ethyleneboronic – –
halogenoboronic – –
siloxyboronic – –
silylboronic – –
– startg. m. f.
α-alkoxyboronic acid esters, asym. synthesis with 1 extra C-atom **60**, 425

–––, functionalized
- startg. m. f.
 α-halogenoboronic acid esters, functionalized, asym. synthesis with 1 extra C-atom **60**, 365
Boronic acids (s.a. Suzuki coupling)
–, 1,4-addition, asym., Rh-catalyzed with – **60**, 452 (s.a. Reviews section)
- special s.
 arylboronic acids
 ethyleneboronic acids
- startg. m. f.
 ketones **60**, 457
 piperazinones, 3-subst. **60**, 416
1,2-Boronyl migration (s.a. Hydroboration-1,2-boronyl migration)
α-(Borylamino)boronic acid esters
- from
 aldimines **60**, 207
Bridgehead ammonium salts, quaternary 29, 970s60
Bromate, diacetoxy- s. Ammonium diacetoxybromate...
Bromine
- as reactant **60**, 176
- as reagent **60**, 108
–, scavengers **3**, 3s60
N-Bromoacetamide
- as reagent **32**, 413s60
Bromotrichloromethane
- as reagent **59**, 86s60
Bruylants reaction, modified 60, 467
***tert*-Butoxyformic anhydride**
–, cyclocarbonylation with – **60**, 135
***tert*-Butyl carbamate**
- as reactant **60**, 330
***tert*-Butyldimethylsiloxymalononitrile**
- as carbonyl equivalent, zwitterionic **60**, 466
***tert*-Butyl hydroperoxide**
- as reagent **60**, 61, 65
N-(*tert*-Butylimino)benzenesulfinyl chloride
- as reagent **60**, 477
***tert*-Butyl isocyanide**
- as reactant **60**, 158
***tert*-Butyl nitrite**
- as reagent **60**, 386

Calcium bis(tetrahydridoborate) 60, 328
– hypochlorite **55**, 159s60
Calixarenes, azaaromatic
–, review **7**, 281s60
Cannizzaro reaction, Cr(III)-catalyzed 60, 72
Carbalkoxyamines s. Carbamic acid esters, Carbenzoxyamines, Urethans
C-Carbalkoxylation, regiospecific, superelectrophilic
- of hydrocarbons **60**, 315
- via carbonylation **60**, 315
O-Carbalkoxylation
- special s.
 O-carballyloxylation
 O-carbobenzoxylation

 O-carbo-*tert*-butoxylation
O-Carballyloxylation 23, 237s60
Carbamic acid esters (s.a. Carbobenzoxyamines, Urethans)
- special s.
 benzyl carbamate
 tert-butyl carbamate
 2,7-dienol carbamates
Carbamic acid silyl esters
- as reagent **60**, 164
- startg. m. f.
 enamines **60**, 164
Carbamyl halides
- special s.
 N-benzyl-N-(6-methyl-2-pyridyl)-carbamyl chloride
– –, N,N-disubst.
- startg. m. f.
 carboxylic acid amides, N,N-disubst. **60**, 376
– –, unsatd.
- startg. m. f.
 lactams **60**, 455
Carbamylphosphonium salts
- as intermediates **60**, 376
Carbanion equivalents
- special s.
 acyl carbanion equivalents
Carbapenams, 1,2-di(alkylidene)-58, 306s60
Carbazole-3,4-quinones 24, 755s60
Carbene complexes
–, applications, review **47**, 955s60
- special s.
 glycosylidene complexes
– –, nucleophilic
–, metathesis with –, review **56**, 495s60
Carbenes
- special s.
 rhodium carbenes
Carbobenzoxyamines
- startg. m. f.
 acylamines **60**, 165
–, polymer-based
- as reactant **60**, 165
O-Carbobenzoxylation 23, 237s60
O-Carbo-*tert*-butoxylation 60, 135
Carbocupration-cross-coupling, regiospecific
- of α,β-acetylenesulfoxides **60**, 411
Carbodiimide, polymer-based
- as reagent **13**, 317s60
Carbohydrate chemistry
–, synthetic technologies, review **57**, 99s60
Carbohydrate conjugates
–, synthesis, combinatorial, review **57**, 99s60
– 1,2-O-isopropylidene derivs.
- startg. m. f.
 1,2-*trans*-C-allyl glycosides **60**, 432
– libraries, combinatorial
–, review **57**, 99s60
Carbohydrates
- special s.
 aldoses
 deoxyuloses
 disaccharides
 glycals
 glycos...
 oligosaccharides
 pentasaccharides

 trisaccharides
- startg. m. f.
 synthons, chiral, review **45**, 579s60
–, O-tritylation-O-acylation-O-silylation **60**, 90
Carbohydrates, branched 45, 368s60
Carbolithiation, regiostereospecific, Fe(III)-catalyzed 60, 258
Carbometalation
- of acetylene and ethylene derivs., heteroatom-functionalized, review **47**, 639s60
- special s.
 carbocupration
 carbolithiation
 carbopalladation
 carboplatination
Carbon dioxide
- startg. m. f.
 cyclopentadienones **60**, 367
Carbon dioxide, supercritical
–, β-aryl- from α,β-ethylene-ketones in – **39**, 640s60
–, dechlorination of PCBs in – **33**, 76s60
–, extraction with – **60**, 33
–, oxidation of alcohols in – **55**, 133s60
–, radical carbonylation in – **60**, 269
–, Wacker oxidation in – **56**, 75s60
– –, –/dimethylformamide
- as medium **23**, 139s60
Carbon disulfide
- as reactant **60**, 160
Carbonic acid esters (s.a. O-Carbalkoxylation)
- special s.
 2-acetylenecarbonic acid esters
 bis(trichloromethyl) carbonate
 diol monocarbonates
 ethylenecarbonic acid esters
 glycol carbonates
– – –, polymer-based
–, prepn. **49**, 168s60
- special s.
 p-nitrophenyl carbonate, polymer-based
– – –, sym.
- from
 alcohols **60**, 135
Carbonium ions
- special s.
 alkoxycarbenium ions
Carbon monoxide
–, N-formylation, Au-catalyzed with – **60**, 132
Carbon monoxide-¹⁴C 43, 808s60
Carbon oxide sulfide
- as reagent **60**, 189
Carbon tetrabromide
- as reagent **60**, 81, 318
– –/aluminum bromide **60**, 315
Carbon tetrahalides
- as reagent **29**, 518s60
Carbonylation (s.a. Aminocarbonylation..., Cyclocarbonylation, Oxypalladation, intramolecular-carbonylation, Radical carbonylation, Ring expansion, carbonylative, Silylformylation)
–, α-aryl-α,β-ethylenecarboxylic acid esters from styrenes **60**, 317
–, arylthiolic acid esters from diaryliodonium salts **55**, 324s60

Carbonylation
-, C-carbalkoxylation, regiospecific, superelectrophilic **60**, 315
-, 2-cyclopentenones from 2-ethylenecarbonic acids esters **60**, 343
-, α,β-ethylene-β-halogeno-α-(organoseleno)carboxylic acid esters from 1-acetylene-1-selenides **60**, 301
-, α,β-ethylenehydroxamic acid esters from α,β-ethyleneiodides **43**, 808s60
- of 1-halogenolethers **60**, 175
-, 1,2,4-oxadiazoles, 5-aryl- from amidoximes **60**, 401
- via ruthenium η³-allyl complexes **60**, 343
-, oxidative, Au-catalyzed **60**, 132
-, Rh-catalyzed
- at carbon atoms, satd. **60**, 293
Carbonylation-nitrogenation, Pd-catalyzed 60, 396
Carbonyl compds.
- special s.
acetylenecarbonyl compds.
arylcarbonyl -
dienecarbonyl -
ethylenecarbonyl -
halogenocarbonyl -
hydroxycarbonyl -
ketocarbonyl-
nitrocarbonyl -
Carbonyl equivalent, zwitterionic 60, 466
Carbonyl ylids, cyclic
- as intermediates **60**, 481
Carbopalladation
- special s.
acylpalladation
Carbopalladation, intramolecular (s.a. Michael addition-intramolecular carbopalladation)
-**, intramolecular-1,4-distannylation**
- of bis(allenes) **60**, 300
-**, intramolecular-1,4-silylstannylation**
- of bis(allenes) **60**, 300
Carbopalladation-cycloisomerization, regiospecific 60, 297
Carboplatination, intramolecular-solvation, regiospecific
- of enynes **60**, 304
β-(o-Carboxyaryl)carboxylic acid esters
- from phthalides and O-silyl O-alkyl keteneacetals **60**, 440
Carboxylic acid allyl esters s. Acoxy-2-ethylenes
Carboxylic acid amides (s.a. Acylamin...)
- from
carboxylic acid halides, polymer-based synthesis **60**, 159, 165
nitriles **60**, 167
- special s.
allenecarboxylic acid amides
aminocarboxylic - -
arylcarboxylic - -
azidocarboxylic - -
carboxylic acid anilides
ketocarboxylic acid amides
- startg. m. f.
amines **60**, 38, 40
glutarimides **60**, 320

- - -, N,N-disubst.
- from
carbamyl halides, N,N-disubst. **60**, 376
- startg. m. f.
aldehydes **60**, 41
- - -, N-unsubst.
- from
carboxylic acids **28**, 144s60
Carboxylic acid anilides
- special s.
ethylenecarboxylic acid anilides
Carboxylic acid anhydrides
- special s.
ethylenecarboxylic acid anhydrides
- startg. m. f.
carboxylic acid esters **60**, 98
α-siloxymalononitriles **60**, 333
Carboxylic acid aryl esters
- from
carboxylic acids **60**, 82
oxalic acid aryl esters **60**, 82
Carboxylic acid esters (s.a. Acoxy..., Carbalkoxy...)
- from
acetals **60**, 98
carboxylic acid anhydrides **60**, 98
- - halides, polymer-based synthesis **60**, 158
- - hydrazides **60**, 87
- acids (with alcohols), preferential conversion **60**, 81
- special s.
acetylenecarboxylic acid esters
acylaminocarboxylic - -
alkoxycarboxylic - -
(alkylideneamino)carboxylic - -
aminocarboxylic - -
arylcarboxylic - -
azidocarboxylic - -
carboxylic - aryl -
cyanocarboxylic - -
diazocarboxylic - -
dicarboxylic - -
epoxycarboxylic - -
ethylenecarboxylic - -
formic - -
halogenocarboxylic - -
hydroxycarboxylic - -
ketocarboxylic - -
nitrocarboxylic - -
(organoseleno)carboxylic - -
siloxyaminocarboxylic - -
silylcarboxylic - -
sulfamidocarboxylic - -
- startg. m. f.
alcohols prim. **60**, 38
-, sec., synthesis **60**, 328
1,1-alkoxyacoxy compds., asym. reduction **60**, 20
2-azetidinones, polymer-based synthesis **60**, 322
carboxylic acids (under microwaves without solvent) **60**, 8
cyclopropanols, trans-2-silyl- **60**, 271
4-ethylenealcohols, with 4 extra C-atoms **60**, 328
Carboxylic acid esters, chiral 60, 191
- - -, tert., cyclic **60**, 315
Carboxylic acid fluorides
- as intermediates **60**, 142
Carboxylic acid halides
- from

1,1,1-trihalides **60**, 92
-, scavengers **9**, 524s60; **60**, 154
- special s.
arylcarboxylic acid halides
carboxylic acid fluorides
- startg. m. f.
α-(benzotriazol-1-yl)ketones, with 1 extra C-atom **60**, 414
carboxylic acid amides, polymer-based synthesis **60**, 159, 165
- - esters, - - **60**, 158
- acids, with 1 extra C-atom **60**, 414
α-halogenocarboxylic acid esters, asym. conversion **60**, 184
Carboxylic acid hydrazides
- startg. m. f.
carboxylic acid esters **60**, 87
- acids **60**, 87
Carboxylic acids
- from
aldehydes **60**, 57
carboxylic acid esters (under microwaves without solvent) **60**, 8
- - halides, with 1 extra C-atom **60**, 414
- - hydrazides **60**, 87
α,β-ethylenecarboxylic acids **60**, 37
2-imidazolidones, 1-acyl- **43**, 162s60
- special s.
acylaminocarboxylic acids
allenecarboxylic -
aminocarboxylic -
arylcarboxylic -
dicarboxylic -
ethylenecarboxylic -
halogenocarboxylic -
hydroxycarboxylic -
pyridiniocarboxylic -
- startg. m. f.
alcohols, prim. **60**, 38
amidinium salts, vinylogous **30**, 634s60
carboxylic acid aryl esters **60**, 82
- - esters (with alcohols), preferential conversion **60**, 81
dithiocarboxylic - - (with mercaptans or alcohols) **60**, 197
hydroxamic - - **60**, 142
thiolic - - (with mercaptans) **60**, 195
Carboxylic acids, fluorous-tagged
- as reactant **12**, 455s60
Carboxylic acid thioamides (s.a. N-Thioacylation)
- special s.
alkoxycarboxylic acid thioamides
Catalysis, asym. (s.a. Synthesis, asym.)
- with
Cinchona alkaloids as ligands/catalysts, review **47**, 114s60
ligands, bidentate, axially chiral, review **52**, 170s60
reagents, combinatorial- and evolution-based, review **41**, 737s60
transition metal imidazolinylidene and triazolinylidene complexes, review **56**, 495s60
Catalyst libraries
-, screening methods, review **19**, 33s60
Catalyst screening, combinatorial
-, review [of homogeneous catalysts] **19**, 33s60
Catecholborane
- as reagent **49**, 836s60
Cellulose

- as support in aq. phase catalysis
  42, 354s60
Cephams, 3-allyl-3-hydroxy-, chiral
  40, 567s60
Cerium(IV) ammonium... s. Ammonium cerium(IV)...
Cerium(III) chloride 21, 9s60; **51**, 2s60; **60**, 180, 479
Cesium carbonate 60, 115, 158, 313, 318, 371, 395
– fluoride **60**, 219, 224
– hydroxide **60**, 152
– pivalate **60**, 397
Choroacetonitrile
– as reactant **60**, 146
Chloroacetylamines
– as intermediates **60**, 146
N-Chlorodiisopropylamine
– as reagent **21**, 606s60
N-(4-Chloro-1,2,3-dithiazol-5-ylidene)-1,1-halogenohydrazones
– from
  tetrazoles **60**, 205
– startg. m. f.
  1,3,4-thiadiazoles, 2-cyano- **60**, 205
Chloroform
– as reactant **60**, 225, 364
N-Chloroformylation
– with
  bis(trichloromethyl) carbonate **16**, 493s60
  trichloromethyl chloroformate **16**, 493s60
***m*-Chloroperoxybenzoic acid**
– as reagent **60**, 50
Chlorosulfonyl isocyanate
– as reactant **60**, 163
4-Chromanones, 2,2-dimethyl-
–, synthesis, review **44**, 976s60
Chromans, 2,4-dialkoxy-
– from *o*-hydroxyaldehydes **60**, 341
Chromate, chloro- s. Benzyltriphenylphosphonium chlorochromate
3-Chromene-4-carboxamides
  43, 808s60
3-Chromenes
– from
  α,β-ethyleneboronic acids **60**, 427
  o-hydroxyaldehydes **60**, 427
3-Chromenes, 3-nitro-
– from
  o-hydroxyaldehydes **60**, 324
  1,1-nitroethylene derivs. **60**, 324
Chromium alkoxycarbene complexes
–, prepn. **43**, 860s60
– startg. m. f.
  δ-alkoxy-γ-ketocarbonyl compds.
  60, 437
  α-(2-alkoxyvinyl)lactones **60**, 417
  benzofurans **60**, 418
– alkynyl(alkoxy)carbene complexes
– startg. m. f.
  2(5*H*)-furanone ring, 5-α-alkoxy-, 4,5-condensed **60**, 492
– alkynyloxycarbene complexes
– startg. m. f.
  2(5*H*)-furanone ring, 4,5-condensed **60**, 492
–(0) carbonyl complexes, organo-
– special s.
  arenedicarbonylchromium(0) complexes

–(III) chloride/lithium tetrahydridoaluminate **60**, 42
– 2,4-diene(alkoxy)carbene complexes
– startg. m. f.
  2,4,6-cyclooctatrienones, 4-alkoxy- **60**, 421
– α,β-ethylene(alkoxy)carbene complexes
– startg. m. f.
  bicyclo[3.n.1]alk-2-en-(n+6)-ones, asym. induction **60**, 419
  Δ³-pyrrolines **60**, 434
–(V) imide complexes, trihalogeno- **48**, 423s60
–(II) nitrate **51**, 76s60
–(III) perchlorate **60**, 72
–(III) salen complex, chiral **60**, 447
4-Chromones, 2-vinyl-
– startg. m. f.
  xanthones **60**, 323
Cinchona alkaloids
– as catalysts and ligands for asym. synthesis, review **47**, 114s60
Cinchonidinium chloride, N-benzyl-, polymer-based
– as reagent **54**, 394s60
Cinchonidinium fluoroborates
– special s.
  N-fluorocinchonidinium fluoroborate
Cinnamic acid esters
– from
  siloxycyclopropanes, 1-alkoxy- and triarylantimony diacetates **60**, 461
Cinnamic acids
– startg. m. f.
  α-amino-β-arylcarboxylic acids, asym. conversion **60**, 126
Cinnolines
– from
  o-acetylenetriazenes **58**, 187s60
Claisen rearrangement, Ir-catalyzed
– of 4-oxa-1,7-dienes **60**, 313
– –, Yb(III)-catalyzed
– of α-allyloxy-α,β-ethylenecarboxylic acid esters **60**, 311
Claisen-Schmidt condensation, polymer-based 55, 327s59
Cobalt/charcoal 58, 281s60
Cobalt(II) acetate 53, 162s60; **60**, 282
Cobalt carbonyl 57, 300s60; **60**, 241, 387
– carbonyl complexation
– of acetylene derivs. with asym. 1,5-hydride shift **60**, 46
– carbonyl complexes
– special s.
  acetylenedicobalt hexacarbonyl complexes
–(II) chloride **43**, 597s60
– complexes
  [η⁵-(4-hydroxybutyrylcyclopentadiene)(1,5-cyclooctadiene)cobalt(I) **31**, 658s60
–(II) 1,1,1,5,5,5-hexafluoroacetoacetonate **47**, 133s60
–(II) phosphine complexes, polymer-based **26**, 463s60; **47**, 162s60
–(III) salen complexes, dendritic, chiral **53**, 59s60
Combinatorial chemistry s.a. Catalyst screening, combinatorial, Compound libraries, Polymer-based...
Combinatorial synthesis

– of
  carbohydrates and conjugates, review **57**, 99s60
  catalysts, chiral, review **41**, 737s60
3-Component reaction
–, 1,5-benzothiazepines, 2,3-dihydro-, 2-aryl- by – **60**, 403
–, furans, tetrahydro-, 3-alkylidene- by – **60**, 402
–, Mannich reaction, asym., amine-catalyzed by – **60**, 336
–, piperazinones by – **60**, 416
–, Δ²-pyrrolines by – **60**, 434
–, α-siloxycarboxamides from oxo compds. by – **60**, 466
– special s.
  radical 3-component reaction
Compound libraries
– of
  carbohydrates, review **57**, 99s60
  pyridines, 1,4-dihydro- **52**, 352s60
  tetrazolo[1,5-*a*]piperazin-6-ones **60**, 338
Copper(II) acetate 40, 286s60; **60**, 201
–(II) acetoacetonate **60**, 361
– complexes
  bis(acetonitrile)[α,α'-bis[((N)-methyl-2-pyridyl)ethylamino]-2-fluoro-*m*-xylene]dicopper(I) **31**, 719s60
  tetrakis(acetonitrile)copper(I) fluoroborate **60**, 379
–(I) cyanide **60**, 264
–(II) formate (s.a. Nickel(II) copper(II) formate)
–(I) hexafluorophosphate **60**, 246
–(I) halides
  – bromide **60**, 228, 411
  – chloride **27**, 530s60; **60**, 86, 271
  – iodide **60**, 174, 204, 374, 394, 403, 460
–(II) halides
  – bromide **60**, 177
  – chloride **60**, 177, 301
– hydride complex **40**, 22s60
–(II) nitrate **10**, 262s60
–(II) –/silica **47**, 146s60
–(I) polypyrazolylborate complexes
– as reagent, review **23**, 819s60
–(II) sulfate (s.a. Potassium permanganate/copper(II) sulfate)
–(II) –/alumina **60**, 404
–(I) thiophene-2-carboxylate **60**, 457
–(I) trifluoromethanesulfonate **24**, 555s60
–(II) –**29**, 752s60; **35**, 655s60; **60**, 227, 247, 260, 261, 327
Coumarins
– from
  α,β-acetylenecarboxylic acid aryl esters **19**, 788s60
  acetylene derivs., carbonylation **60**, 390
  o-halogenophenols, – **60**, 390
Cross-metathesis (s.a. under Ring-closing metathesis..., and Ring-opening metathesis...)
– of acetylene derivs. with kinetic resolution **56**, 295s60
Crown ethers s.a. Thia-crown ethers
Crown ethers, chiral
– special s.
  1,1'-binaphthol crown ether, chiral

18-Crown-6 polyether
– as reagent **60**, 423
Cumene hydroperoxide
– as reagent **60**, 97
Cumulenes
– from
heterocyclics, 5-membered, flash vacuum pyrolysis, review **27**, 675s60
– special s.
heterocumulenes
Cuprates, organo-
–, 1,4-addition with –, review **46**, 641s60
– special s.
lithium dialkylcuprates
Cyanamides, N,N-disubst.
– from
amidoximes **60**, 131
Cyanides s. Acylcyanides, Nitriles
Cyanoallenes
– from
2-acetylenecarbonic acid esters **60**, 347
α-Cyanocarboxylic acid esters
–, fluorination, asym. **60**, 186
(1E,3Z)-3-Cyano-1,3-dienephosphonic acid esters
– from
N-sulfonylimines, with 4 extra C-atoms **60**, 349
2-Cyano-1,4-dienes 60, 373
Cyanohydrin acetates
–, resolution, kinetic **28**, 13s60
Cyanohydrins
– startg. m. f.
nitriles **31**, 49s60
α-Cyano-α-isocyanophosphonic acid esters 30, 43s60
α-Cyanoketones
– startg. m. f.
furans, 3-amino- **60**, 339
Cyanosulfine 49, 507s60
Cyanuric chloride
– as reagent **26**, 806s60
Cyclimmonium salts
– special s.
N-acylimmonium salts, cyclic
Cycloaddition, 1,3-dipolar (s.a.
[2+2+2]-Cycloaddition-1,3-dipolar cycloaddition, [3+2]-Cycloaddition)
– in micelles, aq. **60**, 363
– with
alkoxy-2-ethylenes, asym. induction, review **16**, 888s60
aziridines, 1-sulfonyl- **60**, 265
–, –, asym., amine-catalyzed
– of nitrones **60**, 278
– to α,β-ethylenealdehydes **60**, 278
–, –, –, Lewis acid-catalyzed
– with trimethylsilyldiazomethane **60**, 429
–, –, intramolecular
– with azomethinium ylids, non-stabilized **49**, 388s60
–, –, –, asym.
– with nitrilimines **17**, 871s60
–, –, oxidative
– with pyridinium methylids **60**, 474
–, –, regiospecific
– with nitrile ylid equivalents, under microwaves **60**, 319
–, –, regiostereospecific
– of nitriles to ethylene derivs., review **16**, 888s60

[2+2]-Cycloaddition, ortho-, photochemical
– to benzenes, triplet, review **19**, 764s60
–, regiospecific, Rh-catalyzed
– of acetylene derivs. to norbornenes, 5-subst. **60**, 288
[3+2]-Cycloaddition, regiospecific
– with 2-ketoiodonium ylids **60**, 361
–, regiostereospecific, Ga(III)-catalyzed **60**, 434
[4+2]-Cycloaddition (s.a. Diels-Alder..., and under Diene synthesis in Vol. **1-50**)
–, asym., regiospecific, Pd-catalyzed **60**, 299
[5+2]-Cycloaddition, regiospecific, Rh-catalyzed 60, 292
[2+2+2]-Cycloaddition
– via zirconacyclopentadienes **60**, 274
–, asym., catalytic **60**, 261
[2+2+2]-Cycloaddition-1,3-dipolar cycloaddition, stereospecific 60, 294
Cyclobutanes
– special s.
alkylidenecyclobutanes
Cyclobutanols, 1-(alk-1-ynyl)-
– startg. m. f.
2-alkylidenecyclopentanones, synthesis **60**, 392
–, 2-vinyl- **55**, 478s60
Cyclobutanone O-acyloximes
– startg. m. f.
β,γ-ethylenenitriles **60**, 171
Cyclobutanones
– from
cyclopropanols, 1-α-alkoxy- **60**, 480
– startg. m. f.
cyclopentanones **60**, 407
Cyclocarbonylation (s.a. Cyclohydrocarbonylation)
–, coumarins via – **60**, 390
–, 9-fluorenones by – **60**, 397
– with *tert*-butoxyformic anhydride **60**, 135
Cyclodextrins, per-O-benzyl-
–, O-debenzylation, regiospecific **60**, 7
1,4-Cycloheptadienes
acetylene derivs. **60**, 292
vinylcyclopropanes **60**, 292
4-Cycloheptenone ring, 2-siloxy- 39, 883s60
2,4-Cyclohexadienones
– special s.
hexachloro-2,4-cyclohexadienone
2,5-Cyclohexadienones
– special s.
tetrabromo-2,5-cyclohexadienone
–, 3-silyl-
– startg. m. f.
2-cyclopentenones, 4-alkylidene- **60**, 493
Cyclohexenes
– special s.
vinylcyclohexenes
Cyclohexenes, 3-aryl- 20, 533s60
2-Cyclohexenones, 4-hydroxymethyl-, chiral 53, 288s60
Cyclohexylamine
– as reagent **40**, 475s60; **50**, 415s60
Cyclohydrocarbonylation
–, ring expansion via – **60**, 253
Cycloisomerization (s.a. OCΩ, NCΩ,

CCΩ, Carbopalladation-cycloisomerization)
– of 4-oxa-1,8-enynes **59**, 322
–, asym.
– of 1,6-enynes **60**, 307
–, Au(III)-catalyzed **60**, 74
1,5-Cyclononadien-3-ol carbamates
– from
9-halogeno-2,7-dienol carbamates, asym. conversion **60**, 483
2,4,6-Cyclooctatrienones, 4-alkoxy-
– from
chromium 2,4-diene(alkoxy)carbene complexes and acetylene derivs., terminal **60**, 421
Cyclooctene oxides
–, ring opening, transannular, asym. **60**, 309
4-Cyclooctenones
– from
δ-cyclopropyl-γ,δ-ethylenealdehydes **60**, 306
Cyclopentadienes, subst.
–, synthesis, review **29**, 932s60
Cyclopentadienones
– from
1,4-dihalogeno-1,3-dienes **60**, 367
Cyclopentanes, 2-homoallyl-1-(stannylmethylene)- 49, 669s60
Cyclopentanols, 2-alkynyl- 55, 478s60
Cyclopentanone-3-carboxylic acid esters, 2-alkylidene- 41, 862s60
Cyclopentanones
– from
cyclobutanones **60**, 407
– special s.
alkylidenecyclopentanones
–, 3-alkoxy- **60**, 407
–, 3-alkoxy-2-methylene- **57**, 456s60
–, 2-benzylidene- **60**, 392
Cyclopentenes
– from
1,5(6)-dienes **38**, 706s60
1,6-dienes **60**, 308
Cyclopent-2-en-1-ols, 4-amino-, chiral 4, 3s60
2-Cyclopentenones (s.a. Pauson Khand...)
– from
2-ethylenecarbonic acid esters and ethylene derivs., carbonylation **60**, 343
–, 4-alkylidene-
– from
2,5-cyclohexadienones, 3-silyl- **60**, 493
Cyclopentylcarbonyl ring
– from
ethyleneepoxides **51**, 318s60
Cyclopropanation (s.a. Simmons-Smith...)
– using copper(I) polypyrazolylborate complexes, review **23**, 819s60
Cyclopropaneboronic acid esters, chiral 41, 797s60
***trans*-Cyclopropanecarbonyl compds.**
– from
α-(benzothiazol-2-ylthio)carbonyl compds. **60**, 410
ethylene derivs., electron-deficient **60**, 410
Cyclopropanecarboxylic acid esters
– from
ethylene derivs., electron-deficient, asym. conversion **60**, 355

pyridinioacetic acid esters, – – **60**, 355
Cyclopropanecarboxylic – –, 2,2-dicyano-, chiral 60, 355
cis-**Cyclopropane-1,2-diol O-derivs.**
– from
α-diketones **60**, 233
Cyclopropanes
– from
ethylene derivs., electron-deficient **60**, 426
telluronium salts **60**, 426
– special s.
alkylidenecyclopropanes
siloxycyclopropanes
vinylcyclopropanes
Cyclopropanes, cyano- 60, 171
Cyclopropanes, α-styryl- 43, 806s60
–, **1,1,2-trihalogeno-**
– startg. m. f.
acetylene derivs., review **42**, 979s60
Cyclopropanols, 1-α-alkoxy-
– startg. m. f.
cyclobutanones **60**, 480
–, **1-ω-hydroxy-**
– from
lactones **60**, 384
–, *trans*-**2-silyl-**
– from
carboxylic acid esters **60**, 270
Cyclopropenes
– from
acetylene derivs., Cu(I)-catalysis, review **23**, 819s60
–, **1-chloro-3,3-difluoro-**
– from
difluoromethylene compds. **60**, 364
δ-Cyclopropyl-γ,δ-ethylenealdehydes
– startg. m. f.
4-cyclooctenones **60**, 306
3-Cyclopropyl-2-ethylenesilanes 51, 423s60
Cyclopropyl ketimines
– startg. m. f.
2-pyridones, 3,4-dihydro- **60**, 286

Darzens-Claisen condensation
– in water **60**, 362
N-Dealkylation
– special s.
N-debenzylation
N-de-*tert*-butylation
–, oxidative **16**, 28s59
O-Dealkylation
– special s.
O-debenzylation
O-Debenzylation, on resin 1, 13s60
–, **regiospecific**
– of
per-O-benzylcarbohydrates **60**, 7
per-O-benzylcyclodextrins **60**, 7
N-De-*tert*-butylation 60, 158
O-De-*tert*-butyldimethylsilylation, superbase-catalyzed 60, 5
N-Decarbalkoxylation
– special s.
N-decarballyloxylation

N-decarbomethoxylation
N-Decarballyloxylation, electrocatalytic 60, 16
N-Decarbomethoxylation, selective 60, 17
Dendralenes 17, 968s60
2-Deoxyaldoses
– from
glycals **60**, 64
Deoxygenation (s.a. Radical deoxygenation)
– of
amine N-oxides **60**, 499
N-oxides, cyclic **60**, 499
2-Deoxyglycosides
– special s.
2-amino-2-deoxyglycosides
2-Deoxy-C-α-glycosides
– special s.
2-amino-2-deoxy-C-α-glycosides
2-Deoxyglycosylamines
– special s.
2-nitro-2-deoxyglycosylamines
5-Deoxynucleosides
– special s.
5-halogeno-5-deoxynucleosides
2'-Deoxynucleosides, 2'-C-carboxymethyl-
–, nucleoside base exchange in – **60**, 121
1-Deoxy-2-uloses
– special s.
1-iodo-1-deoxy-2-uloses
Deracemization (s.a. Resolution, kinetic, dynamic)
– of 2-ethylenealcohols **53**, 500s60
–, review **58**, 49s60
O-Desilylation
– special s.
O-de-*tert*-butyldimethylsilylation
Dess-Martin periodinane s. 1,2-Benziodoxol-3-one, 1,1,1-triacetoxy-
N-Desulfonylation
– special s.
N-detosylation
O-Desulfonylation
– special s.
O-detosylation
Desymmetrization
– by O-acylation, asym. with enolesters, update **44**, 214s60
–, 1,2-halogenhydrins from epoxides with – **60**, 173
– of
1,3-diol acetates **28**, 13s60
pyridine-3,5-dicarboxylic acid esters, 1,4-dihydro- **23**, 13s60
–, Sharpless-type epoxidation with – **60**, 67
–, **isotopic**
– as a stereo probe, review **57**, 48s60
N-Detosylation 60, 3
O-Detosylation 60, 3
1-Deuterioalcohols, prim.
– from
aldehydes, reduction, asym. **60**, 28
α-Deuteriobenzylalcohols, chiral
– induction, asym. with –, **60**, 230
1,2-Diacoxy compds. s. Glycol esters
ω,ω'-Di(acylamino)-α-diketones, sym. 60, 251
Di-1-adamantyl-*n*-butylphosphine
– as reagent **57**, 380s60

syn-α,β-**Dialkoxyketones**
– from
acetals, with 2 extra C-atoms **60**, 271
1,2-Dialkylidenecyclobutanes
– from
allenes, dimerization, regiospecific **60**, 285
N,N-Diallylammonium salts, quaternary
– startg. m. f.
quinolines **60**, 359
α,α'-**Diallyl coupling, intramolecular, asym. 60**, 483
(Diallylhydridosiloxy)-3-ethylenes
– startg. m. f.
7-ethylene-1,3,5-triols, via silylformylation **60**, 451
Diamidophosphate
– as reactant **60**, 52
Diamines, chiral
–, mono-N-arylation **52**, 171s60
–, –, bicyclic **40**, 477s60
1,2-Diamines
– startg. m. f.
piperazinones, synthesis **60**, 416
–, chiral, C₂-symmetric **4**, 643s60
–, N-protected differentially
– from
ethylene derivs. **60**, 125
o-**Diamines, unsym., chiral 52**, 171s60
o-**Diamine N,N,N',N'-tetraanions**
– as intermediates **60**, 358
2,3-Diaminoalcohols 50, 151s60
α,β-**Diaminocarboxylic acid esters**
– special s.
α,β-diaminohydrocinnamic acid esters
anti-α,β-**Diaminohydrocinnamic – –, N-protected 60**, 125
α,β-**Diaminonitriles, chiral 17**, 405s60
N,C-Dianions
– as intermediates **60**, 320
O,C-Dianions
– special s.
magnesium O,C-dianions, chloro-
Diarylcarbinols s. Benzhydrols
Diaryl coupling
– special s.
Negishi diaryl coupling
Stille – –
Suzuki – –
Diaryliodonium salts
– startg. m. f.
1,2,4-oxadiazoles, 5-aryl-, carbonylation **60**, 401
– –, **cyclic**
–, review **47**, 933s60
Diaryl ketones 60, 389
Diarylmethanes
– from
benzhydrols **60**, 45
Diaryls
– special s.
dihydroxydiaryls
halogenodiaryls
hydroxydiaryls
Diaryls, functionalized 60, 464
Diaryls, sym.
– from
halides, ar. **60**, 378
Diaryl sulfides
– from
arylthio-2-acetylenes **60**, 204
halides, ar. **60**, 204

1,8-Diazabicyclo[5.4.0]undec-7-ene
– as reagent **44**, 819s60; **60**, 225, 349, 352, 374, 428, 491
1,3,2-Diazaboronidine, 1,3-dimethyl-
– as reagent **50**, 405s60
α-Diazocarboxylic acid esters
– special s.
ethyl diazoacetate
Diazo compds.
– startg. m. f.
O-alkyllactims **60**, 89
α-Diazoketones
– special s.
acetylene-α-diazoketones
diazomethyl β-acylaminoketones
Diazomethanes
– special s.
aryldiazomethanes
Diazomethyl β-acylaminoketones
– startg. m. f.
4-pyridones, 2,3-dihydro-, 5-hydroxy- **60**, 481
Diazomethyltrimethylsilane
– as reactant **60**, 243, 348, 429
1,4-Diazoniabicyclo[2.2.2]octanes
– special s.
N-fluoro-1,4-diazoniabicyclo[2.2.2]-octane…
α-Diazo-α-nitrocarbonyl compds. 20, 271s60
Diazonium o-benzenedisulfonimide salts
– startg. m. f.
thioethers, ar. **41**, 511s60
Diazonium fluoroborates
– startg. m. f.
arylboronic acids **60**, 218
Diazosilanes
– special s.
diazomethyltrimethylsilane
Dibenzo[b,d]thiophene S-oxide
– as reagent **60**, 54
Dibenz[c,e]oxepin-5-ones 31, 964s60
Dibenzyl hydrogen phosphate
– as reactant **60**, 54
Diborane 60, 60
Diboranes
– special s.
alkoxydiboranes
1,2-Di(boronic acid esters) 60, 207
– special s.
1,2-ethylene-1,2-di(boronic acid esters)
Dibromomethylsilanes
– as reactant **60**, 375
Di-tert-butyl disulfide
– as reagent **60**, 268
Di-tert-butyl hyponitrite
– as reagent **60**, 189
(1S,2R)-N,N-Dibutylnorephedrine
– as reagent **60**, 229
2-(Di-tert-butylphosphino)biphenyl
– as ligand **31**, 592s60; **60**, 157
α-Dicarboxylic-… s. Malonic…
β-Dicarboxylic-… s. Succinic…
δ-Dicarboxylic-… s. Adipic…
Dicarboxylic acid esters
– startg. m. f.
dicarboxylic acid monoesters **60**, 6
Dicarboxylic acid imides
– special s.
glutarimides
Dicarboxylic acid monoesters

– from
dicarboxylic acid esters **60**, 6
Dichloroborane 21, 174s60
2,3-Dichloro-5,6-dicyanoquinone
– as reagent **60**, 101
1,3-Dichloro-5,5-dimethylhydantoin
– as reagent **60**, 166
Dichloroisocyanuric acid sodium salt
– as reagent **46**, 397s60; **60**, 166
N,N-Dichloro-o-nitrobenzene-sulfonamide
– as reactant **24**, 555s60; **60**, 125
Dichromate s. Benzyltriphenyl-phosphonium dichromate
Dicyclohexylborinyl triflate
– as reagent **60**, 235
2-(Dicyclohexylphosphino)-2'-dimethylaminobiphenyl
– as reagent **60**, 395
Diels-Alder reaction (s.a. [4+2]-Cyclo-addition, Hetero-Diels-Alder reaction, Pauson-Khand reaction-Diels-Alder reaction, Retro-Diels-Alder reaction, and under Diene synthesis in Vol. 1-50)
– with
enamines, in situ-generated **60**, 323
α-methylenealdehydes, – – **60**, 254
tungsten 2-benzopyran-3-ylidene complexes **60**, 472
– –, asym., Cr(III)-catalyzed
– with
1-amino-3-siloxy-1,3-dienes, **60**, 447
α,β-ethylenealdehydes **60**, 447
– –, –, polymer-based
– with anthracenes, 9-α-alkoxy- **27**, 694s60
– –, intramolecular (s.a. N-Allylation-intramolecular Diels-Alder reaction, Ene-yne metathesis-intramolecular-Diels-Alder reaction, Hetero-Diels-Alder reaction, intramolecular)
–, pyridines, 2-β-carboxy- via – **60**, 314
– with α-(O-acyloximino)nitriles **60**, 314
– –, –, regiospecific, Au-catalyzed **60**, 310
– –, –, regiostereospecific
– with 1,3-dienesulfonic acid amides **46**, 696s60
– –, ionic, endo-selective
– with α,β-ethyleneacetals **60**, 266
– –, polymer-based
–, review **36**, 667s60
– –, transannular
– use in total synthesis, review **43**, 944s60
2,4-Dienals
– special s.
hydroxy-2,4-dienals
1,3-Dienamines
– special s.
3-siloxy-1,3-dienamines
(1,3-Diene)acetals, cyclic
– startg. m. f.
alkoxy-3-ethylenes, isocyclic **57**, 321s60
2,n-Dienecarbonyl compds.
–, radical ring closure **60**, 276
(E,E)-2,4-Dienecarboxylic acid esters 36, 932s60
2,4-Dienecarboxylic acids 28, 852s60
1,3-Dienephosphonic acid esters
– special s.
cyano-1,3-dienephosphonic acid esters

Dienepolyols, chiral 58, 371s60
1,3-Dienes 46, 429s60
– special s.
acoxy-1,3-dienes
alkoxy-1,3-dienes
halogeno-1,3-dienes
hydroxy-1,3-dienes
silyl-1,3-dienes
sulfonyl-1,3-dienes
– startg. m. f.
β,γ-ethylene-δ-stannylketones **60**, 283
4-vinylcyclohexenes, 3-alkylidene-, asym. conversion **60**, 299
(E,E)-1,3-Dienes
– from
acetylene derivs. **51**, 423s60
1,3-Dienes, exocyclic
– from
bis(allenes) **60**, 300
2-halogen-1,n-enynes **60**, 487
1,4-Dienes
– special s.
cyano-1,4-dienes
1,5-Dienes
– special s.
arylthio-1,5-dienes
1,6-Dienes
–, radical ring closure, double, thiyl-mediated **60**, 268
– startg. m. f.
cyclopentenes **60**, 308
cis-3-thiabicyclo[3.3.0]octanes **60**, 268
1,7-Dienes
– special s.
oxa-1,7-dienes
1,3-Dienesulfonic acid amides
–, Diels-Alder reaction, intramolecular, regiospecific with – **46**, 676s60
(1,3-Dienol)ate equivalents
– special s.
–, 3-ethyleneepoxides as – **60**, 431
2,7-Dienol carbamates
– special s.
halogeno-2,7-dienol carbamates
(1,3-Dien)olethers
– special s.
silyl(1,3-dien)olethers
1,4-Dien-3-ols
–, desymmetrization by epoxidation **60**, 61
1,6-Dien-4-ols
– from
3-ethyleneepoxides and potassium 2-ethylene(trifluoro)borates **60**, 431
(E)-2,4-Dienones 60, 1
1,5-Dien-4-ones
– special s.
3-aza-1,5-dien-4-ones
1,3-Dien-5-ynes
– startg. m. f.
benzofurans **60**, 418
Diethoxymethyltri-n-butylstannane
– as reactant **60**, 425
Diethylamine
– as reagent **60**, 40
Diethyl bromomalonate
– as reagent **60**, 117
Difluoromethylene compds.
– startg. m. f.
cyclopropenes, 1-chloro-3,3-difluoro- **60**, 364
α,α-Difluorophosphonic acid esters
–, radical synthesis **38**, 904s60
Di-N-formylation 60, 132

1,1-Dihalides
- special s.
  o-bis(dihalogenomethyl)arenes
  ethylene-1,1-dihalides
1,2-Dihalides
- special s.
  1,2-ethylene-1,2-dihalides
1,3-Dihalides, mixed
- special s.
  1,3-fluoroiodides
o-Dihalides
- startg. m. f.
  indoles **60**, 395
1,1-Dihalogen-1,3-enynes
- special s.
  3-(dihalogenomethylene)-1,4-diynes
1,4-Dihalogeno-1,3-dienes
- startg. m. f.
  cyclopentadienones **60**, 367
γ,γ-Dihalogeno-β-lactones, chiral
50, 397s60
Dihalogenomethylene compds.
- special s.
  difluoromethylene compds.
- startg. m. f.
  1-acetylenephosphonic acid esters
  60, 214
  1,3-diynes **60**, 394
  1,3-diynes, sym. **60**, 394
3-(Dihalogenomethylene)-1,4-diynes
- startg. m. f.
  1,3,5-triynes **60**, 482
1,1,1-Dihalogenosilanes
- special s.
  dibromomethylsilanes
N,N-Dihalogenosulfonic acid amides
- special s.
  N,N-dichloro-o-nitrobenzene-
  sulfonamide
α,β-Dihydroxycarboxylic acid esters
- special s.
  acetylene-α,β-dihydroxycarboxylic
  acid esters
β,γ-Dihydroxycarboxylic – – 41, 888s60
**anti-α,β-Dihydroxycarboxylic acids,
chiral 60**, 235
**o,o′-Dihydroxydiaryls, sym., hindered
31**, 719s60
α,β-Dihydroxyketones, masked s. 1,3-
  Dioxolan-2-ones, 4-(1-aryloxyvinyl)-
**cis-Dihydroxylation, heterogeneous
60**, 71
Diisobutyl(pyrrol-1-yl)alane
- as reagent **51**, 23s60
(+)-Diisopinocampheyl(methoxy)borane
- as reagent **60**, 221
Diisopropylamine
- as reagent **60**, 142
(−)-Diisopropyl tartrate
- as reagent **60**, 61
β,δ-Diketocarboxylic acid esters
- startg. m. f.
  δ-hydroxy-β-ketocarboxylic acid
  esters, asym. reduction **60**, 22
α-Diketones (s.a. α,β-Ethylene-
  α-hydroxyketones)
- from
  1,2-ethylene-1,2-dihalides **60**, 93
- special s.
  benzils
  di(acylamino)-α-diketones
- startg. m. f.

cis-cyclopropane-1,2-diol O-derivs.
  60, 233
α-hydroxyketones **60**, 25
α-Diketones, sym.
- from
  acylcyanides **60**, 468
β-Diketones
- startg. m. f.
  α,β-ethylene-β-halogenoketones
  60, 182
  pyrroles **60**, 476
β-Diketones, sym.
- from
  carboxylic acid halides **55**, 471s60
1,5-Diketones
–, carbocyclization, intramolecular,
  review **11**, 744s60
Dilithium tetramethylzincate
- as reactant **60**, 381
**Dilithium trialkyl(stannyl)manganates
43**, 517s60
**Di-π-methane rearrangement
25**, 549s60
**(S)-2-Dimethylamino-2′-dicyclohexyl-
phosphino-1,1′-binaphthyl**
- as reagent **59**, 447s60
**(1R,2R)-1-Dimethylamino-2-(2-
methoxyphenoxy)-1,2-diphenylethane**
- as reagent **60**, 191
4-Dimethylaminopyridine
- as reagent **60**, 20, 75, 90, 135, 137
Dimethyldioxirane
- as reagent **60**, 62
N,N-Dimethylhydrazine
- as reactant **60**, 155
N,O-Dimethylhydroxylamine
- as reactant **60**, 142
**Dimethyl(methylene)ammonium
chloride**
–, Mannich reaction, double with –
  60, 337
Dimethyl sulfate
- as reagent **60**, 491
Dimethyl sulfide
- as reagent **60**, 188
Dinitriles
- special s.
  malononitriles
**Dinucleoside phosphoromonoamidates
43**, 83s60
Diol esters
- from
  acetals, cyclic **14**, 344s59
1,4-Diol monocarbonates
- special s.
  acetylene-1,4-diol monocarbonates
Diol monoesters
- from
  acetals, cyclic **16**, 342s60
1,3-Diol monoesters s. Aldol-Tishchenko
  reaction
Diol monoethers
- from
  acetals, cyclic, reduction, regiospecific
  60, 26
1,3-Diol monoethers
- special s.
  4-acetylene-1,3-diol 1-(monobenzyl
  ethers)
anti-1,3-Diol monoethers 57, 22s60
Diols
- from

diketones, reduction, asym. **58**, 17s60
hydroxyketones **56**, 30s60
- special s.
  cyclopropanols, 1-ω-hydroxy-
  enediols
Diols, chiral
- as reagent **49**, 898s60
1,2-Diols s. Glycols
1,3-Diols
- special s.
  ene-1,3-diols
syn-1,3-Diols
- from
  enoxy(hydrido)silanes and aldehydes
  60, 448
1,4-Diols
- special s.
  acetylene-1,4-diols
  1,1,4,4-tetraphenylbutane-1,4-diol
1,5-Diols
- special s.
  methylene-1,5-diols
Diosphenols 60, 2
**2,7-Dioxabicyclo[4.3.0]non-4-enes,
8-vinyl- 60**, 475
1,3-Dioxane-4,6-diones, 5-acyl-
- startg. m. f.
  2-azetidinone ring, 3-acyl-, asym.
  induction **60**, 473
1,4-Dioxan-2-ones
–, aldol condensation, asym. with –
  60, 235
**1,3,2-Dioxaphosphorinane 2-sulfides,
2-acylthio-**
–, thioacylation with – **60**, 200
1,6-Dioxaspiro[3.4]octanes 47, 576s60
1,3-Dioxins
- startg. m. f.
  α-methylenealdehydes, in situ-
  generated **60**, 254
**α-(1,3-Dioxin-4-yl)-β,γ-ethylene-
boronic acid esters**
- as reactant **60**, 428
Dioxiranes
- special s.
  dimethyldioxirane
Dioxiranes, chiral, carbohydrate-based
- as reagent **60**, 68
–, polymer-based
- as reagent **60**, 67
1,3-Dioxolanes
- from
  epoxides **50**, 55s60
- startg. m. f.
  glycol monoethers, synthesis, asym.
  60, 222
–, 2-vinyl- **54**, 374s60
1,3-Dioxolane-2-thiones
- startg. m. f.
  2-azidoalcohols **52**, 175s60
1,3-Dioxolan-2-ones
- from
  glycols **60**, 135
–, 4-(1-aryloxyvinyl)-
- from
  2-acetylene-1,4-diol monocarbonates
  60, 84
**Dipeptide amides, N-(o-hydroxy-
benzylidene)-**
- as reagent **60**, 248
Dipeptides
- special s.
  N-(o-phosphinoarylidene)dipeptides

Dipeptides, cyclic, chiral
– as reagent 50, 54s60
Diphenoquinones 31, 719s60
o-Diphenylphosphinophenol
– as reagent 31, 592s60
4-(Diphenylphosphinyl)phenylphosphonic acid
– as reagent 27, 871s60
Diphenyl phosphorazidate
– as reagent 60, 124
Diphenyl-2-pyridylphosphine
– as reagent 31, 592s60
Diphenylsilane
– as reagent 60, 39
Diphenyl sulfoxide
– as reagent 60, 80
Di(phosphine oxides)
– special s.
 di(phosphinyl)...
Di(phosphines)
– special s.
 bis(diphenylphosphino)...
 bis(ditolylphosphino)...
 xanthenes, bis(diarylphosphino)-, guanidinium-modified
3,3′-Di(phosphinyl)-1,1′-bi-2-naphthols, chiral 60, 212
1,1-Diphosphonic acid esters
– special s.
 1-hydroxy-1,1-diphosphonic acid esters
o-Di(propargyl)arenes
– startg. m. f.
 anthracene ring, 9,10-dihydro- 60, 274
C-Disaccharides, pseudo-aza, (1→2)-linked 43, 249s60
Diselenides
– special s.
 bis(4-perfluorohexylphenyl) diselenide
– startg. m. f.
 selenosulfides 60, 190
1,1-Di(silanes)
– special s.
 ethylene-1,1-di(silanes)
Disilazanes
– special s.
 hexamethyldisilazane
 tetramethyldisilazane
1,3-Disiloxy-1,3-dienes
– special s.
 alkoxy-1,3-disiloxy-1,3-dienes
– startg. m. f.
 furans, tetrahydro-, (E)-2-acylene- 60, 441
N,N-Disiloxyenamines
–, 2,2′-disubstitution, electrophilic, sequential 60, 446
– startg. m. f.
 1,3-nitroethers 60, 446
Dispersions, aq. colloidal
–, Mannich reaction in – 60, 340
1,4-Distannylation (s.a. Carbopalladation, intramolecular-1,4-distannylation)
Disulfides
– special s.
 aminodisulfides
 di-tert-butyl disulfide
 glycosyl disulfides
Disulfides, cyclic
– from
 mercaptothioethers, polymer-based synthesis 60, 188
Di(sulfonylamines), chiral

– as reagent 44, 565s60
Disulfoxides, C_2-symmetric 51, 46s60
2,4-Dithia-6,8-diazabicyclo[3.2.2]-nonanes
–, ring expansion via sulfur group migration 60, 206
1,4-Dithiafulvenes
– from
 1,4-dithiines 60, 192
1,2,3-Dithiazoles, 5-imino-
– special s.
 N-(4-chloro-1,2,3-dithiazol-5-ylidene)-1,1-halogenohydrazones
1,4-Dithiins
– startg. m. f.
 1,4-dithiafulvenes 60, 192
Dithiocarbamic acid aryl esters 48, 510s60
Dithiocarboxylic acid esters (s.a. S-Thioacylation)
– from
 carboxylic acids (with mercaptans or alcohols) 60, 197
1,3-Di(thiocyanates)
– from
 1,2-dithiolane 1-oxides 60, 198
1,2-Dithiolane 1-oxides
– startg. m. f.
 1,3-di(thiocyanates) 60, 198
1,3-Dithiolanes (s.a. under Protection)
– from
 oxo compds. 60, 194
–, 2-(α,β-ethyleneacylene)-
– startg. m. f.
 4-thiopyrones, 2,3-dihydro-, 6-(vinylthio)-
Dithiophosphonic acid esters
– from
 ethylene derivs. 60, 217
Dithiophosphonyl radicals
–, radical addition with – 60, 217
Diynes
–, Pauson-Khand reaction-Diels-Alder reaction with – 60, 280
1,3-Diynes
– from
 acetylene derivs., terminal and dihalogenomethylene compds. 60, 394
1,3-Diynes, sym.
– from
 dihalogenomethylene compds. 60, 394
1,4-Diynes
– special s.
 3-(dihalogenomethylene)-1,4-diynes
1,5-Diynes
– special s.
 3-ene-1,5-diynes
Diynols
–, addition of H_2S and thiols to –, review 12, 653s60
Dodecylbenzenesulfonic acid
– as reagent 60, 340
Dötz reaction
–, benzofurans via – 60, 418
Dötz reaction, interrupted 60, 417
Dötz-type reaction
–, 2,4,6-cyclooctatrienones, 4-alkoxy-by – 60, 421

Electrolysis 29, 518s60; 51, 204s60
–, generation of alkoxycarbenium ions under – 60, 422
Electrolysis, catalytic 60, 16
Enamines
– from
 acetylene derivs., polymer-based synthesis 60, 124
 carbamic acid silyl esters 60, 164
 ketones 60, 164
–, hydroboration, asym. 19, 188s60
– special s.
 α/β-amino-α,β-ethylen...
 1,3-dienamines
 N-siloxyenamines
– startg. m. f.
 amines 60, 32
 indoles 60, 395
 naphthalenes 60, 472
 pyridine ring 24, 836s60
 2-pyridones, 3,4-dihydro-, polymer-based synthesis 60, 124
Enamines, in situ-generated
–, Diels-Alder reaction with – 60, 323
Enamines, isocyclic
– startg. m. f.
 bicyclo[3.n.1]alk-2-en-(n+6)-ones, asym. induction 60, 419
Enamines, polymer-based
–, reactions 32, 287s60
Enantiomeric excess
–, NMR determination, review 5, 66s60
trans-3-Ene-1,2-diol 2-monoethers 60, 59
4-Ene-1,2-diol 1-monoethers, chiral 45, 382s60
– 2-monoethers, – 44, 776s60
2-Ene-1,n-diols
– startg. m. f.
 2-vinyl-O-heterocyclics 20, 200s59
5-Ene-1,3-diols, chiral 45, 382s60
3-Ene-1,5-diynes
– startg. m. f.
 benzofurans 60, 418
Enegermanes
– special s.
 β-germylstyrenes
Ene reaction, regiostereospecific, hydroxyl-directed
– with nitroso compds. 60, 118
Eneselenides
–, ring opening metathesis-cross-metathesis with – 60, 290
– special s.
 2-alkylseleno-1,3-enynes
 α,β-ethylene-β-halogeno-α-(organoseleno)...
 halogeneneselenides
(E)-Eneselenides
– startg. m. f.
 (E)-enestannanes 60, 216
Enesilanes
– from
 trialkylmanganates, with 1 extra C-atom 60, 375
– special s.
 trimethyl(vinyl)silane
– startg. m. f.
 ethylene derivs., terminal 60, 44
Enesilanols
– special s.
 alkoxyenesilanols

Enesilicon hydrides
– special s.
 enesilicon hydrides
Enestannanes
– special s.
 germylenestannanes
(E)-Enestannanes
– from
 (E)-eneselenides **60**, 216
Ene synthesis s. Ene reaction
Ene-yne metathesis, intramolecular-Diels-Alder reaction 60, 289
Enolate equivalents
– special s.
 (1,3-dienol)ate equivalents
 zinc enolate –
Enolates
– special s.
 alkali metal enolates
 aluminum –
 boron –
 titanium(IV) –
 zinc –
(Z)-Enol borinates
– as intermediates **32**, 614s60
Enolesters
–, hydroformylation, asym., regiospecific **60**, 291
–, syntheses with – via alkali metal enolates, review **51**, 429s60
Enolethers
– from
 mercaptals, polymer-based synthesis **60**, 113
– special s.
 α-alkoxy-α,β-ethylen....
 alkoxyvinyl...
 halogenenolethers
 o-hydroxy-β-styryl ethers
– startg. m. f.
 ethylene derivs., reduction **60**, 213
 naphthalenes **60**, 472
 pyrrole ring **60**, 372
 (E)-vinylzirconium compds. **60**, 213
Enolethers, cyclic
– from
 ketones, cyclic **60**, 89
Enolethers, metalated
– special s.
 lithioenolethers
Enol phosphates
– from
 enol phosphites **60**, 104
Enol phosphites
– startg. m. f.
 enol phosphates **60**, 104
Enols
– special s.
 α,β-ethylene-α-hydroxy...
Enol triflates
– from
 ketones **60**, 49
Enones s. α,β-Ethyleneketones
Enoxy(hydrido)silanes
– startg. m. f.
 syn-1,3-diols **60**, 448
Enoxysilanes (s.a. under Aldol-type..., Michael-type...)
– special s.
 enoxy(hydrido)silanes
– startg. m. f.
 β-aminoketones **50**, 551s60

α-arylcarbonyl compds. **60**, 461
γ,δ-ethyleneketones **60**, 453
α-fluoroketones, asym. conversion **60**, 186
–, syntheses with – via alkali metal enolates, review **51**, 429s60
Enoxystannanes
–, Michael-type addition, catalytic with – **60**, 449
(Z)-2-En-4-ynamines
– startg. m. f.
 pyrroles **60**, 129
Enynes
–, carboplatination, intramolecular-solvation, regiospecific **60**, 304
– special s.
 halogenenynes
1,3-Enynes
– from
 α,β-ethylenehalides **60**, 460
 lithium alk-1-ynyl(trialkoxyborates) **60**, 460
– special s.
 5-acoxy-3-en-1-ynes
 2-(alkylseleno)-1,3-enynes
 amino-1,3-enynes
 1,3-dien-5-ynes
 1,1-dihalogen-1,3-enynes
 2-en-4-ynamines
1,6-Enynes
–, cycloisomerization, asym. **60**, 307
1,5-Enyn-4-ols, chiral 45, 382s60
Enzymatic reactions (s.a. Biotransformations)
– in plasticized glass phases **49**, 195s60
– with hydrophobic polymers as co-solvent **60**, 24
Enzymes
 aminoacylase **44**, 214s60
 ammonia lyase **60**, 126
 benzoylformate decarboxylase **26**, 675s60
 dehydrogenase **60**, 22
 esterase, fibre-immobilized **41**, 118s60
 lipase **60**, 9, 195
 –, hydrophobic sol-gel-entrapped **28**, 13s60
 –, immobilized **60**, 23
 –, – on HPLC stationary phases **28**, 13s60
 –, inorganic-supported **46**, 171s60
 lipase-on-Celite **44**, 314s60
 phosphotriesterase **60**, 10
 subtilisin mutants **53**, 144s60
Enzymes, artificial
–, design and synthesis, review **58**, 499s60
Epoxidation (s.a. Epoxides from ethylene derivs.)
Epoxidation, asym. (s. Sharpless-type epoxidation)
– of *cis*-ethylene derivs. **60**, 68
– in fluorous 2-phase medium **60**, 69
–, –, V(V)-mediated
– of 2-ethylenealcohols **60**, 65
–, Cu(I)-catalyzed
–, review **23**, 819s60
–, heterogeneous, dioxirane-mediated **60**, 67
Epoxides (s. under Oxido compds. in Vol. 1-50)
– from
 ethylene derivs. (s.a. Epoxidation) **60**, 70

 glycol sulfates, cyclic **60**, 112
– special s.
 ethyleneepoxides
 propylene oxide
– startg. m. f.
 2-(arylamino)alcohols **60**, 120, 122
 furans, tetrahydro-, (E)-2-acylene- **60**, 441
 1,2-halogenhydrins, desymmetrization **60**, 173
 β-hydroxyaldehydes, hydroformylation **60**, 241
 α-hydroxyketones **60**, 58
Epoxides, bicyclic
– special s.
 cyclooctene oxides
α,β-Epoxyaldehydes
– startg. m. f.
 α-halogenoglycols, synthesis, regiosteriospecific **60**, 234
α,β-Epoxycarboxylic... s. Glycidic...
2,3-Epoxyiodides, chiral 21, 606s60
α,β-Epoxyketones
– by Darzens condensation in water **60**, 362
– from
 α,β-ethyleneketones **60**, 63
– startg. m. f.
 α,β-ethylene-α-hydroxyketones **60**, 2
α,β-Epoxyketones, chiral 52, 389s60
Esterification... s. Carboxylic acid esters from...
Ethers
– special s.
 alkoxy...
 aminoethers
 diol monoethers
 enolethers
 ethyleneethers
 nitroethers
 trityl ethers
Ethers, cyclic (s.a. O-Heterocyclics)
Ethyl (carbethoxyimino)acetate
– as reactant **60**, 246
Ethyl diazoacetate
– as reactant **60**, 357
Ethyl dichlorovanadate
– as Lewis acid and oxidant **60**, 263
– as reagent **60**, 381
Ethyldiisopropylamine
– as reagent **60**, 49, 252, 309, 392
α,β-Ethyleneacetals
–, Diels-Alder reaction, ionic, *endo*-selective with – **60**, 266
2-Ethylenealcohols
–, deracemization **53**, 500s60
–, epoxidation, asym. **60**, 65
– from
 acetylene derivs., regiospecific conversion **60**, 284
 aldehydes, – – **60**, 284
 α,β-ethylenealdehydes, reduction, asym. **60**, 24
 2-ethyleneselenides, via [2.3]-sigmatropic rearrangement **60**, 97
– special s.
 1,4-dien-3-ols
 1,6-dien-4-ols
 α,β-ethylene-α-(hydroxymethyl)...
 β'-hydroxy-α-methylene...
 2-methylenealcohols

2-Ethylenealcohols
- startg. m. f.
  aldehydes, asym. isomerization **60**, 73
  ethylene derivs., synthesis **51**, 372s**60**
  α,β-ethyleneoxo compds. **60**, 107
  (E)-2-ethylenetosylamines, double bond shift **60**, 130
2-Ethylene-*prim*-alcohols, chiral 29, 36s**60**
3-Ethylenealcohols
-, epoxidation, asym. **55**, 64s**60**
- from
  aldehydes, synthesis, regiostereospecific **60**, 302
  allenes, –, - **60**, 302
  2-ethylene(trichloro)silanes, –, asym. **49**, 898s**60**
- special s.
  3-aryl-3-ethylenealcohols
  1,6-dien-4-ols
  1-perfluoroalkyl-3-ethylenealcohols
- startg. m. f.
  β-hydroxyketones, ring contraction **27**, 162s**60**
(E)-3-Ethylenealcohols
- from
  3-acetylenealcohols, synthesis **60**, 406
4-Ethylenealcohols
- from
  carboxylic acid esters, with 4 extra C-atoms **60**, 328
β,γ-Ethylenealdehyde equivalents
-, 3-ethyleneepoxides as – **60**, 431
α,β-Ethylenealdehydes (s.a.
  α,β-Ethyleneoxo compds.)
-, cycloaddition, 1,3-dipolar, asym., amine-catalyzed with – **60**, 278
-, Diels-Alder reaction, asym., Cr(III)-catalyzed with – **60**, 447
- special s.
  acoxy-α,β-ethylenealdehydes
  2,4-dienals
  α-methylenealdehydes
  2,4,6-trienals
- startg. m. f.
  2-ethylenealcohols, reduction, asym. **60**, 24
γ,δ-Ethylenealdehydes
- from
  4-oxa-1,7-dienes **60**, 313
- special s.
  cyclopropyl-γ,δ-ethylenealdehydes
Ethyleneamines, macrocyclic, N-protected 20, 685s**60**
(E)-Ethyleneamines, N-protected
- from
  α-amino-N-sulfonylhydrazones, cyclic, N-protected, synthesis **60**, 356
2-Ethyleneamines
- special s.
  2-en-4-ynamines
- startg. m. f.
  3-azabicyclo[3.3.0]oct-1(8)-ene-2,7-diones **60**, 387
2-Ethyleneamines, cyclic
- from
  ω-alleneamines **60**, 287
  ethylene derivs., electron-deficient **60**, 287
2-Ethyleneamines, N-protected
-, Heck arylation, regiospecific **60**, 344
2-Ethylene-*prim*-amines

- special s.
  aryl-2-ethylene-*prim*-amines
2-Ethylene-*tert*-amines
- from α-*tert*-aminonitriles and vinylmagnesium halides **60**, 467
-, Heck arylation, regiospecific **60**, 344
- special s.
  β-*tert*-amino-α-methyleneketones
3-Ethyleneamines, N-protected
- from
  (E)-2-ethylenesilanes and aldimines, N-protected, *in situ*-generated, asym. induction **60**, 430
-, –, chiral **60**, 330
o-Ethyleneamines
- from
  Δ³-isothiazoline 2,2-dioxide ring **60**, 489
Ethyleneazides
- startg. m. f.
  amines, sec., cyclic **60**, 172
3-Ethyleneazides
- special s.
  aryl-3-ethyleneazides
2-Ethyleneboranes (s.a. Allylboration)
α,β-Ethyleneboronic acid esters
- special s.
  α-alkylidene-γ-silylboronic acid esters
β,γ-Ethyleneboronic acid esters
- special s.
  β-alkylidene-γ-silylboronic acid esters
  α-(1,3-dioxin-4-yl)-β,γ-ethyleneboronic – –
α,β-Ethyleneboronic acids
- special s.
- startg. m. f.
  3-chromenes **60**, 427
  α,β-ethyleneketones **60**, 457
2-Ethylenecarbonic acid esters
- special s.
  allyl ethyl carbonate
- startg. m. f.
  2-cyclopentenones, carbonylation **60**, 343
  γ,δ-ethyleneketones **60**, 453
Ethylenecarbonyl compds.
- special s.
  dienecarbonyl compds.
α,β-Ethylenecarbonyl compds.
- startg. m. f.
  δ-alkoxy-γ-ketocarbonyl compds. **60**, 437
  γ-ketocarbonyl compds., synthesis via radical carbonylation **60**, 269
  Δ²-pyrazoline-5-carbonyl compds., 1-acetyl-, asym. cycloaddition **60**, 429
(E)-α,β-Ethylenecarbonyl compds.
- from
  β-hydroxycarbonyl compds. **60**, 479
α,β-Ethylenecarboxylic acid amides
- startg. m. f.
  β-azidocarboxylic acid amides, asym. conversion **60**, 127
(Z)-α,β-Ethylenecarboxylic acid amides
- by Peterson olefination **28**, 856s**60**
Ethylenecarboxylic acid anhydrides
- startg. m. f.
  2*H*-1,4-benzoxazines, lactam-condensed **60**, 78
α,β-Ethylenecarboxylic acid anhydrides
- startg. m. f.

2-pyridones, 3,4-dihydro-, polymer-based synthesis **60**, 124
α,β-Ethylenecarboxylic acid esters
-, Michael addition, asym. to – **60**, 191
- special s.
  acylamino-α,β-ethylenecarboxylic acid esters
  alkoxy-α,β-ethylenecarboxylic – –
  aryl-α,β-ethylenecarboxylic – –
  cinnamic – –
  α,β-ethylene-α-(halogenomethyl)-carboxylic – –
  α,β-ethylene-α-(hydroxymethyl)-carboxylic – –
  hetaryl-α,β-ethylenecarboxylic – –
  α-methylenecarboxylic – –
- startg. m. f.
  glutarimides **60**, 320
(E)-α,β-Ethylenecarboxylic acid esters 60, 479
(E)-β,γ-Ethylenecarboxylic – –
- from
  α-allenecarboxylic acid esters **60**, 325
α,β-Ethylenecarboxylic acids
- special s.
  cinnamic acids
- startg. m. f.
  carboxylic acids **60**, 37
  2-pyridones, 3,4-dihydro-, polymer-based synthesis **60**, 124
β,γ-Ethylenecarboxylic acids
- special s.
  amino-β,γ-ethylenecarboxylic acids
Ethylene derivs. (s.a. C-Allylation, Horner..., Peterson..., Wittig...)
-, cycloaddition, 1,3-dipolar with nitrile oxides, review **16**, 888s**60**
-, fluorination with iodoarene difluorides, review **36**, 470s**60**
- from
  enolethers, reduction **60**, 213
  2-ethylenealcohols, synthesis **51**, 372s**60**
  β,γ-ethylenehalides, with double bond shift **60**, 42
  1,1-iodosilanes **60**, 496
  (E)-vinylzirconium compds. **60**, 213
-, hydrogermylation **43**, 517s**60**
-, radical addition to –, review of controlling factors **39**, 646s**60**
-, radical hydrosilylation **60**, 209
-, ring closure with ω-phthalimidoalcohols **60**, 255
- special s.
  acoxyethylenes
  alkoxyethylenes
  dienes
  nitroethylene derivs.
  polyfluoroethylene derivs.
  siloxyethylenes
  styrenes
  trienes
- startg. m. f.
  α-aminoketones, cyclic, carbonylation **60**, 293
  α-amino-γ-lactones **60**, 332
  benzofurans, 2,3-dihydro-, 5-subst. **60**, 334
  2-cyclopentenones, carbonylation **60**, 343
  1,2-diamines, N-protected differentially **60**, 125

dithiophosphonic acid esters **60**, 217
furan ring, 2,3-dihydro- **60**, 361
glycols s.a. Dihydroxylation
β-hydroxynitriles, regiostereospecific
  conversion **60**, 272
ketones, *o*-subst. **60**, 360
2-oxazolidones, asym. conversion
  60, 166
Δ²-pyrazolines, in aq. micelles or water
  60, 363
pyrroles **60**, 352
pyrrolidines, 1-sulfonyl- **60**, 265
silanes, synthesis **60**, 273
α-sulfonyloxyketones **60**, 62
(E)-Ethylene derivs.
– from
  oxo compds., aza-Horner synthesis
    60, 424
  phosphine N-carbalkoxyimines **60**, 424
– –, N-, O-, and S-heterocyclic, 8- to 10-
  membered
–, synthesis and planar-chiral properties,
  review **49**, 985s60
cis-**Ethylene derivs.**
–, epoxidation, asym. **60**, 68
Ethylene derivs, bicyclic
–, ring-opening metathesis-cross-
  metathesis **60**, 290
– special s.
  norbornenes
– –, electron-deficient
–, Friedel-Crafts alkylation, asym.,
  catalytic with – **60**, 260
– startg. m. f.
  trans-cyclopropanecarbonyl compds.
    60, 410
  cyclopropanecarboxylic acid esters,
    asym. conversion **60**, 355
  cyclopropanes **60**, 426
  2-ethyleneamines, cyclic **60**, 287
  furans, tetrahydro-, 3-alkylidene-,
    synthesis **60**, 402
  β-hydroxyacetals, cyclic **60**, 282
  indolizines **60**, 474
  4-oxa-1,8-dienes **60**, 346
– –, O- or N-functionalized
– from
  acetylene derivs., O- or N-
    functionalized, synthesis,
    regiostereospecific **60**, 258
– –, heteroatom-functionalized
–, carbometalation, review **47**, 639s60
– –, terminal
– from
  enesilanes, 1-subst. **60**, 44
– startg. m. f.
  aldehydes **46**, 116s60
– –, tetrasubst. **60**, 258
–, hydroboration-1,2-boronyl migration,
  syntheses via – **60**, 264
Ethylenediaminetetraacetic acid
– as reagent **60**, 87
1,2-Ethylene-1,2-di(boronic acid esters)
  60, 207
1,2-Ethylene-1,1-dihalides s.
  Dihalogenomethylene compds.
1,2-Ethylene-1,2-dihalides
– startg. m. f.
  α-diketones **60**, 93
2-Ethylene-1,1-di(silanes) 12, 931s60
Ethyleneepoxides
– startg. m. f.

cyclopentylcarbinol ring **51**, 318s60
3-Ethyleneepoxides
– as (1,3-dienol)ate equivalents **60**, 431
– as β,γ-ethylenealdehyde – **60**, 431
–, ring opening, nucleophilic, Rh-
  catalyzed **60**, 59
– startg. m. f.
  1,6-dien-4-ols **60**, 431
2-Ethyleneethers, cyclic (s.a. 2-Vinyl-
  O-heterocyclics)
–, –, chiral **48**, 988s60
α,β-Ethylenefluorides
–, cyclopropanation, asym. **23**, 819s60
–, ozonolysis **32**, 180s60
δ,ε-Ethylenefluorides
– from
  acetylenestannanes **60**, 458
  1,3-fluoroiodides **60**, 458
**α,β-Ethylene-α-formamidocarboxylic
  acid esters**
– from
  ketones, with 2 extra C-atoms **60**, 476
α,β-Ethylenehalides
– startg. m. f.
  1,3-enynes **60**, 460
β,γ-Ethylenehalides
–, Baylis-Hillman-type reaction with –
    60, 373
– special s.
  allyl bromide
  9-halogeno-2,7-dienol carbamates
– startg. m. f.
  allylarenes **60**, 386
  ethylene derivs., with double bond shift
    60, 42
  furans, 2,5-dihydro-, 3-allyl- **60**, 400
δ,ε-Ethylenehalides
– special s.
  δ,ε-ethylenefluorides
1,2-Ethylene-1,3-halogenhydrins
– special s.
  α,β-ethylene-β-halogeno-β'-hydroxy...
3-Ethylene-2,1-halogenhydrins
  46, 429s60
**α,β-Ethylene-α-halogenocarboxylic
  acid esters**
–, reactions with nucleophiles, review
    47, 863s60
**(E)-α,β-Ethylene-α-halogeno-
  carboxylic – –**
– from
  aldehydes, with 2 extra C-atoms
    60, 423
**α,β-Ethylene-β-halogeno-β'-
  hydroxycarbonyl compds.**
– from
  α,β-acetylenecarbonyl compds.
    60, 275
α,β-Ethylene-β-halogenoketones
– from
  β-diketones **60**, 182
**2-Ethylene-α-halogenolactolides
  33**, 477s60
**(Z)-α,β-Ethylene-α-(halogenomethyl)-
  carboxylic acid esters**
– startg. m. f.
  α-methylenecarboxylic acid esters,
    asym. synthesis **60**, 379
**α,β-Ethylene-β-halogeno-α-(organo-
  seleno)carboxylic – –**
– from

1-acetylene-1-selenides, carbonylation,
  stereospecific **60**, 301
**(E)-α,β-Ethylene-γ-halogenosulfoxides
  27**, 555s60
3-Ethylenehydrazines
– from
  hydrazones, with 3 extra C-atoms
    60, 249
α,β-Ethylenehydroxamic acid esters
– by carbonylation **43**, 808s60
anti-**2-Ethylene-2'-hydroxyhydrazines**
– from
  α-vinyloxyhydrazones, asym. induction
    60, 494
α,β-Ethylene-hydroxy...
  s.a. Hydroxy-α-methylene…
α,β-Ethylene-α-hydroxyketones (s.a.
  α-Diketones)
– from
  α,β-epoxyketones **60**, 2
(Z)-α,β-Ethylene-β'-hydroxyketones
– from
  2-acetylenealcohols **60**, 277
β,γ-Ethylene-ε-hydroxyketones 60, 283
3-Ethylenehydroxylamines
– from
  nitrones, with 3 extra C-atoms **60**, 249
threo-**3-Ethylene-2-hydroxylamino-
  alcohols 60**, 118
3-Ethylene-2-hydroxymercaptans
– startg. m. f.
  thiophenes, tetrahydro-, 3-acyl-, asym.
    induction **60**, 331
**(E)-α,β-Ethylene-α-(hydroxymethyl)-
  carboxylic acid esters**
– from
  β-hydroxy-α-methylenecarboxylic
    acid esters **60**, 76
**2-Ethylene-1,1-hydroxysilanes
  44**, 545s60
**α,β-Ethylene-β'-hydroxy-β-silyl-
  carboxylic acid esters, chiral
  52**, 310s60
**γ,δ-Ethylene-β-hydroxysulfoxides,
  chiral 8**, 272s60
**β,γ-Ethylene-γ-iodophosphonic acid
  esters 47**, 658s60
β,γ-Ethylene-α-ketocarboxylic – –
– special s.
  aryl-β,γ-ethylene-α-ketocarboxylic
    acid esters
**γ,δ-Ethylene-β-ketocarboxylic – –
  10**, 633s60
α,β-Ethyleneketones (s.a. α,β-Ethylene-
  oxo compds.)
– from
  1,3-enynes **60**, 66
  α,β-ethyleneboronic acids **60**, 457
–, Michael addition, asym., proline-
  catalyzed to – **60**, 267
– special s.
  amino-α,β-ethyleneketones
  2,4-dienones
  α,β-ethylene-β-halogenoketones
– startg. m. f.
  α,β-epoxyketones **60**, 63
  β,γ-ethyleneketones, synthesis **60**, 329
  3-ketothioethers **60**, 199
α,β-Ethyleneketones, cyclic
–, 1,4-addition-alkylation **60**, 263
(E)-α,β-Ethyleneketones 60, 479
– from
  ketones, dehydrogenation **60**, 477

β,γ-Ethyleneketones
- from
  aldehydes, synthesis 60, 329
  α,β-ethyleneketones, – 60, 329
γ,δ-Ethyleneketones
- from
  enoxysilanes 60, 453
  2-ethylenecarbonic acid esters 60, 453
- special s.
  γ-methyleneketones
α,β-Ethylenenitriles
- from
  β-hydroxynitriles 60, 478
(E)-α,β-Ethylenenitriles 60, 478
β,γ-Ethylenenitriles
- from
  cyclobutanone O-acyloximes 60, 171
- special s.
  β-alkylidenenitriles
Ethylenenitro... s.a. Nitroethylene...
α,β-Ethylene(nitro)carboxylic acid esters
- startg. m. f.
  β-arylaminocarboxylic acid esters, cyclic 60, 170
α,β-Ethyleneoximes
- startg. m. f.
  pyridines via Michael addition 60, 342
(E)-α,β-Ethyleneoximes
- from
  Δ²-isoxazolines 60, 1
α,β-Ethyleneoxo compds.
- from
  2-ethylenealcohols 60, 107
α,β-Ethylenephosphine oxides
-, hydrogenation 19, 60s60
α,β-Ethylenephosphonic acid esters
- special s.
  1,3-dienephosphonic acid esters
β,γ-Ethylenephosphonic – –
-, γ-acylation 60, 1
α,β-Ethylene-α-(2-pyridyloxy)ketones
- as intermediates 60, 2
2-Ethyleneselenides
- startg. m. f.
  2-ethylenealcohols, via [2.3]-sigmatropic rearrangement 60, 97
2-Ethylenesilanes
-, Heck arylation 27, 871s60
- special s.
  cyclopropyl-2-ethylenesilanes
  (diallylhydridosiloxy)-3-ethylenes
- startg. m. f.
  alkoxy-3-ethylenes 60, 422
  anti-3-ethylene-2-silylalcohols, asym. conversion 60, 221
(E)-2-Ethylenesilanes
- startg. m. f.
  3-ethyleneamines, N-protected, asym. induction 60, 430
  pyrrolidines, 3-silyl-, –, – – 60, 330
3-Ethylenesilanes
- from
  acetylene derivs., with 2 extra C-atoms 60, 270
anti-3-Ethylene-2-silylalcohols
- from
  aldehydes and 2-ethylenesilanes, asym. conversion 60, 221
- via allylboration, asym. 60, 221
Ethylene-γ-silylboronic acid esters

s.a. Alkylidene-γ-silylboronic acid esters
β,γ-Ethylene-δ-silylketones 51, 423s60
β,γ-Ethylene-β-stannylketones
- from
  acylstannanes 60, 283
  allenes 60, 283
β,γ-Ethylene-δ-stannylketones
- from
  acylstannanes 60, 283
  1,3-dienes 60, 283
(E)-α,β-Ethylene-α-sulfinylphosphonic acid esters
- from
  N-sulfonylimines, with 1 extra C-atom 60, 349
α,β-Ethylenesulfones
- special s.
  aryltelluro-α,β-ethylenesulfones
β,γ-Ethylenesulfones
- special s.
  nitro-β,γ-ethylenesulfones
α,β-Ethylenesulfonic acid amides
- special s.
  1,3-dienesulfonic acid amides
2-Ethylenesulfonylamines
- special s.
  2-ethylenetosylamines
α,β-Ethylene-β-sulfonyloxyiodides
s. 2-Iodoenol sulfonates
α,β-Ethylenesulfoxides
- special s.
  α-methylenesulfoxides
- startg. m. f.
  oxo compds. 60, 96
β,γ-Ethylenesulfoximines, S-chiral 33, 806s60
(E)-2-Ethylenetosylamines
- from
  2-ethylenealcohols, double bond shift 60, 130
2-Ethylene-N-tosylurethans
- as intermediates 60, 130
7-Ethylene-1,3,5-triols
- from
  (diallylhydridosiloxy)-3-ethylenes, via silylformylation 60, 451
N-(2-Ethylene)urethans
-, 1,4-addition, asym., regiospecific with – 60, 259
Ethyl glyoxylate
- as reactant 60, 227
Ethyl isocyanoacetate
- as reactant 60, 476
Ethynylcarbinols
- startg. m. f.
  1,5-benzothiazepines, 2,3-dihydro-, 2-aryl- 60, 403
  furans, tetrahydro-, 3-alkylidene-, synthesis 60, 402
-, substitution, regiospecific, Nicholas-type 60, 83

Ferrocene-1,1'-dicarboxamides
-, o,o'-disubstitution, asym. 49, 536s60
Ferrocenes, aryl-

- from
  halides, ar. 60, 459
Ferrocenium tetrakis[3,5-bis(trifluoromethyl)phenyl]borate
- as Lewis acid 1, 419s60
Ferroceneylmercury compds.
- special s.
  bis(ferrocenyl)mercury
Ferrocenylphosphines
- special s.
  1,2,3,4,5-pentaphenyl-1'-(di-tert-butylphosphino)ferrocene
-, chiral
- as reagent 60, 299
Flash vacuum pyrolysis
-, cumulenes via –, review 27, 675s60
2-Fluoramines, chiral 23, 570s60
Fluorenes, 9-alkylidene-
- from
  arylacetylenes and halides, ar. 60, 391
9-Fluorenones
- from
  o-halogenodiaryls, carbonylation, 60, 397
Fluorides, ar.
- from
  amines, ar., prim. 60, 183
Fluorination
- of ethylene derivs. with iodoarene difluorides, review 36, 470s60
-, electrochemical
-, review 42, 462s60
α-Fluorination, asym.
- of
  α-cyanocarboxylic acid esters 60, 186
  β-ketocarboxylic – – 60, 179, 186
  ketones 60, 186
  ketones, cyclic 60, 186
Fluorine compds., organo-
-, preparation from HF or HF-base molten salts, review 49, 419s60
β-Fluoroamidinium salts, vinylogous
-, review 37, 807s60
Fluoroboric acid 60, 115
α-Fluorocarboxylic acid esters 32, 180s60
N-Fluorocinchonidinium fluoroborate
- as reagent 60, 186
2-Fluoro-2-deoxyglycosyl azides 49, 407s60
N-Fluoro-1,4-diazoniabicyclo[2.2.2]-octane bis(fluoroborate), N'-chloromethyl-
- as reagent 60, 179
Fluoroform
- as reactant 60, 224
1,3-Fluoroiodides
- startg. m. f.
  δ,ε-ethylenefluorides 60, 458
α-Fluoroketones
- from
  enoxysilanes, asym. conversion 60, 186
N-Fluoroquininium fluoroborate, dihydro-, O-(4-chlorobenzoyl)-
- as reagent 60, 186
Fluorous 2-phase medium
-, epoxidation, asym. in – 60, 69
-, Stille coupling in – 60, 462
Fluorous reagents
- special s.
  (S)-1,1'-bi-2-naphthol, 4,4',6,6'-tetrakis(perfluorooctyl)-

bis(4-perfluorohexylphenyl) diselenide
carboxylic acids, lightly fluorous
manganese(III) salen complexes,
  chiral, fluorous
palladium complexes, fluorous
platinum complexes, –
ytterbium(III) tris(perfluoroalkanesulfonyl)methides
Formaldehyde
– as reactant **60**, 337
Formamidinium salts, vinylogous 60, 43
– startg. m. f.
  pyridines **60**, 351
α-**Formamidinonitriles, chiral 5**, 301s60
α-**Formamidocarboxylic acid esters**
– special s.
  ethylene-α-formamidocarboxylic acid
  esters
Formic acid
– as H-donor **60**, 37
Formic acid-D2 60, 28
N-Formylation
– special s.
  di-N-formylation
N-Formylation, Au-catalyzed
– with carbon monoxide **60**, 132
**Friedel-Crafts alkylation, asym.,
  catalytic**
– with ethylene derivs., electron-
  deficient **60**, 260
– α-*prim*-aminoacylation
– with 2,5-oxazolidiones, retention of
  chirality **60**, 470
– hydroxyalkylation, asym. **60**, 227
– reactions, Cu(II)-catalyzed **60**, 327
Fulvenoids (s.a. Dithiafulvenes)
Fungus 60, 24
**Furan-2-carboxylic acid esters,
  3-amino- 60**, 339
**Furan-4-carboxylic – –, 2,3-dihydro-,
  chiral 47**, 705s60
**2(5*H*)-Furanone ring, 5-α-alkoxy-,
  4,5-condensed**
– from
  chromium alkynyl(alkoxy)carbene
  complexes **60**, 492
– –, **4,5-condensed**
– from
  chromium alkynyloxycarbene
  complexes **60**, 492
2(3*H*)-Furanones, 5-aryl-
– from
  furans, 2-siloxy- and triarylantimony
  diacetates **60**, 461
2(5*H*)-Furanones, 5-acoxy-
– from
  acetylene derivs. **60**, 316
**2(5*H*)-Furanones, 5-alkylidene-
  33**, 477s60
–, **5-(carbalkoxyene)-3-hydroxy-**
– from
  1-alkoxy-1,3-disiloxy-1,3-dienes
  60, 445
–, **5-α-halogeno- 33**, 477s60
Furan ring
– from
  acetylene derivs. **60**, 361
Furan ring, 2,3-dihydro-
– from
  ethylene derivs. **60**, 361
Furans
– from

β,γ-acetyleneketones **60**, 74
α-alleneketones **60**, 74
ketones **60**, 374
Furans, 3-acyl- 60, 374
–, **3-amino-**
– from
α-cyanoketones **60**, 339
–, **3-cyano-, 2,5-disubst. 60**, 374
–, **2,5-dihydro-, 3-allyl-**
– from
2-allenealcohols and β,γ-ethylene-
  bromides **60**, 400
–, **2-siloxy-**
– startg. m. f.
2(3*H*)-furanones, 5-aryl- **60**, 461
–, **tetrahydro-**
– by alkoxyl radical ring closure,
  stereospecific, review **52**, 125s60
–, **tetrahydro-, (E)-2-acylene-**
– from
1,3-disiloxy-1,3-dienes and epoxides
  60, 441
–, –, **3-alkylidene-**
– from
  ethynylcarbinols and ethylene derivs.,
  electron-deficient, synthesis **60**, 402
–, –, **3-allylidene- 60**, 402
–, –, **3-arylidene- 60**, 402
–, –, **3,3-difluoro-, chiral 29**, 970s60
–, –, **3-hydroxy-4-iodo- 33**, 447s60
–, –, **3-iodo- 33**, 477s60
–, –, **3-β-keto- 60**, 276
–, –, **3-methylene-4-vinyl-, chiral
  49**, 699s60; **60**, 307
Furoxans
– from
α-nitrooximes **60**, 48
Furylacetylenes
– startg. m. f.
  phenols, condensed **60**, 310
**Furylcarbinols, tetrahydro-, 4-arylthio-
  44**, 238s60

Gallium trichloride 60, 27, 434
Germanes
– special s.
  enegermanes
Germanethiols
– from
  germanium hydrides, organo- **60**, 189
Germanium hydrides, organo-
– startg. m. f.
  germanethiols **60**, 189
Germylation
– special s.
  hydrogermylation
(Z)-2-Germylenestannanes
– as intermediates **43**, 517s60
(E)-β-Germylstyrenes 39, 887s60
Glass phases, plasticized 49, 195s60
Glutaric acid diamides, chiral 42, 638s60
Glutarimides
– from
  carboxylic acid amides and α,β-
  ethylenecarboxylic acid esters
  60, 320

– special s.
  2-sulfonylglutarimides
Glycals
– startg. m. f.
  2-deoxyaldoses **60**, 64
Glycidic acid esters, chiral 60, 112
Glycidic acids
– startg. m. f.
  β-azido-α-hydroxycarboxylic acids
  60, 119
Glycin(at)es
– special s.
  arylglycin(at)es
Glycoconjugates
–, syntheses, enzymatic and one-pot,
  review **17**, 169s60
Glycol carbonates
– from
  glycols **60**, 135
– –, cyclic s. 1,3-Dioxolan-2-ones
Glycol esters, sym.
– from
  aldehydes **60**, 240
Glycol ethers
– special s.
  α,β-dialkoxy...
Glycol ethers, sym.
– from
  aldehydes **60**, 239
Glycol monoethers
– from
  1,3-dioxolanes, synthesis, asym. **60**, 222
– special s.
  cyclopropanols, 1-α-alkoxy-
  3-ene-1,2-diol monoethers
Glycols (s.a. Pinacol...)
– from
  ethylene derivs. s.a. Dihydroxylation
– special s.
  3-amino-1,2-diols
  α,β-dihydroxy...
  halogenoglycols
– startg. m. f.
  1,3-dioxolan-2-ones **60**, 135
  glycol carbonates **60**, 135
  oxo compds., dehydration (of 1,1-
  disubst. glycols) **60**, 111
Glycols, long-chain, chiral 53, 4s60
Glycol sulfates, cyclic
– startg. m. f.
  epoxides **60**, 112
  1,2-halogenhydrins **60**, 112
Glycolurils
– special s.
  2,4,6,8-tetraiodoglycoluril
Glycopeptides
– with carbohydrate and peptide-
  recognition motifs, review
  17, 169s60
Glycopeptides, artificial
–, synthesis, review **17**, 169s60
**Glycosidation, dehydrative, iterative
  60**, 80
Glycosidation, intramolecular
–, review **59**, 124s60
–, –, **oxidative 39**, 228s60
Glycosides
– from
  glycosyl disulfides **60**, 95
  glycosyl 2-pyridyl sulfones **60**, 94
  trichloroacetimidoyl glycosides **60**, 99
  trifluoroacetimidoyl – **60**, 103

Glycosides
- special s.
  deoxyglycosides
  oligoglycosides
  selenoglycosides
  silyl glycosides
  thioglycosides
  trichloroacetimidoyl glycosides
  trifluoroacetimidoyl –
β-Glycosides
- from
  N-(alkylthio)phthalimides **39**, 189s60
C-Glycosides
- by
  radical 1,4-addition **60**, 442
  Wittig synthesis **29**, 861s60
- special s.
  C-allyl glycosides
  deoxy-C-glycosides
  C-(1-hydroxyalkyl) glycosides
  C-(ketoalkyl) glycosides
  C-pentasaccharides
- via transition metal glycosylidene
  complexes, review **44**, 782s60
C-α-Glycosides
- from
  α-selenoglycosides **60**, 442
1,2-trans-C-Glycosides
- from
  α-glycosyl iodides **60**, 237
  silyl glycosides **60**, 237
Glycosylamines
- special s.
  deoxyglycosylamines
**C-(Glycosylamino)-α-aminocarboxylic
acids, anomerically-linked or fused
–**, synthesis, review **17**, 169s60
Glycosyl azides
- special s.
  2-fluoro-2-deoxyglycosyl azides
Glycosyl(chloro)sulfonium chlorides
- as intermediates **60**, 185
Glycosyl disulfides
- from
  S-glycosyl methanethiolsulfonates
  60, 95
- startg. m. f.
  glycosides **60**, 95
Glycosyl halides
- from
  aldoses **60**, 181
  thioglycosides **60**, 185
- special s.
  glycosyl iodides
Glycosylidene complexes
- special s.
  transition metal glycosylidene
  complexes
α-Glycosyl iodides
- from
  silyl glycosides **60**, 237
- startg. m. f.
  1,2-trans-C-glycosides **60**, 237
S-Glycosyl methanethiolsulfonates
- startg. m. f.
  glycosyl disulfides **60**, 95
1-Glycosyloxyallenes
- as reactant **58**, 465s60
Glycosyl phosphates
- from
  aldoses **60**, 54
Glycosyl 2-pyridyl sulfones

- startg. m. f.
  glycosides **60**, 94
S-Glycosylsulfenamides
- from
  glycosyl thiolacetates **60**, 117
Glycosyl thiolacetates
- startg. m. f.
  S-glycosylsulfenamides **60**, 117
Glycosyl transfer, stereocontrolled
- with glycosyl donors, unprotected,
  review **59**, 124s60
Glycuronides 39, 189s60
Glyoxylic acid
- as reactant **60**, 415, 416
Glyoxylic acid esters
- special s.
  arylglyoxylic acid esters
  ethyl glyoxylate
  methyl –
Gold complexes
  chloro(triphenylphosphino)gold(I)
  60, 132
Gold trichloride 60, 74, 310
Grignard compds.
  s.a. Magnesium halides, organo-
Guanidines
- from
  amines **60**, 162
- (2 different molecules) **60**, 147
- (2 – –), polymer-based synthesis
  60, 161
  isothiocyanates, – – **60**, 161
  N-thioacyltriazenes, – – **60**, 161
  thioureas **60**, 162
- special s.
  arginine...
Guanidines, polymer-based 60, 147
Guanines, 9-subst. 9, 512s60

Halides (s.a. under Replacement)
- special s.
  acetylenehalides
  benzyl halides
  dihalides
  ethylenehalides
  fluorides
  iodides
  sulfonyloxyhalides
  trihalides
- startg. m. f.
  amidines, ar. **60**, 158
  amines, tert., sym. **60**, 150
  hydrazones, dimethyl- **60**, 155
  silanes, synthesis **60**, 273
Halides, ar.
- special s.
  fluorides, ar.
- startg. m. f.
  acylophenones, with 2 extra C-atoms
  60, 405
  amines, ar. (from chlorides) **60**, 156
  arylacetylenes **60**, 398, 460
  arylcarboxylic acid amides, N-unsubst.,
  carbonylation-nitrogenation **60**, 396
  2-aryl-3-ethyleneazides **60**, 298
  aryl ketones **60**, 389

1-aryllactones **60**, 366
aryloxysilanes **53**, 86s60
arylphosphonous acids **60**, 215
1,5-benzothiazepines, 2,3-dihydro-,
  2-aryl- **60**, 403
diaryls, sym. **60**, 378
diaryl sulfides **60**, 204
ferrocenes, aryl- **60**, 459
fluorenes, 9-alkylidene- **60**, 391
ketimines, ar. **60**, 389
methylarenes **60**, 381
triarylamines (from 2 different
  molecules) **60**, 157
Halides, prim.
- from
  alcohols, prim. **60**, 182
α-Halogenacetals
–, C-α-alkylation, intramolecular,
  stereospecific with – **60**, 485
Halogenalcohols s. Halogenhydrins
1,2-Halogenamines, cyclic
- from
  ethylene-N-halogenamines **27**, 530s60
Halogenation
- special s.
  fluorination
Halogenation, ar. 55, 183s60
- special s.
  iodination, ar.
α-Halogenation, asym.
- of β-ketocarboxylic acid esters **60**, 179
α-Halogenation-esterification, asym.
  60, 184
(Z)-1-Halogeneneselenides
- from
  1-acetylene-1-selenides **60**, 174
–, Sonogashira coupling with – **60**, 174
- startg. m. f.
  (Z)-2-(alkylseleno)-1,3-enynes **60**, 174
2-Halogeneneselenides
- special s.
  α,β-ethylene-β-halogeno-α-(organoseleno)carboxylic acid esters
1-Halogenenolethers
–, carbonylation **60**, 175
- from
  alkoxyacetylenes, stereospecific
  conversion **60**, 175
–, Sonogashira coupling with – **60**, 175
–, Stille coupling with – **60**, 175
2-Halogen-1,n-enynes
- startg. m. f.
  1,3-dienes, exocyclic **60**, 487
Halogenhydrins
- special s.
  trihalogenalcohols
1,2-Halogenhydrins
- from
  epoxides, desymmetrization **60**, 173
  glycol sulfates, cyclic **60**, 112
- special s.
  3-ethylene-2,1-halogenhydrins
  1,2-iodohydrins
1,3-Halogenhydrins
- special s.
  ethylene-1,3-halogenhydrins
  β'-halogeno-β-hydroxy...
α-Halogenoboronic acid esters
- special s.
  azido-α-halogenoboronic acid esters
– – –, **functionalized**
- from

boronic acid esters, functionalized,
asym. synthesis with 1 extra C-atom
60, 365
α-Halogenocarbonyl compds.
– startg. m. f.
benzo[b]thiophene-2-carbonyl compds.
60, 368
α-Halogenocarboxylic acid esters
– from
carboxylic acid halides, asym.
conversion **60**, 184
– special s.
ethylene-α-halogenocarboxylic acid
esters
β-Halogenocarboxylic acid esters
– special s.
α,β-ethylene-α-(halogenomethyl)-
carboxylic acid esters
o-Halogenocinnamyloxysilanes
– startg. m. f.
isochromenes, 1-vinyl- **60**, 370
5-Halogeno-5-deoxynucleosides
21, 606s60
o-Halogenodiaryls
– startg. m. f.
9-fluorenones, carbonylation **60**, 397
Halogeno-1,3-dienes
– special s.
dihalogeno-1,3-dienes
9-Halogeno-2,7-dienol carbamates
– startg. m. f.
1,5-cyclononadien-3-ol carbamates,
asym. conversion **60**, 483
Halogenoen... s. Halogenen...
Halogenoformic acid esters
– special s.
acetylenehalogenoformic acid esters
α-Halogenoglycols
– from
α,β-epoxyaldehydes, synthesis,
regiostereospecific **60**, 234
1,1-Halogenohydrazones
– special s.
N-(4-chloro-1,2,3-dithiazol-5-ylidene)-
1,1-halogenohydrazones
β-Halogeno-β′-hydroxycarbonyl
compds.
– special s.
ethylene-β-halogeno-β′-hydroxy-
carbonyl compds.
β′-Halogeno-β-hydroxyketones
–, Baylis-Hillman reaction, accelerated
via – **60**, 275
γ-Halogeno-β-hydroxyketones, chiral
29, 36s60
Halogenoketones
– from
alcohols, tert., cyclic **60**, 176
α-Halogenoketones
– special s.
α-fluoroketones
– startg. m. f.
indoles, 3-amino-1,2-diacyl- **60**, 371
β-Halogenoketones
– special s.
ethylene-β-halogenoketones
Halogenolactamization, oxidative
60, 177
α-Halogenolactolides
– special s.
2-ethylene-α-halogenolactolides
Halogenolactols

– special s.
iodolactols
o-(Halogenomethyl)arylmagnesium
bromides
–, synthesis via – **60**, 377
(E)-α-Halogenomethylene-β-
hydroxyketones 60, 275
α-Halogenonitriles
– special s.
chloroacetonitrile
β-Halogeno-α-(organoseleno)-
carboxylic acid esters
– special s.
ethylene-β-halogeno-α-(organoseleno)carboxylic acid esters
α-Halogenooxo compds.
– startg. m. f.
pyrrole ring **60**, 372
o-Halogenophenols
– startg. m. f.
coumarins, carbonylation **60**, 390
Halogenosilanes
– as reactant **60**, 273
– special s.
dihalogenosilanes
methyltrichlorosilane
trimethylsilyl halides
– startg. m. f.
silanes, synthesis **60**, 273
1,1-Halogenosilanes
– special s.
1,1-iodosilanes
α-Halogeno-β-siloxyboronic acid
esters, chiral
Halogenosulfonic acid amides
– special s.
dihalogenosulfonic acid amides
Halogenosulfonium salts
– special s.
chlorodiphenylsulfonium chloride
glycosyl(chloro)sulfonium chlorides
α-Halogeno(sulfonyl)acetals
– startg. m. f.
β-sulfonyllactolides **60**, 485
α-Halogenosulfonyl derivs.
–, electrophilic and radical reactions,
review **20**, 685s60
γ-Halogenosulfoxides
– special s.
ethylene-γ-halogenosulfoxides
Heck arylation
– in
ionic liquids **60**, 405
micelles aq. **27**, 871s60
2-phase media under thermomorphic
conditions **60**, 399
– –, desilylative
–, allylations by – **27**, 871s60
– –, double, intramolecular, asym.
60, 488
– –, **intramolecular** s.a. N-Arylation-
intramolecular Heck arylation
– –, polymer-based
– of azoles **57**, 376s60
– –, regiospecific
– of
2-ethyleneamines, N-protected **60**, 344
2-ethylene-tert-amines **60**, 344
Heck-type arylation, Rh-catalyzed
– of aldimines **60**, 389
Henry reaction, catalytic, solvent-less
60, 226

***cis*-β-Hetaryl-α,β-ethylenecarboxylic**
acid esters
– from
α,β-acetylenecarboxylic acid esters,
1,4-addition **60**, 295
Heterocumulenes
–, annulation with azoles and azolines,
review **27**, 675s60
–, heavy-atom
–, review **32**, 928s60
Heterocyclics
– from
oximes **44**, 241s60
–, metalation (of [di]azines), review
42, 597s60
– via nitrene insertion, intramolecular,
review **52**, 274s60
Heterocyclics, 5-membered
– startg. m. f.
cumulenes, review **27**, 675s60
Bi-Heterocyclics
–, review **16**, 698s60
N-Heterocyclics
– special s.
alkylthio-N-heterocyclics
O-Heterocyclics (s.a. Ethers, cyclic)
– special s.
vinyl-O-heterocyclics
O,N-Heterocyclics, phthalimidine-
condensed 60, 255
Sb-Heterocyclics
–, review **16**, 698s60
Hetero-Diels-Alder reaction
– with
imines, review **33**, 627s60
3-siloxy-1,3-dienamines **60**, 439
– –, **intramolecular**
– with o-quinone mono(N-acylimines)
60, 78
– –, regiospecific
– with thiophene 1,1-dioxides, 2,5-
dihydro-, 3-arylthio- **60**, 409
– reaction-allylboration
– with asym. induction **60**, 420
Hexachloro-2,4-cyclohexadienone
– as reagent **60**, 184
Hexafluoroisopropanol
– as solvent **60**, 120
Hexamethyldisilazane
– as reagent **60**, 121
–, O-silylation with – **60**, 55
High pressure hydrogenation, asym.,
homogeneous 60, 30
Homoallyl... s. 3-Ethylene...,
γ,δ-Ethylene...
Homobarrellenes 22, 761s60
Horner synthesis (s.a. Aza-Horner
synthesis)
– of
(E)-α-bromo-α,β-ethylenecarboxylic
acid esters **60**, 423
(E)-2,4-dienones **60**, 1
Hydantoins
– special s.
1,3-dichloro-5,5-dimethylhydantoin
Hydrazines
–, scavengers **1**, 391s60
– special s.
N,N-dimethylhydrazine
ethylenehydrazines
hydroxyhydrazines
–, protected
–, N-arylation **40**, 286s60

Hydrazinoalcohols s. Hydroxyhydrazines
β-Hydrazinosulfones, chiral 49, 264s60
Hydrazones
– from
  azides (N,N-dimethylhydrazones) **60,** 155
  halides (–) **60,** 155
– special s.
  halogenohydrazones
  sulfonylaminohydrazones
  sulfonylhydrazones
  vinyloxyhydrazones
– startg. m. f.
  3-ethylenehydrazines, with 3 extra C-atoms **60,** 249
  oxo compds. (from dimethylhydrazones) **60,** 88
Hydroacylation, intramolecular, Rh-catalyzed 60, 306
Hydroacylation-*o*-alkylation, chelation-assisted 60, 360
Hydroamination, asym., regiostereospecific
– of styrenes **60,** 128
Hydroboration
– at low temp. **60,** 60
Hydroboration-1,2-boronyl migration
– of ethylene derivs., tetrasubst., syntheses via – **60,** 264
Hydroboration-intramolecular Suzuki coupling 49, 836s60
Hydrocarbon groups, quaternary 22, 742s60
– –, –, chiral **43,** 607s60; **48,** 772s60
– –, –, vicinal, chiral **60,** 488
Hydrocarbons
–, C-carbalkoxylation, regiospecific **60,** 315
– special s.
  arenes
  diarylmethanes
  methyl and methylene groups
Hydroformylation
–, β-hydroxyaldehydes via – **60,** 241
–, isomerizing **52,** 296s60
–, regiospecific
– of enolesters **60,** 291
Hydrogenation (s.a. Transfer-hydrogenation)
Hydrogenation, asym.
– in ionic liquid/water 2-phase medium **60,** 33
– of ketones with chirally-modified Ni-catalysts, review **36,** 36s60
–, –, homogeneous (s.a. High pressure hydrogenation, asym., homogeneous) **60,** 29, 36
– via asym. activation-deactivation with 2 chiral auxiliaries **60,** 30
– with ligands, chiral, monodentate **60,** 34
–, heterogeneous
– in protic media **60,** 35
1,5-Hydrogen atom migration 60, 86
Hydrogen bromide
–, controlled release *in situ* **60,** 11, 175
Hydrogen fluoride
–, preparation of organofluorine compds. with –, review **49,** 419s60
– fluoride-base molten salts
–, preparation of organofluorine compds. with –, review **49,** 419s60
Hydrogen iodide 60, 43

–, generation, *in situ* **5,** 63s60
Hydrogen peroxide (s.a. Urea-hydrogen peroxide) **60,** 47, 57, 58, 108
Hydrogermylation
– of ethylene derivs. **43,** 517s60
Hydroperoxides
– special s.
  tert-butyl hydroperoxide
  cumene –
Hydrosilylation (s.a. Radical hydrosilylation)
–, intramolecular-cross-coupling, dual-catalyzed **60,** 406
Hydroxamic acid esters
– from
  carboxylic acids **60,** 142
– special s.
  ethylenehydroxamic acid esters
Hydroxamic acids
– special s.
  aminohydroxamic acids
  hydroxyhydroxamic –
Hydroximinohalides
– startg. m. f.
  nitrile oxides, under neutral conditions **60,** 385
β-Hydroxyacetals, cyclic
– from
  acetals, cyclic **60,** 282
  ethylene derivs., electron-deficient **60,** 282
β-Hydroxyaldehydes
– from
  epoxides, hydroformylation **60,** 241
o-**Hydroxyaldehydes**
– startg. m. f.
  chromans, 2,4-dialkoxy- **60,** 341
  3-chromenes **60,** 427
  –, 3-nitro- **60,** 324
α-Hydroxyalkoximes, chiral 29, 36s60
1,2-*trans*-C-(1-Hydroxyalkyl) glycosides 60, 237
o-**Hydroxyarylacetylenes**
– startg. m. f.
  benzofuran-3-carboxylic acid esters, carbonylation **60,** 318
β-Hydroxycarbonyl compds. (s.a. Aldol..., Reformatskii...)
– special s.
  halogeno-β-hydroxycarbonyl compds.
– startg. m. f.
  (E)-α,β-ethylenecarbonyl compds. **60,** 479
ξ-Hydroxycarbonyl compds. 54, 212s60
α-Hydroxycarboxylic acid amides
– special s.
  amino-α-hydroxycarboxylic acid amides
Hydroxycarboxylic acid esters
– special s.
  dihydroxycarboxylic acid esters
α-Hydroxycarboxylic acid esters
– from
  α-ketoaldehydes **60,** 72
  O-silyl O-alkyl keteneacetals **56,** 106s60
– special s.
  aryl-α-hydroxycarboxylic acid esters
– startg. m. f.
  α-ketocarboxylic acid esters, 2-phase medium **60,** 108
β-Hydroxycarboxylic acid esters (s.a.

Aldol..., Reformatskii...)
– from
  β-ketocarboxylic acid esters, reduction, asym. **60,** 29
– special s.
  alkoxy-β-hydroxycarboxylic acid esters
  ethylene-α-(hydroxymethyl)carboxylic – –
δ-Hydroxycarboxylic acid esters
– special s.
  allene-δ-hydroxycarboxylic acid esters
Hydroxycarboxylic acids
– special s.
  dihydroxycarboxylic acids
α-Hydroxycarboxylic acids
– special s.
  azido-α-hydroxycarboxylic acids
– –, chiral
– as reagent **60,** 238
β-Hydroxycarboxylic acids 47, 576s60
– special s.
  amino-β-hydroxycarboxylic acids
– –, chiral **41,** 37s60; **59,** 45s60
γ-Hydroxycarboxylic acids 47, 576s60
o-α-**Hydroxydiaryls 56,** 490s60
(E,E)-7-Hydroxy-2,4-dienals 60, 428
2-α-Hydroxy-1,3-dienes 54, 320s60
1-Hydroxy-1,1-di(phosphonic acid esters)
–, review **28,** 584s60
2-Hydroxyhydrazines
– special s.
  ethylene-2-hydroxyhydrazines
–, cyclic **46,** 612s60
o-**Hydroxyhydroxamic acids**
– startg. m. f.
  1,2-benzisoxazol-3(2*H*)-ones **60,** 48
δ-Hydroxy-β-ketocarboxylic acid esters
– special s.
  β,γ-diketocarboxylic acid esters, asym. reduction **60,** 22
α-Hydroxyketones
– from
  α-diketones **60,** 25
  epoxides **60,** 58
– special s.
  acylamino-α-hydroxyketones
  amino-α-hydroxyketones
  α,β-dihydroxyketones
  ethylene-α-hydroxyketones
β-Hydroxyketones (s.a. Aldol...)
– from
  3-ethylenealcohols, ring contraction **27,** 162s60
– special s.
  alkoxy-β-hydroxyketones
  ethylene-β-hydroxyketones
  halogeno-β-hydroxyketones
  halogenomethylene-β-hydroxyketones
β-Hydroxyketones, ar.
–, resolution, kinetic **44,** 214s60
ε-Hydroxyketones
– special s.
  ethylene-ε-hydroxyketones
Hydroxylamines
– from
  amines **49,** 85s60
  oximes **60,** 32
– special s.
  acetylenehydroxylamines

N-benzylhydroxylamine
  ethylenehydroxylamines
2-Hydroxylaminoalcohols
 – special s.
  ethylene-2-hydroxylaminoalcohols
Hydroxylation, biochemical
 –, docking and protecting groups in –,
  review **45**, 56s60
2-Hydroxymercaptans
 – special s.
  ethylene-2-hydroxymercaptans
β-Hydroxy-α-methylene...
  s.a. Baylis-Hillman
β-Hydroxy-α-methylenecarboxylic
  acid esters
 – startg. m. f.
  (E)-α,β-ethylene-α-(hydroxymethyl)-
  carboxylic acid esters **60**, 76
γ-Hydroxy-α-methylenecarboxylic acid
  esters
 –, resolution, kinetic **44**, 214s60
γ-Hydroxy-α-methylenesulfoxides,
  chiral **40**, 567s60
Hydroxynitriles
 – startg. m. f.
  lactams **20**, 266s60
α-Hydroxynitriles s. Cyanohydrins
β-Hydroxynitriles
 – from
  ethylene derivs., regiostereoselective
  conversion **60**, 272
 – startg. m. f.
  α,β-ethylenenitriles **60**, 478
β-Hydroxynitriles, chiral **29**, 36s60
N-Hydroxypeptides
 – by 1,5-O→N-acyl group migration
  14, 400s60
α-Hydroxyphosphonic acid esters
 – special s.
  amino-α-hydroxyphosphonic acid
  esters
 – – –, acyclic and macrocyclic
 –, synthesis and properties, review
  41, 556s60
 – – –, chiral **24**, 387s60
N-Hydroxyphthalimide
 – as reagent **60**, 134, 282
1,1-Hydroxysilanes
 – special s.
  ethylene-1,1-hydroxysilanes
2-Hydroxysilanes
 – special s.
  3-ethylene-2-silylalcohols
3-Hydroxysilanes
 – from
  aldehydes, with 2 extra C-atoms
  60, 270
β'-Hydroxy-β-silylcarboxylic acid
  esters
 – special s.
  ethylene-β'-hydroxy-β-silylcarboxylic
  acid esters
o-Hydroxy-β-styryl ethers
 – startg. m. f.
  benzofurans **60**, 113
N-Hydroxysuccinimide esters, polymer-
  based
 –, N-acylation, polymer-based with –
  32, 317s60
β-Hydroxysulfoxides
 – special s.
  ethylene-β-hydroxysulfoxides

p-Hydroxysulfoxides
 – startg. m. f.
  benzofurans, 2,3-dihydro-, 5-subst.
  60, 334
3-Hydroxythioethers
 – special s.
  acetylene-3-hydroxythioethers
Hyponitrous acid esters
 – special s.
  di-*tert*-butyl hyponitrite
Hypophosphorous acid/iodine **60**, 45
Hypophosphorous acid salts
 – special s.
  anilinium hypophosphite
Hypoxanthines
 – startg. m. f.
  adenines **60**, 141

Imidazole
 – as reagent **60**, 90
Imidazole ring
 – from
  aldehydes **46**, 321s59
Imidazoles
 –, N-arylation **42**, 397s60; **53**, 162s60
 –, deuteration **39**, 555s60
Imidazolidines
 – startg. m. f.
  Δ²-imidazolinium salts **32**, 413s60
4-Imidazolidone hyperchlorate,
  5(S)-benzyl-2,2,3-trimethyl-
 – as reagent **60**, 278
2-Imidazolidones, 1-acyl-
 – startg. m. f.
  carboxylic acids **43**, 162s60
 –, 1-(N-alkylideneglycyl)-
 –, C-α-alkylation, asym. **27**, 843s60
β-(2-Imidazolidon-1-yl)ketones, chiral
  60, 443
Δ²-Imidazolinium chloride, 2-chloro-
  1,3-dimethyl-
 – as reagent **58**, 80s60; **60**, 182
Δ²-Imidazolinium salts
 – from
  imidazolidines **32**, 413s60
Imidazolinylidene complexes
 – special s.
  transition metal imidazolinylidene
  complexes
Imidazolium salts
 – as ionic liquids **60**, 136, 405, 438
Δ²-5-Imidazolones, 2-amino-
 –, synthesis, polymer-based **43**, 316s60
Imidazol-2-ylidene hydrochloride, 1,3-
  bis(2,6-diisopropylphenyl)-
 – as reagent **57**, 439s60
Imidazo[1,2-*a*]pyridines **60**, 352
Imines (s.a. Azomethines)
 –, radical addition to –, review **50**, 382s60
 – special s.
  phosphinylimines
  sulfinylimines
  sulfonylimines
Iminoesters, cyclic
 – special s.
  O-alkyllactims

Iminophosphoric acid amides, cyclic,
  polymer-based
 – as reagent **60**, 154, 159, 184
Iminosulfinic acid halides
 – special s.
  N-*tert*-butyliminobenzenesulfinyl
  chloride
Iminyl radicals
 – from
  ketoxime xanthates **49**, 370s60
Indenes
 – from
  aldehydes **60**, 352
  benzotriazoles, 1-benzyl- **60**, 352
 –, synthesis, review **29**, 932s60
Indenes, 1-acylamino- **60**, 352
Indium **43**, 699s60; **60**, 14, 249, 329, 468
 –/ammonium chloride **43**, 60s60
 –(III) bromide **60**, 194
 –(III) chloride **55**, 337s60; **58**, 342s60;
  60, 169, 266, 329, 332, 435
 –(III) compds., organo-
 – special s.
  allylindium(III) compds.
 –(I) halides **43**, 699s60
 – hydride complexes **54**, 27s60
 –(III) iodide/silica gel **60**, 8
Indoles (s.a. Nenitzescu)
 – by substitution, nucleophilic, ar., review
  41, 694s60
 – from
  o-dihalides **60**, 395
  enamines **60**, 395
 –, 3-amino-1,2-diacyl-
 – from
  o-acylaminonitriles and α-
  halogenoketones **60**, 371
 –, 2-α-amino-1-sulfonyl-, chiral
  44, 815s60
 –, 2,3-diaryl-
 – by Suzuki coupling, double **53**, 473s60
 –, tetracyclic, N-condensed **40**, 475s60
Indolizines
 – from
  ethylene derivs., electron-deficient
  60, 474
  pyridinioacetic acids **60**, 474
Induction, asym. (s.a. Synthesis, asym.)
 – via 1,5-hydride shift **60**, 46
 – with α-deuteriobenzylalcohols, chiral
  60, 230
Induction, asym., isotopic **60**, 230
Iodic acid **47**, 440s60
Iodides
 – from
  alcohols **60**, 180
Iodination, ar.
 – of arenes, deactivated **60**, 178
Iodine (s.a. Hypophosphorous acid/iodine,
  Triphenylphosphine/–)
 – as catalyst **60**, 55
 – as reagent **25**, 649s60; **60**, 104, 143,
  341
Iodine compds., organo-, hypervalent
 –, synthetic use in radical reactions,
  review **47**, 262s60
Iodine difluorides
 –, fluorination of ethylene derivs. with –,
  review **36**, 470s60
Iodine monochloride **17**, 630s60;
  42, 200s60
1-Iodo-1-deoxy-2-uloses **60**, 220

(E)-2-Iodoenol sulfonates 55, 183s60
1,2-Iodohydrins, cyclic
– special s.
  acylamino-1,2-iodohydrins, cyclic
α'-Iodolactols
– from
  lactones, with 1 extra C-atom 60, 220
Iodomethylzinc aroxides
–, Simmons-Smith-type reaction,
  stereospecific with – 60, 380
Iodonium salts
– special s.
  diaryliodonium salts
Iodonium ylids
– special s.
  ketoiodonium ylids
γ-Iodophosphonic acid esters 47, 658s60
1,1-Iodosilanes
– startg. m. f.
  ethylene derivs. 60, 496
Iodosobenzene
– as reagent 60, 69, 107
N-Iodosuccinimide
– as reagent 51, 205s60; 60, 64, 95
Ionic liquids
–, alcohols, prim. from aldehydes 60, 21
–, Balz-Schiemann reaction 60, 183
–, benzofurans, 2,3-dihydro- from allyl
  phenolethers 52, 334s60
–, *trans*-cyclopropanecarbonyl compds.
  from ethylene derivs. 60, 410
–, diaryls, sym. from halides, ar.
  27, 870s60
–, Heck arylation 60, 405
–, Negishi diaryl synthesis 38, 836s60
–, nitration, ar. 60, 136
–, resolution, kinetic of alcohols
  44, 214s60
–, Wittig synthesis 60, 438
Ionic liquid-water 2-phase medium
  60, 33
Iridium(I) amidophosphine-phosphinite
  complexes 46, 47s60
–(I) complexes
  chloro(cyclooctadiene)iridium(I) dimer
    60, 303, 313
  cyclooctadiene(β-diketonato)iridium(I)
    complexes 39, 555s60
–(I) –, water-soluble 60, 110
–(I) –/montmorillonite 46, 47s60
Iron 60, 170
Iron π-allyl complexes, acyclic
–, syntheses, stoichiometric with –, review
  55, 35s60
– carbonyl
  diiron nonacarbonyl 60, 115
–(II) chloride 24, 228s60; 60, 208
–(III) – 46, 327s60; 60, 92, 155, 342
–(III) –, polymer-based 38, 100s60
– complexes
  (1,3,5-cycloheptatriene)(1,5-cyclo-
    octadiene)iron(0) 42, 676s60
  μ-oxodiiron(III) complex 26, 463s60
–(III) perchlorate 50, 115s60
–(III) trifluoroacetate 50, 55s60;
  55, 205s60
Isobenzopyrylium-4-olates, 1-alkoxy-
–, cycloaddition, asym. with – 43, 943s60
Isochromenes, 1-vinyl-
– from
  o-halogenocinnamyloxysilanes and
    methyl ketones 60, 370

Isocyanates
– special s.
  sulfonylisocyanates
– startg. m. f.
  phthalimidines 60, 377
  quinolines, 2,4-diamino- (from 3
    molecules) 60, 463
α-Isocyanocarboxylic acid esters
– special s.
  ethyl isocyanoacetate
– startg. m. f.
  tetrazolo[1,5-*a*]piperazin-6-ones
    60, 338
α-Isocyanophosphonic – –
– special s.
  cyano-α-isocyanophosphonic acid
    esters
Isoflavanones, chiral 10, 688s60
Isomerization (s. Cycloisomerization,
  Migration, Rearrangement)
Isonitriles (s.a. Isocyano... and under
  Ugi...)
– special s.
  tert-butyl isocyanide
Isopropyl dichlorovanadate 50, 75s60
Isoquinoline ring, 1,2,3,4-tetrahydro-
–, closure by intramolecular ring opening
  of 1,3-oxazines, tetrahydro- with
    asym. induction 60, 484
Isoquinolinium salts, 3,4-dihydro-, chiral
– as reagent 51, 65s60
(–)-α-Isosparteine
– as reagent 60, 309
Δ³-Isothiazoline 2,2-dioxide ring
– startg. m. f.
  o-ethyleneamines 60, 489
Isothiazolo[4,3-*b*]pyridine 2,2-dioxides,
  1,3-dihydro-
– startg. m. f.
  pyridines, 3-amino-2-vinyl- 60, 489
Isothiocyanates
– from
  2*H*-1,2,3-triazolium-1-imides 60, 160
– startg. m. f.
  guanidines, polymer-based synthesis
    60, 161
  thiazoles, 5-amino- 60, 352
Isoxazoles, polysubst.
–, synthesis, polymer-based 43, 316s60
Isoxazolidine-4-carboxaldehydes, chiral
  60, 278
Isoxazolidines, 5-alkoxy-, chiral
  54, 296s60
5-Isoxazolidones
– from
  nitrones and O-silyl O-alkyl ketene-
    acetals, asym. induction 60, 433
  2-pyrrolidones, N-(α,β-ethyleneacyl)-,
    asym. synthesis 60, 149
–, *anti*-3-α-alkoxy-, chiral 60, 433
Δ²-Isoxazolines
– startg. m. f.
  (E)-α,β-ethyleneoximes 60, 1
Δ⁴-Isoxazolines
– from
  2-acetylenehydroxylamines 60, 75

Ketals
– special s.
  acetone dimethyl acetal
Ketene acetals
– special s.
  O-silyl O-alkyl keteneacetals
Ketene diethyl acetal
– as reactant 60, 261
Ketene equivalents
– special s.
  silylketene equivalents
Ketene mercaptals, cyclic
– special s.
  ketoketene mercaptals, cyclic
Ketenes
–, α-halogenation, asym. via – 60, 184
Ketimines
– special s.
  cyclopropyl ketimines
– startg. m. f.
  (Z)-β-amino-α,β-ethyleneketones
    60, 353
Ketimines, ar.
– from
  halides, ar. 60, 389
α-Ketoaldehydes
– startg. m. f.
  α-hydroxycarboxylic acid esters 60, 72
C-β-(β-Ketoalkyl) glycosides
– from
  aldoses 60, 465
α-Ketocarbonyl compds.
– startg. m. f.
  pyrans, tetrahydro-, 2,2,4,4-tetra-
    alkoxy-, 6-acyl-, asym. conversion
    60, 261
γ-Ketocarbonyl compds.
– from
  α,β-ethylenecarbonyl compds.,
    synthesis via radical carbonylation
    60, 269
  alkoxy-γ-ketocarbonyl compds.
β-Ketocarboxylic acid amides
– special s.
  acetoacetic acid amides
Ketocarboxylic acid esters
– startg. m. f.
  1-aryllactones 60, 366
α-Ketocarboxylic acid esters
– from
  α-hydroxycarboxylic acid esters,
    2-phase medium 60, 108
– special s.
  aryl-α-ketocarboxylic acid esters
  ethylene-α-ketocarboxylic – –
β-Ketocarboxylic – –
–, α-fluorination, asym. 60, 179, 186
–, α-halogenation, – 60, 179
– special s.
  β,δ-diketocarboxylic acid esters
  ethylene-β-ketocarboxylic –
  δ-hydroxy-β-ketocarboxylic – –
– startg. m. f.
  β-hydroxycarboxylic acid esters,
    reduction, asym. 60, 29
2-Ketoiodonium ylids
–, [3+2]-cycloaddition, regiospecific
  with – 60, 361
α-Ketoketene mercaptals, cyclic
– special s.
  1,3-dithiolanes, 2-(α,β-ethylene-
    acylene)-

Ketones (s.a. C-Acylation, Carbonyl compds., Hydroacylation, Oxo compds.)
–, α-fluorination, asym. **60**, 186
– from
  acetylene derivs. **60**, 66
  – –, via thioenolethers **7**, 217s60
  alcohols, sec. s. under Oxo compds.
  boronic acids **60**, 457
  methylene groups **60**, 109
  thiolic acid esters **60**, 457
–, radical C-α-alkylation, redoxidative, co-catalytic (with ethylene derivs.) **60**, 279
–, reduction s. under Oxo compds.
– special s.
  acetyleneketones
  acylaminoketones
  alkoxyketones
  alleneketones
  aminoketones
  aryl ketones
  arylketones
  azidoketones
  (benzotriazol-1-yl)ketones
  cyanoketones
  diazoketones
  diketones
  epoxyketones
  ethyleneketones
  halogenoketones
  hydroxyketones
  (2-imidazolidon-1-yl)ketones
  nitroketones
  (pyridyloxy)ketones
  silylketones
  stannylketones
  sulfonyloxyketones
– startg. m. f.
  acoxy compds., asym. reduction **60**, 23
  β-aminoketones, N-protected, asym. conversion **60**, 336
  benzene ring **60**, 464
  1*H*-1,5-benzodiazepines, 2,3-dihydro- **60**, 358
  enamines **60**, 164
  enol triflates **60**, 49
  α,β-ethylene-α-formamidocarboxylic acid esters, with 2 extra C-atoms **60**, 476
  (E)-α,β-ethyleneketones, dehydrogenation **60**, 477
  furans **60**, 374
  pyridines **60**, 351
  – (from 2 molecules) **60**, 335
  xanthones **60**, 323
Ketones, cyclic
–, fluorination, asym. **60**, 186
– special s.
  aminoketones, cyclic
– startg. m. f.
  enolethers, cyclic **60**, 89
–, *o*-subst.
– from
  aldehydes, ar. and ethylene derivs. **60**, 360
α-Ketonitriles s. Acylcyanides
β-Ketosulfoxides, chiral 24, 100s60; **39**, 83s60
3-Ketothioethers
– from
  α,β-ethyleneketones **60**, 199

N-(γ-Keto)thiolactams
– startg. m. f.
  4-pyridone ring, 2,3-dihydro- **60**, 491
Ketoxime xanthates
– startg. m. f.
  iminyl radicals **49**, 370s60
Kharasch acoxylation, heterogeneous 14, 211s60
Knoevenagel condensation (s.a. Aza-Knoevenagel condensation)
– with aldoses, in aq. medium **60**, 465
– –, uncatalyzed **15**, 571s60
Kulinkovich reaction, modified 60, 384

Lactams
– from
  azidocarboxylic acid esters **52**, 13s60
  carbamyl halides, unsatd. **60**, 455
  hydroxynitriles **20**, 266s60
– special s.
  N-acyllactams
  halogenolactam...
– startg. m. f.
  O-alkyllactims **60**, 89
Lactams, N-protected 53, 473s60
β-Lactams s. 2-Azetidinones
γ-Lactams s. 2-Pyrrolidones
Lactolides
– special s.
  aryl lactolides
  halogenolactolides
  sulfonyllactolides
γ-Lactolides
– special s.
  alkylidene-γ-lactolides
δ-Lactolides
– special s.
  pyran-2-yl ethers, tetrahydro-
Lactols
– special s.
  halogenolactols
–, benzo-condensed, tricyclic **60**, 366
–, bicyclic
– startg. m. f.
  lactones **21**, 177s60
Lactones
– from
  lactols, bicyclic **21**, 177s60
– special s.
  (2-alkoxyvinyl)lactones
  aryllactones
  vinyllactones
– startg. m. f.
  cyclopropanols, 1-ω-hydroxy- **60**, 384
  α-iodolactols, with 1 extra C-atom **60**, 220
Lactones, chiral 33, 865s60
β-Lactones
– special s.
  alk-1-ynyl-β-lactones
  dihalogeno-β-lactones
γ-Lactones
– special s.
  alkoxyamino-γ-lactones
  alkylidene-γ-lactones
  amino-γ-lactones

methylene-γ-lactones
spiro-γ-lactones
–, anomerically fused
  – as intermediates **60**, 121
–, α,γ-disubst., chiral **60**, 46
Lanthanides
–, reductive chemistry, review **42**, 993s60
Lanthanum(III) chloride 55, 337s60
Lanthanum triisopropoxide 60, 333
Lead/alumina 39, 555s60
–(II) chloride **58**, 371s60
– tetraacetate **7**, 761s60
–(II) triflate/1,1'-binaphthol crown ether, chiral **60**, 444
Lewis acid catalysis
– under bidentate chelation **60**, 27
Libraries of compds. s. Catalyst libraries, Compound –
Ligands, chiral
–, exchange, review **55**, 297s60
Lipases s. under Enzymes
Lithiation s. Metalation
1-Lithioenolethers
– as acyl carbanion equivalents **60**, 175
Lithium 60, 396
Lithium/N,N'-dimethylethylenediamine/ *tert*-butylamine
– as reductant **5**, 32s60
– **alkoxides**
  – *tert*-butoxide **60**, 156
– **alk-1-ynyl(trialkoxy)borates**
–, cross-coupling, Pd-catalyzed with –
  60, 460
– startg. m. f.
  arylacetylenes **60**, 460
  1,3-enynes **60**, 460
–, Suzuki-Miyaura coupling via – **60**, 398
– **amides**
  – bis(trimethylsilyl)amide **60**, 245, 322, 369, 370, 485
  – N,N,N-trimethyl-2-aminoethylamide **60**, 368
–, N-nitrosation **10**, 251s60
– **aryl(trimethoxy)borates**
–, 1,4-addition, asym. with – **55**, 452s60
– **bis(methylenecyclopropyl)cuprate**
– as reactant **41**, 638s60
– **bromide 55**, 183s60; **56**, 73s60; **60**, 112, 130
– **chloride 55**, 183s60; **60**, 174, 264
– **compds., organo-** (s.a. Carbolithiation)
  sec-butyllithium/TMEDA **60**, 223
  tert-butyllithium **60**, 367, 484
– **cyanide 60**, 326
– **dialkylcuprates**
  – as reactant **60**, 325, 326
– **hypochlorite 35**, 182s60; **51**, 132s60
– **nitrate 39**, 189s60
– **perchlorate 21**, 739s60
– **silylethynolates**
  – as silylketene equivalents, nucleophilic **60**, 243
– **tetrahydridoaluminate/Cp*₂TiCl₂ 44**, 48s60
– **tetrakis(pentafluorophenyl)borate**
  – as reagent **60**, 79
– **trialkylmagnesiates**
  – as reagent **60**, 378
– **trialkylzincates, unsym.**
–, 1,4-addition-alkylation with – **60**, 263
– **tri-*n*-butylmanganate**
  – as reagent **43**, 517s60

Lithium trifluoromethanesulfonate
46, 659s60; **60,** 79
– ynolates
– special s.
lithium silylethynolates

Macrocyclization, transannular
49, 836s60
Magnesiates s. Lithium trialkylmagnesiates
Magnesium
Rieke magnesium **60,** 208
Magnesium bromide 60, 26, 234
– chloride **43,** 402s60; **60,** 208, 375
– O,C-dianions, chloro-
– as intermediates **60,** 478
– halides, organo- (s.a. Grignard...)
methylmagnesium chloride **60,** 478
isopropylmagnesium bromide **60,** 377
–, preparation, transition metal-catalyzed
60, 208
– special s.
arylmagnesium bromides
vinylmagnesium bromide
– oxide **60,** 79, 168
– perchlorate **60,** 149
Maleic anhydrides
– from
acetylenedicobalt hexacarbonyl
complexes **60,** 436
Malonic acid esters
– special s.
acetylenemalonic acid esters
Malononitriles
–, C-arylation, Ni-catalyzed **60,** 388
– special s.
siloxymalononitriles
(R)-Mandelic acid
– as reagent **60,** 238
Mandelic acid esters
– from
arenes, asym. conversion **60,** 227
Manganacyclics
– as intermediates **60,** 375
Manganates, organo-
– special s.
dilithium trialkyl(stannyl)manganates
lithium tri-n-butylmanganate
trialkylmanganates
Manganese(III) acetate 42, 636s60;
60, 279, 316
Manganese complexes
5,10,15,20-tetrakis(pentafluorophenyl)-
porphyrinatomanganese(III)
chloride **56,** 162s60
– dioxide **60,** 474
– –/bentonite **18,** 302s60
–(III) salen complexes, chiral, fluorous
60, 69
–(III) – –, silica-supported, chiral
46, 106s60
Mannich reaction
– in dispersions, aq. colloidal **60,** 340
– –, asym., amine-catalyzed
– by 3-component reaction **60,** 336
– –, double

–, β-tert-amino-α-methyleneketones by –
60, 337
– with dimethyl(methylene)ammonium
chloride **60,** 337
Mannich-type reaction
–, alternative **60,** 382
– –, asym.
– with α-aminosulfones, N-protected,
with 2 extra C-atoms **60,** 408
Mercaptals (s.a. Transthioacetalation)
– from
aldehydes **60,** 194
– startg. m. f.
enolethers, polymer-based synthesis
60, 113
Mercaptans
– from
alcohols, with inversion **60,** 196
trifluoroacetic acid esters, – – **60,** 196
– special s.
aminomercaptans
arylmercaptans
hydroxymercaptans
o-**Mercaptoaldehydes**
– as intermediates **60,** 368
Mercaptothioethers
– startg. m. f.
disulfides, cyclic, polymer-based
synthesis **60,** 188
Mercury chloride 13, 366s60
Mercury compds., organo-
– special s.
ferrocenylmercury compds.
Mercury perchlorate 60, 172
Mercury triflate 11, 224s60
Metalation (s.a. Carbometalation)
– of
N-allenyl-2-imidazolidones and
-2-oxazolidones **31,** 806s60
azines and diazines, review **42,** 597s60
p-**Metalation**
– of arylcarboxylic acid amides **60,** 223
Metals, highly reactive
–, preparation, review **50,** 514s60
Metathesis (s.a. Cross-metathesis, Ene-
yne metathesis, Ring-closing
metathesis, Ring-opening metathesis...,
also under Interchange in Vol. **1-50**)
– using metal carbene complexes,
nucleophilic, review **56,** 495s60
Methanesulfonic acid esters
–, Friedel-Crafts alkylation with – **60,** 327
– special s.
benzyl mesylate
Methoxyallene
– as reactant **60,** 271
p-**Methoxybenzylalcohols**
– startg. m. f.
oxo compds. **60,** 114
p-**Methoxybenzyl trityl ether**
–, O-tritylation with – **60,** 101
p-**Methoxyphenyl benzenethiosulfinate**
– as reagent **54,** 92s60
o-**Methoxyphenylphosphonic acid**
diamides, bicyclic, chiral
– as reagent **60,** 173
Methylarenes
– from
halides, ar. **60,** 381
2-Methylene-sec-alcohols 60, 97
α-**Methylenealdehydes,** in situ-
generated

–, Diels-Alder reaction with – **60,** 254
– from
1,3-dioxins **60,** 254
α-**Methylenecarboxylic acid esters**
– from
(Z)-α,β-ethylene-α-(halogenomethyl)-
carboxylic acid esters, synthesis,
asym. **60,** 379
– special s.
hydroxy-α-methylenecarboxylic acid
esters
Methylene chloride
– as reactant **60,** 365
3-Methylene-1,5-diols 47, 576s60
Methylene groups
– startg. m. f.
ketones **60,** 109
Methylene iodide
– as reactant **60,** 220, 380
α- or γ-**Methyleneketones**
– special s.
amino-α- or -γ-methyleneketones
α-**Methylene-γ-lactones, chiral**
44, 214s60
α-**Methylenesulfoxides**
– special s.
hydroxy-α-methylenesulfoxides
Methyl glyoxylate
– as reactant **60,** 332
Methyl groups
– startg. m. f.
aldehydes **60,** 109
Methyl ketones
– startg. m. f.
β-tert-amino-α-methyleneketones
60, 337
isochromenes, 1-vinyl- **60,** 370
Methyl methoxyacetate
– as reactant **60,** 252
N-Methylmorpholine
– as reagent **60,** 71
– **N-oxide, polymer-based**
– as reagent **58,** 286s60
Methyltrichlorosilane
– as reagent **60,** 17
Micelles, aq.
–, cycloaddition, 1,3-dipolar in – **60,** 363
Michael addition (s.a. Anti-Michael
addition, Aldol condensation-Michael
addition, 1,4-Addition)
–, pyridines via – **60,** 342
Michael addition, asym.
– to α,β-ethylenecarboxylic acid esters
60, 191
– with
aryl mercaptans **60,** 191
phosphorous acid diesters, cyclic
60, 210
– –, –, catalytic
–, review **59,** 206s60
– –, –, proline-catalyzed
– of nitro compds., aliphatic **60,** 267
– to α,β-ethyleneketones **60,** 267
– –, double
– of dicarboxylic acids, tethered to α,β-
acetyleneketones, review **29,** 665s60
– –, intramolecular, reductive **60,** 170
– –, iron(III) chloride-catalyzed
–, review **46,** 667s60
– –, tandem
–, review **59,** 407s60
– –, vinylogous, iron(III) chloride-

catalyzed
-, review **46**, 667s60
Michael addition-intramolecular N-acylation 60, 320
- **addition-intramolecular carbopalladation, stereospecific 60**, 402
Michael-type addition, catalytic
- to 1,1-nitroethylene derivs. **60**, 449
- with enoxystannanes **60**, 449
Microwave irradiation
-, N-arylation (in water) **60**, 153
-, cleavage of tetrahydropyran-2-yl ethers **59**, 9s60
-, dehalogenation **52**, 38s60
-, C-dimethylaminomethylenation **26**, 247s60
-, 1,3-dioxolanes, 2-vinyl- from enol triflates **54**, 374s60
-, C-glycosides by Wittig synthesis **29**, 861s60
-, Heck arylation in aq. media **27**, 871s60
-, - - of 2-ethyleneamines, N-protected **60**, 344
-, hydrostannylation-Stille-coupling **56**, 307s60
-, 2-oxazolidones from N-carbo-*tert*-butoxyaziridines **59**, 123s60
-, pyridine ring from enamines **24**, 836s60
-, tetrazoles from nitriles **13**, 371s60
-, triaryl phosphates, sym. from phenols **13**, 156s60
- -, solventless
-, N-allylation-intramolecular Diels-Alder reaction **59**, 335s60
-, carboxylic acid thioamides from amides **34**, 525s60
-, β-ketocarboxamides from esters **12**, 453s60
-, nitriles from aldehydes **55**, 146s60
-, nitriles, ar. from bromides, ar. **29**, 845s60
-, Δ²-oxazolines from aldehydes **60**, 319
-, O-tetrahydropyran-2-ylation **56**, 73s60
-, 1,3,4-thiadiazoles, 2-acylamino- **12**, 690s60
- - -, solid-supported
-, N-aminomethylation **19**, 419s60
-, aryloxo compds. from benzylalcohols **42**, 236s60
-, carboxylic acid amides, N-subst. from acids **19**, 419s60
-, - - acids from esters **60**, 8
-, 1,3-diynes, sym. from acetylene derivs., terminal **16**, 780s60
-, flavones from *o'*-hydroxychalcones **2**, 288s60
-, imidazoles from α-diketones **23**, 423s60
-, nitriles from hydrazones **22**, 408s60
-, N-oxidation **49**, 85s60
-, oximes, cleavage **47**, 146s60
-, protection of carbonyl groups as 1,3-oxathiolanes **60**, 202
-, *p*-quinones from *p*-quinols **18**, 302s60
-, Suzuki diaryl coupling **37**, 902s60
Mitsunobu reaction-Thorpe cyclization 60, 339
Molybdate s. Ammonium molybdate
Molybdate, tetrathio- s. Benzyltriethylammonium tetrathiomolybdate
Mono-N-alkylation, partial and preferential 60, 152

Monothiolphosphoric acid esters
- startg. m. f.
  nitriles **60**, 413
Montmorillonite KSF 57, 102s60; **60**, 202
Morpholine/silica gel 4, 702s60
Morpholines, N-carbalkoxy- 50, 257s60

Naphthalenes
- from
  o-acetyleneketones **60**, 472
  o-diynes **48**, 665s60
  enamines **60**, 472
  enolethers **60**, 472
Negishi diaryl coupling
- in ionic liquids **38**, 836s60
Nenitzescu indole synthesis, polymer-based 8, 782s60
Nicholas-type substitution, regiospecific
- of ethynylcarbinols **60**, 83
Nickel, activated 20, 356s60
Nickel, chirally-modified
-, hydrogenation, asym. of ketones over
-, review **36**, 36s60
Nickel/carbon 60, 156
Nickel(II) acetate 51, 76s60
-**(II) acetoacetonate 42**, 676s60; **52**, 297s60
-**(II) -, polybenzimidazole-based 47**, 113s60
- **complexes**
  bis(cyclooctadiene)nickel(0) **60**, 283, 284, 285
  [1,2-bis((2R,5R)-2,5-dimethylphospholan-1-yl)benzene]nickel(II) diiodide **34**, 668s60
  dibromo[1,2-bis(diphenylphosphino)ethane]nickel(II) **52**, 255s60
  dibromobis(triphenylphosphine)nickel(II)/zinc **60**, 388
  dichlorobis(triphenylphosphine)nickel(II) **27**, 870s60; **60**, 376
  tris(2,2'-bipyridyl)nickel(II) fluoroborate **60**, 16
  tris(ethylenediamine)nickel thiosulfate **60**, 88
-**(II) copper(II) formate 60**, 144
-**(II) iodide 50**, 531s60
- **peroxide 25**, 649s60
o-**Nitramines**
- startg. m. f.
  benzimidazoles, polymer-based synthesis **60**, 139
Nitration
- via alkyl radicals **60**, 134
Nitration, ar.
- in ionic liquids **60**, 136
Nitrenes
- as intermediates in heterocyclic ring closure, review **52**, 274s60
Nitric acid/bentonite 18, 302s60
Nitrile oxides
-, cycloaddition, 1,3-dipolar, stereospecific with ethylene derivs.,
  review **16**, 888s60

- from
  hydroximinohalides, under neutral conditions **60**, 385
Nitriles
- from
  aldehydes **60**, 143, 144
  aldoximes **60**, 169
  amines, prim., oxidation, aerobic **60**, 167
  cyanohydrins **31**, 49s60
  monothiolphosphoric acid esters **60**, 413
- special s.
  acoxynitriles
  (O-acyloximino)nitriles
  (acylthio)nitriles
  aminonitriles
  cyano...
  dinitriles
  epoxynitriles
  ethylenenitriles
  halogenonitriles
  siloxynitriles
  sulfinylaminonitriles
- startg. m. f.
  carboxylic acid amides **60**, 167
  oxazole ring **60**, 361
Nitrile ylid equivalents
-, cycloaddition, 1,3-dipolar, regiospecific under microwaves with –
  60, 319
Nitrilimines
-, cycloaddition, 1,3-dipolar with – in aq. micelles
-, -, -, intramolecular, asym. with –
  17, 871s60
2-Nitroalcohols s.a. Henry reaction
anti-**2-Nitroalcohols**
- from
  α-nitroketones **60**, 19
o-**Nitroazides**
- startg. m. f.
  1*H*-1,5-benzodiazepines, 2,5-dihydro- **60**, 358
α-**Nitrocarbonyl compds.**
- special s.
  diazo-α-nitrocarbonyl compds.
Nitrocarboxylic acid esters
- special s.
  ethylene(nitro)carboxylic acid esters
Nitro compds.
- special s.
  acoxynitro compds.
Nitro compds., aliphatic
-, β,β'-disubstitution, electrophilic, sequential **60**, 446
-, Michael addition, asym., proline-catalyzed with – **60**, 267
Nitro compds., ar.
- from
  amines, ar., prim. **60**, 47
- startg. m. f.
  amines, ar., prim. **60**, 15
2-Nitro-2-deoxyglycosylamines 43, 455s60
1,3-Nitroethers
- from
  N,N-disiloxyenamines and acetals **60**, 446
Nitroethylene... s.a. Ethylenenitro...

1,1-Nitroethylene derivs.
-, 1,4-addition, asym. to – **60**, 259
-, -, -, Rh-catalyzed to – **60**, 452
-, epoxidation, asym. **52**, 68s60
-, Michael addition, asym. of azomethines to – **43**, 607s60
-, Michael-type addition, catalytic to – **60**, 449
- startg. m. f.
  α-*tert*-aminooximes **60**, 140
  benzene ring (from 3 molecules) **60**, 354
  3-chromenes, 3-nitro- **60**, 324
  β-nitrophosphonic acid esters, asym. conversion **60**, 210
(E)-α-Nitro-β,γ-ethylenesulfones
- from
  aldehydes, with 1 extra C-atom **60**, 321
Nitrogenation s. Carbonylation-nitrogenation
Nitrogen dioxide
- as reactant **60**, 134
- monoxide **39**, 171s60; **47**, 470s60
α-Nitroketones
- startg. m. f.
  anti-2-nitroalcohols **60**, 19
Nitrones
-, cycloaddition, 1,3-dipolar, asym., amine-catalyzed with – **60**, 278
- startg. m. f.
  3-ethylenehydroxylamines, with 3 extra C-atoms **60**, 249
  5-isoxazolidones, asym. induction **60**, 433
α-Nitrooximes
- startg. m. f.
  furoxans **60**, 48
***p*-Nitrophenyl carbonate, polymer-based**
- as reactant **60**, 159
β-Nitrophosphonic acid esters
- from
  1,1-nitroethylene derivs., asym. conversion **60**, 210
- acids, chiral **60**, 210
N-Nitrosation
- of amines, sec. **60**, 116
Nitroso compds.
-, ene reaction, regiostereospecific, hydroxyl-directed with – **60**, 118
α-Nitrosulfones
- special s.
  α-nitro-β,γ-ethylenesulfones
  phenylsulfonylnitromethane
Nitrosyl chloride
-, generation *in situ* **60**, 116
Nitrous acid esters
- special s.
  tert-butyl nitrite
Norbornenes, 5-subst.
-, [2+2]-cycloaddition, regiospecific, Ru-catalyzed with – **60**, 288
Nucleic acids
- special s.
  peptidyl nucleic acids
Nucleoside bases
-, exchange **60**, 121
Nucleoside H-dithiophosphonates **52**, 208s60
- 3'-monothiolphosphonates **31**, 520s60
Nucleosides
-, modification, selective, biocatalytic, review **17**, 169s60

-, O-phosphorylation **60**, 52
- special s.
  deoxynucleosides
-, carbocyclic
-, modification, selective, biocatalytic, review **17**, 169s60
C-Nucleosides
-, modification, selective, biocatalytic, review **17**, 169s60
Nucleoside thiolphosphates, 2',3'-cyclic **38**, 86s60
- thiolthionophosphonates **31**, 520s60

Oligoglycosides, β-(1→6)-linked
-, synthesis, polymer-based **39**, 189s60
Oligosaccharide synthesis, polymer-based 20, 387s60; **57**, 99s60 (review)
- –, –, automated
-, review **57**, 99s60
Oppenauer-type oxidation
- in aq. media **60**, 110
α-(Organoseleno)carboxylic acid esters
- special s.
  halogeno-α-(organoseleno)carboxylic acid esters
Osmate s.a. Potassium osmate
Osmate, silica-supported 60, 71
Osmium carbonyls
  triosmium dodecacarbonyl **60**, 40
4-Oxa-5-azabicyclo[4.4.0]dec-8-en-3-ones, 6-cyano-
- as intermediates **60**, 314
2-Oxa-8-azabicyclo[3.3.0]octanes, 4-alkylidene- 38, 965s60
2-Oxabicyclo[2.2.2]oct-5-en-3-ones, 8-arylseleno-
- startg. m. f.
  2-oxabicyclo[3.2.1]oct-6-en-3-ones **60**, 495
2-Oxabicyclo[3.2.1]oct-6-en-3-ones
- from
  2-oxabicyclo[2.2.2]oct-5-en-3-ones, 8-arylseleno- **60**, 495
- startg. m. f.
  cis-2-oxabicyclo[3.3.0]oct-7-en-3-ones **60**, 495
***cis*-2-Oxabicyclo[3.3.0]oct-7-en-3-ones 60**, 495
9-Oxa-10-borabicyclo[3.3.2]decane, 10-methyl-
- as reactant **46**, 899s60
1,2,4-Oxadiazoles, 5-aryl-
- from
  amidoximes and diaryliodonium salts, carbonylation **60**, 401
4-Oxa-1,7-dienes
-, Claisen rearrangement, Ir-catalyzed **60**, 313
- startg. m. f.
  γ,δ-ethyleneldehydes **60**, 313
4-Oxa-1,8-dienes
- from
  ethylene derivs., electron-deficient **60**, 346
4-Oxa-1,8-enynes
-, cycloisomerization **59**, 322

Oxalic acid aryl esters
- startg. m. f.
  carboxylic acid aryl esters **60**, 82
Oxalyl chloride
- as reactant **60**, 445
Oxamic acid esters
- from
  acetoacetic acid amides **60**, 100
1,3-Oxathianes, 2-phosphino-, bicyclic, chiral
- as reagent **60**, 281
1,3-Oxathiolanes (s.a. under Protection)
-, 5-vinyl-
- as intermediates **60**, 331
1,3-Oxazine-2,6-diones
-, radical 1,4-addition, intramolecular, stereospecific to – **60**, 486
4H-1,3-Oxazines, 5,6-dihydro-
- special s.
  bis(4H-1,3-oxazines, 5,6-dihydro-), phosphino-
1,3-Oxazines, tetrahydro-
-, ring opening, intramolecular with 1,2,3,4-tetrahydroisoquinoline ring closure, asym. induction **60**, 484
Oxazole ring
- from
  nitriles **60**, 361
Oxazoles
- from
  α-formamidoacetals **12**, 136s60
Oxazoles, 5-alkoxy-
- from
  α-(alkylideneamino)carboxylic acid esters **60**, 106
-, 5-(carbalkoxyoxy)-
-, N-acylation, asym. with – **60**, 138
2,5-Oxazolidiones
-, Friedel-Crafts α-*prim*-aminoacylation with – **60**, 470
2-Oxazolidones
- from
  ethylene derivs., asym. conversion **60**, 166
  urethans, – induction **60**, 168
-, 4-aryl-, chiral **60**, 166
-, 3-carbo-*tert*-butoxy-
- from
  2-aminoalcohols **60**, 135
-, (Z)-4-(α-chloroalkylidene)- **59**, 195s60
γ-(2-Oxazolidon-4-ylidene)oxo compds., N-protected 60, 296
Δ²-Oxazolines
- from
  aldehydes, under microwaves **60**, 319
- special s.
  bis(Δ²-oxazolines)
Δ²-Oxazolines, 2-acyl- 50, 73s60
-, 2-[2-(arylseleno)ferrocenyl]-, chiral
- as reagent **48**, 772s60
-, phosphino-
- as reagent **46**, 738s60
-, 2-(*o*-phosphinoaryl)-, chiral
- as reagent **56**, 295s60
-, 2-(2-pyridyl)-, polymer-based, chiral
- as reagent **48**, 772s60
Δ²-5-Oxazolones
- startg. m. f.
  α-acylaminocarboxylic acid esters, asym. conversion **50**, 54s60
Oxepans, 3-bromo- 21, 606s60

Oxepans, 2-α-hydroxy- 54, 71s60
2-Oxetanones s. β-Lactones
N-Oxides, cyclic
–, deoxygenation 60, 499
Oxido compds. s. Epoxides in Vol. 51-60
Oximes
– special s.
  aldoximes
  alkoxyoximes
  aminooximes
  ethyleneoximes
  nitrooximes
– startg. m. f.
  acylamines 60, 14
  heterocyclics, review 44, 241s60
  hydroxylamines 60, 32
Oximes, O-acyl- s. O-Acyloximes
–, O-alkyl- s. Alkoximes
Oxindoles
– special s.
  bi(oxindoles)
Oxindoles, 3-alkylidene- 60, 455
α'-(Oxindol-3-yl)ketones, α-aminomethyl-α,β-ethylene- 53, 429s60
α-Oxocarboxylic acid esters
– startg. m. f.
  α-aryl-α-hydroxycarboxylic acid
  esters, asym. conversion 60, 227
Oxo compds (s.a. Aldehydes, Carbonyl
  compds., Ketones)
– from
  alcohols (s.a. OCⅡH) 60, 107, 114
  – (aldehydes) 60, 105
  – (ketones) 60, 163
  α,β-ethylenesulfoxides 60, 96
  glycols, dehydration (of 1,1-disubst.
    glycols) 60, 111
  hydrazones, N,N-dimethyl- 60, 88
  p-methoxybenzylalcohols 60, 114
– special s.
  aryloxo compds.
  ethyleneoxo –
  halogenooxo –
– startg. m. f.
  ω-acylaminoketones 60, 251
  alcohols, reduction (s.a. HCⅡOC)
  – (prim. alcohols), – in ionic liquids
    60, 21
  – (sec. –), –, asym., enzymatic 60, 24
  – (– –), –, – with modified Ni-catalysts,
    review 36, 36s60
  – (– –), synthesis, asym. 42, 616s60
    (update); 44, 565s60 (update);
    60, 299 (without solvent), 230
  (E)-alkoximes 60, 151
  1,3-dithiolanes 60, 194
  (E)-ethylene derivs., aza-Horner
    synthesis 60, 424
  α-siloxycarboxylic acid amides, asym.
    induction with 1 extra C-atom
    60, 466
  2-siloxy-1,1,1-trifluorides 60, 224
  thiophenes, tetrahydro-, 3-acyl-, asym.
    induction 60, 331
  2,2,2-trihalogenoalcohols, with 1 extra
    C-atom 60, 225
Oxyamination, asym.
– special s.
  Sharpless oxyamination, asym.
Oxypalladation, intramolecular-
  carbonylation 60, 318
Ozonolysis

– of α,β-ethylenefluorides 32, 180s60

σ-Palladacyclics, 5-membered, cationic
– as intermediates 60, 344
Palladation s.a. Aminopalladation,
  Carbopalladation..., Oxypalladation
Palladium 60, 13
  nanoparticles 1, 13s60; 37, 902s60
Palladium/silica 47, 113s60
Palladium(II) acetate 19, 788s60;
  35, 549s60; 46, 738s60; 47, 234s60;
  53, 429s60; 54, 244s60; 56, 75s60,
  295s60; 57, 317s60; 60, 130, 214, 295,
  296, 297, 344, 390, 391, 392, 405, 454,
  455
–(II) alkoxides, organo-
– special s.
  allenylpalladium(II) alkoxides
Palladium-catalyzed reactions
– in a 2-phase medium with soluble
  polymer-based catalysts 60, 399
Palladium(II) chloride 51, 171s60;
  54, 466s60; 60, 37, 129, 158, 301, 317,
  400, 401, 461
– complexes
  η³-allyl(chloro)(tricyclohexyl-
    phosphine)palladium(II) 60, 308
  bis(acetonitrile)[(S)-2,2'-bis(di-p-
    tolylphosphino)-1,1'-binaphthyl]-
    palladium(II) hexafluoroantimonate
    44, 568s60
  bis(acetonitrile)dichloropalladium(II)
    57, 456s60
  bis(allylpalladium chloride) 60, 145
  bis(benzonitrile)dichloropalladium(II)
    27, 851s60; 60, 174
  bis(dibenzylideneacetone)palladium(0)
    52, 128s60; 53, 86s60; 60, 211, 298,
    406
  [(R)-2,2'-bis(diphenylphosphino)-1,1'-
    binaphthyl]palladium bistriflate
    60, 128
  [2,6-bis(diphenylphosphinoxy)-
    4-methylphenyl]palladium(II)
    trifluoroacetate 56, 445s60
  bis(3-methylbenzothiazolin-2-
    ylidene)palladium(II) diiodide
    60, 405
  bis(tricyclohexylphosphine)pal-
    ladium(0) 60, 397
  2-[1-(tert-butylthio)ethyl]phenylpal-
    ladium(II) chloride dimer 56, 445s60
  trans-di(μ-acetato)bis[o-(di-o-tolyl-
    phosphino)benzyl]dipalladium(II)
    60, 457
  di-μ-chlorobis(benzophenone oxime-
    6-C,N)dipalladium(II) 57, 416s60
  dichlorobis(diethyl sulfide)pal-
    ladium(II) 56, 445s60
  dichloro[1,1'-bis(diphenylphosphino)-
    ferrocene]palladium(II) 60, 218
  dichlorobis(triphenylphosphine)pal-
    ladium(II) 27, 851s60; 32, 820s60;
    60, 175, 204, 302, 402, 403, 404, 488
  (phenanthroline)methylpalladium(II)
    chloride 38, 706s60

tetrakis(triphenylphosphine)pal-
  ladium(0) 29, 854s60; 56, 307s60;
  60, 215, 300, 346, 347, 398, 412, 458,
  459, 460
tris(dibenzylideneacetone)dipalladium
  52, 305s60; 59, 447s60; 60, 84, 157,
  171, 299, 345, 370, 394, 395, 396,
  457
Palladium complexes, fluorous
  dichlorobis(triarylphosphine)pal-
    ladium(II), fluorous 60, 462
– –, supported
  bis(pyridyl)palladium(II) dichloride,
    polysiloxane-supported 42, 676s60
  dichloropalladium(II), polymer-based
    37, 902s60
  (2-pyridylaldimine)palladium(II),
    silica-supported 46, 799s60
–(II) hydroxide/carbon 60, 15
–(II) iodide/thiourea 60, 318
– phosphine complexes, polymer-based,
  soluble
–, reactions with – in a 2-phase medium
  60, 399
–(II) residues
–, removal on work-up 53, 473s60
–(II) trifluoroacetate 49, 699s60;
  60, 297
Pauson-Khand reaction, asym.
– using P,S-ligands 60, 281
– –, –, Ir-catalyzed 60, 303
– –, intramolecular (s.a. C-Allylation-
  intramolecular Pauson-Khand reaction,
  Aminocarbonylation-intramolecular
  – –)
– reaction, intramolecular-Diels-Alder
  reaction
– with diynes 60, 280
Penams, 6-acyl-, chiral 60, 473
1,2,3,4,5-Pentaphenyl-1'-(di-tert-
  butylphosphino)ferrocene
– as ligand 52, 128s60; 53, 86s60
Pentasaccharides, ethylene-bridged
–, synthesis, iterative 18, 912s60
–, linear 44, 191s60
Peptide isosteres 45, 209s60
Peptides
– special s.
  dipeptides
  glycopeptides
  N-hydroxypeptides
  polypeptides
Peptides, cyclic
– special s.
  dipeptides, cyclic
–, α-fluoroalkylated 45, 209s60
–, natural, bioactive, unusual
–, synthesis, review 19, 33s60
–, β-turn
– as reagent 60, 27
Peptide synthesis 52, 164s60
– in solution, review 19, 33s60
– with 2-(3-oxidobenzotriazol-1-yl)-
  1,1,3,3-tetramethyluronium
  fluoroborate, polymer-based
  28, 144s60
– –, contemporary
–, review 19, 33s60
– –, solid-phase
– –, update 19, 33s60
– using 4-(4,6-dimethoxy-1,3,5-triazin-2-
  yl)-4-methylmorpholinium chloride
  41, 344s60

Peptide synthesis
- with carboxylic acids, fluorescent 19, 33s60

Peptidyl aldehydes, C-terminal 19, 33s60
- α,β-epoxyketones 19, 33s60
- α,β-ethylenesulfones 19, 33s60
- nucleic acids
-, review 17, 169s60
- sulfonamides 19, 33s60
- thioacids, C-terminal 41, 505s60

Perfluoro... s.a. Polyfluoro...

1-Perfluoroalkyl-3-ethylenealcohols, chiral 33, 865s60

(Perfluoroalkyl)ethylenes, terminal
-, hydroboration 21, 174s60

Perfluorobutanesulfonic acid
- as reagent 28, 827s60

Permanganate s. Benzyltriethylammonium permanganate

Peroxymonosulfate s. Benzyltriphenylphosphonium peroxymonosulfate, Potassium –

Persulfate s. Tetra-n-butylammonium persulfate

Petasis reaction, polymer-based 48, 856s60

Phase-boundary catalysis 36, 235s60

Phase transfer catalyst
-, tetra-n-octylammonium bromide as – 60, 151

Phenacyl esters, 2,5-dimethyl-
-, cleavage 29, 2s60

3(4H)-Phenanthrenones, 1,2-dihydro-
- from
p-alkoxystilbenes 60, 497

Phenanthridines, 6-amino- 60, 463

1,10-Phenanthrolines, chiral 2, 651s60

Phenolesters s.a. Carboxylic acid aryl esters

Phenolethers s.a. Aryloxy...

Phenols
-, identification 37, 152s60
- special s.
halogenophenols
o-hydroxy...
2,4,6-trichlorophenol

Phenols, condensed
- from
furylacetylenes 60, 310

Phenyl iodosoacetate
- as reagent 43, 402s60; 44, 197s60; 48, 299s60; 60, 106, 107, 168, 187

Phenyl iodoso(hydroxy)-p-nitrobenzenesulfonate
- as reagent 26, 325s60

Phenyl iodoso(hydroxy)tosylate
- as reagent 60, 87

Phenyl iodosotrifluoroacetate
- as reagent 47, 262s60 (review)
- -, polymer-based
- as reagent 29, 910s60

Phenylsulfonylnitromethane
- as reactant 60, 321

Phospha-Staudinger reaction
-, review 47, 569s60

Phospha-Wittig synthesis
-, review 47, 569s60

Phosphazene base (P4-t-Bu) 23, 927s60

Phosphine N-carbalkoxyimines
-, aza-Horner synthesis with – 60, 424
- startg. m. f.

(E)-ethylene derivs 60, 424

Phosphine oxides
- special s.
bis(phosphine oxides)
diphenylphosphinyl...
triphenylphosphine oxide

Phosphines
- special s.
aminophosphines
di(phosphines)
tetra(phosphines)

Phosphines, tert.
- special s.
bis(diphenylphosphino...
di-$tert$-butylphosphino...
diphenylphosphino...
1,3-oxathianes, 2-phosphino-, bicyclic
pyrrolidines, (S)-2-diphenylphosphinomethyl-4,4-dibenzyl-1-pivaloyltriarylphosphines
tributylphosphine
tricyclohexylphosphine
tri-2-furylphosphine
triphenylphosphine
tri-n-propylphosphine
tris(2,6-dimethoxyphenyl)phosphine
tris(p-methoxyphenyl)phosphine

Phosphinic acid amides
- special s.
phosphinylamino...

N-(o-Phosphinoarylidene)dipeptides
- as reagent 52, 297s60

Phosphinoferrocenes s. Ferrocenylphosphines

N-[(β-Phosphinylamino)acyl]sultams
- from
N-phosphinylimines, synthesis, asym. 60, 245

1,2-N→C-Phosphinyl group migration 60, 212

1,3-O/S→C-Phosphinyl – –
- with retention of P-chirality 60, 212

N-Phosphinylimines
- startg. m. f.
N-[(β-phosphinylamino)acyl]sultams, synthesis, asym. 60, 245

Phosphonic acid derivs.
- special s.
benzylphosphonic acid derivs.
- – diamides, bicyclic, chiral
- special s.
o-methoxyphenylphosphonic acid diamides, bicyclic, chiral
- – esters
- special s.
acetylenephosphonic acid esters
aminophosphonic – –
diphosphonic – –
ethylenephosphonic – –
hydroxyphosphonic – –
isocyanophosphonic – –
nitrophosphonic – –
sulfinylphosphonic – –
- – monoesters
- special s.
aminophosphonic acid monoesters

Phosphonic acids
- special s.
nitrophosphonic acids

Phosphonium salts
- special s.
carbonylphosphonium salts

tetraphenylphosphonium chloride
- –, polymer-based
- as reagent 60, 203

Phosphonous acid monoesters
- from
phosphonous acids and tetraalkoxysilanes 60, 56

Phosphonous acids
- special s.
arylphosphonous acids
- startg. m. f.
phosphonous acid monoesters 60, 56

Phosphoric acid diesters
- special s.
dibenzyl hydrogen phosphate

Phosphoric acid esters
-, resolution, kinetic 60, 10
- special s.
enol phosphates
glycosyl –

Phosphorodiamidites, bicyclic
- special s.
8-quinolyl phosphorodiamidites, bicyclic

Phosphoromonoamidates
- special s.
dinucleoside phosphoromonoamidates

Phosphoromonohalidates
- special s.
bis(2,4-dichlorophenyl) phosphoromonochloridate

Phosphorous acid esters
- special s.
enol phosphites
- – –, cyclic
-, Michael addition, asym. with – 60, 210

Phosphorus(III) acid amides
-, O-phosphorylation with –, review 42, 91s60

Phosphorus pentasulfide 60, 197

O-Phosphorylation 60, 52
- with phosphorus(III) acid amides, review 42, 91s60
- –, regiospecific
- of aldoses 60, 52

Photooxidation, heterogeneous 60, 91

Phthalans
- from
aldehydes 60, 377

Phthalides
- startg. m. f.
β-(o-carboxyaryl)carboxylic acid esters 60, 440

Phthalimides
- special s.
N-(alkylthio)phthalimides

Phthalimidines
- from
isocyanates 60, 377
- special s.
O,N-heterocyclics, phthalimidine-condensed
-, 3-hydroxy-, N-condensed 60, 255

2-Phthalimidoalcohols, chiral 39, 32s60

ω-Phthalimidoalcohols
-, ring closure with ethylene derivs. 60, 255

Phthalocyanines, acetylenic
-, synthesis review 33, 325s60

3-Picoline, 2-amino-
- as reagent 60, 360

Pinacolborane

– as reagent 53, 481s60
Pinacol esters 60, 240
Pinacol ethers 60, 239
Pinacol-type rearrangement,
  stereospecific
–, thiophenes, tetrahydro-, 3-acyl- via –
  60, 331
Piperazine, *trans*-2,5-dimethyl- 60, 267
Piperazine, (R)-2-hydroxymethyl-
  4-sulfonyl-, polymer-based
– as reagent 25, 15s60
2,3-Piperazinediones, chiral
  49, 300s60
Piperazines
–, N-acylation, regiospecific 58, 154s60
Piperazinones
– by 3-component reaction 60, 416
Piperidines, 1-alkoxy-
  as intermediates 35, 137s60
Piperidin-3-ylcarbinols, N-protected,
  chiral 60, 259
2-Piperidone ring
– from
  3-aza-1,5-dien-4-ones 60, 276
2-Piperidones, chiral 60, 259
Pivalic acid
– as reagent 60, 127
Platination s.a. Carboplatination
Platinum chloride 60, 304
Platinum complexes
  dichloro(1,5-cyclooctadiene)-
    platinum(II) 49, 518s60; 60, 207
  (1,3-divinyl-1,1,3,3-tetramethyl-
    disiloxane)platinum(0) 60, 406
  (ethylene)bis(triphenylphosphine)-
    platinum 60, 210
– –, fluorous
  dichlorobis[tris[4-[2-(perfluorohexyl)-
    ethyl]phenyl]phosphine]platinum(II)
    54, 431s60
Polyamines
–, reactions, selective under metal-
  chelation or -mediation, review
  40, 235s60
Polyarenes
– special s.
  aceanthrylenes
Polyarenes, linearly condensed 60, 274
Poly[4-(N-benzylaminomethyl)styrene]
– as reagent 60, 427
Polyfluoro... s.a. Perfluoro...
Polyfluoroalkoxy(fluoro)sulfuranes
–, review 32, 460s60
Polyfluoroethylene derivs. 60, 496
Polymer-based reactions
– special s.
  N-acylation, polymer-based
  N-arylation, –
  Baylis-Hillman reaction, –
  Diels-Alder –, –
  Heck arylation, –
  Nenitzescu indole synthesis, –
  oligosaccharide synthesis, –
  peptide synthesis, solid-phase
  Petasis reaction, polymer-based
  retro-Diels-Alder reaction, –
  ring closure, –
  O-triflylation, –
  Zincke reaction, –
Polymer-based reagents
– in synthesis, catalytic, review 52, 75s60
– special s.

arenedicarbonylchromium(0)
  complexes, polymer-based
  ammonium diacetoxybromate, –
  – hydroxides, quaternary, –
  azabis(Δ²-oxazolines), –
  1,4-bis(9-O-dihydroquin[id]ine)-
    phthalazine, –
  carbodiimide, –
  cinchonidinium chloride, N-benzyl-, –
  cobalt complexes, –
  –(II) phosphine –, –
  dichloropalladium(II), – under
    Palladium complexes, supported
  iminophosphoric acid amides, cyclic, –
  N-methylmorpholine, –
  nickel(II) acetoacetonate,
    polybenzimidazole-based
  2-(3-oxidobenzotriazol-1-yl)-1,1,3,3-
    tetramethyluronium fluoroborate,
    polymer-based
  palladium phosphine complexes, –
  phenyl iodosotrifluoroacetate, –
  piperazine, (R)-2-hydroxymethyl-4-
    sulfonyl-, –
  rhodium(I) phosphine complexes, –
  ruthenium carbene complexes, –
  scavengers, –
  selenocyanates, –
  triazabicyclo[4.4.0]decene, –
Polymer-based synthesis (s.a. Resin-
  capture-release)
– by resin-to-resin transfer 60, 415
–, monitoring 19, 33s60
– of
  β-aminocarboxylic acid esters from
    aldimines 50, 551s60
  aryl dithiocarbamates 48, 510s60
  arylglycines from arylboronic acid
    esters 60, 415
  2-azetidinones from aldimines 60, 322
  benzofuran-3-carboxylic acid esters
    from *o*-hydroxyarylacetylenes
    60, 318
  2(3H)-benzofuranones from *o*-alkoxy-
    carboxylic acid thioamides 60, 113
  benzofurans from *o*-hydroxy-β-styryl
    ethers 60, 113
  carboxylic acid amides and esters from
    acyl halides 60, 159, 165
  enolethers from mercaptals 60, 113
  glycosides from glycosyl disulfides
    60, 95
  guanidines from amines 60, 161
  heterocyclics from α-sulfonyloxy-
    ketones 60, 62
  N-heterocyclics 43, 316s60
  α-ketocarboxamides from enamines
    20, 502s60
  oligoglycosides 39, 189s60
  oxazolidines, 3-acyl- from 2-(alkylid-
    eneamino)alcohols 17, 216s60
  2-pyridones, 3,4-dihydro- from
    acetylene derivs. 60, 124
–, resin considerations, review 50, 555s60
Polymer linkers, alkoxylamine-type
–, cleavage 51, 9s60
– –, carbamate-type
–, cleavage 27, 110s60
– –, sulfonyloxy-type
–, removal on nucleophilic substitution,
  intramolecular 60, 62
– –, thioester-type 41, 505s60

– –, traceless
–, review 50, 555s60
Polymers, hydrophobic
– as solid organic co-solvent 60, 24
Polymer supports
–, preparation, update 19, 33s60
– –, dendritic and hyperbranched
–, review 50, 555s60
Polymethylhydrosiloxane
– as reagent 56, 146s60
Polyols
– special s.
  dienepolyols
Polypeptides
–, synthesis, catalyzed by living
  polymerization 19, 33s60
Potassium 2-ethylene(trifluoro)borates
– startg. m. f.
  1,6-dien-4-ols 60, 431
– iodide 60, 180
– osmate 60, 166
– permanganate/copper(II) sulfate
–, oxidation, heterogeneous under
  ultrasonication 60, 109
– peroxymonosulfate 45, 120s60; 60, 67,
  68, 87
– –/alumina 22, 408s60
– –/silica gel 49, 85s60
– trisodium diperiodatoargentate(III)
  52, 121s60
Porphyrins, unsym. 29, 853s60
Pressure, high s. High pressure...
L-Proline
– as reagent 60, 267, 336
Propargyl alcohols s.a. Ethynylcarbinols
α-Propargylation
–, furans via – 60, 374
C-Propargylation, asym. 38, 802s60
Propargyl bromide
– as reactant 60, 374
Propylene oxide
– as reagent 60, 214
Protection
– of carbonyl groups as
  1,3-dithiolanes 60, 202
  p-methoxybenzyl alcohols 60, 114
  1,3-oxathiolanes 60, 202
– of carboxyl groups as
  p-methoxybenzyl esters 29, 547s60
  2-phenyl-2-(trimethylsilyl)ethyl –
    33, 8s60
  tris(2,6-diphenylbenzyl)silyl – 60, 4
– of hydroxyl groups as
  p-(trifluoromethyl)benzyl ethers 60, 13
– of *vic*-hydroxyl groups as
  1,3-dioxolanes, 2-(carbo-*tert*-
    butoxymethyl)- 17, 234s60
–, 2-(carbomethoxymethyl)- 17, 234s60
  ethyl orthoformates, cyclic 28, 268s60
– of – –, ar. as
  1,3-benzodioxoles, 2,2-diphenyl-
    28, 268s60
– of phosphoryl hydroxyl groups as
  2-cyanoethyl esters, subst. 17, 169s60
Protective groups
– in biohydroxylation, review 45, 56s60
N-Protective groups, removal (s.a.
  N-Decarbalkoxylation)
– of
  carbo-2,2-bis(2-nitrophenyl)ethoxyl
    44, 23s60
  carbo-6-nitroveratryloxy 44, 23s60

(N-Protective groups, removal
– of)
  carbo-2-phenyl-2-trimethylsilylethoxy
    34, 23s60
  dimethylsulfamyl 27, 18s60
  1-(p-methoxyphenyl)ethyl 49, 330
O-Protective groups, removal
– of
  carballyloxy (selective removal)
    29, 28s60
  p-(trifluoromethyl)benzyl (from
    ethers), reductive removal 60, 13
  tris(2,6-diphenylbenzyl)silyl (from
    esters) 60, 4
Protein ligation, expressed
–, application, review 19, 33s60
Protein synthesis
–, review 19, 33s60
anti-Protonation, asym., catalytic
    60, 191
Pummerer-type rearrangement
–, benzofurans, 2,3-dihydro- via – 60, 334
Purines
– special s.
  adenines
  hypoxanthines
Purines, 2-amino-
–, N⁹-alkylation 60, 154
2H-Pyrans, 3,4-dihydro-, 4-alkylidene-
– from
  3-acetylenealcohols 60, 297
  acetylene derivs., terminal 60, 297
2H-Pyrans, 3,6-dihydro-, 2,6-disubst.
– from
  η³-pyranylmolybdenum complexes,
    dihydro-, 2,6-dialkoxy-, asym.
    synthesis 60, 262
2H-Pyrans, 5,6-dihydro-, 2,3,6-trisubst.,
    chiral 60, 262
Pyrans, tetrahydro-, 6-acyl-2,2,4,4-
    tetraalkoxy-
– from
  α-ketocarbonyl compds., asym.
    conversion 60, 261
–, –, 2-alkoxy-3-arylseleno- 49, 515s60
–, –, 2-homoallyloxy-
– startg. m. f.
  pyrans, tetrahydro-, 2-δ-hydroxy-,
    recyclization 60, 312
–, –, 2-δ-hydroxy-
– from
  pyrans, tetrahydro-, 2-homoallyloxy-,
    recyclization 60, 312
Pyran-2-yl ethers, tetrahydro-
– startg. m. f.
  acoxy compds. 60, 102
η³-Pyranylmolybdenum complexes,
    dihydro-, 2,6-dialkoxy-
– startg. m. f.
  2H-pyrans, 3,6-dihydro-, 2,6-disubst.,
    asym. synthesis 60, 262
N-(2-Pyrazinyl)aldimines
– startg. m. f.
  aryl ketones 60, 389
Pyrazole
– as reagent 60, 70
Pyrazoles
–, N-arylation 8, 563s60
– from
  Δ²-pyrazolines 25, 649s60
Pyrazoles, polysubst.
–, synthesis, polymer-based 43, 316s60

–, 3-silyl- 2, 368s60
Pyrazolidines, 1-acyl-3-allyl-, chiral
    58, 406s60
Δ²-Pyrazoline-5-carbonyl compds.,
    1-acetyl-
– from
  α,β-ethylenecarbonyl compds., asym.
    cycloaddition 60, 429
Δ²-Pyrazolines
–, C₄-alkylation 31, 812s60
– from
  ethylene derivs., in aq. micelles or
    water 60, 363
Pyrid[1,2-a]indoles 60, 352
Pyridine/hydrogen fluoride 60, 4
Pyridine, 1,4-dihydro-, 1-methyl-
– as reagent 60, 492
Pyridine-3-carboxylic acid esters
    60, 342
Pyridine N-oxide
– as reagent 60, 69
– –, 4-phenyl-
– as reagent 60, 107
Pyridine N-oxides
– from
  amidinium salts, vinylogous 25, 581s60
Pyridine ring
– from
  enamines 24, 836s60
Pyridines
– from
  aldehydes 60, 335
  α,β-ethyleneoximes, via Michael
    addition 60, 342
  formamidinium salts, vinylogous
    60, 351
  ketones 60, 351
  – (2 molecules) 60, 335
– special s.
  terpyridyls
–, 3-amino-2-vinyl-
– from
  isothiazolo[4,3-b]pyridine 2,2-dioxides,
    1,3-dihydro- 60, 489
–, 4-amino-, ferrocenyl-, chiral
– as reagent 60, 138
–, 3-aryl 60, 351
–, 2-β-carboxy-
– via Diels-Alder reaction, intra-
    molecular 60, 314
–, 1,2-dihydro-, 2-allyl-1-carballyloxy-
    26, 684s60
–, 1,4-dihydro-
– from
  pyridinium salts 60, 31
–, 2,3-dihydro-, 2-alkylidene-3-imino-
– as intermediates 60, 489
–, library 52, 352s60
–, 1,2,3,4-tetrahydro-, 3-iodo-,
    N-condensed 48, 434s60
–, 1,2,3,6-tetrahydro-, 1-amino-
    6-α-hydroxy-, chiral 60, 420
Pyridinioacetic acid esters
– startg. m. f.
  cyclopropanecarboxylic acid esters,
    asym. conversion 60, 355
– acids
– startg. m. f.
  indolizines 60, 474
α-Pyridiniocarboxylic acids/esters
– special s.
  pyridinioacetic acids/esters

Pyridinium methylids
–, cycloaddition, 1,3-dipolar, oxidative
    with – 60, 474
Pyridinium salts
– startg. m. f.
  pyridines, 1,4-dihydro- 60, 31
2-Pyridone, 6-methyl-
– as reagent 60, 2
2-Pyridone-5-carboxamides,
    2,3-dihydro- 60, 124
4-Pyridone ring, 2,3-dihydro-
– from
  N-(γ-keto)thiolactams 60, 491
2-Pyridones, 3,4-dihydro-
– from
  acetylene derivs., polymer-based
    synthesis 60, 124
  cyclopropyl ketimines 60, 286
  enamines, polymer-based synthesis
    60, 124
  α,β-ethylenecarboxylic acid
    anhydrides, – – 60, 124
  – acids, – – 60, 124
–, synthesis, polymer-based 13, 412s60
–, 5,6-dihydro-, 1-arylsulfonyl-4-
    arylthio-
– from
  thiophene 1,1-dioxides, 2,5-dihydro-,
    3-arylthio- 60, 409
4-Pyridones, 2,3-dihydro-
– from
  aldimines 60, 439
  3-siloxy-1,3-dienamines 60, 439
–, 2,3-dihydro-, 5-hydroxy-
– from
  diazomethyl β-acylaminoketones
    60, 481
α-(2-Pyridyloxy)ketones
– special s.
  ethylene-α-(2-pyridyloxy)ketones
2-Pyridyl sulfones
– special s.
  glycosyl 2-pyridyl sulfones
Pyrimidine ring, 1,2(4)-dihydro-,
    4-imino- 38, 496s60
Pyrimidines, 4-anilino-
–, synthesis, review 26, 817s60
–, 1,4,5,6-tetrahydro-, 2-amino-
    5-hydroxy- 43, 316s60
Pyrimidine-2(1H)-thiones, dihydro-
    55, 337s60
2(1H)-Pyrimidinone-5-carboxylic acid
    esters, 3,4-dihydro-
–, resolution, kinetic 28, 13s60
2(1H)-Pyrimidinones, 3,4-dihydro-
s.a. under Biginelli
Pyrimido[5,4-c]quinoline-2,4(1H,3H)-
    diones, 5-amino- 60, 463
2-Pyrones
– startg. m. f.
  benzene ring 60, 464
–, 3,6-dihydro-
– from
  α-allene-δ-hydroxycarboxylic acid
    esters, via 1,4-addition 60, 325
–, 5,6-dihydro-, 6-acyl-, chiral 60, 261
4-Pyrones, 2,3-dihydro-
– from
  aldehydes 60, 439
  3-siloxy-1,3-dienamines 60, 439
Pyrrole-2-carboxylic acid esters 60, 476
Pyrrole ring

– from
  enolethers 60, 372
  α-halogenooxo compds. 60, 372
Pyrroles
– from
  benzotriazoles, N-α-thioacylamino- 60, 352
  β-diketones 60, 476
  (Z)-2-en-4-ynamines 60, 129
  ethylene derivs. 60, 352
Pyrroles, 1-acyl-
– from
  α-acylaminoketones 60, 490
  Δ³-pyrrolines, 1-acyl-3-sulfonyl- 60, 490
Pyrrolidine
– as reagent 60, 323
–, (S)-2-diphenylphosphinomethyl-4,4-dibenzyl-1-pivaloyl-
– as reagent 60, 247
Pyrrolidines, 4-benzyl-3-methylene-1-tosyl- 55, 454s60
–, 1-carbalkoxy-3-silyl-, chiral 60, 430
–, 2-(2-cyanovinyl)- 33, 640s60
–, 3-silyl-
– from
  (E)-2-ethylenesilanes and acetals, synthesis, asym. 60, 330
–, 1-sulfonyl-
– from
  aziridines, 1-sulfonyl- and ethylene derivs. 60, 265
Pyrrolidinetriones
–, chemistry, review 7, 823s60
Pyrrolidin-3-ols, trans-2-allyl-, N-protected 60, 432
4-Pyrrolidinopyridine
– as reagent 60, 466
3-Pyrrolidone-4-carboxylic acid esters, chiral 33, 918s60
2-Pyrrolidones, 5-acoxy-
– startg. m. f.
  2-pyrrolidones, 5-subst., synthesis stereospecific 60, 435
–, 4-alkoxy-, chiral 11, 905s60
–, (Z)-3-alkylidene- 60, 243
–, 1-amino-, chloro- 40, 493s60
–, 3,4-di(alkylidene)- 58, 273s60
–, N-(α,β-ethyleneacyl)-
– startg. m. f.
  5-isoxazolidones, asym. synthesis 60, 149
–, 3-(1-halogenalkylidene)- 48, 692s60
–, 3-silyl-, N-protected
– from
  aziridines, N-protected 60, 243
–, 5-subst.
– from
  2-pyrrolidones, 5-acoxy-, synthesis, stereospecific 60, 435
Δ¹-Pyrrolines, 5-α-functionalized 49, 370s60
Δ³-Pyrrolines
– from
  aldehydes, aldimines and chromium α,β-ethylene(alkoxy)carbene complexes 60, 434
–, 1-acyl-3-sulfonyl-
– startg. m. f.
  pyrroles, 1-acyl- 60, 490
Δ³-2-Pyrrolones, 4-halogeno-5-hydroxy-

– from
  α-allenecarboxylic acid amides 60, 177
Pyrrolo[3,4-c]pyridazines, hexahydro-, N-sulfonyl- 60, 289
Pyrrolo[2,3-d]pyrimidines 60, 372
Pyrromethanes 21, 948s60

Quaternary C-atoms s. Hydrocarbon groups, quaternary
4(3H)-Quinazolones, 2-amino-
–, synthesis, polymer-based 43, 316s60
Quinidine
– as reagent 50, 397s60
Quini[id]ne, benzyl-
– as reagent 60, 184
Quininium salts, dihydro-
– special s.
  N-fluoroquininium fluoroborate, dihydro-, O-(4-chlorobenzyl)-
Quinoline-2,4-diones 28, 418s60
Quinolines
– by substitution, nucleophilic, ar., review 41, 694s60
– from
  amines, ar., prim. 60, 359
  N,N-diallylammonium salts, quaternary 60, 359
Quinolines, 2,4-diamino-
– from
  benzotriazoles, 1-acyl- 60, 463
  isocyanates (3 molecules) 60, 463
–, 3-formyl-2-halogeno- 22, 826s60
Quinolin-2-ylacetic acid esters, 1,2,3,4-tetrahydro- 60, 170
4(1H)-Quinolones, 2-cyano- 42, 919s60
8-Quinolyl phosphorodiamidites, bicyclic, chiral
– as reagent 48, 772s60
o-Quinone methidimines
– special s.
  pyridines, 2,3-dihydro-, 2-alkylidene-3-imino-
o-Quinone mono-N-acylimines
– from
  acylamines, ar. 60, 77
–, hetero-Diels-Alder reaction, intramolecular with – 60, 78
p-Quinones
– special s.
  acylamino-p-quinones
Quinoxalin-2(1H)-ones, 3,4-dihydro-, 3-subst. 60, 416
Quinoxalin-2-ylacetic acid esters, 1,2,3,4-tetrahydro- 60, 170
Quinuclidine, 3-hydroxy-
– as reagent 39, 593s60

Radical addition
– of dithiophosphonyl radicals 60, 217
– to

ethylene derivs., controlling factors, review 39, 646s60
imines, review 50, 382s60
Radical 1,4-addition
–, C-α-glycosides by – 60, 442
– using tetraphenyldisilane 46, 65s60
– via 1,5-hydrogen atom transfer 43, 958s60
– –, intramolecular, stereospecific
– to 1,3-oxazine-2,6-diones 60, 486
– –, regiospecific, oxidative
– with acetals, cyclic 60, 282
– 1,4-addition-ring closure 60, 382
– C-α-alkylation, redoxidative, cocatalytic
– of ketones (with ethylene derivs.) 60, 279
– 3-component reaction
– with vinylcyclopropanes 57, 257s60
– – –, regiospecific
– of silanes 60, 273
– deoxygenation
– via trifluoroacetates 60, 39
– hydrosilylation
– of ethylene derivs. 60, 209
– reactions
– with α-halogenosulfonyl derivs., review 20, 685s60
– rearrangement, skeletal
– of 2-oxabicyclo[2.2.2]oct-5-en-3-ones, 8-arylseleno- 60, 495
– ring closure
– of 2,n-dienecarbonyl compds. 60, 276
– – –, double, thiyl-mediated
– of 1,6-dienes 60, 268
– – –, regiostereospecific
– of 2-halogen-1,n-enynes 60, 487
Radicals
– special s.
  alkoxyl radicals
  dithiophosphonyl –
  silyl –
Radical vinylation, intramolecular, asym. 60, 494
Rearrangement, [2.3]-sigmatropic
–, 2-ethylenealcohols via – 60, 97
–, skeletal
– of 2-oxabicyclo[3.2.1]oct-6-en-3-ones 60, 495
– special s.
  radical rearrangement, skeletal
Reformatskii-type synthesis, dialkylzinc-mediated
– via zinc enolate equivalents 60, 232
– –, Rh-catalyzed
– via zinc enolates 60, 242
Replacement
– of chlorine by
  hydrogen 60, 43
– of hydroxyl groups by
  chlorine 60, 182
– of – –, N-heterocyclic by
  amino 60, 141
– of tert-hydroxyl groups by
  prim-amino 60, 146
– of sulfonyl groups by
  Si-nucleophiles 60, 443
– of trifluoroacetoxy groups by
  hydrogen 60, 39
Resin-capture-release
–, syntheses via –, review 50, 555s60

Resolution (s.a. **Res** section, and under Stereoisomers in Vol. **1-50**)
- of 1,1'-bi-2-naphthols **60**, 500
Resolution, kinetic
-, considerations, practical, review **46**, 55s60
- of
  alcohols in ionic liquids **44**, 214s60
  amines by asym. N-acylation **60**, 138, 148
  cyanohydrin esters by asym. hydrolysis **28**, 13s60
  cycloalk-2-en-1-yl acetates, 3-nitro- **28**, 13s60
  phosphoric acid esters, P-chiral **60**, 10
- by
  N-acylation, asym. **44**, 314s60
  O-acylation, asym. with enolesters, update **44**, 214s60
  cross-metathesis of alkynes **56**, 295s60
-, -, **dynamic** (s.a. Deracemization)
-, review **58**, 49s60
-, -, **parallel**
- of alcohols, sec. and amines, prim. **60**, 9
-, -, **transition metal-mediated**
-, review **46**, 55s60
Retro-Barbier fragmentation 60, 176
Retro-Diels-Alder reaction, polymer-based 17, 198s60
Rhenium oxides, organo-
  methylrhenium oxide **60**, 70
  methyl[2-(thiolatomethyl)phenylthio]-(triphenylphosphine)rhenium(V) oxide **60**, 499
Rhodium(II) acetate 29, 603s60; **60**, 168, 361, 481
Rhodium(I) 1,1'-binaphthalene-2,2'-diyl phosphites, phosphonites and phosphoromonoamidites, chiral 60, 34
–(**I**) **bis(1,3,2-diazaphospholidine) complexes, chiral 60**, 291
–(**I**) **bis(phosphite) complexes, calixarene-based 4**, 667s60
– **carbenoids**
-, insertion, asym. into C–H bonds, review **47**, 955s60
– **carbonyls**
  dodecacarbonyltetrarhodium **60**, 476
– **complexes**
  acetoacetonato(dicarbonyl)rhodium(I) **60**, 451
  acetoacetonato(diethylene)rhodium(I) **60**, 453
  bis(1,5-cyclooctadiene)rhodium(I) fluoroborate **60**, 450
  [1,2-bis(diphenylphosphino)ethane]-rhodium(I) triflate **60**, 306
  carbonyl(chloro)[1,3-bis(diphenyl-phosphino)propane]rhodium(I) dimer **60**, 345
  chloro[1,4-bis(diphenylphosphino)-butane]rhodium(I) dimer **58**, 311s60
  chloro[(2R,2'R)-bis(diphenyl-phosphino)-(1R,1'R)-dicyclo-pentane]rhodium(I) dimer **60**, 307
  chloro(1,5-cyclooctadiene)rhodium(I) dimer **60**, 293, 389
  chlorotris(triphenylphosphine)-rhodium(I) **60**, 242, 294, 360
  (1,5-cyclooctadiene)bis(methyldi-phenylphosphine)rhodium(I) triflate **60**, 453

(1,5-cyclooctadiene)rhodium(I) tetraphenylborate, zwitterionic **60**, 253
dicarbonyl(chloro)rhodium(I) dimer **60**, 59, 292
–(**I**) **complexes, cationic 22**, 421s60
– –, **organo-**
– special s.
  allylrhodium complexes
–(**I**) **phosphaferrocene complexes, cationic, chiral 60**, 73
–(**I**) **phosphine complexes, dendritic 4**, 667s60
–(**I**) – –, **polymer-based 60**, 35
–(**I**) **phosphine-phosphite complexes, chiral 60**, 36
–(**I**) **phosphine-phosphoromonoamidite complexes, chiral 60**, 36
–(**II**) **trifluoroacetate 60**, 89
–(**II**) **triphenylacetate 38**, 954s60; **60**, 168
Ring-closing metathesis (s.a. under Ring-opening metathesis...)
- using ruthenium carbene complexes, polymer-based, soluble **60**, 498
Ring closure, double, oxidative, stereospecific 60, 78
– –, **fragmentation-type 60**, 193
– –, **polymer-based**
- with traceless release of heterocyclics **60**, 62
Ring expansion, carbonylative, Ru-catalyzed 60, 286
Ring opening, transannular, asym.
- of cyclooctene oxides **60**, 309
Ring-opening metathesis-cross metathesis
- of ethylene derivs., bicyclic **60**, 290
- with eneselenides **60**, 290
Ring-opening metathesis-cross metathesis-ring-closing metathesis, regiostereospecific 60, 475
Ritter reaction
-, amines, prim. from alcohols, tert. via – **60**, 146
- with chloroacetonitrile **60**, 146
Rupe rearrangement 60, 66
Ruthenate, per- s. Tetrapropylammonium perruthenate
Ruthenium/carbon 26, 52s60
– **allenylidene complexes**
- as intermediates **60**, 83
– **η³-allyl complexes**
-, carbonylation via – **60**, 343
– **carbene complexes**
- (arylseleno)carbene complex **60**, 290
- dichloro(imidazolidin-2-ylidene) complexes, polymer-based **54**, 320s60; **58**, 497s60
  Grubbs' complex **60**, 289, 475
– – –, **polymer-based, soluble 60**, 498
– **carbonyls**
  dodecacarbonyltriruthenium **60**, 286
– **complexes**
  carbonyl(6,6'-dichloro-2,2'-bipyridyl)ruthenium(VI) oxo complexes **50**, 53s60
  carbonyl(dihydrido)tris(triphenyl-phosphine)ruthenium(II) **58**, 306s60
  chloro(1,5-cyclooctadiene)(η⁵-penta-methylcyclopentadienyl)-ruthenium(II) **60**, 288

chloro(η⁵-pentamethylcyclopenta-dienyl)bis(triphenylphosphine)-ruthenium(II) **18**, 776s60
(η⁶-*p*-cymene)((1,R,2R)-N-tosyl-1,2-diphenylethylenediamine)-ruthenium(II) chloride **60**, 28
diacetato[(R)-2,2'-bis(di-*p*-tolyl-phosphino)-1,1'-binaphthalene]-ruthenium(II) **60**, 33
dichlorobis(triphenylphosphine)-ruthenium(II) **60**, 359
dichloro(*p*-cymene)ruthenium(II) dimer **60**, 53; **24**, 261s60
dichloro[2,2',6,6'-tetramethoxy-4,4'-bis(diphenylphosphino)-(S)-3,3'-bipyridyl]ruthenium(II) **60**, 29
diruthenium complexes, thiolate-bridged **60**, 83
tricarbonyldichlororuthenium(II) dimer **60**, 343
(1,4,7-trimethyl-1,4,7-triazacyclonon-ane)ruthenium(III) trifluoroacetate **56**, 162s60
tris(acetonitrile)(cyclopentadiene)-ruthenium(II) hexafluorophosphate **60**, 287
-, activation-deactivation, asym. with 2 chiral auxiliaries **60**, 30
–(**III**) **complexes, hydroxyapetite-supported 60**, 167
–(**II**) **formyl complex, cationic 60**, 31
– **hydride complex 60**, 23
– **trichloride 60**, 93
–(**II**) **trifluoroacetate 15**, 261s60

Samarium (s.a. Titanium tetrachloride/samarium)
Samarium complexes
  bis(η⁵-pentamethylcyclopentadienyl)-samarium **50**, 405s60
Samarium diiodide 13, 539s60; **51**, 216s60; **60**, 190, 237, 251
– – /**hexamethylphosphoramide 60**, 437
Samarium(III) triflate 60, 94, 99
Scandium(III) triflate 8, 455s60; **39**, 883s60; **43**, 563s60; **52**, 334s60; **60**, 407, 431
–(**III**) –, **microencapsulated 36**, 101s60
Scavengers
– of
  amines **1**, 391s60; **60**, 159
  bromine **3**, 3s60
  carboxylic acid halides **9**, 524s60; **60**, 154
  hydrazines **1**, 391s60
Scavengers, polymer-based
-, review **8**, 927s60
Schmidt reaction, intramolecular, Hg-promoted 60, 172
Selenides
- special s.
  acetyleneselenides
  eneselenides
  ethyleneselenides
  trifluoromethyl selenides
Selenium dioxide
- as catalyst **60**, 57

Selenocyanates, polymer-based
– as reagent **49**, 210s60
Selenoldithiolphosphoric acid esters, cyclic
– as source of dithiophosphonyl radicals **60**, 217
Selenodithiophosph... s. Selenodithiolphosph...
α-Selenoglycosides
– startg. m. f.
C-α-glycosides **60**, 442
Selenophene 1-oxides
–, chemistry, review **29**, 591s60
Selenosulfides
– from
diselenides **60**, 190
sulfenyl halides **60**, 190
thiolsulfonic acid salts **60**, 190
Selenothiocarboxylic acid ammonium salts
– from
selenothiolic acid 2-(trimethylsilyl)ethyl esters **60**, 18
Selenothiolic acid 2-(trimethylsilyl)ethyl esters
– startg. m. f.
selenothiocarboxylic acid ammonium salts **60**, 18
Sharpless oxyamination, asym., regiospecific
–, 2-oxazolidones, chiral via – **60**, 166
Sharpless-type epoxidation, Zr-mediated
– with desymmetrization **60**, 61
Sialic acids 40, 567s60
O-Sialylation
–, review **59**, 124s60
Silaboration, regiostereospecific
– of alkylidenecyclopropanes **60**, 211
Silanes (s.a. Hydrosilylation)
–, 3-component radical synthesis, regiospecific **60**, 273
– from
ethylene derivs., halogenosilanes and halides, synthesis **60**, 273
– special s.
alkoxysilanes
aryloxysilanes
arylsilanes
diazosilanes
di(silanes)
enesilanes
ethylenesilanes
halogenosilanes
hydroxysilanes
thioacylsilanes
Silanethiols
– from
silicon hydrides, organo- **60**, 189
Silanols
– from
silicon hydrides, organo- **60**, 53
– special s.
enesilanols
Silenes
–, reactivity, review **47**, 625s60
Silica, mesoporous 60, 91
Silica chloride 44, 460s60
Silica gel 60, 337, 495
Silicon hydrides, organo-
– as reagent **60**, 308
– special s.

1-alkoxyenesilicon hydrides
(diallylhydridosiloxy)-3-ethylenes
diphenylsilane
enoxy(hydrido)silanes
1,1,3,3-tetramethyldisilazane
tetraphenyldisilane
triethylsilane
– startg. m. f.
silanethiols **60**, 189
silanols **60**, 53
Silicon tetrachloride 60, 173
Silicotungstic acid 8, 165s60
(Siloxy)acoxy-2-ethylenes
– startg. m. f.
2-vinyl-O-heterocyclics, asym. induction **60**, 115
1,1-Siloxyamines
– special s.
N-[2,2,2-trifluoro-1-(trimethylsiloxy)ethyl]morpholine
N-[2,2,2-trifluoro-1-(trimethylsiloxy)ethyl]piperazine
β-Siloxyboronic acid esters
– special s.
halogeno-β-siloxyboronic acid esters
α-Siloxycarboxylic acid amides
– from
oxo compds. and amines, asym. induction with 1 extra C-atom **60**, 466
Siloxycyclopropanes, 1-alkoxy-
– startg. m. f.
cinnamic acid esters **60**, 461
3-Siloxy-1,3-dienamines
– startg. m. f.
4-pyridones, 2,3-dihydro- **60**, 439
4-pyrones, – **60**, 439
Siloxy-1,3-dienes
– special s.
disiloxy-1,3-dienes
2-Siloxy-1,3-dienes
– special s.
1-amino-3-siloxy-1,3-dienes
N-Siloxyenamines
– special s.
N,N-disiloxyenamines
Siloxyethylene...
s.a. Enoxysilan..., Ethylenesiloxy...
Siloxy-3-ethylenes
– special s.
(diallylhydridosiloxy)-3-ethylenes
1,2-Siloxyhalides
– special s.
α-halogeno-β-siloxy...
syn-β-Siloxylaminocarboxylic acid esters 60, 433
α-Siloxymalononitriles
– from
acyloximes **60**, 333
carboxylic acid anhydrides **60**, 333
– special s.
tert-butyldimethylsiloxymalononitrile
2-Siloxy-1,1,1-trifluorides
– from
oxo compds., with 1 extra C-atom **60**, 224
Silver acetate 60, 294
Silver ate complexes
– special s.
potassium trisodium diperiodatoargentate(III)
– fluoroborate **60**, 467

– hexafluoroantimonate **60**, 307
– nitrate **60**, 161
Silylacetylenes
– from
acetylene derivs., terminal **60**, 219
O-Silyl O-alkyl keteneacetals
– startg. m. f.
β-(o-carboxyaryl)carboxylic acid esters **60**, 440
α-hydroxycarboxylic – – **56**, 106s60
5-isoxazolidones, asym. induction **60**, 433
O-Silylation
– special s.
O-trimethylsilylation
– with simultaneous O-tritylation and O-acylation **60**, 90
O-Silylation, catalytic 60, 55
Silylboranes
– startg. m. f.
silyl radicals **60**, 209
γ-Silylboronic acid esters
– special s.
alkylidene-γ-silylboronic acid esters
β-Silylcarboxylic – –
– special s.
hydroxy-β-silylcarboxylic acid esters
5-Silyl-1,3-dienes 51, 423s60
5-Silyl-(1,3-dien)olethers 51, 423s60
Silyl enol ethers s. Enoxysilanes
Silyl ethers (s.a. Siloxy..., O-Silylation)
– special s.
alkoxysilanes
aryloxysilanes
2-Silylethylation, nucleophilic 60, 270
Silylformylation-allylsilylation, intramolecular, regiospecific 60, 451
Silyl glycosides
– startg. m. f.
1,2-trans-C-glycosides **60**, 237
α-glycosyl iodides **60**, 237
1,2-Silyl group migration 60, 330
Silylketene equivalents, nucleophilic
–, lithium silylethynolates as – **60**, 243
δ-Silylketones
– special s.
ethylene-δ-silylketones
Silyl radicals
– from
silylboranes **60**, 209
1,4-Silylstannylation s. Carbopalladation, intramolecular-1,4-silylstannylation
1,1-Silylthioethers
– special s.
1-benzotriazol-1-yl-1,1-silylthioethers
Simmons-Smith-type reaction, stereospecific
– with iodomethylzinc aroxides **60**, 380
Sodium alkoxides
– tert-butoxide **60**, 157
– azide **60**, 119, 195
– cyanide **60**, 198
– dithionite **60**, 64
– fluoride **60**, 450
– formate **60**, 405
– hydroborates, organo-
– trihydridoacetoxoborate **60**, 116s60
– hydrogen sulfate/silica gel **51**, 104s60
– iodide **21**, 9s60: **31**, 427s60: **60**, 180, 296, 457, 459
– nitrite **60**, 86, 105, 116
– perborate **57**, 60s60

Sodium percarbonate 60, 70
– periodate 60, 93
– sulfite 60, 125
– tetrahydridoaluminate 60, 20
– tetrahydridodoborate 60, 172, 325
– –/tetrakis(triphenylphosphine)-
 palladium(0) 7, 91s60
– –/zirconium tetrachloride 17, 45s60
– tetrakis[3,5-bis(trifluoromethyl)-
 phenyl]borate 60, 308
Solid-phase synthesis s. Polymer-based...
Solid solvents 60, 24
Solid-state Beckmann rearrangement
 11, 217s59
Solid-state oxidation
– of alcohols 46, 237s60
Solvents s.a. Solid solvents
Sonogashira coupling
–, complement 60, 398
– with
 (Z)-1-halogeneneselenides 60, 174
 1-halogenenolethers 60, 175
– –, isomerizing 60, 403
(–)-Sparteine
– as reagent 49, 536s60; 60, 222, 259,
 309, 483
α-Spiro-γ-lactones 27, 162s60
Spiro[pyrrole-3,3′-oxindoles],
 2,3-dihydro- 43, 957s60
Stannanes (s.a. Stille...)
– special s.
 acetylenestannanes
 arylstannanes
 di(stannanes)
 enestannanes
 ethylenestannanes
 silylstann....
1-Stannylacetals
– special s.
 diethoxymethyltri-n-butylstannane
β-Stannylketones
– special s.
 ethylene-β-stannylketones
δ-Stannylketones
– special s.
 ethylene-δ-stannylketones
Staudinger reaction s.a. Phospha-
 Staudinger reaction
Stilbenes
– special s.
 alkoxystilbenes
Stille coupling
–, 2-alkoxy-1,3-dienes by –, with
 inversion 60, 175
– in fluorous 2-phase media 60, 462
– with 1-halogenenolethers 60, 175
Stille diaryl coupling
–, monitoring 34, 862s60
– – –, intramolecular, in situ 45, 583s60
Strecker synthesis, asym. 60, 250
Styrenes
–, 1,2-addition, regiospecific of
 CH-acidic compds. to – 60, 256
– special s.
 sulfonylaminostyrenes
– startg. m. f.
 α-aryl-α,β-ethylenecarboxylic acid
 esters, carbonylation 60, 317
 benzylamines, ar., asym. hydroamination 60, 128
ipso-Substitution
– of sulfoxides, ar. 60, 334

Substitution, nucleophilic, asym.
– at a single prochiral centre 60, 222
–, –, –, sequential 60, 262
Succinimides
– from
 acetylene derivs., biscarbonylation
 54, 312s60
Sugars s. Carbohydrates
Sulfamides
– from
 alcohols and amines 60, 163
o-Sulfamidobenzylalcohols 59, 250s60
α-Sulfamidocarboxylic acid esters
 60, 163
Sulfamoyl chloride
– as reactant 60, 51
Sulfenamides
– special s.
 S-glycosylsulfenamides
Sulfenyl halides
– startg. m. f.
 selenosulfides 60, 190
Sulfines
– special s.
 cyanosulfine
Sulfinic acid amides
– special s.
 sulfinylamino...
Sulfinic acid halides
 s.a. Iminosulfinic acid halides
α-Sulfinylaminonitriles
– from
 N-sulfinylimines, synthesis, asym.
 60, 250
2-Sulfinylamino-1,1,1-trifluorides, chiral
 44, 577s60
N-Sulfinylimines
– startg. m. f.
 α-sulfinylaminonitriles, synthesis, asym.
 60, 250
α-Sulfinylphosphonic acid esters
– special s.
 ethylene-α-sulfinylphosphonic acid
 esters
Sulfones (s.a. Sulfonylation and under
 Replacement)
– from
 sulfonic acid anhydrides 57, 207s60
– acids 57, 207s60
 sulfoximines 60, 50
– special s.
 acylaminosulfones
 allenesulfones
 aminosulfones
 ethylenesulfones
 halogenosulfon...
 hydrazinosulfones
 nitrosulfones
 2-pyridyl sulfones
Sulfones, chiral 60, 50
Sulfonic acid amides (s.a.
 N-Detosylation, N-Sulfonyl...,
 Sulfonylamin...)
–, N-allylation 57, 159s60
– special s.
 ethylenesulfonic acid amides
 halogenosulfonic – –
–, synthesis, combinatorial 48, 348s60
– – –, N-subst.
–, N-dealkylation, oxidative 16, 28s59
Sulfonic acid anhydrides
– startg. m. f.

sulfones 57, 207s60
Sulfonic acid esters (s.a. O-Detosylation,
 O-Sulfonylation, Sulfonyloxy...)
– special s.
 methanesulfonic acid esters
Sulfonic acids
– special s.
 dodecylbenzenesulfonic acid
 perfluorobutanesulfonic acid
– startg. m. f.
 sulfones 57, 207s60
Sulfonic acid tetrazolides
– special s.
 N-2,4,6-triisopropylbenzenesulfonyl-
 tetrazolide
Sulfonimides
– special s.
 trifluoromethanesulfonimide
Sulfonium salts
– special s.
 halogenosulfonium salts
Sulfonylacetals
– special s.
 halogeno(sulfonyl)acetals
O-Sulfonylamidoximes
– as intermediates 60, 131
O-Sulfonylamines (s.a. Sulfonic acid
 amides)
– from
 N-sulfonylimines, synthesis, asym.
 60, 247
– special s.
 di(sulfonylamines)
 ethylenesulfonylamines
γ-Sulfonylaminohydrazones, chiral
 58, 255s60
(E)-β-Sulfonylaminostyrenes 49, 836s60
Sulfonylation, ar. 60, 327
O-Sulfonylation
– special s.
 O-triflylation
2-Sulfonyl-1,3-dienes, cyclic
–, epoxidation, asym., regiospecific
 46, 106s60
N,N′-Sulfonyldiimidazole
– as reagent 25, 190s60
2-Sulfonylglutaramides 60, 320
Sulfonylhydrazones
– special s.
 aldehyde tosylhydrazones
 amino-N-sulfonylhydrazones
N-Sulfonylimines
– startg. m. f.
 amines, prim., synthesis, asym. 60, 247
 aziridines, cis-2-silyl-1-sulfonyl-, asym.
 induction 60, 348
 (1E,3Z)-3-cyano-1,3-dienephosphonic
 acid esters, with 4 extra C-atoms
 60, 349
 (E)-α,β-ethylene-α-sulfinylphosphonic
 – –, – 1 – C-atom 60, 349
 sulfonylamines, synthesis, asym.
 60, 247
Sulfonylisocyanates
– special s.
 N-(arylsulfonyl)isocyanates
Sulfonylketones s. Ketosulfones
β-Sulfonyllactolides
– from
 α-halogeno(sulfonyl)acetals 60, 485
β-Sulfonyloxyhalides
– special s.

2-iodoenol sulfonates
α-Sulfonyloxyketones
– from
 ethylene derivs. **60,** 62
–, reduction, asym. **57,** 20s60
– startg. m. f.
 heterocyclics, polymer-based synthesis
  60, 62
N-Sulfonylurethans
– special s.
 2-ethylene-N-tosylurethans
Sulfoxides
–, desulfuration **45,** 43s60
– from
 thioethers, asym. oxidation,
 V-catalyzed, review **39,** 83s60
– special s.
 acetylenesulfoxides
 diphenyl sulfoxide
 disulfoxides
 ethylenesulfoxides
 hydroxysulfoxides
 ketosulfoxides
Sulfoxides, ar.
–, *ipso*-substitution **60,** 334
Sulfoxides, cyclic
– special s.
 dibenzo[*b,d*]thiophene S-oxides
Sulfoximines
– special s.
 ethylenesulfoximines
– startg. m. f.
 sulfones **60,** 50
Sulfoximines, polymer-based
– as reactant **60,** 50
Sulfoxonium salts
– special s.
 trimethylsulfoxonium iodide
Sulfur 60, 368
Sulfuranes
– special s.
 polyfluoroalkoxy(fluoro)sulfuranes
Sulfuric acid-silica gel 48, 175s60
Sulfuric acid amide esters
– from
 alcohols **60,** 51
– – esters, cyclic
– special s.
 glycol sulfates, cyclic
Sulfur trifluorides, amino-
s. Aminosulfur trifluorides
Sultams
– special s.
 N-acylsultams
Superbases
– special s.
 tetraazaphosphabicycloundecanes
Supercritical fluids
–, reactions in –, review **57,** 106s60
– special s.
 carbon dioxide, supercritical
Superoxide, titanium(IV)-immobilized
 60, 47
Suzuki coupling
– in a 2-phase medium under
 thermomorphic conditions **60,** 399
– –, double
–, indoles, 2,3-diaryl- by – **53,** 473s60
– –, intramolecular (s.a. Hydroboration-
 intramolecular Suzuki coupling)
Suzuki diaryl coupling
– using highly active Pd-catalysts **60,** 456

– with chlorides, ar., non- or de-activated
 60, 456
– – –, heterogeneous
– in a 3-phase medium **37,** 902s60
– under microwaves **37,** 902s60
Suzuki-Miyaura coupling
– via lithium alkynyl(trialkoxy)borates
 60, 398
Synthesis, asym. (s.a. Automultiplication,
 asym., Catalysis, asym., Desymmetrization, Enantiomeric excess,
 Induction, asym., Resolution)
–, effects, non-linear, review **47,** 646s60

Tamao oxidation, improved 50, 407s60;
 58, 300s60
Tantalum pentachloride/silica 60, 122
Tedicyp s. *cis,cis,cis*-1,2,3,4-Tetrakis-
 (diphenylphosphinomethyl)cyclo-
 pentane
Tellurides
–, radical C-alkylation with – **27,** 907s60
Telluronium salts
– startg. m. f.
 cyclopropanes **60,** 426
Terpyridyls 60, 355
Tetraalkoxydiboranes
– special s.
 bis(catecholato)diborane
 bis(pinacolato)diborane
Tetraalkoxysilanes
– startg. m. f.
 phosphonous acid monoesters **60,** 56
2,5,8,9-Tetraaza-1-phosphabicyclo-
 [3.3.3]undecanes, 2,8,9-trialkyl-
– as superbase **60,** 5
**2,4,4,6-Tetrabromo-2,5-cyclohexa-
 dienone**
– as reagent **60,** 184
Tetra-*n*-butylammonium bromide
– as reagent **60,** 410, 449, 499
– chloride
– as reagent **60,** 390, 391, 392
– difluorotriphenylsilicate
– as reagent **44,** 577s60
– dihydrogentrifluoride
– as reagent **51,** 205s60
– fluoride
– as reagent **60,** 44, 406, 448, 496
– hydrogen sulfate
– as reagent **60,** 68
– 1-oxido-1,2-benziodoxol-3-one
– as reagent **60,** 63
– periodate
– as reagent **30,** 244s60
– persulfate
– as reagent **60,** 144
Tetra-*n*-hexylammonium chloride
– as reagent **60,** 363
Tetrahydridoborates s. under specific
 metals and 4-Aza-1-azonia-
 bicyclo[2.2.2]octane tetrahydridoborate, 1-benzyl-
2,4,6,8-Tetraiodoglycoluril
– as reagent **60,** 178
cis,cis,cis-**1,2,3,4-Tetrakis(diphenyl-**

 phosphinomethyl)cyclopentane
– as ligand **60,** 145
2-Tetralone ring closure, photochemical
 60, 497
1,1,3,3-Tetramethyldisilazane
– as reagent **60,** 406
N,N,N′,N′-Tetramethylethylenediamine
– as reagent **60,** 211
Tetramethylguanidine
– as reagent **32,** 820s60; **36,** 877s60
Tetra-*n*-octylammonium bromide
– as catalyst **60,** 151
1,1,4,4-Tetraphenylbutane-1,4-diol
– as reagent **60,** 272
1,1,2,2-Tetraphenyldisilane
– as reagent **46,** 65s60
Tetraphenylphosphonium chloride
– as reagent **60,** 82
Tetra(phosphines)
– special s.
 1,2,3,4-tetrakis(diphenylphosphinomethyl)cyclopentane
Tetra-*n*-propylammonium perruthenate
– as reagent **43,** 219s60 (review)
Tetrathiafulvalenes
–, chemistry of new concepts, review
 16, 820s60
Tetrazoles
– startg. m. f.
 N-(4-chloro-1,2,3-dithiazol-5-ylidene)-
 1,1-halogenohydrazones **60,** 205
 1,3,4-thiadiazoles, 2-cyano- **60,** 205
Tetrazolo[1,5-*a*]piperazin-6-ones
– by Ugi 4-component condensation,
 azide-modified **60,** 338
–, compound libraries **60,** 338
Thallium(I) ethoxide 49, 836s60
cis-**3-Thiabicyclo[3.3.0]octanes**
– from
 1,6-dienes **60,** 268
Thia-crown ethers
– from
 thietanes and thiiranes, review
 31, 592s60
**1,3,4-Thiadiazoles, 2-acylamino-
 12,** 690s60
1,3,4-Thiadiazoles, 2-cyano-
– from
 N-(4-chloro-1,2,3-dithiazol-5-ylidene)-
 1,1-halogenohydrazones **60,** 205
 tetrazoles **60,** 205
Thianaphthenes s. Benzo[*b*]thiophenes
(2Z,6E)-4*H*-1,4-Thiazepin-5-ones
– from
 thiazoles, 2-alk-1-ynyl- **60,** 253
Thiazoles, 2-alk-1-ynyl-
– startg. m. f.
 (2Z,6E)-4*H*-1,4-thiazepin-5-ones
 60, 253
Thiazoles, 5-amino-
– from
 N-α-(benzotriazol-1-yl)aldimines
 60, 352
 isothiocyanates **60,** 352
–, 4-(benzotriazol-1-yl)methyl-
– startg. m. f.
 benzothiazoles **60,** 352
**Thiazolidine-4(R)-carboxylic acid, 5,5-
 dimethyl-**
– as reagent **60,** 336
**2-Thiazolidones, 3-aryl-, conform-
 ationally rigid**
–, C-α-alkylation, asym. **60,** 369

Thiiranes s. Sulfido compds. in Vol. 1-50
Thioacetals s. Mercaptals
Thioacylation
– with 1,3,2-dioxaphosphorinane
  2-sulfides, 2-acylthio- **60**, 200
N-Thioacylation 60, 200
S-Thioacylation, selective 60, 200
Thioacylsilanes
– from
  1-benzotriazol-1-yl-1,1-silylthioethers
  43, 597s60
N-Thioacyltriazenes
– startg. m. f.
  guanidines, polymer-based synthesis
  60, 161
Thioalkylation, ar., asym. 30, 599s60
–, –, Yb-catalyzed **60**, 383
Thioamides s. Carboxylic acid thioamides
Thiocoumarins 34, 525s60
Thiocyanates
– special s.
  di(thiocyanates)
–, trans-S-alkylation **60**, 203
Thioethers (s.a. Alkylthio...,
  Organothio...)
– special s.
  acetylenethioethers
  ketothioethers
  mercaptothioethers
  silylthioethers
  trifluoromethyl thioethers
– startg. m. f.
  sulfoxides, asym. oxidation,
  V-catalyzed, review **39**, 83s60
Thioethers, ar. (s.a. Arylthio...)
– from
  arylboronic acids (with mercaptans)
  60, 201
  diazonium o-benzenedisulfonimide
  salts **41**, 511s60
– special s.
  diaryl sulfides
Thioglycosides
– startg. m. f.
  glycosyl halides **60**, 185
Thiolactams
– special s.
  N-(γ-keto)thiolactams
Thiolcarbamic acid aryl esters
– startg. m. f.
  arenes **3**, 73s60
Thiolic acid esters
– from
  carboxylic acids (with mercaptans)
  60, 195
– special s.
  arylthiolic acid esters
  glycosyl thiolacetates
– startg. m. f.
  ketones **60**, 457
– – –, polymer-based
–, reactions with – **60**, 113
Thiolsulfinic acid esters
– special s.
  4-methoxyphenyl benzenethiosulfinate
Thiolsulfonic acid esters
– special s.
  S-glycosyl methanethiolsulfonates
Thiolsulfonic acid salts
– startg. m. f.
  selenosulfides **60**, 190
Thiophene-2-carboxylic acid amides,

2,3-dihydro- **43**, 943s60
**Thiophene 1,1-dioxides, 2,5-dihydro-,
  3-arylthio-**
– startg. m. f.
  2-pyridones, 5,6-dihydro-,
  1-arylsulfonyl-4-arylthio- **60**, 409
Thiophene 1-oxides
–, chemistry, review **29**, 591s60
Thiophenes, 2,3-dihydro-
– from
  1-acetylene-1-thioethers **60**, 305
–, –, 2-aryl- **60**, 305
–, tetrahydro-, 3-acyl-
– from
  3-ethylene-2-hydroxymercaptans and
  oxo compds., asym. induction **60**, 331
Thiophenol
– as reagent **60**, 494
Thiophenols s. Arylmercaptans
**1,2-N→C-Thiophosphinyl group
  migration 60**, 212
Thiophosphoric acid esters
– special s.
  monothiolphosphoric acid esters
**4-Thiopyrones, 2,3-dihydro-, 6-(vinyl-
  thio)-**
– from
  1,3-dithiolanes, 2-(α,β-ethylene-
  acylene)- **60**, 193
Thioureas
– startg. m. f.
  guanidines **60**, 162
Thorpe cyclization s.a. Mitsunobu
  reaction-Thorpe cyclization
**Tiemann rearrangement, N-alkylative
  60**, 131
Tin 60, 239
Tin(II) bromide 60, 165
–(II) chloride 41, 888s60; **60**, 139, 140,
  359
–(IV) chloride 38, 887s60; **60**, 443
–(IV) compds., organo- s. Stannanes
–(IV) enolates s.a. Enoxystannanes
– oxides, organo-
  bis(tri-n-butyltin) oxide **50**, 71s60;
  60, 385
  dibutyltin oxide **14**, 343s60
–(II) phenylmercaptide 20, 387s60
–(II) triflate 60, 327
Tishchenko reaction (s.a. Aldol-
  Tishchenko reaction)
Titanium(IV) alkoxides
– tetraisopropoxide **51**, 372s60;
  56, 146s60; **60**, 271, 396
– –/ethylmagnesium bromide **60**, 384
– –/isopropylmagnesium chloride
  60, 270
–(IV) –, chiral
–, generation in situ **60**, 238
  2,2-dimethyl-α,α,α′,α′-tetrakis(1-
  naphthyl)-1,3-dioxolane-(4R,5R)-
  dimethanolatotitanium dichloride
  60, 179
–(IV) aroxides, chiral
  di-tert-butoxytitanium(IV) 1,1′-bi-2-
  naphthoxide **60**, 238
  titanium tetraisopropoxide/(R)-1,1′-bi-
  2-naphthol **49**, 898s60; **55**, 434s60
– complexes
  titanocene dichloride/n-butyl-
  magnesium chloride **60**, 273
–(IV) enolates, chloro-

– as intermediates **60**, 252
–(IV) superoxide s. Superoxide,
  titanium(IV)-immobilized
– tetrabromide **60**, 275
– tetrachloride **34**, 896s60; **60**, 19, 102,
  252, 275, 287, 378, 440, 441
– –/diethylamine **60**, 275
– –/nitromethane **7**, 761s60
– –/samarium **8**, 657s60; **13**, 539s60;
  59, 149s60; **60**, 358
– tetraiodide **60**, 25, 271
– trichloride/lithium **60**, 3
p-**Toluenesulfonic acid**
– as reagent **60**, 332, 472
– –/montmorillonite
– as reagent **21**, 948s60
Trans-S-alkylation
– of thiocyanates **60**, 203
Transetherification s.a. 1,2-Alkoxy-
  allylation-trans-etherification
Transfer-hydrogenation
– in aq. alkaline medium **60**, 37
Transfer-hydrogenation, asym.
–, 1-deuterioalcohols, prim. by – **60**, 28
Transfer reactions s.a. Aldol transfer
**Transition metal glycosylidene
  complexes**
– startg. m. f.
  C-glycosides, review **44**, 782s60
– – imidazolinylidene and triazolinylidene complexes, chiral
– in catalysis, asym., review **56**, 495s60
Transthioacetalation 60, 202
Trialkylboranes
– as reductant **60**, 21
Trialkylmanganates
– startg. m. f.
  enesilanes, with 1 extra C-atom **60**, 375
Triarylamines
– from
  amines, ar., prim. **60**, 157
  halides, ar. (2 different molecules)
  60, 157
Triarylantimony diacetates
– startg. m. f.
  α-arylcarbonyl compds. **60**, 461
  cinnamic acid esters **60**, 461
  2(3H)-furanones, 5-aryl- **60**, 461
Triarylphosphine, polymer-based
– as reagent **45**, 220s60
1,5,7-Triazabicyclo[4.4.0]dec-5-ene
– as reagent **39**, 854s60; **60**, 226
–, polymer-based
– as reagent **60**, 226
Triazenes
– special s.
  amidinotriazenes
  thioacyltriazenes
**1,2,4-Triazines, 5-hydroxylamino-
  16**, 409s60
2H-1,2,3-Triazolium-1-imides
– startg. m. f.
  isothiocyanates **60**, 160
Tri-n-butylphosphine
– as reagent **17**, 118s60; **60**, 97, 284, 326
2,4,6-Tri-tert-butylpyridine
– as reagent **60**, 54, 80
Tri-n-butyltin hydride
– as reactant **60**, 216
– as reagent **60**, 26, 27, 217, 276, 442,
  487
Trichloroacetimidoyl glycosides

– startg. m. f.
  glycosides **60**, 99
Trichloromethyl chloroformate
– as reagent **16**, 493s60
2,4,6-Trichlorophenol
– as reagent **60**, 380
Tricyclohexylphosphine
– as reagent **60**, 313
Tricyclo[3.3.0.0^{2,8}]octanes, 1-amino-2-alkoxy- 60, 419
(*all*-E)-2,4,6-Trienals
– from
  aldehydes, with 6 extra C-atoms
  60, 428
1,2,4-Trienes
– from
  5-acoxy-3-en-1-ynes, synthesis with
  remote asym. induction **60**, 326
– startg. m. f.
  4-vinylcyclohexenes, 3-alkylidene-,
  asym. conversion **60**, 299
1,3,9-Trienes, macrocyclic 42, 980s60
Triethylenediamine
– as reagent **60**, 324, 344, 373
Triethylsilane
– as reagent **60**, 40
Triethylsilyl triflate
– as reagent **60**, 95
1,1,1-Trifluorides
– special s.
  2-siloxy-1,1,1-trifluorides
  sulfinylamino-1,1,1-trifluorides
Trifluoroacetic acid
– as reagent **60**, 70, 113, 351, 471, 493
Trifluoroacetic acid esters (s.a.
  Trifluoroacetoxy under Replacement)
– startg. m. f.
  mercaptans, with inversion **60**, 196
Trifluoroacetic anhydride
– as reagent **60**, 54, 80, 136, 196, 334
Trifluoroacetimidoyl glycosides
– startg. m. f.
  glycosides **60**, 103
Trifluoroacetone
– as reagent **57**, 62s60
α,α,α-**Trifluoroacetophenones**
29, 792s60
N-Trifluoroacetylation, preferential
– of amines, ar., prim. **60**, 137
Trifluoroacetyl nitrate
– as reagent **1**, 343s60
Trifluoromethanesulfonic acid
– as reagent **49**, 909s60; **60**, 66, 206
Trifluoromethanesulfonic acid esters
– special s.
  enol triflates
Trifluoromethanesulfonimide
– as reagent **60**, 66
anti-**1-Trifluoromethyl-2-aminoalcohols**
48, 856s60
Trifluoromethylation, nucleophilic
60, 224
Trifluoromethyl selenides 60, 224
α-**Trifluoromethylstilbenes 39**, 887s60
Trifluoromethyl thioethers 60, 224
Trifluoromethyltri-*n*-butylstannane
– as reactant **60**, 219
Trifluoromethyltrimethylsilane
– as reactant **60**, 219
**N-[2,2,2-Trifluoro-1-(trimethylsiloxy)-
  ethyl]morpholine**
– as reactant **60**, 224

**N-[2,2,2-Trifluoro-1-(trimethylsiloxy)-
  ethyl]piperazine**
– as reactant **60**, 224
O-Triflylation, polymer-based 60, 49
Tri-2-furylphosphine
– as reagent **60**, 394, 455, 457
1,1,1-Trihalides
– special s.
  1,1,1-trifluorides
– startg. m. f.
  carboxylic acid halides **60**, 92
2,2,2-Trihalogenalcohols
– from
  oxo compds., with 1 extra C-atom
  60, 225
**N-2,4,6-Triisopropylbenzenesulfonyl-
  tetrazolide**
– as reagent **60**, 52
Triisopropyl borate
– as reagent **60**, 398
O-Trimethylsilylation 60, 55
Trimethylsilyl azide
– as reactant **60**, 123, 127, 298, 338
Trimethylsilyl chloride
– as reagent **60**, 121, 239, 396
– cyanide
– as reagent **60**, 272, 333, 347, 413
Trimethylsilyldiazomethane s.
  Diazomethyltrimethylsilane
Trimethylsilyl halides
– as reactant **60**, 175
Trimethylsilyl iodide
– as reagent **60**, 96
Trimethylsilyl triflate
– as reagent **60**, 76, 103, 121, 312, 445, 446
Trimethylsulfoxonium iodide
– as reactant **60**, 407
Trimethyl(vinyl)silane
– as reactant **60**, 270
1,3,5-Triols
– special s.
  ethylene-1,3,5-triols
Triphenylarsine oxide
– as reagent **52**, 61s60
**Triphenyl borate/(R)-2,2′-diphenyl-
  3,3′-bi-4-phenthrol**
– as reagent **60**, 357
**Triphenylphosphine/diethyl azo-
  dicarboxylate**
– as reagent **60**, 48, 111, 339
Triphenylphosphine/iodine
– as reagent **60**, 141
Triphenylphosphine oxide
– as reagent **60**, 272
Triphenyl phosphite
– as reagent **60**, 253
Tri-*n*-propylphosphine
– as reagent **60**, 389
Trisaccharide synthesis
– in one pot **44**, 191s60
Tris(2,6-dimethoxyphenyl)phosphine
– as reagent **54**, 325s60; **60**, 297
Tris(1-imidazolyl)phosphine
– as reagent **50**, 43s60
Tris(*p*-methoxyphenyl)phosphine
– as reagent **60**, 394
Tris(trimethylsilyl)amine
– as reactant **60**, 224
Tris(trimethylsilyl)silane
– as reagent **37**, 765s60; **41**, 955s60;
  60, 269, 486, 495

Tris(triphenylsilyl) vanadate
– as reagent **60**, 277
**N,N′,N″-Tris[tris(dimethylamino)-
  phosphoranylidene]phosphoric acid
  triamide**
– as reagent **19**, 442s60
O-Tritylation
– with *p*-methoxybenzyl trityl ether
  60, 101
– with simultaneous O-acylation and
  O-silylation **60**, 90
O-Tritylation, preferential
– with benzyl trityl ether **60**, 101
Trityl ethers
– special s.
  benzyl trityl ether
  p-methoxybenzyl – –
**Trityl tetrakis(pentafluorophenyl)-
  borate**
– as reagent **39**, 189s60
1,3,5-Triynes
– from
  3-(dihalogenomethylene)-1,4-diynes
  60, 482
**Tungsten aryl(alkoxy)carbene
  complexes**
– startg. m. f.
  γ-alkoxy-γ-arylboxylic acid esters
  60, 437
**Tungsten 2-benzopyran-3-ylidene
  complexes**
–, Diels-Alder reaction with – **60**, 472
– carbene complexes s.a. Chromium
  carbene complexes
– hexachloride/silica **60**, 116
– pentacarbonyl **60**, 472

**Ugi 4-component condensation, azide-
  modified**
–, tetrazolo[1,5-*a*]piperazin-6-ones by –
  60, 338
Ugi multicomponent condensation
– with isonitriles, review **43**, 700s60
Uracils, 5-fluoro-
–, N-arylation **53**, 162s60
Urea-hydrogen peroxide 56, 106s60
Ureas
– via Lossen-type rearrangement
  58, 80s60
Ureas, cyclic
–, mono-N-acylation **14**, 516s60
Ureas, sym.
– from
  amines, prim. **58**, 133s60
Urethans (s.a. Carbobenzoxyamines,
  N-[De]carbalkoxylation, Carbamic...)
– from
  alcohols and amines, oxidative
  carbonylation **60**, 132
–, N-nitrosation **8**, 364s59
– special s.
  ethyleneurethans
  sulfonylurethans
– startg. m. f.
  2-oxazolidones, asym. induction
  60, 168

(Urethans
- startg. m. f.)
-, synthesis, combinatorial **48**, 348s60
- via Lossen-type rearrangement
58, 80s60
Urethans, polymer-based
- as intermediates **49**, 168s60;
50, 214s60; **60**, 159

Vanadate s. Ethyl dichlorovanadate, Isopropyl -, Tris(triphenylsilyl) vanadate
Vanadium(V) imide complexes, trihalogeno- 48, 423s60
Vanadyl acetoacetonate 60, 97
Vanadyl allenolates
-, aldol-type condensation via - **60**, 277
Vanadyl chloride 60, 240
Vanadyl triisopropoxide 60, 65
Vinyl...
s.a. En..., α,β-Ethylen...
C-Vinylation
- special s.
radical C-vinylation
C-α-Vinylation, intramolecular, Pd-catalyzed 54, 489s60
4-Vinylcyclohexenes, 3-alkylidene-
- from
1,3-dienes and 1,2,4-trienes, asym. conversion **60**, 299
cis-**2-Vinylcyclopropanecarboxylic acid derivs. 60**, 426
Vinylcyclopropanes
-, radical 3-component reaction with -
57, 257s60
- startg. m. f.
1,4-cycloheptadienes **60**, 292
Vinyl halides s. α,β-Ethylenehalides
2-Vinyl-O-heterocyclics
- from
2-ene-1,n-diols **20**, 200s59
(siloxy)acoxy-2-ethylenes, asym. induction **60**, 115
Vinyl ketones s. α,β-Ethyleneketones
α-Vinyllactones
- special s.
α-(2-alkoxyvinyl)lactones
1-Vinyllactones 60, 366
Vinylmagnesium bromide
- as reactant **60**, 328, 467
α-Vinyloxyhydrazones
- startg. m. f.
anti-2-ethylene-2′-hydroxyhydrazines, asym. induction **60**, 494

Vinylsilanes s. Enesilanes
Vinylstannanes s. Enestannanes
(E)-Vinylzirconium compds.
- from
enolethers **60**, 213
- startg. m. f.
ethylene derivs. **60**, 213

Wittig synthesis (s.a. Phospha-Wittig synthesis)
- in ionic liquids **60**, 438

Xanthates
- special s.
ketoxime xanthates
Xanthenes, bis(diarylphosphino)-, guanidinium-modified
- as reagent **4**, 667s60
Xanthones
- from
4-chromones, 2-vinyl- and ketones **60**, 323
Xenon derivs.
-, chemistry, review **33**, 466s60

Yeast 48, 108s60
Yeast, lyophilized 29, 36s60
Ynamines, terminal, cobalt-complexed
-, Pauson-Khand reaction with -
40, 475s60
Ytterbium(III) chloride 55, 81s60
Ytterbium(III) complexes, organo-
- as intermediates **60**, 366
-(III) triflate **43**, 563s60 851s60; 943s60;
44, 610s60 851s60; **49**, 264s60;
59, 220s60; **60**, 94, 366, 383
-(III) tris(perfluoroalkanesulfonyl)-methides **58**, 378s60
Yttrium(III) chloride 11, 215s60
Yttrium complexes, chiral 50, 402s60
- salen complex, chiral **60**, 236
-(III) triflate **11**, 843s60

Zeolite 1, 343s60; **23**, 423s60
Zeolite NaY, modified 36, 235s60
Zinc 60, 240, 412
Zinc, activated
-, prepn., electrochemical **34**, 614s60
Zinc aroxides, organo-
- special s.
iodomethylzinc aroxides
Zincates
- special s.
dilithium tetramethylzincate
lithium trialkylzincates
Zinc bis(tetrahydridoborate) 60, 328
- bromide **60**, 165, 352
- chloride **22**, 742s60; **60**, 116, 365, 425
- compds., dialkyl-
diethylzinc (as reagent) **42**, 676s60;
60, 210, 231, 242, 380
diisopropylzinc (- -) **60**, 232, 264
- as reactant **60**, 229, 230, 247, 248, 379
- compds., organo-
- special s.
bis(iodomethyl)zinc
- enolate equivalents
-, Reformatskii-type synthesis, dialkylzinc-mediated via - **60**, 232
- enolates
- as intermediates **60**, 263
-, Reformatskii-type synthesis, Rh-catalyzed via - **60**, 242
- halides, organo-
- special s.
benzylzinc halides
bis(iodozincio)methane
- iodide **60**, 75
- nitrate/silica **47**, 146s60
- perchlorate **60**, 429
- triflate **50**, 551s60; **51**, 433s60;
54, 71s60
**Zincke reaction, polymer-based
20**, 369s60
Zirconacyclopentadienes
-, [2+2+2]-cycloaddition via - **60**, 274
Zirconium(IV) alkoxides
- tetra-*tert*-butoxide **60**, 271
- tetraisopropoxide **60**, 61, 248
- complexes
bis(2-methoxyethyl)zirconocene dihydride **50**, 428s60
chlorobis(cyclopentadienyl)hydrido-zirconium **60**, 41
zirconocene dichloride **4**, 643s60;
60, 213
- -/n-butyllithium **60**, 274
- compds., organo-
- special s.
vinylzirconium compds.
-(IV) enolates
- as intermediates **45**, 383s60
-(IV) tetrahydridoborate **60**, 38

Supplementary References in Volume 60

No.	Suppl. Ref. Vol. Page

Volume 1

13	60, 7
343	60, 80
391	60, 81
419	60, 105
549	60, 178
569	60, 243
697	60, 127
715	60, 242
775	60, 4

Volume 2

48	60, 9
49	60, 9
75	60, 16
288	60, 64
368	60, 81
403	60, 81
647	60, 178
651	60, 178
692	60, 71

Volume 3

3	60, 102
73	60, 26
568	60, 138
657	60, 71

Volume 4

3	60, 1
214	60, 52
311	60, 73

643	60, 140
667	60, 164 (3)
702	60, 180
768	60, 138
810	60, 248

Volume 5

32	60, 10
63	60, 25
279	60, 81
301	60, 86
355	60, 248
614	60, 248
666	60, 260, 262

Volume 6

339	60, 71
340	60, 71
486	60, 80

Volume 7

91	60, 20
217	60, 45
281	60, 262
384	60, 71
761	60, 176
816	60, 192
823	60, 262
836	60, 202

Volume 8

165	60, 36
272	60, 102
455	60, 82
529	60, 95

563	60, 90, 91
657	60, 117
782	60, 177
823	60, 185, 187
927	60, 262

Volume 9

512	60, 90
524	60, 91
807	60, 191
872	60, 196

Volume 10

251	60, 69
262	60, 71
633	60, 221
688	60, 253

Volume 11

73	60, 19
215	60, 46
224	60, 122
362	60, 70
633	60, 26
660	60, 110
666	60, 114
744	60, 262
843	60, 183
905	60, 220

Volume 12

103	60, 23
136	60, 84
453	60, 81
455	60, 83

Supplementary References

No.	Suppl. Ref. Vol. Page

Volume 12 continued

653	60, 262
686	60, 117
690	60, 119
931	60, 171

Volume 13

156	60, 33
239	60, 50
317	60, 67
359	60, 71
366	60, 95
371	60, 73
412	60, 84
518	60, 97
539	60, 99
681	60, 191
799	60, 27

Volume 14

211	60, 48
343	60, 67
380	60, 71
400	60, 77
516	60, 94
711	60, 262

Volume 15

39	60, 13
261	60, 66
571	60, 177

Volume 16

24	60, 10
28	60, 10
342	60, 68
396	60, 74

409	60, 76
493	60, 78
606	60, 107
681	60, 121
698	60, 262
780	60, 175
820	60, 262
888	60, 262

Volume 17

45	60, 11
118	60, 26
169	60, 4, 262
198	60, 257
216	60, 47
234	60, 38
405	60, 81
630	60, 109
653	60, 74
809	60, 177
871	60, 252
968	60, 254
983	60, 259

Volume 18

26	60, 9
180	60, 37
302	60, 64
776	60, 162
912	60, 225
965	60, 147

Volume 19

33	60, 86 (2), 87, 262
60	60, 20
188	60, 38
201	60, 41
221	60, 38
419	60, 82
442	60, 85
764	60, 152, 262
788	60, 171

954	60, 249

Volume 20

81	60, 27
266	60, 77
271	60, 80
312	60, 84
356	60, 9
369	60, 88
387	60, 99
413	60, 104
450	60, 115
502	60, 138
510	60, 142
511	60, 142
533	60, 170
685	60, 255, 262

Volume 21

9	60, 2
174	60, 38, 39
177	60, 40
606	60, 107
739	60, 156
749	60, 169
948	60, 246

Volume 22

408	60, 100
421	60, 87
737	60, 171
742	60, 173
761	60, 144
826	60, 196
835	60, 147

Volume 23

27	60, 9
66	60, 21
139	60, 35
237	60, 55

No.	Suppl. Ref. Vol. Page

Volume 23 continued

423	60, 82
570	60, 110
819	60, 194, 263
831	60, 198
879	60, 225
927	60, 249
928	60, 250

Volume 24

100	60, 30
228	60, 58
245	60, 62
261	60, 66
387	60, 4
555	60, 102
726	60, 170
755	60, 175
836	60, 194
852	60, 200

Volume 25

15	60, 8
21	60, 11
104	60, 38
190	60, 67
487	60, 128
549	60, 175
581	60, 192
649	60, 98, 248
669	60, 85
701	60, 259

Volume 26

52	60, 21
235	60, 64
247	60, 177
325	60, 79

331	60, 81
386	60, 82
463	60, 66 (2)
675	60, 136
684	60, 141
806	60, 185
817	60, 263
823	60, 187
871	60, 82

Volume 27

18	60, 9
57	60, 21, 23
93	60, 28
110	60, 10
122	60, 37
145	60, 43
146	60, 41
162	60, 46
435	60, 263
530	60, 103
555	60, 102
675	60, 263
694	60, 144
696	60, 144
724	60, 165
761	60, 175, 176
840	60, 200
843	60, 203
851	60, 216
870	60, 209
871	60, 211, 212
907	60, 224

Volume 28

13	60, 5 (2), 263
141	60, 50, 60
144	60, 83
268	60, 6
413	60, 96
418	60, 245
584	60, 263
637	60, 150
827	60, 208

852	60, 226
856	60, 226
869	60, 243

Volume 29

2	60, 3
28	60, 4
36	60, 15
42	60, 19
391	60, 263
518	60, 106
530	60, 111
547	60, 59, 114
553	60, 114
591	60, 263
603	60, 122
665	60, 263
752	60, 180
792	60, 144
845	60, 214
853	60, 220
861	60, 225
910	60, 248
912	60, 256
932	60, 263
959	60, 128
968	60, 250
970	60, 252, 253

Volume 30

22	60, 19
43	60, 197
239	60, 80
244	60, 80
599	60, 207, 208
634	60, 245

Volume 31

49	60, 24
65	60, 27
427	60, 96
520	60, 112

No.	Suppl. Ref. Vol. Page

Volume 31 continued

592	60, 126, 263
616	60, 141
637	60, 169
658	60, 159
719	60, 175
806	60, 199
812	60, 200
905	60, 248
964	60, 253

Volume 32

26	60, 13
180	60, 61
231	60, 134
287	60, 75
317	60, 81
413	60, 98
460	60, 263
614	60, 135
637	60, 200
739a	60, 218
800	60, 201
820	60, 216
928	60, 263

Volume 33

8	60, 7
43	60, 12 (2)
76	60, 26
325	60, 263
466	60, 263
477	60, 104, 105
627	60, 263
640	60, 145
658	60, 158
662	60, 170
806	60, 198
807	60, 148

865	60, 226
879	60, 232
918	60, 248

Volume 34

23	60, 11
88	60, 67
525	60, 115
610	60, 137
614	60, 130
668	60, 170
681	60, 263
693	60, 175
862	60, 240
896	60, 119
999	60, 259

Volume 35

23	60, 19
39	60, 20
57	60, 30
68	60, 116
105	60, 52
137	60, 59
182	60, 101
439	60, 127
549	60, 197
655	60, 248

Volume 36

36	60, 263
101	60, 34
235	60, 42
470	60, 263
539	60, 116
667	60, 263
668	60, 151
754	60, 181
795	60, 198
825	60, 148
838	60, 204
877	60, 227, 232

932	60, 248

Volume 37

127	60, 42
148	60, 50
152	60, 6
234	60, 63
447	60, 9
657	60, 148
765	60, 207
807	60, 263
902	60, 238

Volume 38

24	60, 12
73	60, 31
86	60, 33
91	60, 34
97	60, 37
100	60, 36
238	60, 147
353	60, 85
363	60, 263
496	60, 74
506	60, 115
584	60, 124
668	60, 150
673	60, 148
706	60, 171
723	60, 175
760	60, 154
802	60, 198
836	60, 204
887	60, 234
904	60, 234
954	60, 250
965	60, 252

Volume 39

32	60, 13
33	60, 15

No.	Suppl. Ref. Vol. Page
Volume 39 continued	
83	60, 30, 264
124	60, 44
171	60, 54
189	60, 57 (2)
228	60, 64
450	60, 264
458	60, 105
555	60, 123 (2)
593	60, 130
612	60, 139
640	60, 165
646	60, 264
754	60, 200
854	60, 227
887	60, 241
883	60, 151
Volume 40	
1	60, 3
22	60, 15
81	60, 44
99	60, 51
176	60, 72
186	60, 74
235	60, 264
286	60, 96
424	60, 124
475	60, 158
477	60, 182
493	60, 174
540	60, 124
567	60, 130, 134 (2)
Volume 41	
37	60, 14
118	60, 35, 36
154	60, 48
241	60, 110
286	60, 81
305	60, 77
344	60, 83
352	60, 79
463	60, 105
505	60, 114
511	60, 116
556	60, 120, 264
638	60, 148
694	60, 264
723	60, 183
737	60, 264
797	60, 206
819	60, 209
862	60, 226
888	60, 234
955	60, 252
Volume 42	
3	60, 4
45	60, 21
48	60, 23
91	60, 264
108	60, 35
165	60, 53
200	60, 58
235	60, 64
236	60, 64
354	60, 87
397	60, 96
462	60, 264
597	60, 264
616	60, 131 (2)
636	60, 157
638	60, 147
676	60, 161
695	60, 174
733	60, 180
979	60, 264
980	60, 257
993	60, 264
Volume 43	
45	60, 13
51	60, 17
60	60, 24
83	60, 33
131	60, 47
162	60, 53
213	60, 64
219	60, 264
249	60, 144
298	60, 83
316	60, 83
402	60, 101
456	60, 102
517	60, 122
563	60, 135
597	60, 159
607	60, 143
616	60, 148
628	60, 264
699	60, 190
700	60, 264
770	60, 149
806	60, 214
808	60, 217
841	60, 226
851	60, 231
860	60, 222
943	60, 145, 196, 264
944	60, 264
957	60, 253
958	60, 253
Volume 44	
23	60, 10
48	60, 20
97	60, 35
129	60, 44
191	60, 56
197	60, 57
214	60, 60
238	60, 67

No.	Suppl. Ref. Vol. Page
Volume 44 continued	
241	60, 264
268	60, 37
314	60, 83
403	60, 101
460	60, 115
545	60, 130
565	60, 132
568	60, 136
577	60, 129, 232
602	60, 147
610	60, 151
646	60, 264
776	60, 200, 201
781	60, 216
782	60, 264
815	60, 216
819	60, 219
972	60, 265
976	60, 265
Volume 45	
43	60, 27
56	60, 265
88	60, 59
120	60, 48, 63 (2)
192	60, 82
204	60, 82
209	60, 83
220	60, 89
368	60, 126
382	60, 137
383	60, 137
402	60, 135
413	60, 156
436	60, 173
439	60, 175
528	60, 228
542	60, 265
555	60, 241
577	60, 252
579	60, 265
583	60, 254
Volume 46	
42	60, 18
44	60, 19
47	60, 19
55	60, 265
65	60, 24
90	60, 29
106	60, 43, 44
116	60, 38
123	60, 28
171	60, 54
237	60, 64
321	60, 87
351	60, 91
397	60, 98
429	60, 101
612	60, 142
631	60, 156
641	60, 265
659	60, 156
667	60, 265
696	60, 175
738	60, 189 (2)
767	60, 265
799	60, 217
899	60, 240
971	60, 252
Volume 47	
48	60, 22
111	60, 44
113	60, 44 (2)
114	60, 45, 265
146	60, 53
162	60, 56
243	60, 73
262	60, 265
440	60, 105
470	60, 110
542	60, 122
569	60, 265
576	60, 127
614	60, 140
625	60, 265
639	60, 265
651	60, 151
646	60, 265
658	60, 151
668	60, 149
705	60, 176
715	60, 265
863	60, 265
885	60, 228
897	60, 229
933	60, 265
955	60, 265
Volume 48	
17	60, 9
18	60, 9
48	60, 21
106	60, 35
108	60, 36
134	60, 43
169	60, 52
175	60, 54
299	60, 78
348	60, 92
392	60, 99
423	60, 101
434	60, 106
510	60, 116
600	60, 137
625	60, 140
665	60, 151
681	60, 161
692	60, 165
772	60, 190
802	60, 200
856	60, 222
988	60, 257
Volume 49	
42	60, 15
47	60, 17

No.	Suppl. Ref. Vol. Page
Volume 49 continued	
85	60, 28
168	60, 55
195	60, 61
210	60, 63
264	60, 75
274	60, 77
300	60, 82
312	60, 85
330	60, 10, 191
370	60, 113
388	60, 255
407	60, 103
419	60, 265
507	60, 119
515	60, 121
518	60, 120, 123
536	60, 124
699	60, 152, 171
674	60, 158
805	60, 204
636	60, 253
836	60, 168
849	60, 144
898	60, 231, 232
909	60, 235
932	60, 247
938	60, 249
985	60, 258 (2), 265
Volume 50	
43	60, 32
53	60, 35
54	60, 36
55	60, 36, 102
73	60, 46
75	60, 46
103	60, 57
115	60, 62
123	60, 64

No.	Suppl. Ref. Vol. Page
151	60, 73
198	60, 88
214	60, 91
241	60, 96
257	60, 100
322	60, 116
343	60, 122
382	60, 265
397	60, 203
402	60, 151
405	60, 151
407	60, 155
410	60, 234
415	60, 159
428	60, 154
449	60, 175
514	60, 266
531	60, 136, 220
551	60, 231
555	60, 266
579	60, 253
Volume 51	
2	60, 5
3	60, 2
9	60, 9
17	60, 10
23	60, 13
26	60, 16
32	60, 23
42	60, 26
46	60, 30
62	60, 39
65	60, 42, 43
72	60, 48
76	60, 48
104	60, 59
126	60, 73
132	60, 76
171	60, 93
172	60, 94
204	60, 103
205	60, 107, 109
216	60, 110

No.	Suppl. Ref. Vol. Page
271	60, 131, 132
300	60, 146
318	60, 154
335	60, 167
337	60, 169
372	60, 181
380	60, 177
416	60, 217
423	60, 68, 221
429	60, 266
433	60, 228
453	60, 236
500	60, 260
Volume 52	
13	60, 9
38	60, 26
68	60, 38
75	60, 266
77	60, 47
121	60, 65
125	60, 266
128	60, 67
150	60, 83
164	60, 89
170	60, 266
171	60, 93
175	60, 96
208	60, 112
221	60, 266
247	60, 147
255	60, 131
274	60, 266
297	60, 149
305	60, 166
310	60, 153
334	60, 173
352	60, 177
358	60, 184
389	60, 200
390	60, 203
415	60, 215
422	60, 138
494	60, 258
497	60, 260, 261, 266

No.	Suppl. Ref. Vol. Page

Volume 53

No.	Vol, Page
4	60, 6
22	60, 135
59	60, 37, 266
61	60, 39
82	60, 53
86	60, 56
131	60, 76
144	60, 83
152	60, 88
162	60, 92
228	60, 116
268	60, 134
288	60, 144
291	60, 145
326	60, 167, 266
335	60, 172
336	60, 172
371	60, 249
399	60, 200
429	60, 211
442	60, 206
453	60, 235
470	60, 237
473	60, 239
481	60, 210
500	60, 261

Volume 54

No.	Vol, Page
27	60, 14
30	60, 18 (2)
44	60, 24
58	60, 1
71	60, 45
92	60, 58
96	60, 200
212	60, 221
244	60, 124
261	60, 136
296	60, 150

No.	Vol, Page
312	60, 157
320	60, 162
325	60, 167
374	60, 188
394	60, 197
431	60, 242
466	60, 241
477	60, 245
489	60, 253

Volume 55

No.	Vol, Page
26	60, 17
35	60, 266
64	60, 39
67	60, 41, 42
81	60, 50
113	60, 66
146	60, 83
159	60, 92
161	60, 93
183	60, 104
205	60, 116
239	60, 133
253	60, 142
261	60, 146
277	60, 155
284	60, 156
297	60, 266
299	60, 145
324	60, 214
337	60, 183, 185
357	60, 188
412	60, 217
415	60, 215
421	60, 221
434	60, 232
452	60, 237
454	60, 240
471	60, 204
478	60, 249

Volume 56

No.	Vol, Page
9	60, 7

No.	Vol, Page
13	60, 8
30	60, 16
73	60, 37, 43
75	60, 45
106	60, 59
131	60, 76
146	60, 84
162	60, 80
265	60, 143
275	60, 149
295	60, 166
296	60, 164
307	60, 167
330	60, 177
334	60, 202
445	60, 242
490	60, 174, 256
495	60, 266

Volume 57

No.	Vol, Page
20	60, 14
22	60, 14
44	60, 25
48	60, 266
60	60, 39
62	60, 41
99	60, 266
102	60, 58
106	60, 266
159	60, 89
207	60, 116
238	60, 135
257	60, 145
300	60, 158
304	60, 177
312	60, 179
321	60, 137
376	60, 210
380	60, 212
439	60, 238
456	60, 248
479	60, 252

No.	Suppl. Ref. Vol. Page
Volume 58	
10	*60*, 7
17	*60*, 12
49	*60*, 266
76	*60*, 50
80	*60*, 51
126	*60*, 76
133	*60*, 80
141	*60*, 84
154	*60*, 91
187	*60*, 100
213	*60*, 118
235	*60*, 132
236	*60*, 133
255	*60*, 139
273	*60*, 153
281	*60*, 157
286	*60*, 158
300	*60*, 168
302	*60*, 225
306	*60*, 170
311	*60*, 175
342	*60*, 195 (2)
353	*60*, 198
371	*60*, 204
378	*60*, 183
406	*60*, 233
465	*60*, 243
497	*60*, 258 (2)
499	*60*, 267
Volume 59	
33	*60*, 17
45	*60*, 25
73	*60*, 39
86	*60*, 49
123	*60*, 68
124	*60*, 267
149	*60*, 84
159	*60*, 267
171	*60*, 91
174	*60*, 267
195	*60*, 108
206	*60*, 267
220	*60*, 122
250	*60*, 131
288	*60*, 149
321	*60*, 171
335	*60*, 207
354	*60*, 191
407	*60*, 267
418	*60*, 195
447	*60*, 239
Volume 60	
233	*60*, 267
395	*60*, 267
452	*60*, 267